Introductory Geotechnical Engineering

The environmental effect on the behaviour of the soil–water system is difficult to explain using classical mechanical concepts alone. This book integrates and blends traditional theory with particle-energy-field theory in order to provide a framework for the analysis of soil behaviour under varied environmental conditions.

A complete treatment of geotechnical engineering concepts is given, with an emphasis on environmental factors. Soil properties and classifications are included, as well as issues relating to contaminated land. Both SI and Imperial units are used, and an accompanying website provides example problems and solutions.

Introductory Geotechnical Engineering: An Environmental Perspective explains the "why" and "how" of geotechnical engineering in an environmental context. Students of civil, geotechnical and environmental engineering, and practitioners unfamiliar with the particle-energy-field concept, will find the book's novel approach helps to clarify the complex theory behind geotechnics.

Hsai-Yang Fang is Professor Emeritus at Lehigh University and a Distinguished Fellow at the Global Institute for Energy and Environmental Systems, The University of North Carolina at Charlotte.

John L. Daniels is Assistant Professor of Civil Engineering and Fellow at the Global Institute for Energy and Environmental Systems, The University of North Carolina at Charlotte.

Also available from Taylor & Francis

Craig's Soil Mechanics 7th edition
R.F. Craig
Hb: 0–415–32702–4
Pb: 0–415–32703–2

Taylor & Francis

Applied Analyses in Geotechnics
F. Azizi
Hb: 0–419–25340–8
Pb: 0–419–25350–5

Taylor & Francis

Contaminated Land
T. Cairney
Hb: 0–419–23090–4

Taylor & Francis

Introduction to Geotechnical Processes
J. Woodward
Hb: 0–415–28645–X
Pb: 0–415–28646–8

Taylor & Francis

Soil Mechanics 2nd edition
W. Powrie
Hb: 0–415–31155–1
Pb: 0–415–31156–X

Taylor & Francis

Geotechnical Modelling
D. Muir Wood
Hb: 0–415–34304–6
Pb: 0–419–23730–5

Taylor & Francis

Information and ordering details

For price availability and ordering visit our website **www.tandf.co.uk/builtenvironment**

Alternatively our books are available from all good bookshops.

Introductory Geotechnical Engineering

An environmental perspective

Hsai-Yang Fang and
John L. Daniels

Taylor & Francis
Taylor & Francis Group

LONDON AND NEW YORK

First published 2006
by Taylor & Francis
2 Park Square, Milton Park, Abingdon, Oxon OX14 4RN

Simultaneously published in the USA and Canada
by Taylor & Francis
270 Madison Ave, New York, NY 10016

*Taylor & Francis is an imprint of the Taylor & Francis Group,
an informa business*

© 2006 Taylor & Francis

Typeset in Sabon by
Newgen Imaging Systems (P) Ltd, Chennai, India
Printed and bound in Great Britain by
TJ International Ltd, Padstow, Cornwall

British Library Cataloguing in Publication Data
A catalogue record for this book is available from the British Library

Library of Congress Cataloging in Publication Data
Fang, Hsai-Yang.
    Introductory geotechnical engineering : an environmental
perspective / Hsai-Yang Fang and John Daniels.
        p. cm.
    Includes bibliographical references and index.
      1. Environmental geotechnology. 2. Engineering geology.
    I. Daniels, John, 1974– II. Title.
    TD171.9.F36 2005
    624.1'51–dc22   2004015398

ISBN10: 0–415–30401–6 (hbk)
ISBN10: 0–415–30402–4 (pbk)

ISBN13: 978–0–415–30401–6 (hbk)
ISBN13: 978–0–415–30402–3 (pbk)

Julia S. Fang and Julie K. Daniels
For their encouragement and support

Contents

Figures

Tables

Preface

At the present time, the subject of *geotechnical engineering* stands at a crossroad. One road still dogmatically follows the classical concept developed by K. Terzaghi, and the other adopts a multidisciplinary approach. Motivation for the latter, as the emphasis of this text, is derived from frequently encountered field situations that challenge classical concepts and methods for analyzing soil behavior under varied environmental conditions. Put simply, soil mechanics alone cannot sufficiently explain all soil–water–environment phenomena and soil–structure interactions present in the modern world. While classical concepts will always serve as the "foundation" of geotechnical engineering, adjustments need to be made to evolve the profession into one that can better face increasingly complex situations. To cope with this issue, a compromise approach that incorporates the recently developed *particle-energy-field theory* is introduced in this textbook. In other words, this new textbook is presented in the classical framework with new information blended into it as necessary.

This book is intended to serve as a textbook for the required first year undergraduate geotechnical engineering course. In all cases, essential and conventional information is included in the text. For example, standard soil classification together with identification and classification of contaminated soil are included. Soil properties such as shear strength, soil dynamics, consolidation and settlement, bearing capacity, lateral earth pressures, and slope stability as influenced by both standard and environmental effects are included. Two new chapters, the thermal and electrical properties of soils and cracking–fracture–tensile behavior of soils are added. Experimental data for both laboratory and in situ conditions, together with numerical examples, are also included.

Critical current soil mechanics concepts and methods

Since 1925, the concept of soil mechanics has made rapid strides into being a major discipline in the civil engineering field. When Terzaghi introduced the concept of soil mechanics into the civil engineering field, it became a major subject in instructional curricula. The basic concepts and theorems have been established which greatly improved modern design and construction technology in civil engineering. These approaches are outlined as follows: (a) soil constants such as Atterberg limits (Ch. 2) and specific gravity (Ch. 3) for given soil under any conditions are assumed as constant; (b) constitutive models based on soil's stress–strain relationship (Ch. 10) often fail to accurately describe real soil behavior. In some cases, the assumption may hinge on an individual's preference, not based on the soil behavior; (c) commonly used concepts are the void ratio or porosity (Ch. 3) as indicators of the deformation under load; (d) The water content in the soil mass is mainly based on gravity water (free water), while other types of water in the pore space such as environmental water (Ch. 3) are not included; and (e) flow through a soil mass considers the hydrostatic potential only. Other causes such as thermal, electrical, phase changes are not considered in the analysis and design. The following observations are offered:

1   Current research and instructional efforts in geotechnical engineering places very little effort on the other factors besides load and short-term investigations. Since soil is an interdisciplinary science, not a simple mechanical system, current mechanical approaches may lead down the wrong track;

2   Soil mechanics itself has no unified theory or concepts to analyze all soils under various environmental conditions. For example, the concepts of bearing capacity are based on the plasticity theory. However, when examining the vertical pressure distribution of soil, the elastic theory is applied. For computing settlement, the Terzaghi consolidation theory is used which follows heat conduction concepts. For slope stability analysis, the majority of investigators follow the limit equilibrium (Sec. 12.2) concept but some use limit analysis (Sections 12.2 and 14.8);

3   Heavy emphasis is placed on mathematical manipulation to show how a soil can fit into a mathematical model rather than how mathematics can assist in understanding soil behavior. Most of the constitutional models of soil serve only an academic interest and are not useful for practical applications;

4   Many premature or progressive failures frequently occur. Most of these failures cannot be explained by current concepts or methods. For example, the Terzaghi consolidation theory only considers the load, other factors such as chemical, physicochemical and microbiological factors are not included in this theory and cannot be estimated;

5   Most designs for geotechnical projects hinge on a loading condition. Since loading is not the only controlling factor, design criteria based on the load factor alone do not give the whole picture and neglect an important factor which does control the overall stability of all civil engineering structures – the environmental factor design criteria. The load is an independent parameter, but the ground soil is a dependent variable which fluctuates with local environmental conditions;

6   Over emphasis is placed on mechanical aspects of soil behavior as indicated in Table 1.2 and very little effort is placed on environmental factors as discussed.

Notwithstanding the foregoing discussion, the trend in geotechnical instruction effort has been to consider only the physical and mechanical behavior of soil. In fact, in some institutions soil mechanics courses have become a part of the engineering mechanics discipline, which implies that the fundamental aspects of soil behavior have been ignored. Since soil is extremely sensitive to environmental conditions, its study should encompass areas of soil science, physical chemistry, mineralogy, geology, microbiology, etc. In 1980 the scope of soil mechanics expanded to include rocks, marine sediments, and the title of soil mechanics changed to geotechnology; however, the analysis approach still is dominated by the mechanical energy, that is loading alone. In the text to follow the authors attempt to generalize the soil and rock properties under diverse environmental conditions using *the particle-energy-field theory* allowing environmental conditions to be divided into five basic energy fields, namely mechanical, thermal, electrical, magnetic, radiation, and soil–water behavior within these fields. The mechanical field (loading) of soil behavior is the major part of current soil mechanics and all the other fields are considered due to variable environmental conditions (e.g. temperature, moisture content, pore fluid). Using the general framework created by K. Terzaghi, the new data are presented with detailed explanations and comparisons with existing theories and/or concepts.

Acknowledgements to Prof. Ronald C. Chaney, Prof. Wai-Fah Chen, Prof. Hilary I. Inyang, Prof. Abidin Kaya, Prof. Tae-Hyung Kim, Prof. Ian K. Lee, Prof. Horace Moo-Young, Jr., Prof. Leonardo Zeevaert, Prof. Rajaram Janardhanam, and Prof. Thomas M. Zimmie for their review and suggestions; to AASHO Road Test, National Research Council and Fritz Engineering Laboratory, Lehigh University for permission to use experimental data; to Ms Eleanor Nothelfer for assistance in all phases during preparation of the manuscript and the galley proof readings. Also, many thanks is given to the undergraduate and graduate students at the University of North Carolina at Charlotte who helped with the review and editing, namely, Mr Raghuram Cherukuri, Mr Gautham Das, Mr Nick DeBlasis, Mr Chris Friel, Ms Umamaheshwari Udayasankar, Mr Gabriel Molina, Mr Robie Goins, and Mr Harold Smith.

Hsai-Yang Fang
John L. Daniels
August 2005

Note to instructors

Scope and organization of the text

As stated in the preface, the main purpose of this text is to present geotechnical engineering with a combined approach that is based on a classical framework with new information blended into it as necessary. This is why the particle-energy-field theory is introduced in this textbook. The text contains sixteen chapters and it can be categorized into three groups:

1   Basic concepts of both classical soil mechanics and the proposed particle-energy-field theory are presented. In the analytical procedures, both limit equilibrium and limit analysis techniques are discussed. In addition, two new topics namely thermal–electric–magnetic characteristics and cracking–tensile–fracture of soils are added to traditional soil mechanics. These subjects are primary environmental factors which affect the soil–water system in the environment;

2   Comparisons highlighting the importance of environmental effects on soil and rock as related to various basic soil mechanics concepts such as compaction, consolidation, shear strength, dynamic properties, bearing capacity, and lateral earth pressures;

3   Illustration of these environmental aspects by using various ground improvement methods such as reinforced earth, geosynthetics, anchors, nailing, and pile foundations. Environmental geotechnical problems such as wetlands, marine margins, erosion, soil decontamination as well as antidesertification measures are discussed. Waste control and reuse of wastes is an important subject and presented as a separate chapter.

4   Numerical examples and problems are also provided in each chapter. The book is intended to serve as a standard first year undergraduate textbook. In all cases, core fundamentals are included in the text. For example, standard soil classification together with identification and characterization of contaminated soil are included. Soil properties such as hydraulic conductivity (Ch. 5), compaction (Ch. 7), consolidation, stress distribution and settlement (Ch. 9), shear strength (Ch. 10), soil dynamics (Ch. 11), bearing capacity (Ch. 12), earth pressure (Ch. 13), and earth slope stability (Ch. 14) under both standard and environmental aspects are

presented. Two new chapters are added, given as thermal–electrical characteristics (Ch. 6) and cracking–fracture–tensile behavior of soils (Ch. 8) are also included.

5   In the interest of covering the standard first semester course worth of material, some chapters such as cracking–fracture–tensile behavior of soils (Ch. 8), dynamic properties of soil (Ch. 11), problems in environmental geotechnology (Ch. 16) may be omitted.

Chapter 1

Introduction to geotechnical engineering

1.1 Introduction

Geotechnical engineering is the systematic application of principles and practices which allow construction on, in, or with earthen material. Virtually all civil infrastructure is in direct contact with soil and as such is dependent on the geotechnical properties. Throughout civilization, there has been the need for constructing buildings, roads, dams, bridges, and other structures. Foundation design was historically a trial and error enterprise where no effort was made to quantify or predict soil behavior. A common example of the consequence of this approach is given by the Leaning Tower of Pisa, which prior to recent corrections, was tilted at 5.5° from the vertical due to unanticipated differential settlement. The first rational approach to working with soils came from Charles Coulomb who worked with soils in retaining wall applications for the French army in the latter part of the eighteenth century. A more comprehensive contribution to the field, and what is often noted as the birth of geotechnical engineering, is Karl Terzaghi's 1925 text, named in part "Erdbaumechanik," which may be thought of as the first geotechnical textbook. Still, there were many more whose efforts and work have made the profession what it is today.

Currently, geotechnical engineering has emerged as a well-developed field that interfaces with many other engineers and professionals. Clearly, the work of the geotechnical engineer in estimating settlements and designing foundations is of interest to the structural engineer and the architect in connection with building construction. Similarly, geotechnical work performed to retrieve soil samples and characterize subsurface properties is important for groundwater quality and control where interaction with environmental engineers and hydrogeologists is likely. Other projects for which the services of a geotechnical engineer are needed include designing dams, embankments, landfills, and assessing the stability of slopes. There are many opportunities for geotechnical engineers to find work with private consulting companies as well as state agencies and academia. In short, there will always be a need for understanding and designing with soil.

Although significant advances have been made in geotechnical engineering since the days of Terzaghi, many solutions are at best an approximation, mostly because of the heterogeneous nature of both the soil and prevailing environmental conditions. The word "Environmental" has come to mean many things to different groups. Applied herein, it refers to ambient conditions that are reflected by such variables as temperature, pressure, groundwater composition, microbial population, etc. Soils do not exist in a vacuum, and they are the product of a variety of ongoing physical and chemical weathering phenomena. While some properties remain constant, others are subject to change as a function of mineralogy and environmental conditions.

In addition to being inherently complex, soil is more sensitive to the local environment than other construction materials such as steel or concrete. When soil is combined with water to varying degrees above or below the groundwater table, the result is a multiphase soil–water–gas system. This system may be thought of as a miniature reactor wherein a variety of physical and chemical processes occur within these phases. More details of the relevant reactions and specific properties will be presented in subsequent chapters, however at this point it suffices to note that soil is an engineering material that can change dramatically with time and space. As such, we must make an effort to understand as much as possible about soil and its response to the local environment if we are to make accurate predictions of the engineering behavior during the service life of a particular project.

1.2 Need to study geotechnical engineering from an environmental perspective

In recent years, due to population growth, progressive living standards, and industrial progress, soils that are of good quality (e.g. in terms of strength, compressibility, or permeability) and clean (e.g. free of contamination by metals or organics) are becoming harder to find. Thus, the geotechnical engineer is called upon more frequently to work with sites that would otherwise be rejected because of some deficiency. To work with soils that are physically or chemically deficient requires a broader, environmental perspective.

Geotechnical engineering is actually an interdisciplinary science and one that requires an assessment of mechanical (loading) as well as the response to fluctuations in the local environment. These fluctuations may be summarized as chemical, physico-chemical, and microbiological including such processes as (1) ion exchange reactions (Sec. 4.7) in the soil–water system that can change the arrangement of soil particles; (2) crack formation which fragments the soil surface and arises from an energy imbalance caused by natural variations in moisture or temperature as well as variations in compaction energy during construction. The cracking patterns (Sec. 8.3) have a significant effect on prefailure (Sec. 10.4) characteristics of soil as well as the flow through saturated and unsaturated (Sec. 5.11) fine-grained soils; (3) For a given soil under in situ conditions, the stress–strain behavior can change from elastic to plastic, or from a softening or hardening process, if certain local environmental conditions change; and (4) Bacteria (Sec. 4.12) can influence the character of the pore fluid and can also impact particle contacts through the production of exocellular substances.

In analyzing the soil behavior for practical application at present, most project designs use the test results following American Society for Testing and Materials (ASTM) and American Association of State Highway and Transportation Officials (AASHTO) standards. These standards are important and will be discussed in subsequent chapters. However, many of them are based on controlled conditions at room temperature, often with distilled water or low concentration electrolyte (e.g. $CaSO_4$) as the pore fluid, in part to insure uniformity of results and test repeatability.

Also, many analyses concentrate on loading conditions tested under short-term duration conditions but projected into long-term performance. Since field conditions and the standard control condition are significantly different, many premature or progressive failures are difficult to predict on the basis of controlled tests alone.

1.3 Environmental geotechnology and geoenvironmental engineering

Those new to the field or even rigidly trained in geotechnical engineering may be confused by the "environmental perspective" proposed herein as it relates to other rapidly emerging areas, namely environmental geotechnology and geoenvironmental engineering. In particular, geotechnical engineering was defined at the beginning of the chapter in terms of engineering with soil and soil–structure interaction. An environmental perspective simply interprets and modifies these results in light of the relevant site-specific and time-dependent environmental influences, that is, it attempts to reflect more accurately the actual in situ behavior of soil. This is in contrast to environmental geotechnology or geoenvironmental engineering, which are discussed as appropriate in the text and summarized as follows.

1.3.1 Environmental geotechnology

Environmental geotechnology has been defined as an interdisciplinary science which includes soil and rock and their interaction with various environmental cycles, including the atmosphere, biosphere, hydrosphere, lithosphere, and geomicrobiosphere (Fang, 1986, 1997). The latter includes trees, vegetation, and bacteria as they influence soil behavior. By definition, the emphasis in geotechnology is broad in scope and includes elements of fields beyond civil or geotechnical engineering such as soil science, material science, and geology. Environmental geotechnology has grown quickly since the first international symposium was organized in 1986 at Lehigh University. Environmental geotechnology is not only of relevance to traditional geotechnical problems but also has been expanded to include (a) hazardous/toxic waste control; (b) wetlands, coastal margins, dredging and marine deposits; (c) arid and desert regions; and (d) sensitive ecological and geological environments as well as archaeological science and technologies.

1.3.2 Geoenvironmental engineering

Geoenvironmental engineering may be considered the part of environmental geotechnology that deals with geological, geohydrological, and geotechnical aspects of environmental engineering problems. Common examples relate to the containment and remediation of municipal, hazardous, and nuclear waste in soil and groundwater, including: (a) hazardous/toxic waste controlling systems such as hydraulic barriers and various types of containment systems; (b) various aspects of landfill problems including selection of landfill sites, compaction control, stability analysis, settlement prediction of landfill, and design and construction of barrier, top seal (cover, cap) and bottom seal (liners); (c) geological and hydrogeological considerations of pollution control systems of groundwater aquifers; (d) soil and groundwater remediation technologies including immobilization and in situ treatment such as solidification, stabilization, and vitrification; and (e) utilization of waste materials in civil engineering construction. Some of these aspects will be discussed in Sections 15.5, 15.6, and Chapter 16.

1.4 The particle-energy-field theory

As the foregoing suggests, an analysis of soil behavior indeed requires an environmental perspective. As a basis for this perspective, a new approach entitled *the particle-energy-field theory* is proposed (Fang, 1989, 1997) for a unified approach for analyzing soil behavior under various environmental conditions. The main purpose for developing this theory is to link otherwise unrelated phenomena into one system that reflects in situ conditions.

1.4.1 Assumptions and approaches

The particle-energy-field theory consists of three major components: (a) elementary particles; (b) particle systems; and (c) energy fields. The combination of these three components into one system is called *the particle-energy-field theory*. Basically, the theory combines the concepts of solid state physics and chemistry on one side; organic chemistry, physical chemistry, and microbiology on the other side. Interacting between these two groups is the common denominator known as the "particle." Particles are the fundamental building units of all types of materials including soil, water, gas, and pollutants. In addition, environmental phenomena such as ion exchange reactions, absorption, adsorption, soil–bacteria interaction, etc. which pose difficulties to an approach without an environmental perspective are incorporated in this theory. The particle-energy theory is based on the following assumptions, some of which may require the student to revisit their chemistry text:

1 that the physical world is constructed of particles such as atoms, ions, molecules, macro- and micro-particles;
2 these particles may attract or repel each other depending on their electromagnetic forces and structures;
3 bonding energies such as ionic, covalent, chemical bonding and linkage such as cation, water dipole, dipole-cation control the stress–strain–strength and durability between particles;
4 energies such as kinetic, potential, heat, electrical, magnetic, and radiation are caused by the relative movement of these particles;
5 particle systems can be:

   a   solid state if attraction (A) > repulsion (R)
   b   liquid state if attraction (A) ~ repulsion (R)   (1.1)
   c   gaseous state if attraction (A) < repulsion (R)

1.4.2 Particles, particle systems, and bond energies

1 Elementary particles: Elements are composed of tiny, fundamental particles of matter called *atoms*. Ordinarily atoms are neutral, that is, they do not carry an electrical charge. However, under certain circumstances, atoms can become electrically charged. Such charged atoms are called *ions*. Some elements form positive ions, called *cations*, and some form negative ions, called *anions*. The atom as a basic particle of matter is composed of still smaller particles called *subatomic* particles. The neutron, electron, and proton are classified as subatomic particles. Positive subatomic particles are present in the atom and are called as *protons*. Units of negative charges are known as *electrons*. A third subatomic particle found as a constituent of atoms which carries no electrical charge (neutral) is known as the *neutron*. The sharing or transfer of a pair of electrons binds the atoms together to form a new kind of particle called a *molecule*. Molecules are stable particles and are characteristic chemical particles of many compounds. Table 1.1 presents basic types of particles which serve as building units of matter.

2 Particle systems: Since the physical world consists of three states of matter, solid, liquid, and gas (air), any other elements existing are these in combinations. Basic physics and chemistry indicate energy gradients are the main causes for particle movement from one place to another. Particle motion, whether it is monotonic or dynamic, originates from particle behavior under energies such as potential, kinetic, thermal, electrical, magnetic, etc.

3 Particle Strength and Bonding Energy Between Particles: There are two major types of bonds existing within atoms and molecules comprising soil particles: the *primary bond* and the *secondary bond*. The primary bond is what combines atoms together to form molecules. The secondary bond occurs when the atoms in one molecule or ion bond to another. Bond energies are normally expressed as kcal per mole of bonds. By division through the Avogadro Number (6.025×10^{23}) one obtains the energy per single bond which can be converted into ergs or other appropriate energy units. Finally, by dividing through the length of bond or an appropriate multiple thereof, one can obtain the bond force which when divided by the pertinent

Table 1.1 Basic types of particles which serve as building units of matter

---

A *Subatomic particles*
   (a) Electron (negative charge)
   (b) Proton (positive charge)
   (c) Neutron (neutral)

B *Atom (neutral)*
   Such as Carbon (C), Hydrogen (H), Magnesium (Mg),
   Nitrogen (N), Oxygen (O), Sodium (Na)

C *Ions (charged atom)*
   (a) Cation (positive charge)
       such as Magnesium ion (Mg^{2+}), Sodium ion (Na^+)
   (b) Anion (negative charge)
       such as Chloride ion (Cl^-), Oxide ion (O^{2-})
   (c) Polyatoms ions
       Groups of covalently bonded atoms with varying charges
       such as Carbonate ion (CO_3^{2-}), Hydroxide ion (OH^-), Nitrate ion (NO_3^-)

D *Molecules (neutral)*
   A group of covalently bonded atoms
   such as Ammonia (NH_3), Hydrogen Chloride (HCl), Methane (CH_4)

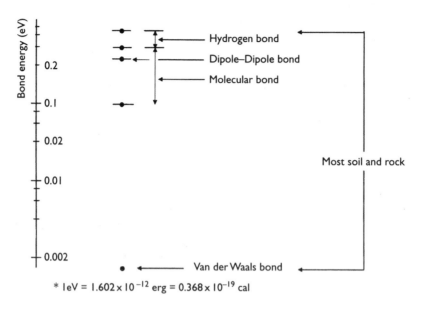

Figure 1.1 Ranges of particle bonding energy for common types of soil and rock.

molecular or ionic cross-section gives the bond strength in force per unit cross-section. By way of example, the bond energy of the secondary dipole–dipole bond in water is 4.84 kcal per mole of water which is about one half of the 9.7 kcal required to evaporate one mole of water at its boiling point under normal atmospheric pressure. The individual bond strength is of the order of 10^4 kg/cm^2 (1.4×10^5 psi). The range of bonding energies for common types of soil and rock is presented in Figure 1.1.

4   *Attractive and repulsive forces between particles:* All clay particles carry an electrical charge. Theoretically, they can carry either a net negative or net positive charge, however, only net negative charges have been measured. When two particles are close to each other in face-to-face arrangement, an attractive force exists between the negatively charged surfaces and the intervening exchangeable cations. If the atoms in an adjacent surface approach each other so closely that their outer electron shells overlap, a net repulsion force results. When the various attractive and repulsive energies are summed algebraically, the net energy of interaction is obtained. Both attractive and repulsive forces are important to the soil–water behavior and their interaction with the environment. The several methods for measuring or computing these forces are discussed and summarized by Pauling (1960) and Low (1968).

1.4.3   Energy, energy charge, energy field and particle energy field

1   *Energy and energy charge:* Energy is the quality possessed by an object that enables it to do work. The source of energy is the energy charge such as E_1 and E_2

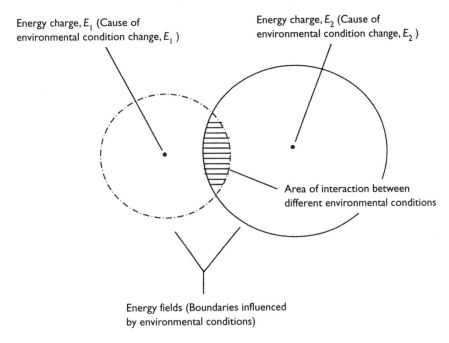

Energy charge, E_1 (Cause of environmental condition change, E_1)

Energy charge, E_2 (Cause of environmental condition change, E_2)

Area of interaction between different environmental conditions

Energy fields (Boundaries influenced by environmental conditions)

Figure 1.2 Relationship between energy charge and energy field.

indicated in Figure 1.2. These energy charges are the impetus for a change in environmental condition. The energy charge can be derived from surface force and body force. The surface force creates an energy source including potential, kinetic, thermal, electrical, magnetic, and radiation, as well as the body force (i.e. gravity). Further discussion on gravity force will be presented in Section 1.8.4.

2 *Energy field:* An energy field is defined as a space in which each energy charge reacts with another energy charge or the boundaries influenced by environmental conditions as illustrated in Figure 1.2. In other words, the energy field is an area of influence in the vicinity of the energy charge and the interaction among the other energy charges. From a geotechnical viewpoint, the energy field is called the *influence area* which is influenced by the energy charges. For example, when driving piles for deep foundations, the energy charge is the drop hit on the pile which is mechanical potential energy. The shaded area indicated in Figure 1.2 is the area of interaction between different environmental conditions, also called the *interaction zone*. Here, the combined influence of both charges is observed.

3 *Particle energy field:* The particle energy field is the collection or assemblage of individual particles in space which interact and exhibit surface and/or body forces. For practical purposes, let the energy fields or particle energy fields be divided into five basic groups, namely (a) mechanical energy field (including the Potential Energy Field, that is, energy of position), kinetic energy field (Energy of motion); (b) thermal energy field; (c) electrical energy field; (d) magnetic energy field; and (e) radiation energy field.

1.4.4 Particle behavior in various energy fields

To evaluate particle behavior in various energy fields, proper laws, theory, or principles are required as indicated in Table 1.2. For example, flow movement due to a hydraulic gradient (Sec. 5.4) will follow Darcy's Law, however, if flow movement is caused by a thermal gradient (Sec. 6.3), then it should follow Fourier's Law, and if it is due to an electric potential (Sec. 6.8), then it should follow Ohm's law. Because environmental conditions change, soil behavior will also change, consequently, the method of interpretation must also change. There are five basic energy fields stated in the Table 1.2. Although each energy field has its own identity with individual characteristics, they are interconnected and may operate simultaneously in the long-term as shown in Table 1.3. Detailed discussions of these effects will be presented in Section 1.9.3 and Chapter 6.

1.5 Particle energy field and environment

Particles are the basic structural units for all materials, however, each particle reacts differently at various energy fields. In other words, particles respond to various environments differently. As indicated in Section 1.3.4, and Figure 1.3, there is a

Table 1.2 Law/theory required for evaluation of particle behavior in various energy fields

Major elements in each energy field	Law or/and theory required for evaluation
A *Mechanical energy field* (potential and kinetic)	
Load, deformation	Hooke's law
Weight, mass	Newton's law
Fluid in motion	Darcy's law
Velocity, acceleration, wave, sound	Laws of motion
B *Thermal energy field*	
Hydration	Laws of thermodynamics
Heat of wetting	Fourier's law
Kinetic dispersive force,	General gas law
Thermal conductivity and resistivity	
Thermoosmosis	
C *Electric energy field*	
Polarization,	Ampere's law
Proton migration	Coulomb's law
Electromotive force	Joule's law
Electric conductivity and resistivity	Ohm's law
Electrophoresis, electroosmosis	
D *Magnetic energy field*	
Electromagnetic	Faraday's law
Ferromagnetism	Lenz's law
Electromagnetic induction	Biot-Savart law
Electromagnetic waves	Gauss's law
E *Radiation energy field*	
Decay process	Atomic physics
Radioactivity, nuclear reactions	Nuclear physics
Fundamental forces	

Table 1.3 Long-term implications of particle energy fields and examples of their interaction

	At construction (initial condition)	After construction (possible long-term effects)			
Example of system change	Structure/ surcharge loading	Fluctuating temperature	Variable soil oxidation/ reduction potential	Variable iron content/ ferromagnetism	Emanating radon gas
Energy field	Mechanical	Thermal	Electrical	Magnetic	Radiation

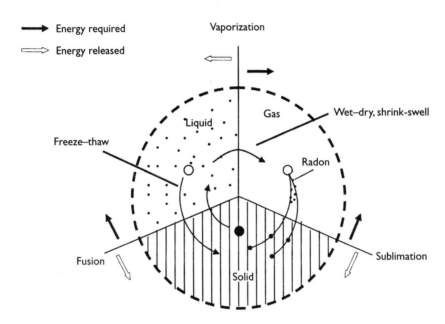

Figure 1.3 State of matter: solid–liquid–gas phases in thermal energy field, indicating wet–dry, freeze–thaw and radon gas relationships.

Source: Fang (1997).

similarity between environmental phenomena and energy fields. Therefore, the environment can also be divided into five environmental zones such as mechanical, thermal, electrical, magnetic, and radiation zones as suggested and discussed by Fang (1992). Further explanations of soil behavior in each energy field or each environmental zone including state of matter and inter-phases are presented as follows.

1.5.1 State of matter in thermal energy field

There is a unique relationship between the state of matter and the thermal energy field. The physical world consists of three major states of matter: solid, liquid, and gas as shown in Figure 1.3. In examining Figure 1.3, there are three basic heating systems which control the change of state of matter, namely (a) heat of fusion (solid to liquid); (b) heat of sublimation (solid to gas); and (c) heat of vaporization (liquid to gas). When a change of state of matter occurs, energy is either required or released as indicated in Figure 1.3. Soils are commonly subjected to wet–dry and freeze–thaw cycles in response to seasonal and diurnal temperature and moisture fluctuations. Further discussions on these phenomena will be presented in Sections 5.2 and 5.10. Among these three heating conditions, the heat of sublimation phenomenon is the most interesting. The common example for this phenomenon is dry-ice and moth-balls. In environmental geotechnology applications, this phenomenon also occurs in the uranium (U)–radium (Ra)–radon (Rn) system (Sec. 16.8).

The soil system is generally in some multiphase state. If the soil is dry and in a vac-uum, it is in a solid state; when it is saturated, it becomes a two-phase system; if soil is partially saturated, it is in a three-phase system involving solid, liquid, and gaseous states. Regardless of the state of matter, the micro-structure is composed of particles. Stress–strain relationships of soil hinge on the bonding behavior of two or more par-ticles. The water content of the soil and flow of water through soil are dependent on the energies between particles. Since particles are the basic structural units for all materials, the particle-energy-field theory can be used for explaining the engineering behavior of soil under various environmental conditions. A conceptual diagram is presented in Figure 1.3 that shows the state of matter changes during cycles of wet–dry (Sec. 4.2), shrink–swell (Sec. 4.4), freeze–thaw (Sec. 6.7) as well as the phe-nomena of radon gas relative to radium (Ra), radon (Rn), and the radon daughter (P0). Further explanations on why and how the state of matter changes in the ther-mal energy field will be presented in Sections 4.2 and 16.10.

1.5.2 Solid–liquid–gas interface

1 *Single-phase interface:* It covers liquid–liquid, solid–solid, and gas–gas. Among these three cases, the liquid–liquid interface occurs most commonly as clean water interacting with polluted water, saltwater intrusion, and oil–water mixtures. In solid–solid interfaces such as dry sand–gravel mixtures, coal, crushed stone, if moisture is present between them, then the single-phase interface becomes the double-phase or even the multiphase interface. Gas–gas (air–air) interfaces can be evaluated by the kinetic molecular theory, however, in many cases, gas particles will be absorbed by a solid such as dust (Sec. 3.11), then the behavior of gas–gas becomes a gas–solid interface. Oil–water interface is more complicated than any other single-phase interface because oil itself lies between liquid–solid–gas form. The degree of consistency of oil itself will affect oil–water interface mechanisms.

2 *Two-Phase Interface:* In the two-phase interface, the characteristics of adsorption (Sec. 4.4) play an important role. Some natural soils such as sandy silts or silty sands have inter-particle contacts joined by moist cohesive (clay) soil to form composite particles. The linkage between two particles is through adsorbed water, water dipole, or dipole–cation–dipole (Sec. 3.6). In many cases, they are only temporary, and once the soil becomes dry, the linkage force between two particles can be dismissed.

3 *Multiphase Interface:* The soil–water interaction is commonly treated as two-phase interface. However, in the natural case, this interaction is a multiphase interface because whether or not soil is saturated or dry, it always contains some gases. Other cases include water-repellent soils (Sec. 3.9), where water movement is in a water-repellent soil and the wetting phenomena is a vapor–liquid–solid interaction. All types of polluted transport in the soil–water layers belong to this group.

Multiphase phenomena also occur in natural environments. Water vapor exists in the soil–water system due to the relative humidity of the air in soils. The pressure of the water vapor in the soil voids increases with temperature. In general, water vapor moves from the warmer zone and condenses in the cooler soil. For example in the summer season, hot weather warms the soil to considerable depth, followed by a cool spell which cools the surface soil rapidly. As a result, appreciable amounts of water vapor move up from the warm soil below and condense in the upper soil layer. Such movement may also occur in the autumn season when the lower soil horizons have not yet cooled to the temperature of the surface soil. Likewise, some moisture may condense onto the soil surface from a warm atmosphere with high humidity.

1.6 Particle behavior under load

The Law of Conservation of Energy states that energy cannot be created or destroyed but rather is transformed from one form to another. We also know from basic physics and chemistry that energy gradients are the main causes for particle movement from one place to another. Particle motion, whether it is monotonic or dynamic, originates from particle behavior under energies such as radiation, heat, electrical, potential, kinetic, etc. Basic types of load used in geotechnical engineering are static (e.g. foundation) and dynamic (e.g. earthquake or vibration) loads. Indeed, the response of soil to these types of loading conditions is of prime importance in geotechnical engineering and remains the focus of this text. However, it should be noted that most mechanical energy field related problems are considered short-term, with the exception of excess pore pressure dissipation (Ch. 9). Moreover, the influence of local environmental conditions is often neglected. Unfortunately, most geotechnical projects occur in nature and, therefore, must be considered as long-term installations constructed outdoors where they will be open to various environmental effects. Further discussions on these aspects will be presented in Chapters 4 and 5.

1.6.1 Particle behavior under mechanical load

1 *Potential load*: Mechanical load or mechanical energy includes both potential and kinetic energies which dominate today's geotechnical engineering concepts and approaches. It is true that mechanical energy plays the most important role relating to the performance of all geotechnical engineering projects as illustrated in Table 1.4. Potential energy derives from some type of loading which includes compaction, consolidation, distortion, bending, crushing, kneading, shearing, and other processes.

2 *Kinetic load*: It is caused by kinetic energy, the energy of motion. Flow through soil or other porous media is a typical case of particle behavior under kinetic load which is characterized by capillarity, hydraulic conductivity, and seepage pressure. Vibrations from heavy equipment such as turbines and construction vehicles as well as seismic activity represent kinetic loads.

1.6.2 Particle dynamics

The basic parameters of particle dynamics are velocity, acceleration, mass, force, work, energy, wave, vibration, etc. In a liquid or gas, compression waves are called *sound waves*. The characteristics of sound waves include the pulse, frequency, and type, that is transverse or longitudinal. When Newtonian mechanics is applied to the motion of a system, it is found that motion can be regarded as a wave motion called *normal modes* of vibration. The frequency of oscillation in a normal mode is termed as the natural frequency of the system. The lowest natural frequency is called the *fundamental frequency*. When the driving frequency is near a natural frequency of the vibrating body, the amplitude of these forces oscillating becomes exceptionally large. It is for this reason that knowledge of the natural frequency of a structure is of particular importance when assessing seismic stability. The large response at a certain driving frequency is called *resonance*. A great variety of particle resonance is possible in natural systems.

In many geotechnical engineering projects, knowledge of the dynamic behavior of soil is needed. Such projects include compaction (Sec. 7.3), dynamic compaction (Sec. 7.8), earthquake loading (Sec. 11.2), wind, wave, current (Sections 11.7–11.8), machine vibration (Sec. 11.10), blasting (Sec. 11.11), pile driving (Sec. 15.12), and many others; likewise, soil–structure or structure–soil interaction problems can be interpreted by dynamic behavior of particles (Sec. 15.3).

1.6.3 Gravitational force

Gravitational force (F_G) is one of the basic forces in nature, and it is always attractive. The law of universal gravitation was discovered by Newton in 1686. It may be stated as: Every particle of matter in the universe attracts every other particle with a force that is directly proportional to the product of the masses of the particles and inversely proportional to the square of the distance between them.

$$F_G = \frac{G\,m_1 m_2}{r^2} \tag{1.2}$$

where F_G = force, m_1, m_2 = masses, r = distance between particles, and G = gravitational constant. The numerical value of the constant, G, depends on the units in which force, mass, and distance are expressed. Since the constant, G, in Equation (1.2) can be found from measurements in the laboratory, the mass of the earth may be computed. From measurements on freely falling bodies, we know that the earth attracts a 1 g mass at its surface with a force of about 980 dynes or 9.8 m/s^2. The gravitational field is a condition in space setup by a mass to which any other mass will react.

1.7 Particle behavior in multimedia energy fields

1.7.1 General discussion

With time after application of a given load, the soil behavior may no longer be controlled by the initial mechanical energy. Changes in the ambient environment as noted by temperature changes, cycles of freezing–thawing or wetting–drying, or pore fluid composition, etc. will change the soil particle characteristics. Depending on the specific change, these fluctuations give rise to the other energy fields, namely the

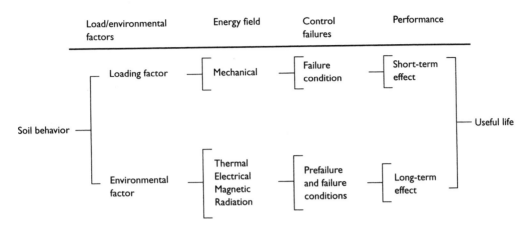

Figure 1.4 Effects of load/environmental factors on useful life of soil.

thermal, electric, and magnetic energy fields. A change from one energy field to another may be initiated by natural or anthropogenic activity. As discussed in Section 1.1, the mechanical energy alone cannot effectively explain all the geotechnical problems the modern world presents, therefore, a combined approach which includes environmental factors is needed. Figure 1.4 presents a flow diagram illustrating the effects of load/environmental factors on the useful life of soil. The useful life of soil is the result of both loading and environmental factors. Some of the relatively important sources affecting the soil–water behavior and their interaction are outlined in the following sections, and detailed discussions will be presented in Chapters 4 and 6.

1.7.2 *Particle behavior in thermal energy field*

The thermal energy field affects soil behavior in several different ways. Perhaps the most obvious occurs when the temperature drops sufficiently to freeze porewater in a soil system. This alone causes a volume expansion of approximately 10% in addition to possible ice lensing (Ch. 6). Other lesser known but significant thermal aspects of soil include

1 the forces produced when water is added to dry or partially saturated soil. Such forces include the kinetic dispersive force (Sec. 4.2.3) and heat of wetting force (Sec. 4.4.5). These forces are referred to as internal environmental forces or stresses;

2 the ability of the soil to retain or dissipate heat, which is dependent on its heat capacity and thermal conductivity. The heat transfer process in the soil is through three basic processes: conduction, convection and radiation, although primarily controlled by conduction;

3 the thermoelectric effect which was discovered by J. T. Seebeck in 1822. This is the phenomena of temperature gradients giving rise to electrical potential. His discovery of a novel method for the direct transfer of heat into electric energy became the phenomenon now known as the *Seebeck* or *thermoelectric effects*. Further discussion on this and related coupling processes with experimental data will be presented in Chapter 6.

1.7.3 Particle behavior under electric and magnetic energy fields

The electric energy field is central to all energy fields, and it plays an important role relating to the basic soil–water behavior. Some fundamental characteristics are outlined as follows and further discussions will be presented in Chapter 6.

1 *Polarization and proton migration*: These phenomena can be used for explaining the soil's stress–strain relationship especially for predicting stress-hardening and stress-softening processes (Sec. 10.9). Also, it can explain the creep behavior or rheological characteristics of soil. Geomorphic process (aging process) (Sec. 4.11) of soil/rock can also be evaluated;

2 *Electrokinetic process*: This process includes electroosmosis and electrophoresis (Ch. 6) for the purpose of ground improvement, subsurface drainage, dewatering, and soil decontamination;

3 *Electroviscous effect*: This effect can be used for explaining the internal cracking of soil mass (Ch. 8) which is related to progressive failure, surface erosion, as well as prediction of landslides potential (Ch. 14);

4 *Magnetic energy field*: The sources of this field are moving charges and electrical currents. Their distribution in a soil system is in a random pattern due to the bombardment of the dispersed particles by molecules of the medium traveling according to *Brownian movement*. When additional electric current is applied into the soil–water system, (as is done, for example, when a site is dewatered or decontaminated using electrokinetics) the particles remain in random motion, but the energy field boundary will change. Because of this, when two or more moving electric charges interact in the system, the thermoelectric energies change into thermal–electric–magnetic energies (Ch. 6).

1.7.4 Particle behavior in radiation energy field

Geotechnical problems interacting with the radiation energy field can be grouped into three general areas: (a) disposal or management of radioactive nuclear wastes; (b) control of radioactive radon gas (Sec. 16.10); and (c) utilization of gamma-rays in nondestructive testing methods. To tackle these problems, we must understand some atomic and nuclear physics including atomic, nuclear, and molecular structures, radioactive decay processes, and soil–rock interaction in the radiation energy field. Further discussions on this aspect will be presented in Section 16.10.

1.8 Justification and application of the particle-energy-field theory

1.8.1 Justification of the particle-energy-field theory

The particle-energy-field theory introduced in this text is mainly applied to geotechnical engineering. In nature, soil is normally composed of solid, liquid, and gaseous phases consisting of soil particles of various sizes, ranging from small boulders (0.3 m) to colloidally dispersed mineral and organic particles (<2 μm);

with the mineral character ranging from that of fragments of igneous, sedimentary, and metamorphic rocks through a wide range of weathering products to clay minerals and hydrous oxides. As discussed in Section 1.3.2, all matter whether solid, liquid, or gas is constructed from various types of particles, therefore, it is logical to use these particles as a common denominator for the evaluation of various problems in geotechnical engineering. Soil and water are very sensitive to local environments such as pollution, more than any other construction material. These chemical substances are also formed from various types of particles.

Explanation of the solid, liquid, and gaseous states of matter by the particle-energy-field theory represents the relationship between the volume of the solid particles and the volumes of the material as a whole in the solid, liquid, and gaseous states. It is a general concept that recognizes that all solid engineering materials are systems of interconnected particles. The behavior of particles that are interconnected depends on (a) sizes, shapes, and mutual arrangement of component particles of a system; and (b) cementing agents or forces acting to hold the particulate component together. Finally, it may be concluded that the particle-energy-field theory is a bridge to link these unrelated groups into a related system as illustrated in Figure 1.5 for practical applications in geotechnical engineering as well as other applications.

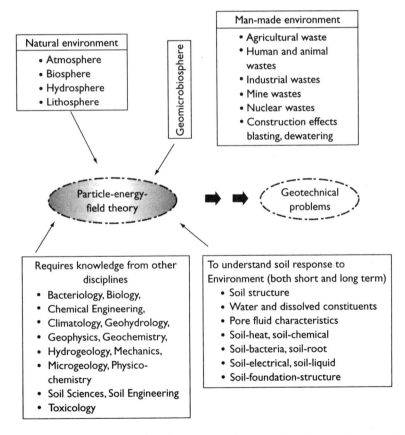

Figure 1.5 Particle-energy-field theory: a bridge to link these unrelated groups into a related system.
Source: Fang (1997).

1.8.2 Applications in geotechnical engineering

While the importance of environmental conditions on soil behavior has become more accepted, these concepts require further exploration and refinement before they will enjoy practical use. Practice requires a working formula or specifications for analysis and design of various projects in order to have service quality and durability of these facilities. Most environmental effects have not been studied enough to establish reliable relationships with soil. Currently, these effects are incorporated into a given design through use of a "Factor of Safety" that is coupled with experience. Typical design procedures also include: (a) careful planning and field investigations; (b) data collection, testing, and history review of material properties to make genetic diagnosis of the projects; and (c) development of localized factor of safety if site proves problematic. A brief discussion for designing for the environment is presented as follows:

1 *Basic considerations*: During planning stage, the following basic items must be considered such as (a) avoid direct pollution intrusion routes; (b) avoid great differences in thermal gradients; and (c) avoid great moisture transmission properties of the different constituent subsurface soil layers.
2 *Genetic diagnosis*: During analyses and design stage, the following items should be evaluated such as (a) mineral structure, (b) sensitivity of material and/or structural elements to environment (c) strength or loading history, and (d) several examples are illustrated in the text in terms of shear strength (Sec. 10.0), landslide analysis (Sec. 14.3) and landfill studies (Sec. 16.9).
3 *Development of localized factor of safety*: The localized factor of safety is a special type of factor of safety, which deals with certain types of soil or site that frequently appear as problematic with higher risk of potential failure. In such a case, the conventional factor of safety must be adjusted to suit the design need. Criteria for localized factor of safety based on genetic diagnosis will be discussed further in Section 12.4.

1.8.3 Identification and classification of geotechnical problems

1 *Identification and classification of parameters*: As discussed in Section 1.2, the analysis and design of geotechnical problems often centers on loading conditions with tests conducted for short durations while the results are assumed to represent long-term performance. However, an applied load is an independent parameter, while soil is a dependent variable that fluctuates with local environmental factors. Table 1.4 summarizes some geotechnical problems related to the various energy fields. The particle-energy-field approach can assist in visualizing a given problem. Details of each case listed in Table 1.4 will also be discussed throughout the text in each of the relevant chapters.
2 *Predicting long-term performance*: In the short-term, mechanical energy controls a large part of geotechnical engineering problems. As time progresses, soil behavior is no longer controlled by mechanical energy alone. Local environments such as temperature changes, freezing–thawing, wetting–drying, pollution intrusion, etc. will change soil particle characteristics. These characteristics, in turn, dictate the resulting soil behavior, as manifested by a possible change in shear strength, compressibility, or hydraulic conductivity.

Table 1.4 Identification of some geotechnical problems based on the particle-energy-field theory

Problems	Energy field	Notes
1 *Hydraulic conductivity* Macro-soil particle Micro-soil particle	Potential Multimedia	Ch. 5
2 *Volume change* Shrinkage Swelling	Thermal Multimedia	Ch. 4
3 *Sorption* Absorption (saturation) Adsorption	Kinetic Multimedia	Ch. 4
4 *Compaction* Dry-side Wet-side	Mechanical Multimedia	Ch. 7
5 *Consolidation* Primary Secondary	Mechanical Multimedia	Ch. 9
6 *Overconsolidated pressure* Caused by load Caused by environment	Mechanical Multimedia	Ch. 9
7 *Stress–strain–time* Stress-softening Stress-hardening Creep phenomena	Multimedia Multimedia Multimedia	Ch. 10
8 *Failure criteria* Prefailure Failure	Multimedia Mechanical	Ch. 10
9 *Friction resistance* Macro-soil particle Micro-soil particle	Mechanical Multimedia	Ch. 10
10 *Liquefaction* Macro-soil particle Micro-soil particle	Mechanical Multimedia	Ch. 11
11 *Earth pressure* Active Passive At rest	Mechanical Multimedia Mechanical	Ch. 13
12 *Landslide* Prefailure phenomena Failure stage	Multimedia Mechanical	Ch. 14

1.8.4 Guide in selection of parameters for correlation study

For the purposes of design, it is often necessary to relate some measurable property to another, perhaps more difficult property to measure. For example, loose relationships exist (Ch. 9) between compressibility and plasticity index. These relationships are useful in part because it is far more time consuming to determine the compressibility than the plasticity index. Moreover, to understand a particular process, and therefore learn how to control it for engineering purposes, it is often necessary to systematically investigate various parameters for correlations. Energy field considerations include the following:

1 Correlation of test results from two or more test methods: To correlate two or more parameters of soil properties or correlate results from two test methods, the natural characteristics of each parameter in various energy fields must be examined. Otherwise, larger variations between two parameters will be expected or can give meaningless results. For example, if one parameter is in the mechanical energy field and the other is in the multimedia energy field, the latter is more sensitive to the environment than the former; therefore, any observed relationship is likely to be inconsistent.

2 Correlation between theoretical and field test results: Field measurements are strongly influenced by the local environment, and most theoretical approaches are based on loading conditions with little consideration for the environment.

3 Correlation between laboratory and field test results: Most laboratory tests follow standards and are performed at room temperature with distilled water as the pore fluid; however, in the in situ condition, local environmental conditions can influence results significantly.

1.9 Soil testing

1.9.1 The importance of soil testing

Most construction materials such as steel and concrete used in civil engineering are well known and well-defined. Thus, except on the research level, any experimentation is generally done for confirmatory or quality control purposes. Soil, however, is a different story. In the first place, the fundamental controlling relations regarding soil behavior under normal conditions are uncertain. The second and equally important difference is that the soil constituency is variable and, except in a few cases, cannot be controlled. For these reasons, the role of experimentation takes on major importance, as it is the only manner of determining soil behavior. These tests are not confirmatory in nature, but are used to determine the actual or postulated soil reaction to environmental conditions for a given condition. Thus, the first and primary importance of a soil test is to solve a particular problem using a particular soil under its own special environmental conditions. While there are standards for field and laboratory testing, it should be noted that each test must be investigated and designed with special regard for the situations indigenous and peculiar to each problem. It is for these reasons that the geotechnical properties of soils are as important as the way they were measured. A soil testing program covers sampling, laboratory testing, field measurements, data collection, and presentation. Figure 1.6 lists various tests for obtaining a variety of soil properties and potential applications. However, in this primary textbook, emphasis is given to the basic principles.

1.9.2 Sampling techniques

Although sampling procedures and in situ testing methods are continually being refined, the basic types of tests have remained unchanged as discussed by Lowe and

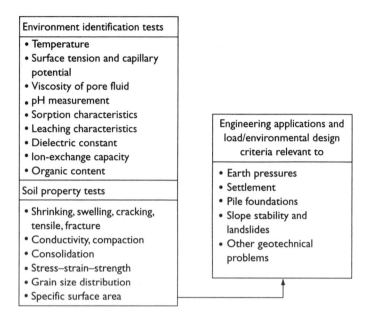

Figure 1.6 Tests with potential applicability in geotechnical design.

Zaccheo (1991). General outlines of each case are presented as follows:

1   *Disturbed sample:*   (a) Soil layers within the first 2–3 m (7–10 ft) of the ground surface can usually be inspected and sampled from test pits. Both high quality "undisturbed" block samples of cohesive soils and disturbed samples of all soils may be obtained. Disturbed samples within this zone may also be obtained by hand auger following ASTM D1452 (ASTM 2003). For explorations below a depth of 3 m (10 ft), it is normally advantageous to drill or bore a hole into the soil. Methods for advancing the hole include washing boring, rotary drilling, and percussion drilling; (b) washing boring is accomplished by pumping water at high velocity through the end of a drill pipe immersed in a cased or uncased hole. Although the soil washed out of the hole during boring cannot be considered of any value for soil properties determination, washing boring is a valuable method of rapidly advancing holes through many soils. It can be conveniently used in conjunction with split spoon sampling as noted in ASTM D1586 (ASTM 2003). The principal disadvantage is the need for an experienced operator to detect changes in soil strata (Sec. 2.3) as the washing boring is advanced; (c) if fine soils or dense granular materials (Sec. 3.3) are encountered, rotary drilling may be used. The principle of operation is similar to washing boring, however the drill rod and cutting bits are rotated during drilling, and pressure is applied on top of the drill to facilitate its movement into the soil. In addition, drilling mud (e.g., bentonitic slurry) is usually used in place of water. Percussion drilling consists of repeatedly dropping the drill rod and cutting bits into the drill hole in order to advance the hole. This method has the disadvantage of introducing repeated dynamic stresses (Sec. 11.1) into the soil which may result in significant soil disturbance; and (d) sampling procedures in deep bore holes may be divided into those that yield disturbed samples and into those that

yield undisturbed samples. A number of different samples are in current use, and the quality of the sample that each one provides can be expressed in terms of the *area ratio*, A_r (Hvorslev, 1949). In other words, the area ratio is an indication of the volume of soil displaced by the sampling spoon (tube).

$$A_r, \% = \frac{D_e^2 - D_i^2}{D_i^2} 100 \tag{1.3}$$

where A_r = area ratio, D_e = external diameter of sampler (tube) that enters the soil during sampling, and D_i = internal diameter of sampler.

2 Soil disturbance: A sampler is considered to cause minimum disturbance if its area ratio determined from Equation (1.3) is less than 20%. In common practice, area ratios of 13% or less are acceptable, but values of 10% are preferred. It is unlikely that perfect sampling will ever become a reality. Even if the problems of physical disturbance of the sample were to be entirely eliminated, stress changes that occur during sampling cannot be avoided. Some of the disturbance created during the tube sampling arises because the soils are not sampled in their true thickness. This is due, in fact, to the adhesion and friction of the soil in contact with the tube. The problem can be minimized by providing the sampler with a piston that closes the lower end of the sampler tube until sampling begins. At this time, the piston is released and permitted to move onto the sample tube at the same rate as the soil.

X-radiography and the computed tomography (CT) techniques are also used to determine the soil disturbance as reflected by internal soil cracks. X-radiography has been shown to be valuable aid in nondestructive examination of sample quality. This technique has been used for examination of the variation of soil density in the sample tube or to evaluate sample disturbance. CT is a relatively new X-ray method and measures point-by-point density values in the cross-sections of an object, thus allowing three-dimensional imaging of the internal structure when successive transverse sections are compared.

3 Undisturbed soil sample: The preparation of an undisturbed soil sample is directly related to the technique used in obtaining the sample. The degree of disturbance during the sampling and preparation of a soil specimen is very important; therefore, proper care must be taken during these processes. It is especially true for cohesionless soils and soft clays. Soil sampling techniques are related to the type of soil encountered, as noted below:

a   *Soft to medium consistency cohesive soils*   A common sampler used in soft to medium consistency cohesive soils is the thin-walled "Shelby Tube" sampler as described in ASTM D1587 (ASTM 2003). The wall thickness of this sampler is usually 1/16–1/8 in. (1.5–3.2 mm). The area ratio (Eq. (1.1)) for these dimensions is about 13%. Thus, reasonably satisfactory samples may be obtained.

b   *Stiff to hard cohesive Soils*   For stiff to hard cohesive soils, the tension sampler has been successful. The Denison Double-Tube Core Barrel Soil Sampler is also useful in hard soils. This sampler contains an outer rotating core barrel fitted with a drilling bit and an inner stationary sample barrel with a sharp cutting edge. Drilling mud is introduced between the inner and outer barrels. This device has also been used in cohesionless soils.

c *Cohesionless soils (sand)* Sampling in cohesionless soils is more difficult than in clays, as it is difficult to remove a contiguous sample that doesn't fall apart. There are two general cases: (i) above the groundwater table; and (ii) below the groundwater table. If the sand is above the water table, soil moisture may provide the soil with sufficient cohesion to permit relatively undisturbed samples to be obtained. If the sand is below the water table, special techniques are required which incorporate some form of core catcher to retain the sample. Core catchers are also used in very soft cohesive soils.

d *Contaminated soil sample* Sampling for contaminated soil is similar to the routine soil except that the test equipment of contaminated soil must be protected.

Figure 1.7 presents the steps for sampling and preparation of a laboratory undisturbed soil test specimen. In examining Figure 1.7, the sampling and preparation steps

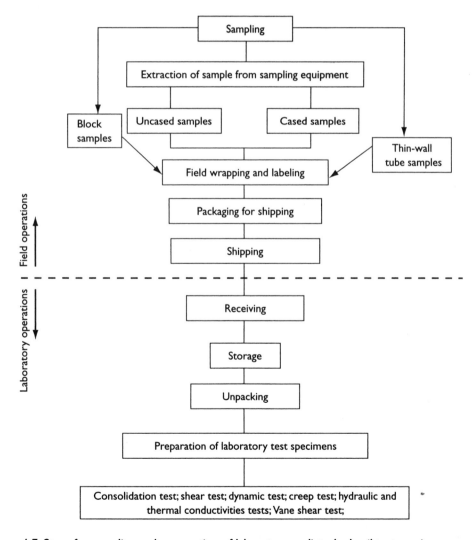

Figure 1.7 Steps for sampling and preparation of laboratory undisturbed soil test specimen.

Figure 1.8 Drill rig in operation (left) with a hollow-stem auger (close-up of auger, right) for use in subsurface exploration.

Source: Photos courtesy of Central Mine Equipment Co., Earth City, MO, Reprinted with permission.

includes both field and laboratory considerations. For contaminated soil samples, additional care, consistent with the specific chemical classification, should be taken. Figure 1.8 shows the picture of a drill rig being fitted with a hollow-stem auger, with a close-up of the auger itself. This type of auger is commonly used, and it allows the sampler or well casing to be driven through the interior (hollow) of the auger itself.

1.9.3 Laboratory soil testing

1 Routine laboratory testing: While details will be presented in appropriate chapters in the text, geotechnical testing is generally directed toward either classification and characterization or determining the engineering properties. Classification is typically based on particle size and consistency while the engineering behavior is defined by an assessment of permeability, compressibility, and strength. Since soil is sensitive to the ambient environment, some additional parameters such as specific surface, pore fluid pH, adsorption coefficients, etc. may also be of relevance. Some of these tests are standardized already by the ASTM and AASHTO or the international equivalent, while others are not.

2 *Testing on contaminated soils:* Preliminary evaluation of contaminated soil may be observed through soil surface cracking patterns, color, odor, and volume change characteristics. Analytical chemistry is usually required to determine the type and concentration of contaminants. Other measurable parameters may be grouped into three categories: (a) basic phenomena such as sorption and dielectric constant; (b) conductivity such as thermal and electric; and (c) loading tests such as tensile and fracture loads. All testing equipment used for testing of contaminated soils must be made of chemical resistant material especially for long-term studies. Various triaxial-permeameters for studying hydraulic conductivity by use of hazardous/toxic pore fluid are discussed in Section 5.4.5.

1.9.4   In situ measurements of soil properties

The properties of some sensitive soil deposits must be determined in situ, on location in as close a state of disturbance or non-disturbance as the respective engineering use may require. Table 1.5 summarizes some commonly used methods or devices at in situ condition. Of practical importance is the Standard Penetration Test (SPT) discussed more in Section 2.6.5.

Table 1.5 In situ measurements on soil-rock properties

Measuring devices	Shear strength	Bearing capacity	Settlement	Earth pressure	Others	Notes
Acoustic emmision	X		X			Ch. 10
Burggraf shear	X					Ch. 10
California bearing ratio	X	X				Ch. 12
Cone penetration test	X					Ch. 10
Cross-hole	X					Ch. 11
Dilatometer	X					Ch. 10
Echo					X	—
LVDT			X	X		Ch. 13
Piezometer			X		X	Ch. 5
Plate load		X	X			Ch. 13
Pressure cell		X	X	X		Ch. 12
Pressuremeter	X					Ch. 10
Settlement rod			X			Ch. 9
Slope inclinometer				X	X	Ch. 13
Standard penetration test		X	X		X	Ch. 2
Thermal needle					X	Ch. 6
Vane shear		X				Ch. 10

1.10   Data collection and presentation

1.10.1   General discussion

The major risk in the construction of any foundation is the uncertainty involved in predicting ground conditions and behavior. Of course, the accuracy of these predictions will improve with increasing effort devoted to the subsurface investigation, but the cost must also be considered. A schematic diagram illustrating the relationship between risk, effort, and cost is presented in Figure 1.9. In examining Figure 1.9, Curve (1) indicates the cost of a site investigation linear with effort. Curve (2) is the cost of risk and presumes that increasing the effort of site investigation will decrease the risk and cost of an unexpected failure. Curve (3) combines both cost and effort and represents a combination task curve. This optimum system should also satisfy the requirements of minimum costs and reduction of the risk to an acceptable level.

A practical evaluation of soil properties must recognize the inherent variability of natural soil deposit (Fig. 2.1). Thus it is important to retain a macroscopic view of the proposed structure and its foundation at all times. The problem of lateral (horizon) and vertical (profile) variation of soil properties, although common to all foundations, is particularly important in the case of deep foundations (Sections 15.12 and 15.13). Here the structural loads must be transferred not only to different areal locations, but also to different vertical levels in the soil profile. The exact properties of the load applied to each level is a function of complex relationships between deep foundation and soil which are not clearly understood.

A frequently used convenient method for handling soil variability involves establishing "average" properties, averaged with respect to both vertical and horizontal variation. How are these averages established? How are the number of borings and the number of samples and measurements in each boring determined? Local experience and economic considerations often play an important role in such determinations. If such considerations outweigh an objective determination of the necessary "level of confidence" for the soil properties required in design, perhaps

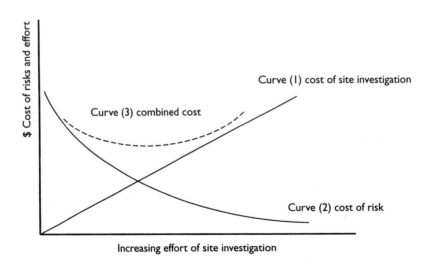

Figure 1.9 Risk and effort relationships for subsurface investigation.

some attempt to incorporate statistical analyses and/or experiment design techniques into the subsoil exploration program may prove fruitful. A brief discussion on data collection and experiment design is presented as follows.

1.10.2 Statistical methods and experiment

Statistical methods are procedures for summarizing observed data and/or for drawing scientific inferences (generalizations) from experimental data. The science of experiment design and analysis is based on mathematical statistics, the study of random variables. Concepts and analysis used in mathematical statistics are drawn from various branches of mathematics such as algebra, geometry, calculus, probability theory, and decision theory.

The method of choosing a sample is called the *design of the experiment*. An experiment is a set of observations on experimental units which have been subjected to treatments within some environment. The experiment and its conclusions must remain ambiguous unless each of the terms is well-defined. Scientific inference must be relative to those environments, treatments, units, and observations that are admissible in the experiment design. The purpose of the experiment must be to obtain average values for various combinations of units, treatments, transducers, and environments and/or to infer how variation in observation is associated with variations among units, treatments, transducers, or environments. If observations are made for all possible combinations of levels (one from each variable), the experiment is called a factorial experiment. If there is repetition of the experiment under various environmental conditions, the experiment is often called a *randomized-block experiment*.

1.10.3 Knowledge-based expert systems

Expert systems are intelligent computer programs that are able to perform an intellectual task in a specific field as a human expert would. Systems are being applied to classification problems such as interpretation and diagnosis, as well as general problems such as planning, analysis, and design. Expert systems can be used as data management systems which facilitate correlation studies, risk analysis, and computer aided design. Information produced with these expert systems includes colorful pictorial displays and/or tabular results at any given stage of interaction. Also, the user can trace back-forth to see what has been done or may interactively alter technical and/or financial criteria and constraints. The significant advantage of the computer integrated systems is that they can lead to a greater degree of unification in the processes across many disciplines. There can be an updating of information and an expansion of capacities within both the human–computer interface as well as in the subsystems to maintain the currency of the overall system at any given time.

1.11 Summary

This chapter served as an introduction to the text and to the field of geotechnical engineering. It should be clear that the behavior of soil is far less straightforward than other construction materials such as steel or concrete. It is in part because of this that geotechnical work remains challenging and exciting. The environmental perspective of this book has been discussed and explained in terms of trying to capture the true behavior of soil. Completely separate from this perspective are the fields of environmental geotechnology or geoenvironmental engineering. Environmental geotechnology was described as a broad interdisciplinary science while geoenvironmental engineering focuses on the hydrogeological and geotechnical of environmental engineering problems.

The particle-energy-field theory is introduced in the text for the purpose of explaining various soil behaviors under different environmental conditions. A brief discussion of the theory including assumptions and approaches summarized in the tabulated and graphical forms are presented. Further discussions and its applications will be made throughout the text.

Another key point is the importance of soil sampling and in situ measurements in soil testing. This is explained in terms of the various soil types, along with in situ instruments commonly used in geotechnical engineering.

PROBLEMS

1.1 Why do ground pollution problems challenge current soil mechanics concepts, and what are the methods for effectively analyzing soil behavior under various environmental conditions?

1.2 What is the particle-energy-field theory? Does this theory have merit? What are the basic concepts, assumptions and limitations?

1.3 Define the terms energy, energy field, and particle energy field. What are the differences between surface energy and body energy?

1.4 Why is mechanical energy considered a short-term process, and why are the chemical and physicochemical energies are considered long-term processes?

1.5 What is the environment? Explain why the air–water–ground soil pollution are interrelated?

1.6 Explain why in situ testing for certain soil deposits are so important and illustrate a practical example for your statement.

1.7 How are soil samples collected, and by what criteria are they judged disturbed or undisturbed? Is there such a thing as an undisturbed sample?

Chapter 2

Nature of soil and rock

2.1 Introduction

Soils are formed from rock as it is acted upon by physical, chemical, and biological forces. The extent to which a parent rock changes to a soil is a function of the rate and overall time of the prevailing reactions and processes. Depending on the viewpoint, there are three basic definitions of soil namely (a) from an engineering viewpoint, soil is any earthy material that can be removed with a spade, shovel or bulldozer and is the product of natural weathering. This soil includes gravel and sand deposits; (b) from a geological viewpoint, soil may be considered as the superficial unconsolidated mantle of disintegrated and decomposed rock material; and (c) from a pedological (soil science) viewpoint, soil is the weathered transformation product of the outermost layer of the solid crust, differentiated into horizons varying in type and amounts of mineral and organic constituents, usually unconsolidated and of various depths.

Soil is truly a unique creation. It differs from the parent rock below in morphology, physical properties, and biological characteristics. The soil mantle of the earth may be termed the "pedosphere" in contact with the atmosphere, the lithosphere, and the hydrosphere. A soil system is a dynamic system subject to temperature, moisture, and biologic cycles and it develops in a certain genetic direction under the influence of climate. The rate of this development is influenced by the parent material, vegetation, and human activity. Coupling the pedologic perspective with the particle-energy-field theory (Ch. 1), soil is constantly under the influence of mechanical, thermal, electric, magnetic, and radiation energies.

2.2 Rocks and their classification

Rocks serve as parent material for natural soil formation. They are also used as ground foundation support and the crushed rock fragments are used as major construction materials. In general, rock classification may be made on the basis of (a) geological origin and genesis, (b) rock mass strength, and (c) weathering and environmental factors.

2.2.1 Rock classification based on geological origin and genesis

The classification of rock based on its geological origin and genesis is the most common rock classification system. Rocks are broadly classified as igneous, sedimentary, and metamorphic. Igneous rocks have solidified from a molten or partly molten siliceous solution. This molten solution is called *magma*. When magma cools and solidifies in direct contact with the atmosphere it is referred to as *extrusive*, while cooling in the subsurface leads to an *intrusive* formation. Sedimentary rocks are naturally consolidated or unconsolidated transported materials. Metamorphic rocks form as a result of subjecting igneous or sedimentary rocks to elevated temperatures and pressures. Igneous rocks comprise about 80% and metamorphic rocks about 15% of the terrestrial and suboceanic earth crust, leaving about 5% for the sedimentary rocks. Common rock examples include granite and basalt (igneous), sandstone and limestone (sedimentary), and schist and gneiss (metamorphic).

2.2.2 *The engineering classification of rock*

Engineering classification of rock are generally made on the basis of strength. The Deere and Miller classification system is based on values of unconfined compression strength (Ch. 10) and modulus of elasticity. This classification applies to intact rock and provides qualitative descriptors according to observed strength and modulus. In terms of strength, intact rock maybe classified as very high strength, high strength, medium strength, low strength, and very low strength when the observed unconfined compressive strength is > 2250, $1125–2250$, $562–1125$, $281–562$, and < 281 kg/cm^2, respectively. Likewise, in terms of modulus, intact rock may be described as very stiff, stiff, medium stiffness, low stiffness, yielding, and highly yielding when the tangent modulus is $8–16$, $4–8$, $2–4$, $1–2$, $0.5–1.0$, and $0.25–0.50 \times 10^5$ kg/cm^2, respectively. In terms of rock types, intrusive igneous rocks (e.g. granite) tend to have high strength and a stiff modulus, while extrusive rocks have a wider range and may be considerably weaker and more plastic. Sedimentary rocks exhibit extreme variability in terms of both strength and modulus. Metamorphic rocks also exhibit a wide range in strength and modulus, although the process of increased temperature and pressure generally increases strength, that is, metamorphic rocks tend to be stronger than their original (pre-metamorphosed) material. Limestone and dolomite are the exception to this rule, as they lose strength after being metamorphosed to marble (Kehew, 1988). Since numerous rock classification systems based on the strength of the rock material have been proposed, the interested reader is referred to a state-of-the-art review of these systems given by Bieniawski (1989).

2.2.3 *Rock classification with environmental considerations*

1 Rock Quality Designation (RQD): An important parameter frequently used for identification and classification of rock mass is the RQD as proposed by Deere (1963). This parameter is a quantitative index based on a core-recovery procedure that incorporates only those pieces of core 100 mm (4 in.) or more in length. The cumulative length of these pieces divided by the total length of the coring run represents the RQD which can range from 0% to 100%. The RQD is considered excellent if near 100%, poor if less than 50% and good or fair if in between. The RQD is a measure of drill-core quality, and it disregards the influence of orientation, continuity, joint tightness, and gauge (infilling). Therefore, the RQD cannot serve as the only parameter for the full description of a rock mass. Because this parameter is easy to use and simple to understand, practicing engineers use this index widely for the preliminary identification and classification of rock mass.

2 *Unified Rock Classification System (URCS):* The URCS is used commonly in the Forest Service of the US Department of Agriculture (Williamson, 1980). The URCS was originally conceived in 1959, and it has been extended and refined since then. The basic elements include four major physical rock properties: (a) degree of weathering, (b) strength of rock mass, (c) discontinuity or directional weakness, and (d) gravity or unit weight. By establishing limiting values of these elements using field tests and observations combined with other geotechnical information, URCS permits a rough estimate of rock performance such as foundation and excavation suitability, slope stability, material use, blasting characteristics, and hydraulic conductivity.

2.2.4 Engineering properties of common rocks

Engineering properties of common rocks are presented in Table 2.1. In particular, Table 2.1 provides typical ranges for strength, modulus of elasticity, and hydraulic conductivity. The wide ranges in values for the engineering properties listed are caused by rock age, depth, test methods, as well as stress history and environmental conditions. Some problematic rocks such as highly weathered rock and clay shale will be discussed further in Section 2.11.

2.3 Soil as a natural system

A soil system may be considered as an assemblage of particles. The behavior of this assemblage is much different than that of the original rock material. In particular, the strength of monolithic materials including rock, but also concrete and steel is governed by the internal bonds of the material itself. In the case of soil, it is the friction and forces which set up between individual particles that dictate its strength, not the individual bonds within a given particle. Soil as a natural, genetic system is composed of (a) solid inorganic and organic particles, (b) an aqueous phase carrying matter in solution or colloidal dispersion, and (c) a gaseous phase of varying composition that is functionally related to biological activity. The aqueous and gaseous phases are usually considered together as pore space or porosity. The porosity varies with time and space, according to different soil layers, depths, and seasons.

Table 2.1 Typical range in selected engineering properties for common, intact rocks

Rock	Unconfined compressive strength (kg/cm^2 or $tons/ft^2$)	Modulus of elasticity (kg/cm^2 or $tons/ft^2$)	Hydraulic conductivity (cm/s (ft/yr))
Limestone and dolomite	500–2500	4–8×10^5	10^{-6} (1)
Granite	1000–2000	6–8×10^5	10^{-10} (10^{-4})
Quartzite	1500–4000	7–8×10^5	10^{-10} (10^{-4})

Source: *Selected data compiled from Freeze and Cherry, 1979 and Kehew, 1988.*

Note
1 kg/cm^2 = 1.02 ton/ft^2.

Theoretically it can assume any value between 0% and 100%, although values ranging from 20% to 50% are common.

2.3.1 Characteristics of the solid phase

Soils may contain a wide array of particle sizes, from clay particles that cannot be seen by the naked eye to large boulders. The particles themselves exhibit a variety of shapes, from smooth and rounded to sharp and angular. The collective distribution of these particles in any given formation is a function of the parent material and subsequent physical and chemical weathering. The size and nature of the solid phase serves as the basis for soil classification as discussed later in this chapter.

2.3.2 Characteristics of liquid and air interfaces

The portion of the soil porosity not filled with water represents the soil-air. Soil-air is in constant exchange with atmosphere and its composition reflects that of the atmosphere except for the concentration of those components that are used up or produced by microbiological activities in the soil. Such substances are oxygen (O), which is used up, and carbon dioxide (CO_2), which is produced. The oxygen content of soil-air decreases as carbon dioxide content increases, since the carbon dioxide is a product of aerobic respiration. It must be noted that natural soils always possess air spaces even if allowed to take in all the water they can. Of course, after a long duration of flooding this air space may be rather small.

2.3.3 Dynamic in situ soil conditions

Soil systems result from climatic forces. These forces derive from daily and seasonal temperature variations, fluctuations in moisture content, the changes in biological potential, and from any other periodic phenomenon that affects the surface layer of the earth. Soils continue to be exposed to the forces that formed them and their properties are in a continuous state of flux. Based on these characteristics, it is clear that any measured soil property may be subject to change. For example, a given soil may be sampled and found to have a low permeability to water. However, depending on in situ variations in groundwater composition, this property may change. As such, absolute descriptions such as "incompressible" or "impervious" have dubious meanings. Further discussion on impervious soil layers as they relate to waste landfills will be presented in Section 16.12.

2.4 Soil texture, strata, profile, and horizon

2.4.1 General discussion

Soils are three-dimensional systems; they have a two-dimensional areal extent and a third depth dimension. Whether they are geological depositions or formed on-site by the interaction of geologic parent material, climatic factors, topography, and living organisms, soils show areal variations and change with depth. Horizontal as well as vertical transition into another soil type may be gradual or abrupt depending on geologic and soil forming factor. A soil may be composed of only one size fraction of narrow range such as beach sand or loess or any number of size fractions in continuous or irregular grading. The size distribution of a soil is called its *texture*. Stones or gravel retained on a US #4 sieve (4.76 mm) are called *coarse aggregate*. Materials passing a #4 sieve are called *fine aggregate*. The fraction that passes the US #200 sieve (0.074 mm) is called *soil fines*. Many different terms and lines of demarcation are used in describing soil particle sizes and details on textural classifications are presented later in this chapter.

The properties of soils are largely influenced by the characteristics of the parent rock. If soil is formed in place by rock weathering it is called a *residual soil*. This is a situation whereby the rate of weathering exceeds the rate of erosion. Soil carried away from the location of rock weathering and deposited elsewhere by gravity, ice, water, or wind is called *transported soil*. Transported soils cover most of the land. Many of them have special geologic names. General relationships between the parent rock and soil types and characteristics are presented in Table 2.2. Some of these soil types will be further discussed in Sections 2.11 and 16.3.

When vertical changes are due to differing geologic processes, the resulting layers are called *strata*, and when they are caused by soil forming factors, the resulting layers are called *horizons*. The set of horizons from the soil surface to the original or physically altered parent rock is called the *profile*. The horizon containing the parent material or substrata is commonly called the *C-horizon*. The top layer which spans from the surface deposit of decaying plant litter to a depth at which the organic matter is completely humified, is called the *A-horizon*. Between the A- and C-horizons lies the B-horizon which is usually a locus of accumulation of material in suspension or colloidal solution washed down from the A-horizon by percolating precipitation. Both the A- and B-horizons develop at the expense of the C-horizon or parent material. If distinct differentiation has taken place in the three primary horizons, they are subdivided into subhorizons and are denoted respectively as A_{oo}, A_o, A_1, A_2, A_3, B_1, B_2, B_3, and C_1, C_2. The theoretical soil profile showing the principal horizons is given by McLerran (HRB, 1957) and Hillel (1998).

Table 2.2 Relationship between parent rock, soil types, and characteristics

A Residual types of soil	
Parent rock	Soil types and characteristics
Igneous and Metamorphic Rocks	Soils are often plastic and expansive
Limestone	Highly plastic soil
Sandstone	Silty sand, sandy clay, silty clay
Shale	High in clay constituents
B Transported types of soil	
Transport mechanism	Types and characteristics forms
Gravity	Colluvial deposits, talus, detritus
Ice	Glacial deposits, till, eskers, kames
Water	Alluvial deposits, (River), lacustrine deposits, (Lake), marine deposits (Ocean)
Wind	Aeolian deposits, dunes, loess, volcanic ash

Figure 2.1 Soil profile showing the various horizons. (Piedmont Residual Soil, North Carolina.)

2.4.2 Soil profile

As indicated in a previous section, the degree of change from a parent material to a soil system is a function of time and of the rate of reaction of the aging processes. The relative maturity of a soil is judged from the development of its characteristic profile or assembly of horizons. The diagnostically important layers are the A_2 and the B_2 horizons. The thickness of the horizons is determined primarily by the permeability of the parent materials; sandy, gravelly, and elastic parent materials develop deep profiles, while solid rock and impervious loose rock develop shallow profiles. Figure 2.1 shows the profile of a Piedmont residual soil, at a site north of Winston-Salem, North Carolina, USA.

2.4.3 Simplified soil profile and horizon system

Over the years, the system of letter designations of the different horizons has been changed and extended several times. The designations shown in Figure 2.2 are termed Master Horizons obtained from the US Department of Agriculture (USDA) Soil Survey Manual (1993). There are 24 further subdivisions within the Master Horizons that are termed Subordinate Distinctions. A complete description of these horizons and their subordinates is given by USDA (1993). Since the Master Horizons system is too extensive to describe here, only the general characteristics of the O, A, E, B, C, and R horizons are summarized.

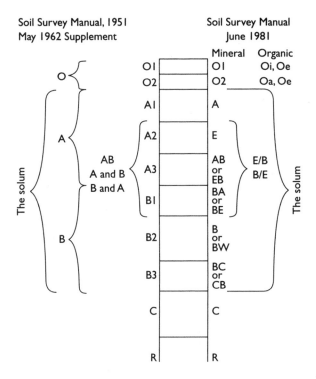

Figure 2.2 A simplified pedalogical soil profile showing the principal horizons.

The O, A, E, and B horizons are layers that have been modified by weathering, while the C-horizon is unaltered by soil-forming processes (Sec. 4.11). The R-horizon, below the other soil layers, is the underlying parent material in its original condition.

1   *O-horizon*: The top layer composed primarily of organic litter, such as leaves, twigs, moss and, lichens, that has been deposited on the surface. This layer, as well as underlying layers, may not exist due to erosion.

2   *A-horizon*: The original top layer of soil having the same color and texture through its depth. It is usually 10–12 in. (25.4–30.5 cm) thick but may range from 2 in. to 2 ft. (5.1–61 cm). The A-horizon is also referred to as the topsoil or surface soil when erosion has not taken place.

3   *E-horizon*: This layer is characterized largely by a loss of silicate clay, iron, aluminum or a combination thereof. It may be lighter than the A- or B-horizon and has less organic material than the A-horizon.

4   *B-horizon*: The soil layer just below the O-, A-or E-horizons that has about the same color and texture throughout its depth. It is usually 10–12 in. (25.4–30.5 cm) thick but may range from 4 in. to 8 ft (10.2–244 cm). In regions of humid or semi-humid climate, the B-horizon is a zone of accumulation in the sense that colloidal material carried in suspension from overlying horizons has lodged in it. The B-horizon is also referred to as the subsoil.

5   *C-horizon*: The soil layer just below the B-horizon having about the same color and texture throughout its depth. It is quite different from the B-horizon. It may

be of indefinite thickness and extend below any elevation. The C-horizon may be clay, silt, sand, gravel, combinations of these soils, or stone. The C-horizon is also referred to as parent material.

6   *R-horizon*: The layer of solid bedrock underlying the C-horizon. It is of indeterminate depth and is in its original condition of formation.

2.5   Soil consistency and indices

2.5.1   Atterberg limits

Soil consistency, in conjunction with its grain size distribution, constitutes the primary basis by which an engineering classification of soil materials is made. The consistency of a fine-grained soil in the remolded condition depends on the water content, which can be measured by Atterberg's consistency system. This system was proposed by the Swedish soil physicist A. Atterberg in 1911. Four consistency states have been recognized, namely liquid, plastic, semisolid, and solid, as shown in Figure 2.3. The liquid limit is the moisture content at which a soil passes from a plastic to a liquid state. The plastic limit is the moisture content at which a soil changes from semisolid to a plastic state. The shrinkage limit is the lower limit of the semisolid state, and also represents the point of minimum volume for the soil, that is, further drying is not accompanied by more shrinkage. These indices correspond to different physical and mechanical characteristics of soil at various water contents. The following section will discuss the use of these indices in geotechnical engineering.

1   *Liquid limit (ω_L, LL)*: The test procedure for determination of the liquid limit has been standardized by ASTM (D423) and AASHTO (T89). This involves filling a dish with soil, placing a groove through the middle of it, and alternately raising and dropping the dish by a fixed distance until the groove closes to 13 mm (0.5 in.). The data are plotted as moisture content (*y*-axis) versus the number of blows (*x*-axis) required to close the groove. The slope of this relationship (*flow curve*) is defined as the *flow index*, (I_F, FI). The moisture content at which 25 blows closes the groove is called the *liquid limit*. The standard procedure specified in ASTM notes two methods, one which requires at least three trials in order to obtain the liquid limit and a one-point method. Another simple and repeatable one-point method for determining the liquid

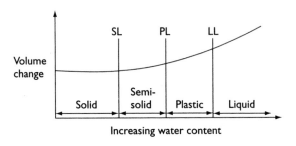

Figure 2.3 Liquid, plastic, and shrinkage limits relative to volume change and moisture content.

limit of soils has been developed (Fang, 1960). This method requires only one trial, provided the blow count is between 17 and 36, and is derived from the definition of the flow index as follows:

$$I_F = \frac{\omega_L - \omega_n}{\log N - \log 25} \tag{2.1}$$

or

$$\omega_L = \frac{N}{(\omega_n + I_F \log 25)} \tag{2.2}$$

where I_F = flow index (slope of flow curve), ω_L = liquid limit, N = number of blows ($17 < N < 36$), and ω_n = moisture content at N blows. The term $[I_F \log 25]$ in Equation (2.2) is called the *moisture correction factor* and is a function of the number of blows, N, and the soil type as reflected in the flow index, I_F. The flow index value increases as the clay content increases. The correction factor has been prepared in the term of a simple chart or table and the flow index can be estimated from the following equation:

$$I_F = 0.36 \, \omega_n - 3 \tag{2.3}$$

The liquid limit values vary from zero for non-plastic soils (e.g. sand-gravel, cohesionless) to higher than 500 for very plastic clay. Also, the composition of the pore fluid influences the results and will be discussed further in Chapters 5 and 7.

2 *Plastic limit (ω_p, PL)*: The test procedure for the plastic limit is standardized by ASTM (D424) and AASHTO (T90). The plastic limit is determined by hand-rolling a thread of fine-grained soil until the diameter is 3.2 mm (0.125 in.). The sample loses moisture as it is handled and rolled, and the process of forming soil threads is repeated until the 3.2 mm thread can no longer be formed without crumbling apart. The corresponding moisture content is the plastic limit. The plastic limit is mainly governed by clay content; hence some silt and sandy soils do not exhibit a plastic limit. Indications are that a significant change in load-carrying capacity of soils occurs at the plastic limit. Load-carrying capacity increases rapidly as the moisture content decreases below the plastic limit.

3 *Plasticity index (I_ω, PI)*: The plasticity index, I_P, is the numerical difference between the liquid limit, ω_L, and plastic limit, ω_p as:

$$I_P = \omega_L - \omega_p \tag{2.4}$$

where I_P = plasticity index, ω_L = liquid limit, and ω_P = plastic limit. The plasticity index represents the moisture range of a soil in which plastic properties dominate soil behavior. When the liquid limit or plastic limit cannot be measured or when the plastic limit is equal to or larger than liquid limit, the plasticity index is termed as *non-plastic*, and recorded as NP.

The liquid and plastic limits are the major part of Atterberg's consistency system and have been widely used in geotechnical engineering since their potential value was first observed in 1926. Comprehensive research relating to the test procedures, apparatus, applications, and their limitations are given by the ASTM Symposium on Atterberg Limits. The fundamental aspects and clay mineralogical effects on liquid limit results are reported by Seed *et al.* (1964) and Vees and Winterkorn (1967). Because these limit values are easily determined and simple to use, they have been used for basic soil classification or for predicting soil behavior such as strength, volume change, hydraulic, and thermal conductivity, or use for correlation of these values to other complicated soil parameters, such as tensile strength (Sec. 8.10), compression index, coefficient of consolidation (Sec. 9.3), cohesion, and internal friction angle (Sec. 10.8).

4 Shrinkage limit (ωₛ, SL): This method is standardized by ASTM (D427) and AASHTO (T92). The shrinkage limit is the moisture content at which further drying will not cause a decrease in volume of the soil mass, but at which an increase in moisture content will cause an increase in the volume of the soil mass. The value can be used as a general index of clay content and will, in general, decrease with increasing in clay content. For example, sands containing some silt and clay have a shrinkage limit of about 12–24, and the shrinkage limit of clays ranges from 4 to 12. Further discussions on the shrinkage limit and related behavior will be presented in Section 4.3. In addition to the Atterberg limits, there are other important indices which have bearing on engineering behavior, as noted in the following section.

2.5.2   Moisture equivalent

1 Field moisture equivalent (FME): The FME of a soil is defined as the minimum moisture content expressed as a percentage of the oven-dried soil at which a drop of water placed on a smooth surface of the soil will not immediately be absorbed by the soil but will spread out over the surface and give it a shiny appearance (ASTM D246) or (AASHTO T93).

2 Centrifuge moisture equivalent (CME): The CME is the moisture content of a soil after a saturated sample is centrifuged for one hour under a force equal to 1000 times the force of gravity. This test (ASTM D425) or (AASHTO T94) is used to assist in structural classification of soils. A value lower than 12 indicates permeable sands and silts while a value greater than 25 indicates impermeable clays with high capillarity. CME values as high as 68 have been observed for soft marine clays from the Gulf of Mexico and 56 from the Gulf of Maine. Both FME and CME are qualitative indicator properties and must be correlated with soil performance in order to have significant meaning. Some useful engineering applications by use of these two parameters include (a) an FME greater than the liquid limit indicates there is the danger for autogenous liquefaction of the soil in the presence of free water (Winterkorn and Fang, 1991); (b) when both FME and CME are more than 30 and if FME is greater than CME, the soil probably expands upon release of a load and should be classified as an expansive soil (PCA, 1992); (c) both the FME and CME tests can be used to predict absorption and adsorption behavior of fine-grained contaminated soil. Further discussions on this aspect will be presented in Section 4.5.

2.5.3 Soil indices

1 *Activity (A)*: Activity was proposed by Skempton (1953) and defined as the ratio of the plasticity index to the clay fraction (% finer than 0.005 mm).

$$A = \frac{I_P}{\% \text{ finer than } 0.005 \text{ mm}}$$ (2.5)

where A = activity and I_P = plasticity index. The activity values range from 0.23 for muscovite to about 6.0 for montmorillonite. This term has use for classifying the nature of the clay components of various soils such as the interrelationships of activity with other soil parameters, including the plasticity index, shrinkage limit, field moisture equivalent, water intake ability, and heat of wetting behavior of various clay minerals.

2 *Liquidity index (I_L, LI)*: Liquidity index also called *relative water content*, was proposed by Terzaghi (1936) and defined as

$$I_L = \frac{\omega_o - \omega_p}{I_P}$$ (2.6)

where I_L = liquidity index, ω_o = natural water content, ω_p = plastic limit, and I_P = plasticity index. Skempton and Northey (1952) reported that the liquidity index decreases when shear and unconfined compressive strength increase. The well defined relationship between liquidity index and sensitivity for all types of soils has indicated (Bjerrum, 1954) that sensitivity increases when the liquidity index increases. This index is also useful for soil classification, for example, when I_L = 1.0 the soil is at the liquid limit and when I_L = 0 the soil is at the plastic limit. A liquidity index value less than 0.4 may imply that the clay is overconsolidated (Fang, 1997). Further discussion on the liquidity index relating as it relates to bearing capacity and residual strength of overconsolidated clays will be presented in Section 10.7.

3 *Toughness index (I_T, TI)*: The toughness index (Casagrande, 1948) is defined as the ratio between plasticity index, I_P, and the flow index, I_F, as shown in Equation 2.7:

$$I_T = \frac{I_P}{I_F}$$ (2.7)

where I_T = toughness index, I_F = flow index (slope of flow curve), and I_P = plasticity index. Recall that the flow index, I_F, is the slope of the flow curve (change in moisture content per number of blows) while the plasticity index is the difference between the liquid and plastic limits. The toughness index is commonly used in soil stabilization to indicate the performance of stabilizing admixtures. Many values range from 0.4 to 1.8. A definite correlation between the toughness index and the tensile strength of compacted soils has been observed (Sec. 8.10).

4 *Consistency Index (I_C, CI)*: The consistency index (ASCE, 1958) is defined as:

$$I_C = \frac{\omega_L - \omega}{I_P}$$ (2.8)

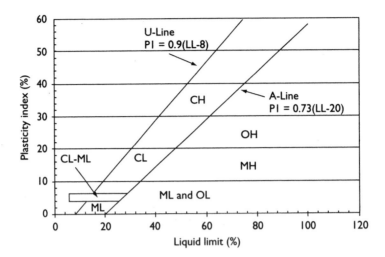

Figure 2.4 The plasticity chart of the Unified soil classification system (D2487, ASTM 2004).
Source: Copyright ASTM INTERNATIONAL. Reprinted with permission.

where ω_L = liquid limit, ω = water content, and I_P = plasticity index. The typical consistency index values range from 0.3 to 0.8 for common silts and clays. This index value has been correlated with the skin friction between soil and piles used in deep foundations. When the consistency index value increases, the skin friction also increases.

   5 *Plasticity angle (β)*: The plasticity angle (β) proposed by McNabb (1979) is based on Casagrande's A-line in the plasticity chart (Fig. 2.4) of the Unified Soil Classification System (Sec. 2.6.3) and can be presented as

$$\beta = \tan^{-1}\frac{I_P}{\omega_L - 20} \tag{2.9}$$

where β = plasticity angle, I_P = plasticity index, and ω_L = liquid limit. The range of plasticity angles varies from 10 to 40 degrees. This angle is a useful parameter for identification and characterization of low plasticity volcanic ash soil.

2.6   Classification systems of soil

Soil classification systems provide a language which quickly communicates information without the necessity of a lengthy description. In order to classify a soil, it is necessary to identify soil parameters with engineering significance. There are numerous soil identification and classification systems existing such as those given as AASHTO (M145), ASTM (D2487), Federal Aviation Administration (FAA), USDA and many others. The three most common soil classifications are the AASHTO, USCS, and USDA systems. The basis of identification systems are the description of the soil by (a) specifying its various components and (b) specifying the proportions of the various components. The proportions are established as ranges that are easily distinguishable by visual means. In addition to the proportion terms which apply only to the soil components, a measure of the gradation within the components is necessary. The overall plasticity index and overall liquidity (Sec. 2.6) are also identifying terms in the description of a soil. The color of the soil can be an important measure of its behavior, and thus becomes an integral part of soil description. Further discussion on soil color is presented in Table 2.10. Soils particles may be described in various terms such as boulders, gravel, sand, silt, and clay. The size limits associated with these terms for the main classification systems are given as Table 2.3.

Table 2.3 Particle size classification according to the USDA, USCS and AASHTO

Particle size (mm)	Unified Soil Classification System	American Association of State Highway Officials Soil Classification	U S Department of Agriculture Soil Classification
<0.001	Colloids	Colloids	Clay
0.001			
0.002	Clay	Clay	
0.005			
0.006	Silt	Silt	Silt
0.008			
0.01			
0.02			
0.03			
0.05			
0.06			
0.08			Very fine sand
0.1	Fine sand	Fine sand	Fine sand
0.25			
0.3			
0.4	Coarse sand	Coarse sand	Coarse sand
0.6			
0.8			
1			Very coarse sand
2			
3	Gravel	Fine gravel	Fine gravel
4			
6			
8			
10		Medium gravel	
20			
30		Coarse gravel	Coarse gravel
40			
60			
80		Boulders	Cobbles
>80			

2.6.1 Visual identification of soils

The visual identification of soils is an important field and laboratory procedure for developing an approximate grain size distribution curve (Fig. 3.1). This curve can be used to evaluate the suitability of a given soil for a particular engineering application. The test procedure includes

1   *Sample size required for visual identification*: (a) if gravel is present, select a representative sample of approximately 0.5 kg (1.0 lb) by the quartering method (ASTM, D421); (b) if no gravel is present, select a representative sample no larger than 0.25 kg (0.5 lb) by weight.
2   *Ocular examination*: Examine the soil by eye and make simple measurements for the following characteristics: (a) color of the whole soil (Table 2.10), preferably moist; (b) odor, to identify between organic and inorganic soils; (c) maximum particle size of gravel or coarse sand; (d) predominating grain shape, that is, water worn, sub-angular, or angular grains; (e) type of rock (Sec. 2.5) or minerals (Sec. 3.9); (f) hardness, soundness or friable condition of rock; (g) constituents such as micro shells, roots, humus, and other foreign matter. Additional information is given in ASTM D2488, entitled "Description of Soils (Visual Manual Procedure)" (ASTM, 2003).

2.6.2 AASHTO classification system (AASHTO, 1988)

The American Association of State Highway and Transportation Officials (AASHTO) soil classification system is derived from the US Bureau of Public Roads (BPR) system of soil classification as illustrated in Table 2.4. They have classified soils in accordance with their performance as subgrade soil beneath highway pavements. There are seven basic groups, A-1 to A-7, although sometimes an organic soil is called out as A-8. The members of each group have similar load bearing values and engineering characteristics under normal traffic conditions. The best soils for road subgrades are classified as A-1, the next best A-2, etc., with the poorest soils classified as A-7. Groups A-1 to A-3 soils possess, in the densified state, an effective sand-size granular skeleton. Groups A-4 to A-7 soils possess no such bearing skeleton and their engineering behavior is governed by water affinity and amount. Group A-2 is subdivided into A-2–4 to A-2–7 subgroups; the last number identifying the type of minus #200 sieve fraction present. Differentiation between the quality within a certain group is made by the *group index*, (I_G, GI). The group index is a function of liquid limit, ω_L, plasticity index, I_P, and the percent passing the #200 sieve, F. Then the group index can be determined by Equation (2.10) or a graphical procedure.

$$I_G = (F - 35)[0.2 + 0.005(\omega_L - 40)] + 0.01 (F - 15) (I_P - 10) \qquad (2.10)$$

where I_G = group index, F = % passing #200 sieve, ω_L = liquid limit, and I_P = plasticity index.

EXAMPLE 2.1
Assume that an A-6 soil has 55% passing a #200 sieve, a liquid limit of 40, and a plasticity index of 25, determine the group index.

Table 2.4 AASHTO soil classification system

General classification	Granular materials (35% or less passing No. 200)							Silty clay materials (more than 35% passing No. 200)			
	A-1		A-2	A-2	A-2	A-2	A-2				A-7[a]
Group classification	A-1-a	A-1-b	A-3	A-2-4	A-2-5	A-2-6	A-2-7	A-4	A-5	A-6	A-7-5 / A-7-6
Sieve analysis Percent passing:											
No. 10	50 max										
No. 40	30 max	50 max	51 min								
No. 200	15 max	25 max	10 max	35 max	35 max	35 max	35 max	36 min	36 min	36 min	36 min
Characteristics of fraction passing No. 40:											
Liquid limit				40 max	41 min	40 max	41 min	40 max	41 min	40 max	41 min
Plasticity index	6 max	6 max	NP	10 max	10 max	11 min	11 min	10 max	10 max	11 min	11 min
Usual types of significant constituent materials	Stone fragments Gravel and sand		Fine sand	Silty or clayey gravel and sand				Silty soils		Clayey soils	
General rating as subgrade	Excellent to good							Fair to poor			

Source: From Manual on Subsurface Investigations, 1988, by the American Association of State Highway and Transportation Officials (AASHTO), Washington DC. Used by permission. AASHTO publications may be purchased from the association's bookstore at 1-800-231-3475 or online at http://bookstore.transportation.org

Note
a Plasticity index of a A-7-5 subgroup is equal to or less than LL minus 30. Plasticity index of A-7-6 subgroup is greater than LL minus 30.

Table 2.5 Subgrade soil classification based on group index

Group index value	Condition of subgrade soil
0	Excellent
0–1	Good
2–4	Fair
5–9	Poor
10–20	Very poor

Source: From Manual on Subsurface Investigations, 1988, by the American Association of State Highway and Transportation Officials (AASHTO), Washington DC. Used by permission. AASHTO publications may be purchased from the association's bookstore at 1-800-231-3475 or online at http://bookstore.transportation.org

SOLUTION
From Equation (2.10)

$$I_G = (55 - 35)[0.2 + 0.005(40 - 40)] + 0.01\,(55 - 15)\,(25 - 10)$$
$$= 4.0 + 6.0 = 10$$

The group index is given in parentheses after soil groups and should be rounded to the nearest whole number. If a negative result is obtained, it should be reported as zero. The AASHTO subgrade soil and soil–aggregate mixture classification is shown in Table 2.5. The values of the group index range from 0 to 20. The smaller the value, the better quality of the soil for highway construction use within that subgroup. General quality of subgrade soil is indicated by the group index.

2.6.3   *Unified soil classification system*

This system grew out of the soil classification and identification system developed by A. Casagrande in 1948. The system was significantly revised in 1983. The essence of the system and its nomenclature is in Table 2.6. The significant changes and revision adopted (ASTM D2487) are included in the following:

Soil classification consists of both a name and a symbol such as CL-lean clay, or sand lean clay, and gravelly lean clay with sand. The names (or symbols) are standardized. These names have a single unique name for each symbol (except for organic silts and clays). In Figure 2.4, the upper limit or "U" line was added to the plasticity chart to aid in the evaluation of test data. This line was recommended by Casagrande as an empirical boundary for natural soils, as noted in D2487 of ASTM (2004).

EXAMPLE 2.2
A soil has liquid limit, $\omega_L = 38$, plasticity index, $I_P = 21$, and 82% passing #200 sieve, use the USCS system to classify this soil.

SOLUTION

For the USCS, when $I_P = 21$ and $\omega_L = 38$, using Figure 2.4, the soil is classified as CL. Also note, CL means the soil is a fine-grained soil (50% or more passes the #200 sieve); silts and clays; inorganic; lean clay.

2.6.4 USDA soil classification system

Comparison among these existing classification methods, the USDA soil classification system is particularly useful in a wide array of applications such as shallow foundations, stability of landfills, design of barriers, wetlands, surface and subsurface drainage systems, and erosion investigations. The USDA soil classification system is based on a system developed by Russian agricultural engineers in 1870 to permit the close study of soils with the same agricultural characteristics. Around 1900 this system was formally adopted by the USDA. Highway engineers found that this system and the resulting valuable soil information could be used in identifying suitable soils. However, this system is limited as a preliminary step in soil investigation since the engineering properties of soil must be determined after it is identified. The USDA system is divided into orders, series, and geographic names. A brief discussion of each category is given as follows:

1 *Orders-zonal, intrazonal and azonal:* In the USDA system, soil is divided into three main orders: zonal, intrazonal, and azonal, depending on the amount of soil profile developed. (a) *Zonal soil:* Mature soils characterized by well differentiated horizons and profiles found where the land is well drained but not too steep; (b) *intrazonal soils:* Those with well-developed characteristics resulting from some influential local environmental factors. Bog soils, peat and saline-alkali soils are typical examples; and (c) *Azonal soils:* Relatively young and reflect to a minimum degree the effects of the environment. They do not have profile development and structure developed from the soil forming processes. Alluvial soils of flood plains and dry sands along large lakes are typical examples.

2 *Great soil groups and soil series:* The USDA systems are subdivided into suborders as noted earlier and then further subdivided into great soil groups on the basis of the combined effect of climate, biological factors, and topography. The essential features for the definition of a soil unit are number, color, texture, structure, thickness, chemical and mineral composition, relative arrangement of the various horizons, and the geology of the parent material (Table 2.7).

Soils within each great soil group are divided into soil series. A soil series comprises all soils that have the same (a) parent material: solid rock (igneous, sedimentary, metamorphic), loose rock (gravel, sands, clays, other sediments); (b) special features of parent material: residual or transported by gravity, ice, water, wind, ice (Table 2.2) or combinations; (c) topographic position: rugged to depressed; (d) natural drainage: excessive to poor; and (e) profile characteristics.

3 *Geographic names:* The different series usually have geographic names indicative of the location where they were first recognized and described such as: Cecil, Hagerstown, Lufkin, Putnam, Wabash, etc. (Table 2.8).

The USDA has also developed a textural soil classification system based on the amount of sand, silt and clay within a given sample. This soil "triangle" is given as Figure 2.5.

Table 2.6 Unified soil classification system (D2487, ASTM 2004)

Criteria for assigning group symbols and group names using laboratory tests				Soil classification[a]	
				Group symbol	Group name[b]
Coarse-grained soils more than 50% retained on No. 200 sieve	Gravels More than 50% of coarse fraction retained on No. 4 sieve	Clean gravels less than 5% fines[c]	Cu ≥ 4 and 1 ≤ Cc ≤ 3[e]	GW	Well graded gravel[f]
			Cu ≥ 4 and 1 ≤ Cc ≤ 3[e]	GP	Poorly graded Gravel[f]
		Gravels with fines more than 12% fines[c]	Fines classify as ML or MH	GM	Silty gravel[f,g,h]
			Fines classify as ML or MH	GC	Clayey gravel[f,g,h]
	Sands 50% or more of coarse fraction passes No. 4 sieve	Clean sands less than 5% fines[d]	Cu ≥ 6 and 1 ≤ Cc ≤ 3[e]	SW	Well-graded sand[i]
			Cu ≥ 6 and 1 ≤ Cc ≤ 3[e]	SP	Poorly graded sand[i]
		Sands with fines more than 12% fines[d]	Fines classify as ML or MH	SM	Silty sand[g,h,i]
			Fines classify as CL or CH	SC	Clayey sand[g,h,i]
Fine-grained soils 50% or more passes the No. 200 sieve	Silts and clays Liquid limit less than 50	Inorganic	PI > 7 and plots on or above "A" line[j]	CL	Lean clay[k,l,m]
			PI < 4 or plots below "A" line[j]	ML	Silt[k,l,m]
		Organic	Liquid limit − oven dried / Liquid limit − not dried < 0.75	OL	Organic clay[k,l,m,n]
				OM	Organic silt[k,l,m,o]
	Silts and clays Liquid greater than 50	Inorganic	PI plots on or above "A" line	CH	Fat clay[k,l,m]
			PI plots below "A" line Cu ≥ 4 and 1 ≤ Cc ≤ 3	MH	Elastic silt[k,l,m]
		Organic	Liquid limit − oven dried / Liquid limit − not dried < 0.75	OH	Organic clay[k,l,m,p]
					Organic silt[k,l,m,q]

| Highly organic soils | Primarily organic matter, dark in color, and organic odor | PT | Peat |

Notes

a Based on the material passing the 3 in. (2.5 cm).

b If field sample contained cobbles or boulders, or both add "with cobbles or boulders, or both" to group name.

c Gravels with 5–12% fines require dual symbols: GW-GM well graded gravel with silt; GW-GC well graded gravel with clay; GP-GM poorly graded gravel with silt; GP-GC poorly graded gravel with clay.

d Sands with 5–12% fines require dual symbols: SW-SM well graded sand with silt; SW-SC well graded sand with clay; SP-SM poorly graded sand with silt; SP-SC poorly graded sand with clay.

e $C_u = D_{60}/D_{10}$ $C_c = (D_{30})^2/D_{10}*D_{60}$.

f If soil contains ≥15% sand, add "with sand" to group name.

g If fines classify as CL-ML, use dual symbol GC-GM or SC-SM.

h If fines are organic, add "with organic fines" to group name.

i If soil contains ≥15% gravel, add "with gravel" to group name.

j If atterberg limits plot in hatched area, soils is a CL-ML, Silty sand.

k If soil contains 15–29% plus No. 200, add "with sand" or "with gravel" whichever is predominant.

l If soil contains ≥30% plus No. 200, predominantly sand, add "sandy" to group name.

m If soil contains ≥30% plus No. 200, Predominantly gravel, add "gravelly" to group name.

n PI ≥4 and plots on or above "A" line.

o PI<4 or plots below "A" line.

p PI plots on or above "A" line.

q PI plots below "A" line.

Table 2.7 USDA soil classification system

Order	Suborder	Great soil groups
Zonal soils	1 Soils of the cold zone 2 Light-coloured soils of arid regions	Tundra soils Desert soils Red Desert soil Sierozem Brown soils Reddish-brown soils
	3 Dark- coloured soils of semiarid, subhumid, and humid grasslands	Chestnut soils Reddish chestnut soils Chernozem soils Prairie soils Reddish prairie soils
	4 Soils of the forest-grassland transition 5 Light-coloured podzolized soils of the timbered regions	Degraded chernozem Noncalcic brown Podzol soils Gray wooded Brown podzolic soils Gray brown podzolic soils Red-yellow podzolic soils[a]
	6 Lateritic soils of forested warm temperature and tropical regions	Reddish-brown lateritic soils[a] Yellowish brown lateritic soils Laterite soils[a]
Intrazonal soils	1 Halomorphic (saline and alkali) soils of imperfectly drianed arid regions and littoral deposits 2 Hydromorphic soils of marches, swamps, seep areas, and flats	Solonchak Solonetz soils Solloth soils Humic-glei soils[a] Alpine meadow soils Bog soils Half-bog soils Low-humic-glei[a] soils Planosols Groundwater podzol soils Groundwater laterite soils
	3 Calcimorphic soils	Brown forest soils Rendzine soils
Azonal soils		Lithosols Regosols Alluvial soils

Source: USDA 1993.

Note
a New or recently modified great soil groups.

2.6.5 *Other soil classification systems*

1 FAA classification system: The FAA classification is based on the soil gradation, soil consistency, soil expansive characteristics, and California Bearing Ratio (CBR) (Sec. 12.7). This system is used mainly for airfield pavement design.

2 The Standard Penetration Test (SPT) method: The SPT is an in situ testing technique, as noted in Chapter 1, and is frequently used in geotechnical engineering for soil classification, estimation of shear strength and bearing capacity. A brief

Table 2.8 Typical SSR and Si/Al ratios for some natural soils and clay minerals

Soil	SSR	Si/Al (Fe) ratio	Sources
Cecil	1.3	0.65	Alabama clay loam
Susquehanna	2.3	1.15	Well oxidized Alabama soil
Putnam	3.2	1.6	Heavy Missouri silt loam
Wabash	3.2	1.6	Missouri alluvia clay
Lufkin	3.8	1.9	Black Belt soil from Alabama
Montmorillonite	5.0	2.5	Wyoming bentonite

Source: Data from Winterkorn, 1955 and Others.

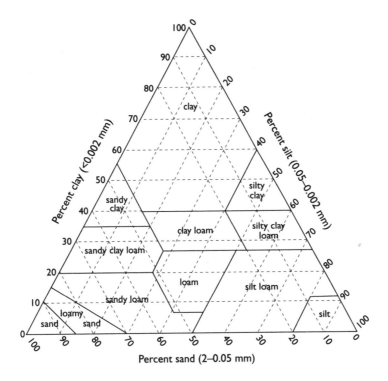

Figure 2.5 USDA textural soil classification system.

Source: Reprinted from Environmental Soil Physics, D. Hillel, p. 64 Copyright (1998), with permission from Elsevier.

description of this method is presented as follows. The test is standardized by ASTM as D1586. It is performed by driving a specified split spoon sampler (I.D. = 3.5 cm; 1.375 in.) into the ground soil with a hammer of certain weight (63.5 kg; 140 lb) dropped freely from a given height (76 cm; 30 in.). The number of hammer blows (N) required to drive the sampler a fixed distance (46 cm; 18 in.; three 15.2 cm; 6 in. increments) is counted and recorded. The SPT provides some indication of the relative density of the soil. The denser the soil, the more difficult it is to penetrate and therefore more blows are required to drive the sampler. The SPT blow count, or N value, is obtained by adding the number of blows for the second and third 6 in. increments and discarding the value for the first increment. If the number of counted blows for three increments were 8, 9, and 11, the N value would be reported as ($9 + 11 = 20$). The N value has units of blows per foot. N values have many uses such as prediction of liquefaction potentials (Sec. 11.4), and estimation of bearing capacity of ground soil (Sec. 12.7) and pile capacity (Sec. 15.12). This technique has the added advantage of producing a sample, albeit disturbed (Ch. 1) that can be used for classification purposes.

It is useful to be able to classify a soil using the three most common systems, namely the AASTHO, USCS, and USDA methods.

EXAMPLE 2.3
For the soil described in Example 2.2, (liquid limit, $\omega_L = 38$, plasticity index, $I_P = 21$, and 82% passing #200 sieve), use the AASHTO table and USDA soil triangle to make a classification. Additional information: 20% sand, 60% silt, and 20% clay.

SOLUTION
Using Table 2.4, when $\omega_L = 38$, and $I_P = 21$, the AASHTO classification is A-6, clayey soil. According to the USDA triangle Figure 2.5, the soil is a silt loam.

2.7   Chemical composition of natural soils

Considering the great variability of the chemical composition of the parent materials from which soils are formed, an equally great variability in soil composition may be expected. The most general chemical soil classification was made by Marbut (1920) by dividing soils into two classes, namely

(a) *Pedalfers*: Soils of humid climates (precipitation > evaporation), therefore, water percolation and elutriation resulting in a relative concentration of iron (Fe) and aluminum (Al) compounds in one of the profile/horizons; and
(b) *Pedocals*: Soils of semiarid and arid climates (precipitation < evaporation) resulting in salts such as calcium carbonate ($CaCO_3$) concentration within or on the surface of the soil profile.

2.7.1   General chemistry of soil

Soil is composed of a variety of chemical compounds. Considering soil as part of the earth's crust, over 98% of the material is composed of the elements oxygen, silicon, aluminum, iron, calcium, sodium, potassium, and magnesium. Combinations of silicon and oxygen in the form of silicates represents the majority of soil minerals.

2.7.2   Soil organic matter

Soil organic matter is a mixture of many different compounds, the more important of which are carbohydrates, proteins, fats, and resins. The absolute amounts and relative proportions of these compounds vary with the climate and macro- and micro-biologic activities in the soil. The organic matter is concentrated in the surface layers and decreases with increasing depth in the soil. It ranges from less than 1% in inorganic solids and sands to almost 100% of the solid matter in peat bogs. Further discussion is presented in Section 2.11.

2.8 Characteristics of granular soils

Granular soils in cold and temperate-humid climates contain predominantly quartz and mixed silicates of various tri-, di-, and monovalent metals. In hot-humid climates, they may contain mechanically strong and water resistant secondary aggregations of hydrogen ions (H^+) and aluminum oxides (Al_2O_3) typical for laterite soils, and in dry climates they may be any kind of minerals.

2.8.1 Cobbles and boulders

Cobbles and small boulders are employed in the construction of road bases and pavements if their mineral composition is such that it provides strength, toughness, and durability sufficient for the service requirements. The size range for cobbles and boulders are listed in Table 2.3. Cobbles and boulders are rounded in shape, otherwise they are called *field stones* or *rock fragments*. The mineral nature of cobbles and boulders is essentially that of the parent rock with a surface film or layer of weathering products, the composition and thickness of which depends on the environmental factors and the time of exposure.

2.8.2 Gravel, sand, and silt

Gravel, sand, and silt may be mechanically comminuted parent rock material or may represent the mechanically and chemically most resistant mineral constituents of the parent rock. The extent to which the mechanical and chemical breakdown takes place depends on the environmental conditions and the length of exposure to them.

1 *Gravel and sand*: There are several types of gravel and sand existing from various sources. Fluvial gravel and sands become more quartzitic the longer the path of transportation. In humid climates, gravel and sands tend to be siliceous and quartzitic, but they may be any type of mineral in dry climates. The typical size ranges for gravel and sand are listed in Table 2.3. Sand generally has a granular appearance in which the individual grain size can be detected. It is free-flowing when in a dry condition. When sand is dry, it will form a cast that falls apart when pressure is released, when moist, it forms a cast which will crumble when lightly touched. There are several types of sand mixtures such as sandy loam, sandy clay loam, and sandy clay are presented dependent on the amount of sand existing in the mixtures. If gravel and sand are used as components of concrete and mortar, gravel is called *coarse aggregate* (CA) and sands are *fine aggregate* (FA).

2 *Silt*: Silt particles often resemble the composition of the parent rock with feldspar, muscovite, and quartz usually well represented. In the silt fraction, the silicon content typically ranges between 21% and 47% and the oxygen content between 42% and 53%. When dry, it may be cloddy and readily pulverizes to a powder with a soft flour-like feel. When dry, it forms a cast which can be handled without breaking, when moist, it forms a cast which can be freely handled. When wet, it puddles easily. There are several types of silt mixtures such as silty loam, and silty clay loam dependent on the amount of silt existing in the mixtures.

2.8.3   Loam (sand-silt-clay mixture)

Loam is a common term used in agricultural and highway engineering. It is a uniform mixture of sand, silt, and clay. It is mellow, has a somewhat gritty feel, yet is fairly smooth and slightly plastic. When dry, it forms a cast which will bear careful handling without breaking, when moist it forms a cast which can be handled freely without breaking. Mechanical composition of loam mixtures ranges, sand 30–50%; silt, 30–50% and clay, 0–20%, with varied names such as sandy loam, sandy clay loam, silty loam, silty clay loam, and clay loam.

2.9   Silica–sesquioxide ratio (SSR) of soil–water system

2.9.1   Definition

The term *silica–sesquioxide ratio* (SSR) proposed in early 1930 by Mattson (1932) and Winterkorn and Baver (1934) recognizes that most clay minerals contain various amounts of SiO_2, Al_2O_3, and Fe_2O_3. Many of the properties of clays are related with the SSR. A formula for estimation of SSR is shown in following equation.

$$SSR = \frac{A}{B + C} \tag{2.11}$$

where SSR = silica–sesquioxide ratio and

$$A = \frac{\% \text{ of } SiO_2}{\text{molecular weight of } SiO_2}$$

$$B = \frac{\% \text{ of } Al_2O_3}{\text{molecular weight of } Al_2O_3}$$

$$C = \frac{\% \text{ of } Fe_2O_3}{\text{molecular weight of } Fe_2O_3}$$

EXAMPLE 2.4
Given: A clay mineral contains 70% of SiO_2, 15% Al_2O_3, 10% Fe_2O_3, and various other constituents. Determine SSR.

SOLUTION
1   From Appendix B-1, the atomic weight of these elements are:

Si = 28.1;   O = 16.0;   Al = 27.0;   Fe = 55.8.

2   Compute molecular weight of

SiO_2 = (28.1) + (16.0) (2) = 60.1,
Al_2O_3 = (27.0) (2) + (16.0) (3) = 102.0,
Fe_2O_3 = (55.8) (2) + (16.0) (3) = 159.6.

Substituting these numerical values into Equation (2.10), then

$A = 70 / 60.1 = 1.16$
$B = 15 / 102.0 = 0.15$
$C = 10 / 159.6 = 0.06$

Then, SSR = 5.52

2.9.2 Relationship of silica–sesquioxide ratio with other soil parameters

The SSR defined in Equation (2.11) is a useful parameter for identification of the characteristics of clay minerals. Table 2.8 presents the typical SSR values for some natural soils and clay minerals. SSR versus activity for various natural soils and clay minerals is presented in Figure 2.6. In examining Figure 2.6, SSR value increases with activity.

2.10 Identification and characterization of contaminated soils

At present, the most commonly adopted soil classification procedures are AASHTO and USCS methods as discussed in Section 2.6. Both methods are based only on particle size and soil consistency limits. Since soil is extremely sensitive to the local environment, some additional parameters such as specific surface, pH in pore fluid, adsorption, dielectric constants, etc. as noted in Chapter 1 may also be of interest. Contaminated soil can be identified by soil surface cracking patterns, colors, odor, volume change, in addition to the relevant analytical chemistry.

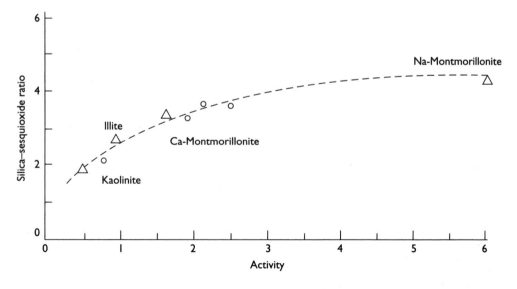

Figure 2.6 Silica–sesquioxide ratio versus activity for some natural soils and clay minerals.

2.10.1 Visual identification of contaminated soil and water

For the design, construction and maintenance of hazardous and toxic waste controlling facilities, or utilization of abandoned landfill sites as building sites, roadways, recreation parks, etc., the condition of a site that is contaminated, or suspected to have contamination, must be known. The general procedure involves a thorough site characterization, including a visual, laboratory, and field-based assessement. Details on site characterization at contaminated sites is given by LaGrega *et al.* (2001) and ASTM D6235 (ASTM, 2003).

1 *Reconnaissance and field investigations*: Table 2.9 lists various components of a field investigation at a contaminated site. The term "brownfield" has been used to indicate an abandoned site that is either contaminated or suspected as having contamination from previous industrial uses.

2 *Identification based on color of soil*: Significant differences in color exist not only between different soils, but even horizons of the same soil. The color may be inherited from the parent material or represent chemical weathering products, or it may be due to organic matter in various amounts and degrees of humification. In many cases, once soil-water is polluted it also produces various colors. One may differentiate between uniform, spotted, streaked, and mottled colors, all of which have physical or chemical significance. For color description, the Munsell notations should be used whenever possible. They take into account (a) hue-dominant spectral (rainbow) color, (b) value-relative lightness of color, and (c) chroma-relative purity of spectral color. Table 2.10 summarizes characteristics of soil related to its color under normal environmental conditions. However, if the ground soil is heavily contaminated, then these colors will change again.

3 *Identification based on odor of soil*: Fresh, wet organic soils usually have a distinctive odor of decomposing organic matter. This odor can be made more noticeable by heating the wet sample. Odor also can be used to identify any natural existing gases. However, some highly poisonous gases existing in nature or as by-products from

Table 2.9 Reconnaissance and field investigations at hazardous waste and Brownfield sites

1 *Characteristics of site*
    (a) Number and condition of abandoned buildings, residential housing, and vehicles
    (b) Dead, dying, and degree of decomposition of trees, shrubs, vegetation, birds, animals, etc.
    (c) Conditions of roads in surrounding areas
    (d) Degree of corrosion in surrounding facilities
2 *Ground soil and water characteristics*
    (a) Soil–water color
    (b) Soil–water-air odors
    (c) Soil erosion features in surrounding areas
    (d) Ground soil cracking patterns
3 *River and stream conditions*
    (a) Flow velocities of surface water
    (b) Color–odor of water
    (c) Water bubbles (approximately number, size, and color)
    (d) Temperature of water
    (e) Degree of turbidity in the water

Table 2.10 Characteristics of soil related to its color

Soil colors	Abbreviations	Soil characteristics
Black	Bk.	Organic soils, or poor drainage
Brown	Br.	Dark brown may be due to dark colored minerals such as: manganese (Mn), titanium (Ti)
Grey	Gr.	Dark gray indicates bad drainage and anaerobic conditions
Red	R.	Presence of nonhydrated hematite or bloodstone.
Yellow	Yr.	Good drainage and aeration
White	Wt.	Preponderance of silica, lime, gypsum, and kaolinite

Source: Fang, 1997.

manufacturing are colorless and odorless such as carbon dioxide (CO_2), carbon monoxide (CO), methane (CH_4), and radon (Rn) gas (Sec. 16.11). Therefore, odor alone cannot be used for identification or classifying polluted air–water–ground soil systems. Odor classification of water by chemical types is standardized by ASTM (D1292). The odor characteristics include sweetness, pungency, smokiness, and rottenness.

4 *Identification based on soil surface cracking pattern*: Cracking patterns and soil color are closely related. They are both affected by chemical contamination, for example by acetic acid, aniline, and carbon tetrachloride. Cracking patterns between a non-polluted (drinking water) and polluted soil pad made from various clay minerals such as bentonite, muscovite and illite have been studied. A significant difference of volumetric change between non-polluted and polluted soil samples is observed. As such, ground surface cracking patterns may be a useful tool for preliminary evaluation of contaminated soil and will be discussed further in Section 8.3.

2.10.2 *Characterization and classification of contaminated soils*

As noted earlier in this chapter, soil classification methods are based on particle size and composition and their interaction with water as observed by volume and consistency changes. Since soil–water interaction is dominated by the total amount of surface present in a sample and since the ratio of surface area to volume (specific surface) increases with decreasing particle size, the importance of these interactions are normally determined by the grain size distribution curve. With this in mind, some suggested procedures for characterizing contaminated soils are given as follows:

1 *Characterization based on pollution sensitivity index:* The pollution sensitivity index (PSI), as proposed by Fang and Mikroudis (1987), has been used for contaminated site characterization. The PSI is based on soil particle surface area and size, recognizing that smaller particles are more subject to physical and chemical interactions. Figure 2.7 shows the relationship between PSI versus particle size. As particle size decreases, PSI increases. A simple classification system for contaminated fine-grained soil is proposed as shown in Table 2.11. In examining Table 2.11, it is indicated that when clay content increases, the pollution sensitivity potential increases significantly.

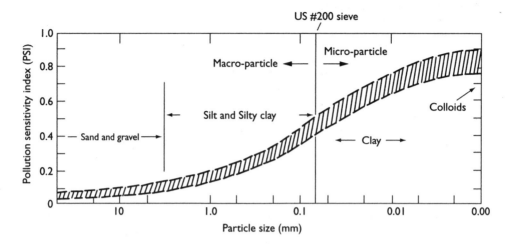

Figure 2.7 Pollution sensitivity index (PSI) relating to soil particle size.

Table 2.11 Identification and characterization of clay based on PSI

PSI	Soil type	Size (cm)	Surface area (cm²/cm³)	Sensitivity
0–2	Gravel	0.2–8.0	15–0.125	Very low
2–4	Sand	0.005–0.2	600–15	Low
4–6	Silt	0.0005–0.005	6000–600	Medium
6–8	Clay	0.0001–0.0005	30,000–6000	High
8–10	Colloids	< 0.0001	> 30,000	Very high

Table 2.12 Identification and characterization of clay based on SSR

SSRᵃ	Base exchange capacity	Adsorption	Swelling	Conductivity	Pollution potential
1–2	Very low	Very low	Very low	High	Very low
2–3	Low	Low	Low	Medium	Low
3–4	Medium	Medium	Medium	Low	Medium
4–5	High	High	High	Very low	High
> 5	Very high	Very high	Very high	Very low	Very high

Note
a SSR computation (see Example 2.4). Typical data of SSR in Table 2.8.

This conclusion should be used with caution, however, given that multiple interactions may result in counter-balancing behavior. Site-specific testing for the parameters of concern (e.g. hydraulic conductivity, shear strength) is recommended.

2   *Characterization based on SSR:* The SSR, as defined in Section 2.9, is a useful indicator of soil chemical composition. The value of SSR can be used for identification or characterization of fine-grained soil behavior and reactivity to various contaminants as shown in Table 2.12.

3 Characterization based on dielectric constant: The dielectric constant is an important measurable parameter of soil surface electrochemistry. It is a function of ion types and concentration and is closely related to the soil behavior as reflected by Atterberg limits, hydraulic conductivity, acidity, etc. It is useful for identification and characterization of contaminated fine-grained soils. The mechanism of dielectric constant and electric conductivity interacts in the solid–liquid medium are explained by Kaya and Fang (1997). Further discussion on these aspects will be presented in Section 6.10.

2.11 Some special types of soil and problematic soils

Because some soils and rocks have special features, they deserve special attention. In this section, we shall discuss the engineering properties of some natural problematic soils and rocks frequently appearing in geotechnical engineering literature. The term called *collapsible soil is* also frequently used as a general term for any soil that has collapsible characteristics such as loess, dispersive clays, residual soil, organic soils, and expansive clay.

2.11.1 Some special types of soil

1 Aeolian deposit: An aeolian deposit (loess) is one whose particles are predominantly of silt size but with a certain amount of fine sand and aggregated clay particles present. The valley loess is typically developed in areas peripheral to those covered by the last ice sheets. Typical loess has a calcium carbonate content that acts as a bonding agent which, though weak, allows the loess to form vertical or even overhanging walls on the banks of streams. In the natural state, loess is characterized by a columnar structure and by its ability to stand unsupported with nearly vertical slopes. It is highly susceptible to erosion. Loess deposits are comparatively pervious but if reworked they become impervious and are very difficult to compact. Since the bond between the particles is due in part to a more or less calcareous (lime) binder, the surface of a loess deposit may settle on saturation.

2 Dispersive clay: Dispersive clay is a fine-grained soil that rapidly erodes, forming tunnels and deep gullies by a process in which the individual clay particles slough off and into suspension when in contact with moving water. In general, these clays have a preponderance of dissolved sodium cations (Na^+) in the porewater, whereas ordinary, erosion-resistance clays have preponderance of dissolved calcium (Ca^{2+}) and magnesium (Mg^{2+}) cations.

3 Expansive clays (swelling soils): Soils that exhibit the greatest volume change when dried or wetted usually possess a considerable percentage of montmorillonite clay or a related three-layer clay mineral. Soils subject to considerable swelling and shrinking do not have a continuous granular skeleton with sufficient interstitial porosity to accommodate the volume changes that accompany changes in moisture content. Major factors affecting swelling behavior include the mineralogy, grain size distribution as well as types and concentrations of electrolytes in solution. Further discussion will be presented in Section 4.4.

4 *Lacustrine deposits (marls and tufa):* Lacustrine deposits are confined to those sediments laid down in lakes and streams associated with glacier or the pleistocene epoch (1.8 million to 8000 years ago). It also includes those sediments deposited in lake basins or in valleys occupied by streams. Lake sediments consists of marls, tufa, clays, silts, sand gravel, iron hydroxides ($Fe(OH)_2$), iron carbonate ($Fe(CO_3)_2$), silicon dioxide (SiO_2), manganese oxide (MnO_2), calcium phosphate ($Ca(PO_4)$), organic matter, and evaporites. Lake marls are a mixture rich in calcium carbonate ($CaCO_3$) content with various impurities that impart an array of colors. Most are gray to white or pale blue but red and black marls are not uncommon. *Tufa* is a limestone deposit that is more or less porous and banded.

5 *Laterite soil:* Laterite soils form under conditions of high temperature and high rainfall where the degree of weathering is intense. Silica (SiO_2) and bases (e.g. Na_2O, K_2O) are leached while iron and aluminum oxides (Fe_2O_3 and Al_2O_3) tend to concentrate in these soils. As such, the clay fractions have a low SSR, low activity, and low base exchange capacity. Laterites have a low content of soluble constituents and of most primary minerals.

6 *Organic soils and pear:* Organic soils are those solid constituents consisting predominantly of plant matter in various stages of decomposition or preservation. They are designated as bog, muskeg, and moor soils with differentiation between peat, muck soils, and coastal marshland soils. Muck implies a higher degree of decomposition of the plant matter and of inter-mixing with mineral soil constituents in contrast to peat soils that have well preserved plant remains. Several types of muck are recognized depending on the source of water, topographic characteristics and types of underlying soil or rock. Peat is organic soil composed of partially decayed deposits of dead vegetation that has been kept anaerobic by almost continuous submergence. It is, therefore, an organic soil with a distinct fibrous texture, and the methods already describe are adequate for its identification. Peat is listed separately here because it is a treacherous soil that will not support any additional sustained load in the form of a fill or a structure without a very considerable reduction in volume, often accompanied by lateral displacement.

7 *Saline-alkali soils:* Saline-alkali soils have more that 15% of their base exchange saturated with Na^+ ions and contain appreciable quantities of soluble salts. The electric conductivity of their saturation extract is greater than 4 milli-mhos per cm at 25°C (77°F) and the pH in the saturated soil solution is usually 8.5 or less. Saline soils (also known as non-alkali soils) contain appreciable amounts of soluble salts which impair crop production. Both saline-alkali and saline soils may be deleterious to contacting concrete structures, especially when they contain appreciable amounts of Na^+ and Mg^{2+} sulfates.

8 *Residual soil and weathered rock:* Residual soil is produced by the in situ decomposition of the underlying rock and the action of the pertinent soil forming factors such as micro-climate, flora, fauna, and geometric features. Chemical breakdown is particularly active in hot humid regions with production and decomposition of large amounts or organic materials. The texture and mineralogy may still reflect the original rock structure with added complication of decreased weathering with increasing depth below the ground surface.

9 *Shale:* Shales predominate among the sedimentary rocks in the earth's crust. Their properties vary from those of "solid" rock that must be blasted for excavation to those of soil-like materials that fall within the engineering definition of soil. *Rock-like*

shale retains its strength and integrity even during repeated exposure to wetting and drying cycles while soil-like shale slakes under these conditions.

10   *Varved clays*: Varved clays are characterized by layers of relatively coarser, clayey silt, and finer silty clay. These soils formed in lakes during the glacial retreat stage where inflowing meltwater deposited particles carried in suspension. The coarser particles settled first, usually during the summer months, followed by deposition of the finer clay particles during the winter months. Many varved clays are sensitive to disturbance.

2.11.2   Natural soil deposits

Certain soil deposits in various regions of the United States and the world have received considerable attention because of their effects on large construction projects or unique properties. Notable examples are presented as follows:

1   *Natural soil deposits in the U.S.*: (a) AASHO Road Test Soil: This is an Illinois yellow-brown silty clay having an A-6 classification. It was a C-horizon material. The soil was quite uniform, a few pebbles and small boulders were found. (b) Boston Blue Clay: Sediments transported by streams of melted glaciers of Pleistocene epoch and deposited in the quiet marine waters of the Boston Basin. The clay was uplifted, submerged and then uplifted again. The clay is often overconsolidated near the surface and normally consolidated at depth. (c) Ottawa Sand: Ottawa sand is white with a uniform size and round shape. It is a natural deposit from Ottawa, Illinois. (d) Wyoming Bentonite: A colloidal clay which has the property of being hydrophillic, or water swelling. It is a hydrous silicate of alumina and may have a liquid limit greater than 500.
2   *Bangkok clay*: Bangkok clays are alluvial deposits that originated from sedimentation at the delta of ancient rivers in the Chao Phraya Plain. The deposits of the more recent ages consist of a series of alluvial clays and cover a large area surrounding the city of Bangkok.
3   *Canadian sensitive clay*: This clay is also sometimes referred to as LeBaie quick clay and has a rock flour consistency that comprises plagioclase, potassium feldspar, amphibole, and calcite. The clay fraction is usually illite with trace amounts of kaolinite and chlorite.
4   *London clay*: Deposited under marine conditions during the Eocene era, about 30 million years ago. Uplift and erosion removed the overlying deposits. Most deposits are overconsolidated.
5   *Mexico clay*: Sediments of volcanic origin deposited in a lake in the valley of Mexico during late Pleistocene epoch. They have both overconsolidated and normally consolidated layers.
6   *Norwegian marine clay*: Sediments transported by river of melted glaciers of Pleistocene epoch and deposited in the sea. Sediments were subsequently uplifted and leached. Surface deposits were dried and weathered. The soil is normally consolidated below the surface crust.
7   *Shanghai soft clay*: The wet and soft alluvial deposits cover large areas of the populated lower Yangtze Valley, China. They contain three distinct layers, the first 3 m (10 ft) is called the shallow strata of the foundation layer. The second

layer is about 20 m (66 ft) deep, gray colored and soft containing organic matter, shells, and some very fine sand at a depth below 18 m (59 ft).

2.12 Summary

The engineering behavior of soil and rock is governed largely by the nature of these materials; their chemical composition, particle shape and size, weathering environment, and mode of deposition. Various classification schemes have been developed that account for some of these factors to assist in evaluating the geotechnical suitability of a particular site. The most commonly used classification systems are those proposed by AASHTO, USCS, and USDA. Natural soil is defined differently according to perspective, that is, from engineering, geology or soil science, although all of these approaches have bearing on geotechnical performance.

PROBLEMS

2.1 Distinguish clearly between soil identification and soil classification systems. To which of these does the Unified Classification System (ASTM D2487–83) belong, and why?

2.2 Define organic soil. What is the general behavior of this type of soil and why is it not suitable as an embankment fill?

2.3 Without the aid of laboratory facilities, how would you (a) identify an organic soil? (b) identify an inorganic silt? (c) distinguish between a silty clay and a highly plastic clay?, and (d) determine whether a damp fine sand was clean or dirty?

2.4 Determine the group index of the following materials: (a) Assume that an A-4 material has 60% passing a #200 sieve, a liquid limit of 25, and plasticity index of 1, (b) assume that an A-7 material has 80% passing a #200 sieve, a liquid limit of 90, and a plasticity index of 50, and (c) assume that an A-2–7 material has 30% passing a #200 sieve, a liquid limit of 50, and plasticity index of 30. (Note that only the plasticity index portion of the formula is to be used).

2.5 What are the three major rock types and how is each formed? What are the major ones produced by sedimentation? What ones are associated with igneous rocks?

2.6 How do the parent rocks form residual and transported soils? Discuss the types of residual transported soils and their usual characteristics.

2.7 What are the advantages and disadvantages for using a soil science classification in geotechnical engineering applications?

2.8 Define SSR and discuss the significance of this ratio with respect to clay mineral and contaminated soil identification.

Chapter 3

Soils and clay minerals

3.1 Introduction

In general, soils can be grouped into two categories, coarse or fine-grained. Coarse soils have a macro-structure, with larger sized soil particles ranging from 0.074 mm (US sieve #200) to 0.3 m. Fine-grained soils, including clay minerals, have a micro-structure that includes soil particles smaller than 0.074 mm as indicated in Table 2.3. Granular soils tend to have a higher strength, bearing capacity, and hydraulic conductivity relative to fine-grained materials. Figure 3.1 presents typical grain size distribution curves (ASTM D421; ASTM D2217) which represent uniform sand, well-graded embankment soil and weathered soil. The values for D_{60}, D_{30}, and D_{10} indicated in the Figure 3.1 will be explained in Section 3.3 and illustrated in Example 3.3.

3.2 Air–water–solid relationships

3.2.1 General discussion

From an engineering viewpoint, soil may be considered in terms of three states; solid (soil minerals), liquid (water), and gas (air). Figure 3.2 shows two conceptual pictures of a soil system, 3.2(a) depicts how a soil system might look naturally, while 3.2(b) represents an idealization useful for engineering calculations. Note that the variables are defined in Figure 3.2(b). Saturated soils ($V_A = W_A = 0$) and completely dry soils ($V_W = W_W = 0$) represent special cases.

Since the weight of the gaseous phase may, for practical purposes, be assumed to equal zero, then

$$W_T = W_\omega + W_S \tag{3.1}$$

$$V_T = V_A + V_\omega + V_S = V_V + V_S \tag{3.2}$$

This basic weight and volume description of a soil system may be used to derive a number of useful parameters and relationships, discussed as follows.

3.2.2 Parameters used in the air–water–solid relationship

1 *Specific gravity*: Specific gravity is the ratio of the weight of a given substance to the weight of an equivalent volume of a standard substance. The specific gravity

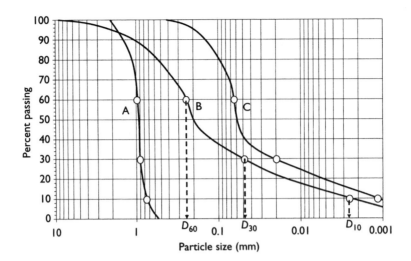

Figure 3.1 Typical types of grain size distribution curves: (A) Uniform sand; (B) Well-graded embankment soil; and (C) Weathered soil. The values for D_{30}, and D_{10} are marked for (B).

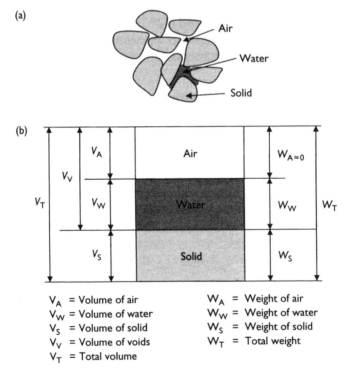

V_A = Volume of air
V_W = Volume of water
V_S = Volume of solid
V_V = Volume of voids
V_T = Total volume

W_A = Weight of air
W_W = Weight of water
W_S = Weight of solid
W_T = Total weight

Figure 3.2 Components of air–water–solid in the soil mass: (a) True soil mass and (b) Idealized soil mass.

therefore represents relative weight and is a dimensionless quantity. In the case of gases the standard substance to which specific gravity refers is air at standard temperature and pressure conditions. In the case of solids and liquids the standard is water at 4°C (39°F). As such, the specific gravity of a solid or liquid indicates how many times heavier it is relative to an equivalent volume of water at 4°C.

2   *Specific gravity of solids, G_s*: The ratio of the weight of the solids to the weight of an equivalent volume of water;

$$G_s = \frac{W_s}{V_s \gamma_\omega} \qquad (3.3)$$

The specific gravity of soil particles is useful not only for sedimentation analysis but also for assessing the volumetric contributions of the different size fractions to a soil system as illustrated in Examples 3.1 to 3.3. Typical values of specific gravity of some common soil minerals are given as follows: illite: range from 2.60 to 2.90; kaolinite: range from 2.50 to 2.61; montmorillonite: range from 2.40 to 2.51; and quartz: range from 2.50 to 2.80.

3   *Specific mass gravity, G_m*: The ratio of the total weight of a given mass of soil to the weight of an equivalent volume of water;

$$G_m = \frac{W}{V \gamma_\omega} \qquad (3.4)$$

4   *Unit weight*: Unit weight is defined as the weight of a given substance per unit of volume. Unit weight must be expressed in correct dimensions. The following expressions of unit weight are important:

5   *Bulk unit weight, γ_b*: The total weight of solids and water (W_T) per unit volume (V_T):

$$\gamma_b = \frac{W_T}{V_T} \qquad (3.5)$$

6   *Unit weight of solids, γ_S*: The weight of solids (W_S) per unit volume of solids (V_S):

$$\gamma_S = \frac{W_S}{V_S} \qquad (3.6)$$

7   *Unit weight of water, γ_ω*: Weight of water per unit volume of water;

$$\gamma_\omega = \frac{W_\omega}{V_\omega} = 62.4 \text{ pcf} = 1.0 \text{ g/cm}^3 \qquad (3.7)$$

8   *Dry unit weight, γ_d*: The dry unit weight of soil is defined as the ratio of weight of dry soil solids to the total volume of soil mass.

$$\gamma_d = \frac{W_S}{V_T} \quad \text{and} \quad W_S = V \gamma_d \qquad (3.8)$$

9   *Submerged unit weight (buoyant unit weight), γ_b'*: The weight of the solids of soil in air minus the weight of water displaced by the solids per unit of volume of soil mass; or the saturated unit weight of soil, minus the unit weight of water;

$$\gamma_b' = \gamma_b - \gamma_\omega \qquad (3.9)$$

Conversion between weight and volume for the aforementioned unit weight expressions involves simple rearrangement:

$$W = \gamma V \quad \text{and} \quad V = \frac{W}{\gamma}$$

10 *Water content (moisture content), ω*: The ratio, expressed as a percentage, of the weight of water in a given soil mass to the weight of solid particles:

$$\omega = \frac{W_\omega}{W_S} \tag{3.10}$$

11 *Void ratio, e*: A void is defined as the volume in a soil mass not occupied by solid mineral matter. The ratio of this volume, which may be occupied by air, water, or other gaseous or liquid material to the volume of solid material is given as

$$e = \frac{V_V}{V_S} \tag{3.11}$$

12 *Porosity, n*: The ratio of the volume of voids of a given soil mass to the total volume of the soil mass:

$$n = \frac{V_V}{V_T} \tag{3.12}$$

Void ratio and porosity both indicate the proportion of void space in a given soil; the one expresses the volume of void space in terms of the volume of solid matter, the other expresses the volume of void space of the total volume of soil mass. They are related as shown in Table 3.1.

13 *Degree of saturation, S*: The degree of saturation is the ratio of the volume of water in a given soil mass to the volume of voids:

$$S = \frac{V_\omega}{V_V} \tag{3.13}$$

3.2.3 Relationship between volume and weight of an idealized soil mass

1 *Relationship between volume and weight:* It is often necessary to establish expressions for the proportions of solid, water, and air in a given soil mass as defined by the terms unit weight, water content, void ratio, etc. This need arises because these terms are useful in describing the engineering characteristics of a given soil. Since all of the defined proportions are expressed in terms of quantities indicating the weight and/or volume of the soil mass or of its components, they are interrelated. A common problem is to express one defined proportion, the value of which is unknown, in terms of other defined proportions, the values of which are known.

2 *Methodology for the solution:* Following the conceptual diagram shown in Figure 3.2, the relationships between volume and weight of an idealized soil mass are

Table 3.1 Summary of soil parameters, definitions, conversion equations units, and ranges

Soil parameters	Definition	Conversion equations	Common units and ranges
Specific gravity, G or G_s	$G = w_s/v_s\gamma_\omega$	$G = Se/\omega$	Dimensionless 2.60–3.20
Moisture or water content, ω	$\omega = W_\omega/W_s \cdot 100$	$\omega = Se/G \cdot 100$ $\omega = (\gamma/\gamma_s - 1)\cdot 100$	Percentage 5–80
Void ratio, e	$e = V_v/V_s$	$e = G/\gamma_0 - 1$	Dimensionless 0.50–3.20
Porosity, n	$n = V_v/V \cdot 100$	$n = (e/1 + e)\cdot 100$ $n = (1 - \gamma_0/G)\cdot 100$	0–100%
Degree of saturation, S	$S = V_w/V_W \cdot 100$	$S = \omega G/e$ $S = \omega\gamma_0/n$	0% (dry soil) to 100% (saturated soil)
Unit weight, γ	$\gamma = W/V$	$\gamma = G + Se/1 + e$	kN/m³, pcf, g/cm³ 16–20 kN/m³
Dry unit weight, γ_0 or γ_d	$\gamma_d = W_s/V$	$\gamma_d = \gamma/100 + \omega\% = G/1 + e$	kN/m³, pcf, g/cm³ 14–18 kN/m³
Saturated unit weight, γ_{sat}	$\gamma_{sat} = W_s + V_s\gamma_\omega/v$	$\gamma_{sat} = G + e/1 + e$	kN/m³, pcf, g/cm³ 16–21 kN/m³
Buoyant unit weight, γ_b	$\gamma'_b = \gamma'_b - \gamma_\omega$ $\gamma_\omega = $ Unit weight of water	$\gamma' = (G - 1)\gamma_\omega/1 + e$ $= G - 1/1 + e$	kN/m³, pcf, g/cm³ 7–10 kN/m³

summarized in Table 3.1, including soil parameters, definitions, conversion equations, units, and ranges. It must be noted that problems in air–water–solid relationships, like all other problems, can be solved in a variety of ways; however, a specific methodology for the solution of these problems has been developed that offers the following advantages: (a) it will generally lead to the most direct solution to a given problem, (b) it minimizes the possibility of mistakes due to improper use of units, and (c) it provides a means for determining rapidly whether or not there are sufficient data for the solution of a problem, or for determining what additional data may be necessary in a given case. The basic methodology is illustrated in Examples 3.1 and 3.2. There are three general types of problems:

1   For problems in which only weights, W, and/or volumes, V, are known, all the necessary blanks must be filled in without assuming any unknown quantities.

2   For problems in which only defined ratios are known, assume an unknown in terms of which one of the known quantities can be expressed as a product. Fill in the necessary blanks using the known quantities and the one unknown.

3   For problems in which defined ratios and weights and/or volumes are known, apply the method of step (1). If a solution is not possible, assume an unknown in terms of which the required quantity can be expressed as a product. Introduce this product into the sketch and obtain an equation for the desired quantity from the equalities of the sketch. If the problem is determined, the unknown can be eliminated or canceled.

EXAMPLE 3.1

Given the total weight, W_T, the total volume, V_T, and the weight of solids, W_S, for a saturated mass of soil, and the specific gravity, G_s determine expressions for water content and void ratio.

SOLUTION

1   Fill in the spaces representing total weight, total volume, and weight of solids as identified in Figure 3.2.
2   The weight of water is determined from the relationship, $W_T = W_S + W_W$; the volume of water is obtained by from Equation (3.7) and the unit weight of water; and the volume of solids is obtained from the relationship, $V_T = V_S + V_V$, since the soil is saturated, the volume of voids is equal to the volume of water. If the soil was not saturated, Equation (3.3) could be used to determine V_S. All the values identified in Figure 3.2 have now been determined.
3   The desired expressions for water content, ω, and void ratio, e, follow directly from their definitions:

$$\omega = \frac{W_T - W_S}{W_S}$$

$$e = \frac{V_T - V_S}{V_S}$$

EXAMPLE 3.2

Given the void ratio, e, and the specific gravity of solids, G_s, for a saturated soil mass, determine expressions for water content (ω) and bulk unit weight (γ_b).

SOLUTION

The solution is provided with the following steps:

1   An unknown quantity, V_S, is introduced.
2   The volume of voids can now be expressed as, eV_S, and the total volume as, V_T $(1 + e)$.
3   The weight of water is obtained from the unit weight of water, the weight of solids from the definition of the specific gravity of solids, and the total weight from their sum.
4   The required expressions for the water content, ω, and mass unit weight, γ_m, can be written directly from their definitions:

$$\omega = \frac{eV_S}{V_S G_S} = \frac{e}{G_S}$$

$$\gamma_m = \frac{V_S (e + G_S)}{V_S(1 + e)} \frac{(e + G_S)}{1 + e}$$

3.3   Geometric relationships of granular soil systems

For practical application, granular soils with macro-structured materials are measured by particle size, shape, and packing characteristics. Often particle shape is described by general terms such as: spherical, rounded, angular, and irregular. Particle size is simply given as differentiation between boulders, cobbles, gravel, and sand. Particle shapes may also be characterized by the closest geometric form, such as *cylindrical, elliptic, cubic, prismatic, plate*, and *needle* shape.

3.3.1   Grain size distribution curves

Some commonly used parameters for the description of granular materials include effective size, uniformity, and concavity coefficients, the fineness modulus and the Santos constant. These are discussed as follows:

1   *Effective size, D_{10}:* The term effective size, D_{10}, as determined from sieve analysis (ASTM D422) is the size of the screen opening that permits 10% of the granular material to pass and retain 90% as indicated in Figure 3.1.

2   *Uniformity coefficient, C_u:* The uniformity coefficient, C_u, is the ratio of the size of screen openings passing 60% to that passing 10%:

$$C_u = \frac{D_{60}}{D_{10}} \tag{3.14}$$

where C_u = uniformity coefficient; D_{60} = diameter at which 60% of the soil is finer; and D_{10} = diameter at which 10% of the soil is finer.

3   *Gradation coefficient, C_G:* This coefficient can be used to indicate the characteristics of grain size distribution curve or the shape of the curve between the D_{10} and D_{60}. The gradation coefficient also is referred to as the *coefficient of concavity, C_c. D_{30}* is the diameter at which 30% of the soil is finer. Other notations are the same as previously stated.

$$C_G = \frac{(D_{30})^2}{(D_{10})(D_{60})} \tag{3.15}$$

4   *The fineness modulus:* The fineness modulus is a measure of gradation developed by Abrams in 1918 and widely used in concrete technology. It is defined as one-hundredth of the sum of the cumulative percentages retained in the sieve analysis when using the US Standard sieve series: 1–1/2″, 3/4″, 3/8″, #4, #8, #16, #30, #50, and #100.

5   *The Santos constant:* The Santos constant is a measure of influence of gradation on consistency properties of soils. It is defined by the formula

$$a = \frac{\Sigma\, y}{100\, n} \tag{3.16}$$

where **a** = Santos constant, y = sum of percentages passing each of a set of n sieves, that with the smallest openings being a #200. (The set of US Standard sieves usually consists of the following size openings: #8, #16, #30, #50, #100, #200.)

EXAMPLE 3.3
Determine the effective size, D_{10}, and the uniformity coefficient, C_u, for the uniformly graded soil (Curve A) and well-graded soil (Curve B) as indicated in Figure 3.1.

SOLUTION

1   From uniformly graded soil of Curve A in Figure 3.1, the

    D_{10} = 0.75 mm, and
    D_{60} = 1.0 mm, then from Equation (3.13), the uniformity coefficient is
    C_u = 1.0/0.75 = 1.33.

2   From the well-graded soil of Curve B in Figure 3.1, the

    D_{10} = 0.025 mm, and
    D_{60} = 0.25 mm, then the uniformity coefficient is
    C_u = 0.25/0.025 = 10.0.

Results from Example 3.3, show that for uniform sand, the uniformity coefficient is 1.33, and the well-graded embankment soil is 10.0. The smaller the value of C_u, the more uniform the material. The lowest theoretical value is 1.0 while values for weathered soils can reach as high as 400.

The finer fraction of a soil sample, that is, the total percent passing the #200 sieve (0.074 mm) has a significant impact on the overall engineering characteristics, including permeability (Ch. 5), frost depth and frost heave (Ch. 6), shear strength (Ch. 10) as well as pollution potential, illustrated in Figure 3.3. In particular, pollution potential may increase with increasing percent of particles passing the #200 sieve because of

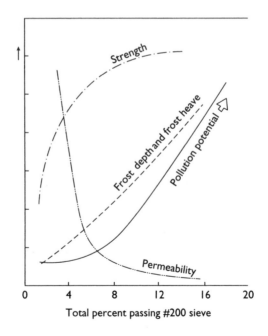

Figure 3.3 Total percent passing #200 sieve relating to soil behavior.

Source: National Crushed Stone Association, 1972 with modification.

increasing surface area available for chemical reactions. The permeability tends to decrease as finer particles fill void spaces. Initially, strength actually increases with increasing percent of particles finer than the #200 sieve because of increased particle to particle contact and friction. Note that Figure 3.3 does not show these relationships beyond 20% passing the #200 sieve and further extrapolation will yield different results. For example, a continued increase in finer particles will usually result in a reduction in strength, particularly as clay minerals with plasticity and lower friction angles affect the behavior. Also, frost heave is more of a concern for silts than it is for sand or clay, and so continued adjustment of the grain size distribution with more clay particles will ultimately reduce the potential for frost heave. Pollution potential and permeability, however, are expected to follow the trends shown, with an asymptotic approach to some ultimate value.

3.3.2 Volumetric determination

The size of granular materials can be estimated by volumetric relationships or by direct measurement. The volumetric relationships in granular systems can be expressed by phase volume, absolute volume, and total volumes. A brief description of each case is presented as follows: (a) *phase volume* is the portion of the total volume contributed by the various solid, liquid, and gaseous components of the system, expressed as a fraction or percent of the bulk volume; (b) *absolute volume* is the actual volume occupied by the various phases, usually expressed in volumetric units such as cubic feet, cubic yards or cubic meters; (c) *total* or *bulk volumes* are the sum of the phase and absolute volumes.

3.3.3 Surface area measurements for granular soils

Another characteristic of soils that has bearing on engineering behavior is the *specific surface area (SSA)*. SSA is defined as the surface area of particles per unit volume or mass and units are often expressed as m^2/m^3 or m^2/g. The surface area of particles may be measured directly or indirectly, and include: (a) Slide calipers for large pieces; (b) smaller particles can be determined by sieve analysis, sedimentation and elutriation methods; and/or (c) adsorption techniques where a known amount of fluid such as water, ethylene glycol monoether or liquid nitrogen is used to coat soil particles. Indirect estimation of surface area from consideration of particle size and shape can be used to explain the relationship of particle size, surface area, and pollution potential. While all methods can be used in granular soils, method (c) is more suited toward fine-grained soils, which will be discussed further in Section 3.6. The surface area of soil particles per unit volume of a solid is a function of the particle size and for particles of spherical and cubic shape can be derived as

Surface area of spherical shape = $3/r$ (3.17a)

Surface area of cubic shape = $6/d$ (3.17b)

where r = radius of sphere and d = edge length of cube. Therefore, values for the amount of surface area per unit solid volume for various soil fractions can be

Table 3.2 Typical specific surface area of various soil types

Soil type/mineral	Specific surface area (m²/g)
Fine gravel	0.0011
Coarse sand	0.0022
Medium sand	0.0045
Fine sand	0.0090
Silt	0.0450
Kaolinite	7–30
Illite	65–100
Montmorillonite	600–800

Source: Thibodeaux, J. L. *Environmental Chemodynamics*, 2nd edition © 1996 Wiley. This material is used by permission of John Wiley & Sons, Inc.

estimated. Typical ranges for the SSA of various soils are given in Table 3.2. This table indicates that SSA increases with decreasing particle size. Moreover, if clay particles are present in a soil, even in small percentages, they contribute the overwhelming portion of the total solid surface area.

3.3.4   *Particle shape measurements*

The strength of an assemblage of soil particles is governed by the ease with which individual grains can be moved relative to one another. For movement (deformation) to occur, the force of friction must be overcome. Naturally, the shape of an individual particle will influence the ease or difficulty with which individual particles will move, that is, consider the difference between flat plates, round spheres, rough surfaces, smooth surfaces, etc. These shapes can be expressed numerically, using terms such as sphericity, volumetric coefficient, elongation and flatness ratios, and shape factor. Brief discussions of each term are presented as follows:

1   *Sphericity, S*: Sphericity was defined by Wadell (1932) as the ratio of the surface area of a sphere of the same volume as the particle or as

$$S = \frac{\text{Surface area of sphere}}{\text{Surface area of particle}} \tag{3.18}$$

2   *Volumetric coefficient (VC)*: Volumetric coefficient, VC, was defined by Joisel (1948) as the ratio of the actual volume of a particle to that of a sphere in which it can just be enclosed as shown in Equation (3.19).

$$\text{VC} = \frac{6V}{a^3} \tag{3.19}$$

where VC = volumetric coefficient, V = volume of particle, and a = largest dimension of the particle. The typical value of VC of various shapes of particle are: (a) *Sphere*: VC = 1.00, (b) *Cubic*: VC = 0.37, (c) *Round gravel*: VC = 0.34, (d) *Angular stone*: VC = 0.22; (e) *Plates*: VC = 0.07, and (f) *Needles*: VC = 0.01.

3   *Elongation ratio, flatness ratio and shape factor*: The particle shape expressed by
elongation ratio, flatness ratio, and shape factor are proposed by Zinng (1935). If
a is the largest, **b** the intermediate, and **c** the smallest dimension of a particle, then,

Elongation ratio, $q = b/a$
Flatness ratio, $p = c/b$ (3.20)
Shape factor, $F = p/q = ca/b^2$

Aschenbrenner (1956) developed an equation by means of which **p** and **q** values can
be converted into sphericity, ψ.

$$\psi = \frac{12.8\,(p^2 q)^{1/3}}{1 + p(1 + q) + 6[1 + p^2]^{1/2}}$$ (3.21)

Lees (1964) modified and proposed a soil particle classification based on the particle
shape and subdivided into four shape groups namely discs, equidimensional, blades,
and rods. Combining their results, a single classification chart was modified
(Winterkorn and Fang, 1991) as shown in Figure 3.4. This chart can be used for characterization of railroad ballast, and aggregates used in asphalt and concrete mixtures.

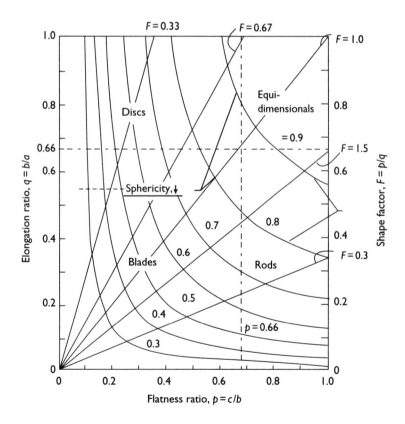

Figure 3.4 Classification of granular soils based on particle shapes.

Source: From Foundation Engineering Handbook, 2nd Edition, Fang, H. Y. ed., p.118, Ch. 3, *Soil Technology and Engineering Properties of Soils*, Winterkorn, H. F. and Fang, H. Y., Copyright (1991) by Van Nostrand Reinhold. With kind permission of Springer Science and Business Media.

3.4   Packings of particles and their primary structure

3.4.1   General discussion

Granular systems are devoid of inter-particle cohesion, and the individual grains are independent of each other except for frictional interaction and geometric constraints to the type of packing. The component particles may be of any size from the finest sand to gravel and cobbles. Primary structure refers to the natural arrangement of the constituent particles of a soil in what may be regarded as a continuous system such as a body of sand or gravel. The theoretically possible types of "continuous, incompressible, uniform packing of identical spheres" are given in Table 3.3 with their porosity range from 26% to 47% (Kezdi, 1964). With sand-sized spheres and rounded sand particles, it is very difficult in practice to get uniform packing with porosity higher than 50% or less than 36% nor is the packing between these limits of a uniform character. Rather, photographs show that in a sand mass with a given bulk porosity, there are domains of regularly packed particles, in various orientations, separated by inter-phases of more loosely and irregularly packed particles. The lower the void ratio, the lesser the volume proportion of the inter-phases and the greater of more orderly and densely packed domains.

3.4.2   Systematic packing of uniform spheres

The systematic arrangement of spheres in porous media was first studied by Slichter in 1899. An understanding of the packing of grains is of importance in many branches of science and technology, such as ceramics, concrete and asphalt technology, crystallography, and geotechnology. A brief discussion is presented as follows:

1   *Square layer arrangement*: Three different systems may be formed by stacking square horizontal layers one above another (Figure 3.5(a)). There are three cases including: (a) *Cubic system*: The cubic system obtained when each sphere in the next horizontal layer has its center vertically above that of the sphere below; (b) *Orthorhombic system*: When the center of the upper sphere is offset a distance R

Table 3.3 Properties of regular packings of uniform spheres

Type	Void ratio	Porosity (%)	Coordination No.	Layer spacing
Cubic	0.91	47.64	6	R 4
Orthorhombic	0.65	39.54	8	R 3[a]
				R 4[b]
Tetragonal/ spheroidal	0.435	30.19	10	R 3
Rhombohedral	0.35	25.95	12	R 2[c]
				R 2/3[d]

Source: Farouki, O. T. and Winterkorn, H. F. 'Mechanical properties of granular systems'. In Highway Research Record No. 52, Highway Research Board. National Research Council, Washington DC, 1964, pp. 10–42. Reproduced with permission of the Transportation Research Board.

Notes
(a) Case 2; (b) Case 4; (c) Case 3; (d) Case 6 (Figure 3.5).

in the direction of one of the rows (R is the radius of the spheres); (c) *Rhombohedral system* (Pyramidal or face-centered cubic): This system is sometimes called pyramidal or face-centered cubic in analogy with crystal lattices when the center of the upper sphere is moved a distance $2R$ in a direction bisecting the angle between two sets of horizontal rows.

2   *Rhombic layer arrangement*: In a similar manner, three types of packing may be formed by stacking simple rhombic layers (Figure 3.5(b)) one above another. (a) *Orthorhombic system*: The spheres of the next rhombic layer are placed in such manner that the center of each sphere lies vertically above the sphere below it; (b) *Tetragonal-spheroidal system*: Each sphere in the next rhombic layer rests in the

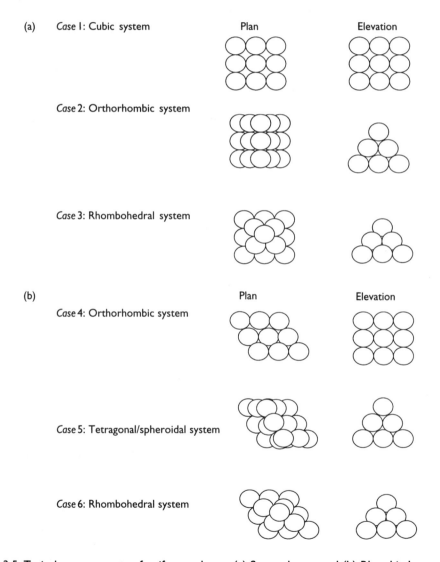

Figure 3.5 Typical arrangements of uniform spheres: (a) Square layers, and (b) Rhombic layers.

Source: Farouki, O. T. and Winterkorn, H. F. 'Mechanical properties of granular systems'. In Highway Research Record No. 52, Highway Research Board. National Research Council, Washington DC, 1964, pp. 10–42. Reproduced with permission of the Transportation Research Board.

cusp between two spheres in the layer below; and (c) *Rhombohedral system*: When each sphere of the next rhombic layer is placed in the hollow formed by three spheres of the lower layer. This case is also referred to as the closely packed hexagonal system and it is also termed tetrahedral.

Some important physical properties of packed uniform spheres are given in Table 3.3. In examining Table 3.3, the coordination number is the number of contacts that a typical sphere makes with its neighbors. The unit cell may be defined as that smallest portion of the system which gives a complete representation of the manner of packing. Note that the rhombohedral system is the densest possible state.

3.4.3 Packing of spheres with different sizes

A theoretical approach to packing of spheres has two general types, one is a systematic packing of spheres of different sizes and the other is a random packing of unequal spheres. Experience indicates that random packing is more representative of a real densely packed system, although systematic models are useful in establishing expected ranges of porosity. Brandt (1955) developed a packing arrangement whereby primary spheres are packed randomly to a porosity, n, and smaller uniform spheres are packed to the same porosity in the voids of the primary system. Still smaller spheres are packed to the same porosity in the remaining voids and so on.

3.5 Mechanical behavior of granular systems

3.5.1 Evaluation of mechanical behavior of granular systems

A primary reason for developing idealized packing arrangements of soil particles is to allow for mathematical modeling of soil behavior. Such models are generally written as computer-based applications and are useful in predicting field performance. A theoretical model for determining the mechanical behavior of a granular system is an arrangement of discrete spheres in direct elastic contact with one another. There are several approaches including contact theory, assessment of the wave velocity, and stress-strain solutions for idealized granular systems. A brief discussion of each approach is presented as follows:

1 *Contact theory:* The classical theory of contact predicts that when two contacting elastic spheres are compressed by a force along their lines of center, there will be a circular area of planar contact. Within the limit of contact theory, there are several classical solutions available as discussed by Timoshenko and Goodier (1951).

2 *Velocity of waves through granular system:* To determine the velocity of compressional waves through a granular system, the grains have been assumed to be in elastic contact with each other as noted in the contact theory. This model considers only limited cases and uses equal spheres as noted in Cases 3 and 6 in Figure 3.5(a) and (b). Theoretical results are in qualitative agreement with experimental data although they predict lower values of velocity wave propagation. This is because only the normal components of the forces at the points of contact are considered, while the tangential components are neglected. Because the tangential

stiffness of a contact has the same order of magnitude as its normal stiffness, it must be accounted for in certain idealized granular systems.

3 *Continuous grading:* Continuous grading yields low porosity mixtures requiring little compactive effort and therefore of great practical importance in soil stabilization and in the making of concrete with hydraulic and bituminous cements. The greater the range from the maximum to the minimum particle size, the lesser the porosity of the system. Representative values for various ratios of D_{max}/d_{min} can be calculated with the equations (Winterkorn, 1970).

(a) For rounded gravel and sand
$$n, \% = 38.5{-}8 \log_{10} D_{max}/d_{min} \qquad (3.22)$$

(b) For crushed stone
$$n, \% = 47.5{-}8 \log_{10} D_{max}/d_{min} \qquad (3.23)$$

where n = porosity, D_{max} = diameter of the largest particle, and d_{min} = diameter of smallest particle. The calculated values are for rounded mixtures and lie between those for maximum and minimum obtainable densities. For most natural materials of relatively narrow gradation, the range of easily obtainable and reproducible porosity lies between 36% and 46%.

4 *Stress–strain solutions for idealized granular systems:* Tangential forces or twisting moments at the contacts between the grains of a granular system cause the load–displacement relations to be nonlinear and inelastic. Therefore, the mechanical response of the system depends not only on the initial loading but also on the history of loading. As such, the stress–strain relations at any point of the system must be expressed as increments of stress related to increments of strain. Further discussion on stress–strain relations of granular soil are presented in Section 10.11.

3.5.2 Experiments on packing characteristics

Packing characteristics of granular materials can be assessed in terms of bulk density. Bulk density can be determined by placing a known mass of soil into a container of known volume (see Equation (3.4)). Table 3.4 presents the bulk density for typical granular materials under three commonly used conditions, loose, packed (dense), and working state.

Table 3.4 Bulk density for some typical granular materials

Type of material	Dense		Loose		Working	
	lb/ft^3	kN/m^3	lb/ft^3	kN/m^3	lb/ft^3	kN/m^3
Aluminum ore	66	10	59	9	60	9
Clay, kaolin	38	6	26	4	30	5
Coal, bituminous	62	10	49	8	52	8
Mica ore	42	7	27	4	33	5
Sand (dry)	108	17	97	15	98	15
Sand (moist)	91	14	61	10	71	11

Confusion may result by using the term "density," when "unit weight" is the actual measured quantity as discussed in Section 3.3. Density generally refers to the mass (measured in grams) while unit weight refers to weight (measured in lbs or N). The two are related, as weight is equal to mass times the acceleration due to gravity (9.81 m/s^2). Bulk density has attained wide engineering usage, especially in the geotechnical construction field. The measured weight includes the solids in the system plus any moisture or gas within the particle or adhered to its surface. The measured volume includes both the solids and the voids between the particles. Thus, bulk density can vary from a maximum value with the particles packed as closely as possible, to a minimum value when there are large void spaces. Factors affecting the packing state including mechanical densification, moisture content, container characteristics, particle shape, surface, size, and distribution have been reviewed and discussed by Fowkes and Fritz (1974). The effect of particle characteristics on soil strength has been reported by Koerner (1970). Soils with more angular particles and lower sphericities have significantly higher angles of shearing resistance.

3.5.3 Winterkorn's macromeritic liquids theory

The *macromeritic liquids theory* developed by Winterkorn in 1953 can be used for evaluation of mechanical behavior of cohesionless granular soil including liquefaction due to earthquake, ocean floor slope stability, river and massive beach erosion as well as concrete mixtures. This interesting theory is based on the following concepts that tie chemistry and geotechnical engineering together:

1 *Relationships for solid and liquid states*: These relationships include the mole volumes of chemical substances at the absolute zero point, the melting and boiling points at atmospheric pressure, as well as the critical temperature as defined for solid and liquid states.
2 *Critical void ratio and melting point*: There exists a critical void ratio (CVR) introduced by Reynolds in 1885 (Casagrande, 1936; Taylor, 1948) to which a denser particulate system must expand for shear to take place and a looser one collapses. While the CVR decreases with increasing confining pressure, the ratio of the volume of a uniformly sized sand at its CVR to that at its greatest density is approximately the same as that of the mole volume of a liquid at its melting point to that at the absolute zero temperature. Accordingly, the CVR can be considered as a volumetrically defined *melting point* of a "macromeritic" system.
3 *Internal friction of macro-particle systems*: For simple molecular liquids, Batschinski (1913) had shown that their viscosity or internal friction at different temperatures could be expressed as a function of the mole volumes at the respective temperatures. This Batschinski-type of formula for the internal friction of macro-particle systems may be written as

$$\tan \phi = \frac{C}{e - e_{min}} \qquad (3.24)$$

Where, tan ϕ = coefficient of internal friction, C = constant; e = actual void ratio (in situ condition), and e_{min} = minimum void ratio of the system.

4   *Verification of the macromeritic shear equation*: Theoretically, Equation (3.23) should apply only for systems having void ratios above the CVR. However, it has been applied to a large body of experimental data and the following points have been summarized (Farouki and Winterkorn, 1964, 1970a): (a) Equation (3.23) yields a good reproduction of experimental data obtained within the same range of confining pressure, that is the same range of stored strain energy; (b) the C value remains relatively constant as long as the same type of shear test (Sec. 10.5) is used; and (c) the e_{min} values generally decrease with increasing strain energy.

5   *Dynamic liquid states*: Granular systems with void ratios above their CVR are in a potentially liquid state, and hence may represent a liquefaction hazard. They may be changed into actual macromeritic liquids, not only in specific shear zones but throughout the system. This occurs through the addition of sufficient kinetic energy (e.g. an earthquake, localized loading) to keep the particles in a translatory and rotary movement that prevents the system from decreasing its void ratio below the CVR.

3.6   Cohesive soil systems

3.6.1   Structure of cohesive soil

A great deal of discussion on granular soil systems has been presented in Sections 3.2 to 3.5. However, cohesive soils are much more complicated than granular soils, in part because the cohesive soil has a large surface area per unit volume and a complex structure. Structural units of cohesive soils (clay) are composed primarily of silica (Si) and alumina (Al) or ferric oxide with varying amounts of other elements. The silicate layer minerals are built up of two fundamental structural units, the silica sheets and the gibbsite (or brucite) sheets. These are essentially oxygen structures that are "glued" together by Si, Al, Mg, and other cations. The clay minerals are formed by various combinations of tetrahedral and octahedral sheets. These structural sheets are presented in Figure 3.6. In examining Figure 3.6, the three types of symbols that represent silica, brucite, and gibbsite sheets are listed. In general, type 1 is commonly used in geotechnical engineering, types 2 and 3 are used in geological and agricultural sciences.

3.6.2   Ionic and electrical structures of clay minerals

Since solid particles react at their surface (both external and internal surfaces of soil particles), special interest is given to clay minerals. This is because their smallness and usually platy, fibrous, and ribbon-like shapes contribute most to the total surface area of a cohesive soil system. The structural arrangement of the oxygen ions is the predominant feature, dictating the extent to which electrical neutralization by available cations is necessary. By way of a chemistry review, oxygen atoms have an oxidation state of -2, and this negative charge must be balanced by positive charges to satisfy electrical neutrality. Silica (Si), for example, has a charge of $+4$. The primary reason for excess negative charge development is *isomorphous substitution*. Isomorphous substitution occurs when cations of similar size, but different charge, substitute for one another. For example, if Si $(+4)$ is replaced by Al $(+3)$ then there is a net

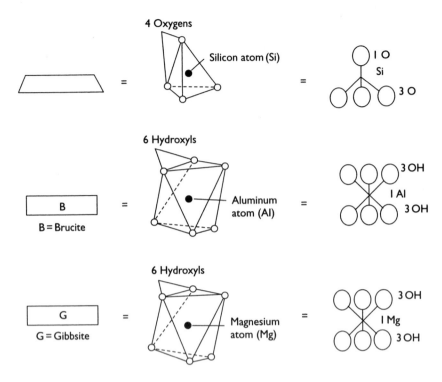

Figure 3.6 Typical structural sheet of cohesive soil: (a) Silicate sheet; (b) Brucite sheet; (c) Gibbsite sheet.

deficiency of 1, or a -1 charge for the overall structure. If other cations do not provide enough positive charges for neutralization within the mineral structure, then the excess negative charges must be neutralized at the mineral surface by adsorption of cations from the environment (pore fluid). This phenomena gives rise to another property known as base, ion or cation exchange capacity (Sec. 4.7). The great variety of possible substitutions of cations within the clay mineral structures and inter-growth of various structural layers leads to a great diversity of actual clay minerals. Isomorphous substitutions for typical clay minerals include aluminum (Al^{+3}) for silicon (Si^{+4}), and magnesium (Mg^{+2}) or iron $(Fe^{+2}$ or $Fe^{+3})$ for aluminum.

3.6.3 Identification and determination of clay minerals

For identification and determination of clay mineral types, there are numerous laboratory techniques such as X-ray diffraction (XRD) analysis, differential thermal analysis, scanning electron microscope, and others. The most common approach in engineering practice is to infer the mineralogy from the Atterberg limits (Ch. 2).

1 *X-ray diffraction analysis*: Clay minerals are crystalline, and their physical structure is defined by a unique geometry. XRD is used for the identification of minerals based on this unique crystal structure. In XRD, characteristic X-rays of a particular wavelength are passed through a crystallographic specimen. The specimen then diffracts these X-rays according to its crystal structure.

2 *Differential thermal analysis*: Differential thermal analysis (DTA) takes advantage of water molecules in clay. It involves the simultaneous heating of a test sample and a thermally inert material at a constant rate of about 10°C/min (18°F/min) to over 1000°C (1800°F) and continuously monitoring the differences in temperature between the sample and the inert material. Differences in temperature between the sample and inert material represent heat-induced reactions in the sample and are characteristic of a given chemical composition.

3 *Scanning electron microscope*: The scanning electron microscope (SEM) can be used for magnifications exceeding 650,000 × with a depth of field that is more than 3000 times greater than for a light microscope. The clay particles themselves and cracking surfaces through soil masses may be viewed directly, and it is an effective method for measuring clay particles.

3.6.4 *Particle size measurement of fine-grained soils*

1 *Sieve analysis:* Routine sieve analysis of soil particles smaller than US #200 sieve (0.074 mm) has been standardized by ASTM (ASTM D1140). The practical sieve size limit is the #400 sieve (0.038 mm) and so this technique is limited for clays which can have much smaller particle sizes.

2 *Hydrometer analysis:* For soils whose particle size is smaller than US #200 sieve, the hydrometer analysis (ASTM, D422) is generally preferred. Soil is mixed with water to form a slurry in a tall glass cylinder. A hydrometer is then placed in the slurry at various time intervals to measure the density of the slurry. As the particles settle out of suspension, the density of the remaining fluid decreases and the hydrometer settles. Naturally, larger particles settle first while smaller particles settle later, and Stokes law is used to relate the size of the particle to the rate at which it settles. The result is a curve similar to Figure 3.1, and in fact many times an overall grain size distribution curve is developed on the basis of both sieves and the hydrometer. Stokes law holds true only for non-hydrated spherical particles that are not so large that steady fall conditions are not attained during the time and distance of fall available and not so small that counter displacement by Brownian movement equals or exceeds the displacement due to gravity. Despite these limitations, the law is useful for soil particles within the range of 0.1 and 0.001 mm.

3 *Surface area measurements:* Surface area measurements for coarse-grained soil have been discussed in Section 3.3. However, for the fine-grained soil presented herein, it must be pointed out that the surface area for fine-grained soil is more significant than coarse-grained soil. For clay particles less than two microns in dimension, the surface area to mass ratio is high. As such, the electric forces at the particle surfaces are relatively strong, influencing the overall behavior. A knowledge of the specific surface area is useful to compute some basic properties, such as electric charge density, particle spacing, ion exchange capacity, and pollution potentials. There are numerous methods available for estimating the specific surface area, however, a simple equation proposed by Sridharan and Rao (1972) is presented as follows:

$$S = \frac{\omega}{M} \frac{N}{10^4} A \, 10^{-16} \qquad (3.25)$$

where S = specific surface area, m²/g, ω = equilibrium moisture content in g̊ water adsorbed per g of soil, N = Avogadro's number (6.025×10^{23}), M = molecular weight of water (18.016 g), and A = area in square Angstroms per water molecule (10.8 Å²). Hence Equation (3.24) can be reduced to:

$$S = 3612 \ \omega \ (m^2/g) \qquad (3.26)$$

4 *Relationship between coarse- and fine-grained soil particles:* The determination of soil particles smaller than US Standard #200 sieve by the hydrometer test is one of the most time-consuming tests associated with routine soil analysis. A set of statistical relationships between coarse-fine-grained soil particles may be established, and estimates of percentage material passing can be computed as:

$$\% \text{ passing } 0.02 \text{ mm} = (\% \text{ passing } \#200) \ (0.967) - 15.8 \qquad (3.27a)$$
$$\% \text{ passing } 0.005 \text{ mm} = (\% \text{ passing } \#200) \ (0.811) - 24.7 \qquad (3.27b)$$

$$\% \text{ passing } 0.002 \text{ mm} = (\% \text{ passing } \#200) \ (0.450) - 12.3 \qquad (3.27c)$$

In view of the time required for thorough laboratory testing and the personal error involved in the results obtained, it is evident that the use of the above statistical approach will reduce the time required to obtain results consistent with laboratory testing methods. The data used for establishing these are based on data obtained from AASHO Road Test and more than 300 sieve analysis (ASTM D1140) and hydrometer (ASTM D422) tests from three borrow pits. However, it must be noted that these types of experimental equations are limited to the geologic formations on which they are based.

3.7 Fundamentals of clay mineralogy

3.7.1 *General discussion*

The concepts of clay mineralogy are necessary to study the behavior of clay and clay–water system. Work began in the field by about 1887 by LeChatelier and Lowenstein in 1909. By 1923–1924, these concepts were well established by Hadding of Sweden and Rinne of Germany. The US Geological Survey (USGS) began studies sometime in 1924 by Ross and Shannon. Engineering applications of clay mineralogy concepts relating to foundation engineering were given by Casagrande (1932) and to the soil stabilization and highway construction materials by Winterkorn and Baver (1934) and Winterkorn (1937). Further applications in various geotechnical problems are presented by Grim (1968). Recent developments are given by Velde (1992) and Mitchell (1993).

A *mineral* is defined as a naturally occurring inorganic substance with an orderly internal arrangement and with chemical composition and physical properties that are either fixed or vary within definite limits. Clay contains various natural minerals, and in most clays these minerals are either kaolinite, illite, montmorillonite, attapulgite, halloysite, or combinations thereof. The basic ionic and electrical structure of the soil minerals and mineral surfaces have been discussed in Section 3.6. The strength of a clay system and the bonds between particles will be further discussed in this section including the basic terminology and concepts of clay mineralogy as relevant to geotechnical engineering.

3.7.2   Clay particle bonds and linkages

There are two major bonds within atoms and molecules comprising clay particles, the primary bond and secondary bond:

1   *Primary bond*: The primary bond is also called the *inter-atomic bond* and is the bond between atoms forming molecules. These bonds are strong and are not broken during conventional engineering works.
2   *Secondary bond*: The secondary bond occurs when atoms in one molecule bond to atoms in another molecule. There are two types of secondary bonds: (a) van der Waals bonds, and (b) Hydrogen bonds. The van der Waals bond is weaker than the hydrogen bond and is attributed to instantaneous imbalances in the electron cloud surrounding an atom. Linkages are a weaker form of association than bonding and occur mainly through adsorbed water, water–dipole or dipole–cation–dipole types of arrangements. Many natural soil deposits consist of (a) water linkage (adsorbed water), (b) water dipole linkage, and (d) dipole–cation–dipole linkage as illustrated in Figure 3.7.

3.7.3   Interaction of clay particles

1   *Electric charge*: All clay particles carry an electrical charge. Theoretically, they can carry either a net negative or net positive charge, however, net negative charges are more common.
2   *Attractive forces*: When two clay particles are close to each other in face-to-face arrangement, an attractive force exists between the negatively charged surfaces and the intervening exchangeable cations.
3   *Repulsive forces*: If the atoms in adjacent surfaces are so close that their outer electron shells overlap, a repulsive force results. When the various attractive and repulsive energies are summed algebraically, the energy of interaction is obtained.

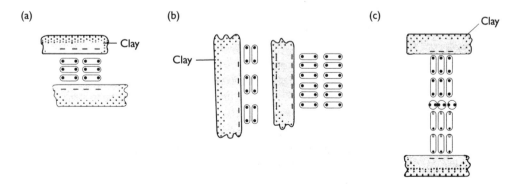

Figure 3.7 Various types of linkage between soil particles: (a) Water dipole linkage; (b) Absorbed water; (c) Dipole–cation–dipole linkage.

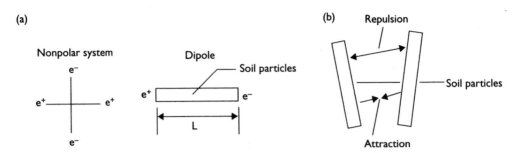

Figure 3.8 Basic characteristics of inter-particle structures: (a) Polar system and dipole; and (b) Repulsion and attraction forces between particles.

Table 3.5 Typical dipole moment of various substances

Substance	Dipole moment (Debye)
Acetic acid	1.74
Acetone	2.90
Aniline	1.55
Benzene	0
Carbon tetrachloride	0
Water	1.89

Some basic characteristics of particle interactions in soil are illustrated in Figure 3.8.

4 *Dipole moment*: If the center of action of the positive charge coincides with the center of action of the negative charge, the system has no dipole moment and is called a *non-polar system*; otherwise, it has a dipole moment and is called a *polar system*. The unit of the dipole moment is a *debye*. A debye is defined as a molecule in the unit that electrical charges separate by 2.1×10^{-11} m. Typical data for various substances are presented in Table 3.5.

3.7.4 *Clay particle structures and arrangement*

1 *Particle arrangement*: Two basic types of particle orientation are frequently considered, the flocculated orientation, which is a edge-to-face arrangement and the dispersive orientation, which is a face-to-face arrangement as illustrated in Figure 3.9. In the natural case, the particle arrangement is generally more random with three-dimensional orientations. Major factors which cause soil to tend toward flocculation include increasing the concentration of the electrolyte, valence of the ion, and temperature, or decreasing the dielectric constant of the pore fluid, size of the hydrated ion, pH value, and anion adsorption. A summary of these factors in relation to double-layer thickness will be discussed further in Section 6.8. Basic properties for common clay minerals are provided in Table 3.6.

2 *Clay structures*: For the idealized clay structures, various researchers have proposed several structures. Barden and Sides (1971) proposed that clay structures

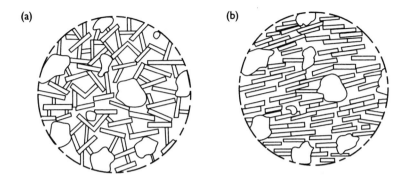

Figure 3.9 Clay particle structure and arrangement: (a) Flocculated structure and (b) Dispersive structure.

Table 3.6 Geotechnical properties of some common clay minerals

Properties	Kaolinite	Illite	Montmorillonite
Specific gravity, G_s	2.61	2.60	2.51
Liquid limit, LL	50–62	95–120	150–170
Plastic limit, PL	33	45–60	55
Plasticity index, PI	20–29	32–67	100–650
Shrinkage limit, SL	29	14–17	6
Activity, A	0.2	0.6	1–6
Compressibility index, C_c	0.2	0.6–1.0	1–3
Friction angle drained, ϕ_d, degree	20–30	20–25	12–20
Particle shape		Platy	
Particle size			
Specific surface, m^2/g	10–20	65–100	50–800
Water adsorption			
Cation exchange capacity, meq/100g	3–15	10–40	80–150
Isomorphous substitution (nature/amount)	Al for Si 1 in 400	Al for Si 1 in 7, Mg, Fe for Al Fe, Al for Mg	Mg for Al 1 in 6
Linkage between sheets	H bonding + secondary valence	Secondary valence + K linkage	Secondary valence + exchangeable ion linkage
Silica sesquioxide ratio	2	2–3	>4

include card-house, honeycomb, dispersed, turbostratic, and stack. Collins and McGown (1974) have proposed microfabric and macrofabric systems for the idealized clay structures including marine clays.

3.8   Clay–water–electrolyte system

The clay micelle includes the solid clay particle itself as well as its sphere of influence in the surrounding water or aqueous solution. Within this sphere of influence, exchangeable ions are in a state of dynamic equilibrium. These ions are attracted to the net negative surface charge of the particle, but they also seek to diffuse away under their own inherent kinetic energy. Water existing in the soil–water system can be divided into two groups: free water and environmental water. *Free water* (gravity water) can be removed by gravity force or can be determined by a standard laboratory oven-dry procedure (ASTM D2216). *Environmental or Adsorbed waters* are under the influence of electrochemical forces induced by the clay particles. This water has been given different names in connection with proposed mechanisms. Mechanisms for water adsorption include hydrogen bonding, ion hydration, attraction by osmosis, and dipole attraction. Environmental waters are strongly influenced by local environmental conditions, especially for contaminated fine-grained soils.

3.9 Clay minerals

3.9.1 Clay minerals and representative groups

Of the approximately 2000 known minerals, the following are the most significant. They are grouped into two categories: bulky (approximately equidimensional), and flaky or platey (one axis being much less than the other two axes):

1 *Bulky grained soil Minerals*: Bulky grained soil minerals include the following types: (a) Quartz, (b) Feldspar, (c) Carbonates, (d) ferromagnesian minerals, (e) Oxide minerals, (f) Sulfide minerals, and (g) Sulfate minerals.
2 *Flaky or platy grained soil minerals*: These soil minerals consist almost entirely of the clay mineral groups. They are hydrous aluminum silicates with magnesium or iron replacing all or most of the aluminum in some minerals and with alkalies or alkaline earth present in some others.

As mentioned in Section 3.7, most fine-grained soil contains kaolinite, illite, montmorillonite or combinations thereof. The representative minerals of each group are listed as follows: (1) *Kaolin group*: (a) Anauxite, (b) Dickite, (c) Halloysite, (d) Kaolinite, (e) Metahalloysite, and (f) Nacrite; (2) *Illite group*: (a) Attapulgite; (b) Hydrous mica; (c) Illite; and (d) Sarospatite; and (3) *Montmorillonite group*: (a) Beidellite, (b) Bentonite, (c) Hectorite, (d) Montronite, (e) Pyrophillite, (f) Saponite, and (g) Talc.

3.9.2 Clay mineral structures

Conceptual representations of some common clay minerals are illustrated in Figure 3.10. Detailed descriptions of these major minerals are presented as follows:

1 *Kaolinite group*: Members of this group are considered 1:1 minerals because there is 1 tetrahedral sheet (indicated by $\triangle$) for every octahedral sheet (indicated by $\rule{1cm}{0.4cm}$) as illustrated in Figure 3.10(a). Within the octahedral layer there are generally aluminum atoms, and within the tetrahedral layer the cations are silicon. The chemical formula of kaolinite group is $(OH)_8Si_4Al_4O_{10}$.

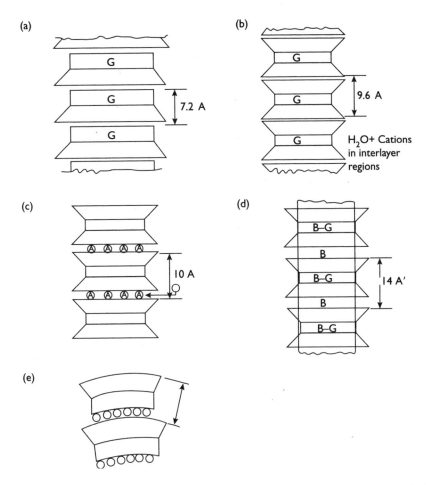

Figure 3.10 Diagrammatic sketch of the structures of some common clay minerals: (a) Kaolinite, (b) Illite, (c) Montmorillonite, (d) Chlorites, and (e) Halloysite.

2   *Illite group*: Illite has a 2:1 structure and consists of a gibbsite sheet between two silica tetrahedral sheets, with the layers bounded together by potassium cations in the interlayer region (Figure 3.10(b)). The chemical formula of illite is $(K,H_2O)_2Si_8(Al,Mg,Fe)_{4.6}O_{20}(OH)_4$. The octahedral cations are aluminum, magnesium or iron, and the tetrahedral cations can be aluminum or silicon.

3   *Montmorillonite group*: Montmorillonite is a 2:1 mineral and consists of two sheets of silica tetrahedra on either side of a gibbsite sheet (Figure 3.10(c)). Hence, for a montmorillonite that has not experienced any cation substitution or exchange, the chemical formula would be $(OH)_4Si_8Al_4O_{20}(H_2O)_n$, where n = number of layers. Sodium (Na) montmorillonite is a common form of the clay within bentonite, an expansive clay often specified in conjunction with drilling muds, borehole sealing, and waste containment.

4   *Chlorites*: Chlorites have a 2:1:1 structure and the basic layer consists of two tetrahedral sheets bonding a brucite or gibbsite sheet. They are basically similar

to illite except that an organized octahedral sheet in chlorites replaces the area otherwise populated by potassium ions. A conceptual representation of chlorite is presented as Figure 3.10(d).

5 *Halloysite*: Halloysite is a form of kaolinite in which water is held between structural units in the basal plane, and consists of crystals of hollow cylinders wherein the diameter is about 1/5 to 1/10 of the length (Figure 3.10(e)). Halloysite can be divided into hydrated and dehydrated halloysites. Hydrated halloysite (4H$_2$O) contains a mono-molecular layer of water within its mineral layers and when this water is completely removed by drying the result is a dehydrated halloysite (2H$_2$O).

3.9.3 Geotechnical properties of clay minerals

Clay contains various natural minerals. In most clays these minerals are either kaolinite, illite, montmorillonite, chlorite, halloysite or combinations thereof. Engineering properties of interest for these major minerals are summarized in Table 3.6.

3.10 Homoionic, pure, and man-made soils

3.10.1 Homoionic (pure) soils

1 *General discussion*: If a soil contains only one type of ion, it is called a *homoionic or pure soil*. For engineering applications, the relationship between clay fraction and the engineering behavior of soils depends not only on its quantity and the physico-chemical properties but also on the relative amounts and characteristics of the other soil constituents, including the aqueous and gaseous phases with which the clay particles interact. In order to illustrate how clay content affects the engineering properties of a natural soil, a series of experiments were carried out at Princeton University. A part of experimental data is condensed and summarized as shown in Table 3.7. The purpose of the homoionic modification is to (a) Aid in the recognition, understanding and separation of the component factors that produce the engineering properties of clays and clay soils as normally defined by activity, plasticity, etc.; (b) Provide certain physical properties of pure clays and their homoionic variants which may be directly useful for a particular engineering purpose; and (c) Define probable ranges of particular engineering properties in cases where specific data are not available.

2 *Method of preparation of pure (homoionic) soil specimen*: The method for preparation of the homoionic modifications from a natural soil were presented in batches using the method described by Scheffer and Schachtschabel in 1959 as cited and modified by Vees and Winterkorn (1967). A brief description of the procedures is presented First, a 10 l (2.6 gal) aqueous solution was made containing 10 times the amount of cations required for base exchange in the form of a soluble salt. The salt was the nitrate in the case of thorium, a sulfate in the case of aluminum, and the acetates for the other modifications. These solutions were placed in the bowl of an industrial model Hobart mixer which along with its paddle, had previously been coated with a chemically inert synthetic resin film. The powdered clay was added slowly while the mixer was running, and the resulting dispersion was mixed for two days. Then it was placed in a container and left at rest for seven days.

Subsequently, the supernatant clear solution was decanted and the dispersion or paste was concentrated and washed either in a filter press or by means of large Buchner funnels connected to a vacuum. The filtrate was frequently tested for anion content, and the process was considered completed when no anion presence could be detected by a pertinent chemical method. For the washing process, about 25 1 (6.6 gal) of distilled water is required when the Buchner funnels were used and half that amount in the case of the filter press.

3   *Physical properties of homoionic soils*: Physical properties of some natural and homoionic soils are presented in Table 3.7 and Figure 3.11. The data indicate that homoionic modifications could significantly change the physical properties of a given natural soil.

3.10.2 Man-made soils

The concept and approach for man-made soil was proposed in 1992 (Fang, 1997). Basically, the approach hinges on controlling the ion substitution or manipulating (adding or removing) ions in the soil element to meet engineering needs.

1   Three basic steps are proposed for making such man-made soils: (a) evaluation of soil genetics, (b) selection of proper homoionic soil types, and (c) estimation of the percentage of homoionic soil to be included in the man-made soil.

Table 3.7 Physical properties of some common natural and homoionic soils

Soil type	Natural soils	Homoionic soils							
		H	Na	K	Mg	Ca	Ba	Al	Fe
Putnam subsoil									
Plasticity Index	17	16	32	18	22	21	19	15	16
% Clay (<0.005 mm)	33								
Activity	0.52	0.48	1.00	0.55	0.67	0.64	0.58	0.45	0.48
Loess pampaneo subsoil									
Plasticity Index	45	27	37	20	34	29	34	32	26
% Clay (<0.005 mm)	50								
Activity	0.90	0.54	0.74	0.40	0.68	0.58	0.68	0.64	0.52
Marshall subsoil									
Plasticity Index	28	21	48	18	36	30	28	21	17
% Clay (<0.005 mm)	37								
Activity	0.76	0.57	1.30	0.49	0.97	0.81	0.76	0.57	0.46
Loess pampaneo topsoil									
Plasticity Index	3	6	12	—	12	12	9	7	6
% Clay (<0.005 mm)	27								
Activity	0.48	0.22	0.44	—	0.44	0.44	0.33	0.26	0.22
Cecil subsoil									
Plasticity Index	37	34	30	38	34	32	36	31	36
% Clay (<0.005 mm)	76								
Activity	0.49	0.45	0.39	0.50	0.45	0.42	0.47	0.41	0.47

Source: From *Foundation Engineering Handbook*, 2nd Edition, Fang, H.Y. ed., Ch. 3, *Soil Technology and Engineering Properties of Soils*, Winterkorn, H. F. and Fang, H. Y., Copyright (1991) by Van Nostrand Reinhold. With kind permission of Springer Science and Business Media.

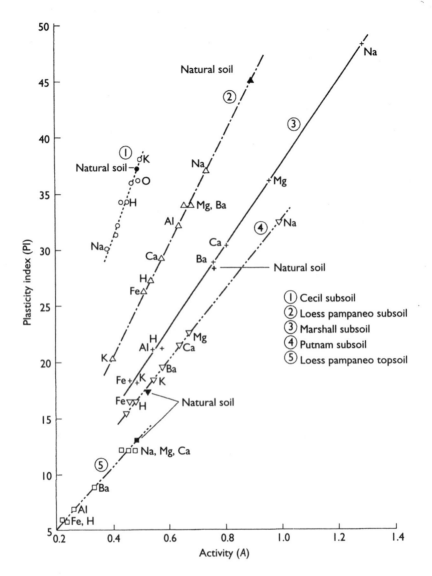

Figure 3.11 Comparison of plasticity index versus activity between natural and homoionic soil.

Source: Data from Table 3.8.

2   The procedures for assessing a natural soil include an evaluation of in situ characteristics of a natural soil such as activity, plasticity index and silica–sequioxide ratio (SSR).

3   The selection of proper ion type(s) is based on the properties of a homoionic soil relative to project requirements. These properties include soil classification, hydraulic conductivity, stress–strain relationships, and compressibility.

4   The use and study of such man-made soils can be used in various practical areas such as (a) Increasing the effectiveness of ground improvement techniques as well

as liners for waste containment, (b) Gaining a better understanding of the underlying mechanisms responsible for landslides, soil erosion, and progressive failures, and (c) Understanding more about long-term soil behavior and soil–water interaction in the environment as well as soil–pollution interactions.

3.11 Summary

Soil may be broadly categorized as coarse- or fine-grained. Coarse soils tend to have a higher strength, bearing capacity, and hydraulic conductivity as compared to fine-grained soils. Fine-grained soils, particularly those with clay minerals have a much greater specific surface area and are more susceptible to changes in the local environment.

The air–water–solid relationships are useful for making engineering calculations and deriving various parameters. Important and interrelated parameters include bulk density, dry density, void ratio, porosity, water content, and degree of saturation.

Granular soils may adopt an array of packing configurations, according to individual particle shape and size. These packing configurations can be idealized in an effort to establish a range within which field soils are likely to exist. Engineering behavior is affected by these configurations, as is obvious where strength comparisons are made between dense and loose deposits.

The behavior of a clay mineral is controlled by its mineralogy. The main mineral types include kaolinite, illite and montmorillonite, although many more exist and with differing combinations as discussed in the chapter. A key consideration with clays is the isomorphous substitution which typically occurs, giving rise to a net-negative surface charge. These charged surfaces have direct bearing on engineering and environmental behavior.

PROBLEMS

3.1 Use a phase diagram (Figure 3.2) to find each of the following relationships in terms of the given quantities (Note: the unit weight of water is always considered to be a known quantity). (a) Given e, find n; (b) Given G_s and e, find moisture content, ω, for a fully saturated soil.

3.2 A soil is at a void ratio of 0.9 with a specific gravity of the solid particles of 2.70. (a) Can the water content be determined from the information given? (b) If the water content cannot be determined, can upper and lower limits be determined? If so, what are they?

3.3 An earth dam is to be constructed by the hydraulic fill method. The mass unit weight of the mixtures of soil and water issuing from the discharge line of the dredge is 84.0 pcf (13.2 kN/m³). In its natural state, the soil is found to be only partially saturated with a dry unit weight = 102.0 pcf (16.0 kN/m³) and a specific gravity of solid, $G_s = 2.72$. How many ft³ of pumping will be required for each ft³ of excavation?

3.4 What are the advantages of plotting grain-size distribution curves for soils on a semi-logarithmic plot as opposed to an arithmetic one?

3.5 Why would a graded, dirty, sand and gravel be is superior to a pea gravel or crushed stone for the purpose of backfilling the footing drains around a building or a house?

3.6 Explain the difference between flocculation and dispersive structures.

3.7 Using the symbols for gibbsite, brucite and the silica tetrahedron, sketch the commonly accepted structure for kaolinite and illite.

3.8 Define the homoionic modification soil. What is man-made soil? Does it have practical value?

Chapter 4

Soil–water interaction in the environment

4.1 Introduction

4.1.1 Importance of soil–water interaction

The amount of water existing in the soil mass will significantly influence the engineering behavior of soil. In fact, Karl Terzaghi has said in effect, that there would be no need for soil mechanics if not for water. This is because the presence of water affects the state of stress within a soil mass. The water content also has bearing on potential volume change, progressive failure, densification, shear strength, and settlement. The mechanism of soil–water interaction is complex and its behavior is not only dependent on soil types, but is also related to the current and past environmental conditions and stress histories. The specific nature of soil–water interaction varies according to the corresponding energy field(s). Types of interaction and their characteristics occurring in various energy fields include: (1) *Thermal energy field*: water intake ability, absorption and saturation, heat of wetting, thermoosmosis, volume change, fracture cracking, shrinkage and shrinkage cracking, and thermal cracking; and (2) *Electric and electromagnetic energy fields*: adsorption, electroosmotic, electrophoresis, ion exchange reaction, redox reaction, and swelling.

4.1.2 Soil acidity in aqueous phase

All wet soils contain hydrogen (H^+) ions in their pore spaces since water itself dissociates into H^+ and OH^- ions. Dissociation of H_2O into H^+ and OH^- ions is a function of temperature. Soil acidity also varies with the season and is normally determined by measuring the liquid portion of a 1:1 (mass ratio) mixture of soil and distilled water. Acidity actually includes all chemical species which serve to reduce the pH, just as alkalinity refers to all chemical species which serve to increase pH. The pH itself is defined as the negative logarithm of the H^+ ion concentration as indicated in the following equation.

$$pH = -\log [H_3O^+] \tag{4.1}$$

The term $[H_3O^+]$ in Equation (4.1) represents the concentration of hydronium ion in moles per liter and is often simply written as the hydrogen ion $[H^+]$. Accordingly, a pH of 7 indicates neutrality and a value less than 7 denotes acidity. A special instrument called a *pH meter* can be used to measure the pH of an aqueous solution. pH values versus H^+ ion concentration for various types of solution is presented in Figure 4.1.

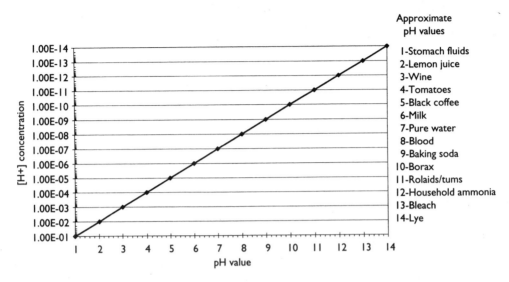

Figure 4.1 pH values versus H$^+$ ion concentration for various types of solution.

The acidity in the solution increases with decreasing pH value. The pH of water and other liquids can be tested by ASTM Standard procedure (ASTM D1293–84). Soil–water acidity is important to the geotechnical engineer because of its corrosive effect on foundation structures and construction materials; on the other hand, acidity is sometimes desirable because of its catalytic effect on certain reactions employed in soil stabilization for ground improvement (Sec. 15.3).

4.2 Mechanisms and reactions of soil–water interaction

4.2.1 Role of surface electrochemistry in soil–water interaction

Basically, the interaction between a liquid (water) and a solid (soil) can occur only at the solid's surface. Therefore, the role of surface electrochemistry is important, and it must be evaluated. The basic concept of electrochemistry was established by Faraday in 1834, who discovered the fundamental law of electrolysis. However, the application of this concept in engineering is a relatively new interdisciplinary subject. It is a combination of chemistry, soil science, and engineering and their interactions with electricity and environment. Progress in electrochemistry also hinges on the development of other related subjects such as clay mineralogy (Sec. 3.7) and ion exchange capacity of soil (Sec. 4.6).

4.2.2 Mechanisms of soil–water interaction

1 Winterkorn (1942), using a physicochemical concept to explain the mechanics of reacting water with a dry cohesive clay, postulated that a dry soil system is held together by a remaining adhesive water film. The cementing power of these films is a function of their own physical and chemical characteristics and of the particles

they hold together. The cementing causes include (a) the surface tension of the soil moisture; (b) chains of oriented dipoles linking a positive charge on one particle with a negative one on the neighboring particles; and (c) formation of an electric field by ions dissociated from the particles by water dipoles. Winterkorn concluded that two phenomena must be considered and analyzed: First, the penetration of water into the soil mass; and second, the action of the water on the cementing films resulting in a lowering and possible destruction of the cohesion of the soil.

2   Terzaghi (1943) using a mechanical (kinetic energy) approach to explain that when a dried soil is rapidly immersed in water, the outer portions of the soil become saturated and air is trapped in the inner portions. The pressure in the air produces a tension in the solid skeleton that is likely to cause failure of soil in tension. This process is known as slaking. Terzaghi claimed that it is responsible for the breaking up and ultimate sloughing off of unprotected clay slopes. Other explanations on soil–water interaction are examined and reviewed by Yong and Warkentin (1966), Mitchell (1976, 1993), and Fredlund and Rahardjo (1993) and many others.

4.2.3   Phenomena and/or reactions during soil–water interaction

1   *Phenomena observed during soil–water interaction*: As indicated in Section 4.1.1, there are numerous phenomena that occur when water and soil interact. Some of the relatively important interactions are presented as follows: (a) shrinkage and swelling, (b) water intake ability, (c) sorption (absorption and adsorption), (d) conductivity (Sec. 5.4), and (e) soil cracking (Sec. 8.2).

2   *Reactions obtained during soil–water interaction*: Reactions from soil–water interactions include (a) heat of wetting, (b) ion exchange reaction, and (c) redox reaction. In many cases, the phenomena and reactions that occur during soil–water interaction are closely related. Ion exchange and redox reactions are typical examples. In this section the focus is on shrinkage, swelling, sorption, absorption, and adsorption.

4.3   Structures and properties of water and water substances

4.3.1   Water structures

Water is a polar liquid with a bi-lateral electric structure. The water molecule consists of one oxygen atom (O) and two hydrogen atoms (H_2). The distance between the center of the oxygen atoms and the center of each of the hydrogen atoms is 0.9 Å (9×10^{-11} m). These molecules possess electric polarity, that is, the centers of the negative and positive charges within these molecules do not coincide. The angle formed by the two hydrogen atoms is 105°.

4.3.2   Properties of water and water substances

1   *Density and specific weight*: The *density of water* is defined as mass per unit volume, and in the foot–pound–second (fps) system is expressed in slugs per cubic foot.

Since mass is equivalent to weight divided by the acceleration of gravity, the density of a substance is equal to its specific weight divided by the acceleration of gravity. The density of water at 4°C is 62.4 lb/ft³ or 1.94 slugs/ft³. In the metric system, it is equal to 1 g/cm³ or 9.807 kN/m³ when expressed in terms of mass or weight, respectively.

2 *Absolute, dynamic, and kinematic viscosities:* The *viscosity* of a fluid is defined as the ratio of shear stress to the rate of shear strain. The shear stress is the shear force divided by its corresponding area while the rate of shear strain is the change in velocity, divided by the shearing distance. The viscosity determined under such conditions is called *absolute viscosity* or *dynamic viscosity, u.* Experimentally, this may be performed through measuring the torque which develops when a rod or cylinder immersed in the fluid of interest is rotated. It is expressed in pound-second per square foot (lb-s/ft²) or slugs per foot-second. In the metric system, it may be expressed in dyne-second per square centimeter, gram-second per square centimeter, or in *poise.* The term poise is an honor of Poiseuille, a French scientist. The centi-poise (cP) or 0.01 poise is a common unit and water has an absolute viscosity of about 1 cP at 20°C (68°F).

The term *kinematic viscosity, (v),* is the ratio of the absolute viscosity divided by the fluid density. The units for kinematic viscosity can be expressed in terms of feet squared per second, or centimeters squared per second. In honor of Sir George Stokes, an English scientist, 1 cm²/s is called a *stoke.*

Glycerine or glycerine–water mixtures are sometimes used in laboratory soil testing for confined lateral pressure in the triaxial shear test (Sec. 10.6) or as a viscous fluid in laboratory (physical) models of groundwater movement. The relationship between viscosity and temperature of glycerine or glycerine–water mixtures is shown in Figure 4.2.

3 *Surface tension of water:* The molecules on the surface of a liquid are attracted to each other, and this creates a tensile force that may be considered as acting across any line in the surface of the liquid. The intensity of the molecular attraction per unit length along any line in the surface of the fluid is called the *surface tension.* There are several methods for measuring surface tension of water or liquid as suggested by ASTM (D1590). The surface tension of liquid is influenced by the nature of the liquid itself, the type of adjacent fluid (which may be air), and the temperature. Typical surface tension results for various substances at various inter-phases are: air–acetone at 20°C = 23.7 dynes/cm; air–glycerin at 20°C = 63.4 dynes/cm; air–mercury at 15°C = 487.0 dynes/cm; air–water at 20°C = 72.8 dynes/cm; and benzene–water at 20°C = 35.0 dynes/cm.

4.3.3 *Solutions, compounds, and mixtures*

1 *Solution, solvent, and solute:* A *solution* is a mixture of substances which dissolve in one another to become a single-phase systems. Solutions can be unsaturated, saturated, and supersaturated depending on the amount of solute that is dissolved in the solvent. A *solvent* is a substance, usually a liquid, in which another substance called the *solute* is dissolved.

2 *Compound and mixture:* Compounds have definite properties and a homogeneous composition. For example, water (H_2O) is composed of the chemical elements hydrogen (H) and oxygen (O). This compound always contains hydrogen and oxygen and any sample will contain 8 g of oxygen to every 1 g of hydrogen. Sugar and water mixtures or salt and water mixtures can contain variable

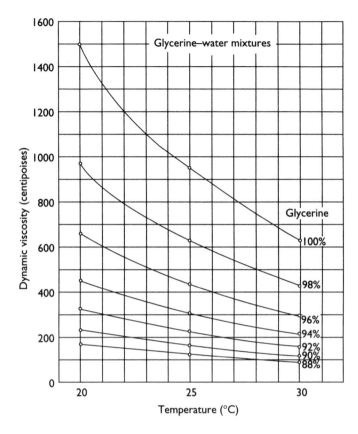

Figure 4.2 Viscosity of glycerine–water mixture versus temperature.

amounts of sugar in water or salt in water. When such variable compositions occur, they are called *mixtures*.

3   *Units of solution, compound, and mixture*: There are several units commonly used to indicate the concentration of the solutions, compounds, and mixtures. (a) *Mole* (mol): A mole is the amount of a substance in grams that is equal to its molecular weight; (b) *Molarity*: Molarity is defined as the number of moles of solute per volume of solution in liters; (c) *Parts per million* (ppm): Solutes found in very low concentrations in solutions are expressed in terms of the number of milligrams of solute per kilogram of solution, or the number of milligrams of per liter of solution; and (d) *Equivalent per million* (epm): A unit chemical equivalent weight of solute per million unit weights of solution (ASTM D1129).

4.4   Shrinkage, swelling, and heat of wetting of soils

4.4.1   *Shrinkage characteristics and mechanism*

Shrinkage is one of the major causes for volume change associated with variations of water content in soil. Haines (1923) and Hogentogler (1937) have shown that when soil decreases its water content, the volume decreases. A typical relationship between

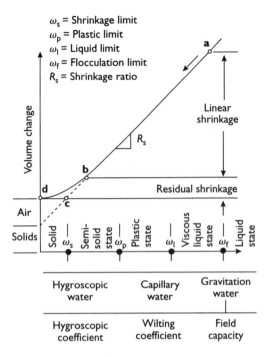

Figure 4.3 Correlation of soil volume change and consistency.

Source: Based on Haines (1923) and Hogentogler (1937a).

volume change and soil consistency is illustrated in Figure 4.3. The change from points **a** to **b** is called *normal shrinkage*, as the volume changes linearly with a decrease in water content. This linear shrinkage is due to surface tension forces in the capillary moisture. When the soil color changes, a small amount of volume change, termed *residual shrinkage* or *curvilinear shrinkage*, occurs between points **b** and **d**. The water content may be reduced to point **c** at which further reductions in water content do not result in a volume change. This point is termed the *shrinkage limit* (Sec. 2.5) and generally decreases with increases in clay content. In addition to the clay content, other factors affecting the shrinkage characteristics of soil include dry process, soil particle orientation, unit weight, and grain size distribution.

The slope of the straight-line portion (from points **a** to **b**) of the volume change and soil consistency curve in Figure 4.3 is called the *shrinkage ratio*, (R_s). This value changes within a narrow range. For example, the R_s value for muck soil is 1.42, black cotton soil from India is 2.1, and silty clay from the AASHO Road Test, Ottawa, Illinois is 1.95. The shrinkage limit of soil can be directly measured as described in ASTM (D427). It is also frequently estimated from the liquid limit and plasticity index as given by Casagrande in Equation (4.2). Volumetric shrinkage and linear shrinkage can also be computed from shrinkage factors as suggested by ASTM and presented in Equations (4.3) and (4.4).

$$SL = LL - PI\left(\frac{43.5 + LL}{46.5 + PI}\right) \tag{4.2}$$

4.4.2 Shrinkage types and relations with other soil constants

1 *Volumetric shrinkage*: The volumetric shrinkage of soil is the decrease in volume, expressed as a percentage of the soil mass, when the water content is reduced from a given percentage to the shrinkage limit. The relationship between volumetric shrinkage and shrinkage limit is shown in Equation (4.3).

$$V_s = (\omega_1 - \omega_s) \, R_s \tag{4.3}$$

where V_s = volumetric shrinkage (%), ω_1 = given percentage of water content, ω_s = shrinkage limit, and R_s = shrinkage ratio (Fig. 4.3).

2 *Linear shrinkage*: Linear shrinkage is defined as the one-dimensional decrease in soil mass expressed as a percentage of the original dimension, when the water content is reduced from a given value to the shrinkage limit (ASTM D427). The relationship between linear and volumetric shrinkage is shown in Equation (4.3) or by means of the curve shown in Figure 4.4.

$$L_s = 100\left(1 - \sqrt[3]{\frac{100}{V_s + 100}}\right) \tag{4.4}$$

where L_s = linear shrinkage (%), and V_s = volumetric shrinkage (%). The relation between linear shrinkage and shrink-swell potential is presented in Figure 4.5. The shrink-swell potential increases with increases in linear shrinkage as reported by Ring (1966).

3 *Factors affecting shrinkage of soil*: Shrinkage-induced volume change is often the result of fluctuations in moisture content. In addition to water content, other factors that affect shrinkage characteristics include (a) clay content, (b) drying process, (c) soil particle orientation, (d) unit weight of soil, and (d) grain size distribution. Rao (1979) reported on experimental results of kaolinite and montmorillonite, in which shrinkage limits are affected by changes in fabric, initial water content and consolidation pressure. Soil cracks caused by shrinkage and other mechanisms are discussed further in Section 8.2.

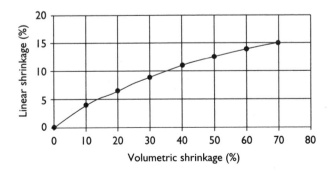

Figure 4.4 Curve for determining linear shrinkage.

Source: D427, ASTM 2004. Copyright ASTM INTERNATIONAL. Reprinted with permission.

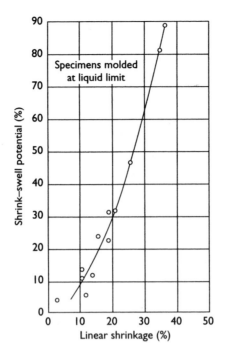

Figure 4.5 Relation of linear shrinkage to shrink–swell potential.

Source: Ring, G. W. III Shrink–swell potential of soils. In Highway Research Record No. 119, Highway Research Board. National Research Council, Washington, DC, 1966, pp. 1–11. Reproduced with permission of the Transportation Research Board.

4.4.3   Swelling characteristics and mechanism

1   *General discussion*: When soil loses its water content, the volume decreases or shrinks. However, if water is added to the dry soil, the volume of soil increases and this phenomenon is called *swelling* or *expansion*. Swelling and shrinkage involve different energy fields. Shrinkage is in the thermal energy field, but swelling is in the multimedia energy field as illustrated in Table 1.4 and Figure 4.6. Also, the swell shrinkage is not a reversible process. In other word swell is not equal to shrinkage.

2   *Swelling mechanism*: The swelling process is actually more complicated than the mere addition of water to a dry or unsaturated soil system. Soil swelling is a spontaneous process that occurs when there is a decrease in free energy or an increase in system entropy in the presence of moisture. The rate of swelling is directly proportional to the available free energy, with the following consequences: (a) the rate of swelling decreases with time, and (b) the swelling pressure, as a measure of free energy, also decreases. The rate of swelling is also controlled by the amount of water available and ease with which water is imbibed, as dictated by the hydraulic conductivity.

4.4.4   Swelling measurements and identification

1 *Swelling measurements:* There are numerous methods for making swelling measurements. The existing methods can be divided into two major groups; direct and indirect measurements. Direct measurements consider one-dimensional volume

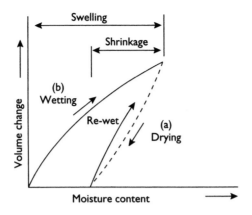

Figure 4.6 Characteristics of swell–shrinkage processes.

Table 4.1 Bureau of reclamation method for identification of swelling soils

Data from index tests			Estimation of probable expansion percent total volume change (dry to saturated condition)	Degree of expansion
Colloid Content (percent minus 0.001 min)	Plasticity index	Shrinkage Limit (%)		
>28	>35	<11	>30	Very high
20–31	25–41	7–12	20–30	High
13–23	15–28	10–16	10–20	Medium
<15	<18	>15	<10	Low

Source: USBR, 1974.

change in samples prepared with standard compaction (Sec. 7.2), consolidation (Sec. 9.3), or CBR tests (Sec. 12.9.1). Indirect methods use a measurement of related soil properties such as mineralogy, Atterberg limits, activity (Sec. 2.5), centrifuge moisture equivalent (Sec. 2.5), and dielectric dispersion (Sec. 6.10).

2   *Swelling identification methods:* (a) Single index method: There are numerous methods for identification of swelling or expansive soil, the US Bureau of Reclamation method (USBR, 1974) is one of most commonly used as shown in Table 4.1.

A term called the expansion index, I_E, has been proposed by Fernando *et al.* (1975) in assessing swelling on the basis of dielectric dispersion. The classification of expansive soils based on magnitude of dielectric dispersion is presented in Table 4.2. The dielectric dispersion can be determined by experimental data provided by Fernando *et al.* (1975) and Anderson and Lade (1981). Based on these experimental data and others, an empirical equation relating the expansion and plasticity indexes is given by Equation (4.5) (Fang, 1997):

$$I_E = 2.72\, I_P \tag{4.5}$$

where I_E = expansion index (Table 4.3), and I_P = plasticity index (Sec. 2.6).

Table 4.2 Classification of expansive soils based on magnitude of dielectric dispersion

Magnitude of dielectric dispersion	Expansion index, I_E	Expansive potential
1–10	1–20	Very low
11–25	21–50	Low
26–45	51–90	Medium
46–65	91–130	High
Above 66	Above 130	Very high

Source: Fernando, J., Smith, R., and Arulanandan, K. New approach to determination of expansion index, *Journal of the Geotechnical Engineering Division. Proceedings of ASCE*, v. 101, no. GT9, pp. 1003–1008 © 1975 ASCE. Reproduced by permission of the American Society of Civil Engineers.

3   *Factors affecting the characteristics of swelling soils:* Relationships between swelling and soil type or other parameters have been studied by many researchers. In general, factors that influence the volume changes of swelling clays are (a) the soil particle composition and soil structure, (b) the amount, mineral character, and exchange ion of clay fraction, (c) types and concentration of electrolytes in the soil solution, (d) amount and electrolyte content of soil–water, and (e) initial moisture content and unit weight of soil.

In addition, Dakshanamurthy (1979) reported that maximum axial and radial swelling decrease exponentially with increasing stress in the normal direction. This relationship is dependent on the principal stress ratio, that is, the ratio of major principal stress to minor principal stress in the triaxial shear apparatus. The swelling ratio is the ratio of axial swelling to radial swelling and is found to decrease with the time allowed for swelling, reaching an equilibrium value termed the *equilibrium swelling ratio*. This equilibrium value is a function of boundary loading conditions.

4.4.5   Heat of wetting

1   *Mechanism of heat of wetting:* When water contacts dry or partially saturated soil, heat is produced. The heat of wetting is an evolution of heat from hygroscopic materials that occurs upon contact with water and is analogous to exothermic chemical reactions. In this case, water molecules are reacting with the solid particle surface. This phenomenon is of particular relevance to fine-grained soil, especially the colloidal fraction and the quantity of heat evolved increases with particle fineness. Colloidal soil materials give comparatively high heats of wetting though varying according to type of mineral and pore fluid. These relationships have bearing on precipitation/dissolution and adsorption/exchange reactions and as such may be significant to contaminant transport.

2   *Factors affecting heat of wetting:* The smaller the soil particle the greater the heat of wetting produced upon contact with water. In general, sand produces less heat than silt and silt is less than clay. Typical ranges of heat of wetting of soils are (a) sand: 0.5–2.0 cal/g, (b) silt: 2.0–5.4 cal/g, and (c) clay: >5.4 cal/g. Experimental data comparing heat of wetting versus activity of natural and homoionic of Cecil and Marshall soils (Table 2.10) is presented in Figure 4.7.

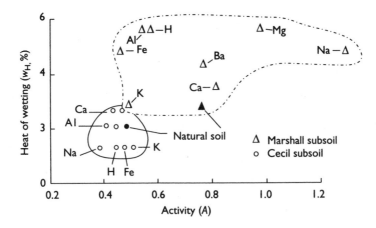

Figure 4.7 Heat of wetting versus activity, A, between natural and homoionic soils.

Source: From *Foundation Engineering Handbook*, 2nd Edition, H. Y. Fang, ed., ch. 3, *Soil Technology and Engineering Properties of Soils*, Winterkorn, H. F. and Fang, H. Y., Copyright (1991) by Van Nostrand Reinhold, with kind permission of Springer Science and Business Media.

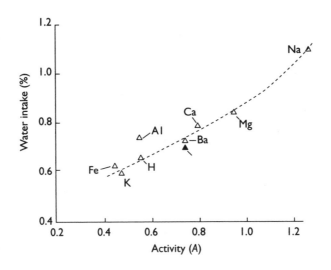

Figure 4.8 Water intake ability versus activity, A, between natural and homoionic soils.

Source: From *Foundation Engineering Handbook*, 2nd Edition, H. Y. Fang, ed., ch. 3, Soil Technology and Engineering Properties of Soils, Winterkorn, H. F. and Fang, H. Y., Copyright (1991) by Van Nostrand Reinhold, with kind permission of Springer Science and Business Media.

4.5   Water intake ability and sorption

4.5.1   *Water intake ability*

A soil's *water intake ability* is defined as the maximum amount of water that it can absorb at a given time. Figure 4.8 presents water intake ability versus activity, A, for natural and homoionic preparations of Putnam soil (Heavy Missouri silt loam).

4.5.2 Sorption characteristics

1 General discussion: Soil–water interaction in the environment may involve sorption, absorption and adsorption reactions. These reactions are influenced by soil types, particle size, bonding energy between particles and these factors are all influenced by local environments such as temperature, pore fluids, etc. However, in geotechnical engineering, these terms are not clearly defined and in most cases, terms such as water content, degree of saturation and water absorbing capacity are used to describe soil–water behavior. However, these terms tell only part of the story, and in order for us to understand soil–water interaction, it is necessary to further examine the characteristics of sorption, absorption, and adsorption phenomena.

2 Sorption mechanisms: When water is added to dry or partially saturated soil, the water will be absorbed by soil. This simple statement covers three complex processes: sorption, absorption, and adsorption.

Table 4.3 illustrates these three water intake processes. In examining Table 4.3 in terms of the particle energy field approach (Ch. 1), absorption is a form of mechanical energy. However, adsorption involves several forms (multimedia energy) and is controlled by physical and physicochemical processes. The term sorption is a general term which covers both absorption and adsorption processes. One example commonly given to differentiate between adsorption and absorption is the case of a sponge which will *absorb* water, as opposed to charcoal, which will *adsorb* acetic acid. Alternatively, adsorption is typically considered as occurring at a surface, as opposed in absorption which represents penetration into the soil matrix. From these simple discussions, we learn the following significant points relative to practical applications in geotechnical engineering: (a) for two given soil samples at the same water content, two different types of behavior may result. These differences will depend on the flow path direction and sorption characteristics; (b) in situ water content of soil includes both absorption and adsorption waters. The relative amount of these waters depends on the soil type and local environmental conditions. In general, fine-grained soils are more sensitive to the local environment than cohesionless soil. In particular, cohesive soil is more likely to have both types of water than cohesionless soil; and (c) the degree of saturation alone cannot be used to indicate the water absorbing capacity of soil.

3 Water sorption: Vees and Winterkorn (1967) presented a series of laboratory experiments with homoionic clays as shown in Figure 4.9. Figure 4.9 shows the

Table 4.3 Sorption, absorption, and adsorption relating to water types in the soil–water system

Environmental water		
1 Solid solution (at solid–water interface)		
2 Double layer water		
3 Hydrated water or/and hydration water	Adsorption	
4 Osmotic water		Sorption
5 Oriented water (at air–water interface)		
6 Free water	Absorption	
7 Gravity water		

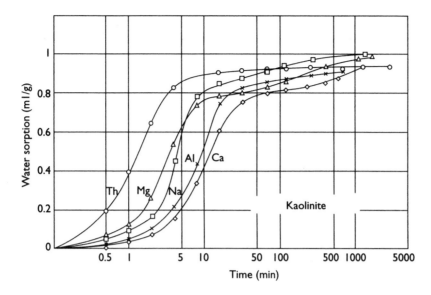

Figure 4.9 Water sorption as a function of type of exchangeable ions and time for kaolinite clay.

Source: Vees, E. and Winterkorn, H. F. Engineering properties of several pure clays as functions of mineral type, exchange ions, and phase composition. In Highway Research Record No. 209, Highway Research Board. National Research Council, Washington, DC, 1967, pp. 55–65. Reproduced with permission of the Transportation Research Board.

water sorption of kaolinite powder as a function of the type of exchange ion and time. Th-clay (Thorium) shows the highest initial sorption rate and the lowest ultimate intake. This is to be expected from its flocculated structure and its low water affinity. For the other ions the picture is not so clear-cut. This is due to the interaction between absorption and adsorption processes with respect to the types of ion exchange reaction in the soil. Considering the relatively low base exchange capacity of the kaolinite, the observable ion effect appears to be markedly influenced by the type of secondary structure induced by the different cations rather than by the individual water affinities of the ions themselves.

Of particular interest is the closeness in rate and final volumes in the water sorption by Na- and Mg-kaolinite. It also illustrates, to some extent, how increasing the cation valence reduces the affinity of water to the clay surface. As such, the rate of water intake and the ultimate volume increase with increasing exchange ion valency.

Table 4.4 presents liquid sorption of oven-dry clays for three clay minerals with water, H_2O and dimethyl sulfoxide $(CH_3)_2SO$ (DMSO). DMSO is a commercial solvent chemically related to acetone $(CH_3)_2CO$. It is a colorless liquid, completely miscible with water and extremely hygroscopic. Properties of special interest in geotechnical engineering are its solvent power for both water and many organic substances and its very rapid diffusion through hydrophilic systems. In examining Table 4.4, for all clays except kaolinite, the final sorption data were higher for water than for DMSO. The respective ratios were 7.2 for bentonite, 1.45 for attapulgite, and 0.7 for kaolinite. In all cases, except for attapulgite with DMSO, the sorption values were higher than the corresponding liquid limit (Sec. 2.5) values, indicating that slight

Table 4.4 Liquid sorption of oven-dry clays

Time	Kaolinite		Attapulgite		Bentonite	
	H_2O	DMSO	H_2O	DMSO	H_2O	DMSO
I h	0.782	0.958	2.640	1.890	0.304	1.334
I day	0.868	1.040	2.928	2.128	0.880	1.480
6 days	1.024	1.454	3.330	2.304	1.714	1.516
Final	1.930	1.460	3.340	2.910	11+	1.520
H_2O/DMSO ratio	0.7		1.45		7.2	

Source: Andrews, R. E., Gawarkiewicz, J. J., and Winterkorn, H. F. Comparison of the interaction of three clay minerals with water, dimethyl sulfoxide, and dimethyl formamide. In Highway Research Record No. 209, Highway Research Board. National Research Council, Washington, DC, 1967, pp. 66–78. Reproduced with permission of the Transportation Research Board.

Notes
Unit of liquid sorption = ml/g clay.
H_2O = Water; DMSO = Dimethyl sulfoxide $(CH_3)_2SO$ (DMSO).

Table 4.5 Time required to absorb maximum amount of water

Material	Maximum potential for absorbed water, % of dry weight	Time required to absorb maximum amount of water
Kaolin	90	20 min
Mica	125	25 sec
Ca-Montmorillonite	300	40 min
Na-Montmorillonite	700	600 min (10 h)

disturbance would put the saturated systems into the liquid state. In the case of attapulgite the sorption value at any given time was higher for water than for DMSO, while in the case of bentonite the amount of liquid sorbed was for the first few days higher for DMSO than for water. The sorption values for kaolinite, the same as the corresponding liquid limit values, indicate a greater affinity for DMSO than for water. Experiments such as these clearly indicate the importance of soil type, exchangeable ions, and pore fluid on sorption phenomenon which in turn may influence engineering behavior.

4.5.3 Saturation and absorption phenomena

In many cases, the term absorption is used rather loosely. The degree of saturation is commonly used as an indication of the extent of water absorption in soil. Water contributing to the degree of saturation may however involve both absorption and adsorption phenomena. Water in the absorbed state is easily influenced by gravity and evaporation. The characteristics of adsorption phenomena are complex, and are discussed in Section 4.6. Table 4.5 illustrates the potential for water absorption exhibited by various types of clay. The water affinity of clay minerals is a function of temperature also, and decreases with increasing temperature.

4.6 Adsorption phenomena

Adsorption is caused by multimedia energy fields which include physical and physico-chemical processes. All solids tend to adsorb gases and solutes with which they are in contact. However, in order for the adsorption to be appreciable, it is necessary that the adsorbent have a large surface area. The amount of gas or solute adsorbed by a solid depends on (a) the nature of the adsorbent, (b) the nature of the substance being adsorbed, (c) the surface area of the adsorbent, (d) the temperature, and (e) the pressure in the soil–water system. In general, a decrease in the temperature or an increase in the pressure increases the amount of adsorption.

4.6.1 Types of adsorption

There are two types of adsorption, defined loosely as physical adsorption and chemical or activated adsorption.

1 *Physical adsorption*: Physical adsorption is characterized by low heats of adsorption, in which the surface is gradually coated with a layer of molecules held in place by van der Waals forces. Since van der Waals forces have weak binding energy, physical adsorption cannot significantly affect their chemical properties of adsorbed molecules and is relatively transient.

2 *Chemical or activated adsorption*: Chemical adsorption is much stronger than physical adsorption, and is characterized by heats of adsorption between 10 and 100 kcal per mole of gas. This is comparable in magnitude to the heat of formation of a chemical compound. Physical adsorption is common at low temperature, and chemical adsorption is common at high temperature, however, physical adsorption at low temperatures may pass into chemical adsorption as the temperature is increased.

4.6.2 Adsorption pressure in the soil–water system

Large adsorption forces exerted on water molecules by the surfaces of solid soil particles act similar to externally applied pressures; that is, they may liquefy solid water or solidify liquid water. In fact, adsorption forces of the order of magnitude that solidify water at 45°C at a pressure of 12,000 kg/cm^2 are not at all uncommon on the surfaces of soil particles (Winterkorn, 1955). These forces may in turn influence the physical properties of cohesive soils, including the shear strength and compressibility. The water layers next to the solid particles are under high adsorption pressures, which may be larger than 25,000 kg/cm^2. The adsorption forces decrease exponentially to about 50 kg/cm^2 at the so-called *hygroscopic moisture content*, and then more slowly to zero for the water content at which the soil–water system behaves essentially as a liquid (at liquid limit stage).

4.6.3 Adsorption measurements

The concepts of adsorption have been widely used in environmental science, engineering as well as organic, surface, and colloidal chemistry. Winterkorn and Baver used the concept adsorption for evaluation of soil stabilization and characterization of highway materials in the early 1930s (Winterkorn and Baver, 1934). More recently the same concept has been applied to radioactive radon gas (Sec. 16.11) and acid rain interaction with soil. There are various methods for measuring sorption, absorption, and adsorption phenomena. These methods are time-consuming procedures. It is suggested that using existing standard ASTM procedures, the Field Moisture Equivalent (FME) and Centrifuge Moisture Equivalent (CME) can indicate absorption and adsorption phenomena for soils. A brief discussion and justification of how these procedures can be used is given as follows:

1   *Absorption relating to the field moisture equivalent:* The FME of a soil is defined as the minimum moisture content, expressed as a percentage of the oven-dried soil, at which a drop of water placed on a smooth surface of the soil will not immediately be absorbed by the soil but will spread out over the surface and give it a shiny appearance. Both absorption and FME, are short-term mechanical processes for characterization of the water intake ability of soil.

2   *Adsorption relating to the centrifuge moisture equivalent:* The CME is the moisture content of a soil after a saturated sample is centrifuged for one hour under a force equal to 1000 times the force of gravity. The CME is used to assist in structural classification of soils. A value lower than 12 indicates permeable sands and silts while values greater than 25 indicate impermeable clays with high capillarity. When FME and CME are both more than 30, and if FME is greater than CME, the soil probably expands upon release of a load and should be classified as an expansive soil. In addition, when FME is greater than the liquid limit, there is the potential for liquefaction in the presence of free water. Both adsorption and CME are long-term in comparison with absorption process and they cover some degree of physicochemical interaction between soil and water.

4.6.4   *Factors affecting adsorption characteristics*

From an environmental viewpoint, three major factors affect the characteristics of water adsorption: (a) Particle size, (b) Pore fluid characteristics, and (c) Organic content. The adsorption potential generally increases with decreasing particle size, since smaller particles have a greater surface area available for reaction when considered on a unit mass or volume basis. The effect of pore fluid as reflected by pH value on adsorption. Using pH as an indicator of pore fluid characteristics, it has been observed that increasing pH leads to higher water adsorption, up to a pH of 8, beyond which the behavior is less pronounced. Adsorption also tends to be higher and stronger with increasing organic content.

4.7   Ion exchange capacity and ion exchange reactions

4.7.1   *General discussion*

Thompson in 1850 and Way in 1850–1852 discovered the phenomenon of ion exchange. It was shown that clay minerals have the property of absorbing certain cations and anions and retaining these ions in an exchangeable state. The quantitative studies made by Gedroiz (1912–1913) and Hissink (1922–1932) called this exchange phenomenon a *base exchange* or *ion exchange*. Winterkorn (1937) was the first to introduce the significance of ion exchange in highway construction and soil stabilization. There are two types of ion exchange; namely, cation exchange and anion exchange. A great deal of information is available regarding cation exchange; however, very little data are found on anion exchange. Anion exchange tends to occur less in part because most soils have a net-negative surface charge and also because it is difficult to isolate these reactions for measurement. From the geotechnical engineering point of view, ion-exchange reactions gives us two important phenomena. The first is that a reaction can cause changes in the soil–water structure, from dispersive to flocculative structures or vice-versa; and second the reaction can change the water composition, for example, change hard water to soft water by the removal of calcium ion (Ca^{2+}) and magnesium ion (Mg^{2+}). Common ions found in soil–water–pollutant systems are: (1) Soil: Ca^{2+}, Mg^{2+} H^+, K^+, NH_4^+, Na^+, SO_4^{2-}, Cl^-, PO_4^{3-}, NO_3^-; (2) Water: Cl^-, SO_4^{2-}, Ca^{2+}, H^+, K^+, CO_3^{2-}, Na^+, HCO_3^-, Mg^{2+}; and (3) Pollutant: As, F, Pb, Cr, Hg, Cl, I, Br, NH_3, H_2S, HNO_3, H_3PO_4, $KBrO_3$, KBr.

4.7.2 Ion exchange capacity of soil

Ion exchange capacity (IEC) of soil is defined by US Department of Agriculture (USDA, 1938) as the maximum quantity of ions that the soil is capable of adsorbing from a neutral solution, expressed in terms of milli-equivalents per 100 g of soil (meq/100 g). Ion exchange capacity of some clay minerals and natural soils are: kaolin = 3–5 meq/100 g, illite = 10–40 meq/100 g, and montmorillonite = 80–150 meq/100 g. Numerous factors affect the ion exchange capacity including, particle size, temperature, concentration of pore fluid, and other local environmental conditions.

4.7.3 Ion exchange capacity relating to other soil parameters

Figure 4.10 shows the relationship between ion exchange capacity, silica sesquioxide ratio, and several soil types.

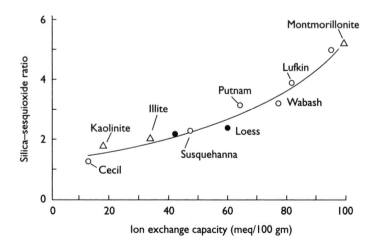

Figure 4.10 Relationship between silica–sesquioxide ratio (SSR) and ion exchange capacity of several natural clays.

Based on various published data, a simple empirical equation has been developed (Fang, 1997) as shown in Equation (4.6). In examining the Equation (4.6), the ion exchange capacity is about half of the plasticity index (Sec. 2.5) for common non-swelling and non-contaminated soils.

$$IEC = 0.50 \, I_P \tag{4.6}$$

where IEC = ion exchange capacity, meq 100 g, and I_P = plasticity index (ASTM D424).

4.7.4 Anion exchange

Anion exchange is a more complex phenomenon than cation exchange and as such there is very little data for soils. However, some comments from various researchers are briefly stated herein. Both cations and anions are involved in reactions of soils with electrolytes. As the pH increases, negative charges are released. Anions are repulsed by these additional negative charges and thus retention is diminished. The maximum negative charge from anion exchange in many soils is developed after they are phosphated and adjusted to a pH equal to 8.2.

4.7.5 Ion exchange reactions

The ion exchange reaction mechanism may be illustrated by considering the use of synthetic silicate exchangers (zeolites) in the softening of hard waters. Usually sodium (Na) is the exchangeable element of an active zeolite (Z). Hard water generally has high concentrations of soluble calcium (Ca) and magnesium (Mg), and when these ions pass through the softening material, the calcium and magnesium are absorbed while sodium is released into solution as shown in the following reaction:

$$Na_2Z + CaSO_4 = CaZ + Na_2SO_4 \tag{4.7}$$

where Na_2Z = sodium zeolite, and $CaSO_4$ = calcium sulfate. The calcium ion may also remain hydrated in solution by itself. CaZ is the calcium zeolite, and Na_2SO_4 is sodium sulfate. Calcium is a very active displacing agent, so that in dilute solution it displaces the sodium of the zeolite. Numerous ion exchange examples are available in standard chemical textbooks.

4.8 Osmotic and reversed osmotic phenomena

4.8.1 Osmotic phenomena

Osmosis is a phenomenon involving solutions separated by a membrane. The membrane acts as a barrier between the solutions and has the property of allowing certain types of molecules to pass through, while preventing the passage of other species in solution. In other words, when the concentration of a solution at one point is different from that at another point, there is a tendency for the more dilute liquid to diffuse into the region of higher concentration. A classic demonstration of osmotic phenomena from basic physics is conducted a U-tube as shown in Figure 4.11(a), in

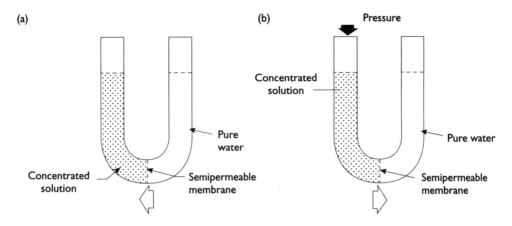

Figure 4.11 Osmotic and reversed osmotic phenomena: (a) osmotic phenomena and (b) reversed osmotic phenomena.

which a semi-permeable membrane has been placed to separate the U-tube as indicated by a vertical dotted line. If a salt solution is placed on the left side of the membrane, and pure water is placed on the right side, then the pure water will enter the solution by passing through the membrane. The only way to prevent this movement is to apply pressure as shown in Figure 4.11(b). Movement of the solute in the opposite direction is inhibited by the membrane. This one-directional movement is referred to as *osmotic phenomena*.

The U-tube analogy is used to describe the attraction of water toward soil particles which are surrounded by a higher concentration of ions (generally cations). The cations are attracted to the net negative surface charge of the soil particles. This negative surface charge results from isomorphic substitution, as discussed in Sec. 3.6.

4.8.2 Reversed osmotic phenomena

As discussed above, osmosis is the spontaneous passage of a liquid from a dilute to a more concentrated solution across an ideal semi-permeable membrane that allows passage of the liquid but not of dissolved solids. Reverse osmosis, shown in Figure 4.11(b), is just the opposite in which the natural osmotic flow is reversed through the application of an external pressure. The amount of pressure applied to the concentrated solution is sufficient to overcome the natural osmotic pressure.

4.9 Soil–water–air interaction in the environment

4.9.1 General discussion

Soil–water interaction is complex as discussed in Sections 4.1 and 4.2. If polluted water is involved, the complication will significantly increase. In this section additional comments are added with respect to soil–water interaction in the environment. For evaluation of the system, we must treat the soil element in three-phases, solid, liquid, and gas. The liquid portion can be divided into two parts, gravity water and environmental water (Sec. 3.8). Gravity water is not under the influence of dissolved ions in the diffuse double-layer or existing solid–liquid and liquid–gas interfaces. Environmental waters are strongly influenced by local environmental conditions such as pore fluid character, temperature, and pressure. This region is the most likely place for ion exchange and bacterial activities to occur.

4.9.2 Water transport in the vapor phases

Dry clay minerals establish different equilibria when brought in contact with water in gaseous or liquid phases. Water in the gaseous phase consists mainly of the individual H_2O molecules, while in the liquid phase water may, in many respects, be considered as one coherent mass. Water may move into, through and out of soil in the solid, liquid, or vapor phase. Vapor phase movement may be due to gradients in relative humidity of the soil–air caused by differences in temperature, moisture content, and in the water affinity of the internal soil surface, to passing air currents, or to the pumping of air into and out of soil as a result of changing barometric pressure in the atmosphere or biosphere. Vapor movement in soil by diffusion alone is normally not very important; movement by convection, however, may acquire great significance under certain special conditions.

4.9.3 Water movement in vadose zone

Water movement in the vadose zone (unsaturated soil zone) is described by a three-phase system of solid, liquid, and gas as discussed previously. A significant feature of the vadose zone is that moisture exists in a state of less than atmospheric pressure. Movement of water is always in the direction from high to low potential, and in unsaturated soil this means that water will travel from a less negative (wetter) to a more negative (drier) pressure zone. This explains the suction effect that dry soils have when exposed to water.

4.10 Sensitivity of soil to environment

4.10.1 Characteristics of sensitivity of soil to environment

The sensitivity of soil to the environment is hinged not only to the local environment, but is also influenced by naturally inherited mineral structure, such as particle size, bonding characteristics between particles, ion exchange capacity, etc. (a) smaller soil particles have larger surface area per unit mass volume. The larger the surface area, the more chance for a soil particle to interact with environment; and (b) the weaker the bonding energy between the particles or higher the ion exchange capacity, the higher the sensitivity of the particles to the environment. For example, montmorillonite clay is potentially more sensitive to the environment than kaolinite or illite clays, because the montmorillonite clay has larger surface areas, weaker bonding energy, and higher ion exchange capacity (Table 3.6). In this section only the load and pore fluid are discussed.

4.10.2 Sensitivity of soil to load

The sensitivity of soil to load has been reported by Fang (1997). The sensitivity of soil to load is related the amount of contact area between soil particles. Essentially, increasing the contact or surface area of particles per unit volume leads to greater opportunity for the soil to react with the environment. Consequently, the type of test will influence soil–environment interaction. It has been reported that tensile tests (Sec. 8.10) give a greater contact area during the loading process than the direct shear (Sec. 10.5.2) and unconfined compression test (Sec. 10.7). In other words, the tensile test is more sensitive to the local environment.

4.10.3 Sensitivity of soil to pore fluid

In conventional geotechnical engineering, it is assumed that soil parameters such as specific gravity (G_s), liquid limit (ω_L), plastic limit (ω_P) are constant for a given soil under any environmental conditions. However, results in various publications indicate that these parameters are not a constant for a given soil and change when the pore fluid or other environmental conditions change. Andrews *et al.* (1967) compared the interaction of three clay minerals with three pore fluids. The pore fluids were water, dimethyl sulfoxide (DMSO) and dimethyl formamide (DMF). The clay minerals used were kaolinite, attapulgite, and Na-montmorillonite. Both DMSO and DMF are acidic solvents. Typical test results on effect on Atterberg limits (Sec. 2.6) of soil are summarized in Table 4.6.

The effect of pore fluid pH on the grain size characteristics is shown in Table 4.7. It is observed that the pore fluid affects the grain size distribution curves significantly when soil particle size is smaller than 0.074 mm (US #200 sieve). Note from the table that the grain size corresponding to 90% passing is the same at all pH levels. As the pH increases, the grain size distribution becomes skewed toward a finer distribution. For example, at a pH of 11, the D_{10} is found to be less than 0.003 mm, while for a pH of 3 it is approximately 0.015 mm. In essence, increasing the pH leads to greater inter-particle repulsion (and thus smaller particles) while low-pH

Table 4.6 Effect of pore fluids on Atterberg limits

Clay minerals	Pore fluid	ω_L	ω_P	I_P	ω_S
Kaolinite	H_2O	62	33	29	29
	DMSO	105	50	55	—
Attapulgite	H_2O	291	110	181	80
	DMSO	309	155	154	—
Na-bentonite	H_2O	506	55	451	6
	DMSO	140	80	60	—

Source: Andrews, R. E., Gawarkiewicz, J. J., and Winterkorn, H. F. Comparison of the interaction of three clay minerals with water, dimethyl sulfoxide and dimethyl formamide. In Highway Research Record No. 209, Highway Research Board. National Research Council, Washington, DC, 1967, pp. 66–78. Reproduced with permission of the Transportation Research Board.

Notes
ω_L = liquid limit; ω_S = shrinkage limit; H_2O = water;
ω_P = Plastic limit; I_P = Plasticity index; DMSO = Dimethyl sulfoxide.

Table 4.7 Influence of pH on grain size characteristics

Percent passing	Grain size, D (mm) as a function of pH		
	pH = 3	pH = 7	pH = 11
90	0.4	0.4	0.4
60	0.03	0.025	0.02
30	0.021	0.015	0.005
10	0.015	0.011	<0.003

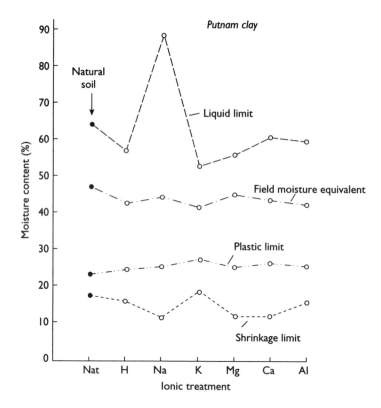

Figure 4.12 Effect of ionic treatment on Putnam soil.

Source: Winterkorn, H. F. and Moorman, R. B. B. A study of changes in physical properties of Putnam soil induced by ionic substitution. In *Proceedings of the 21st Annual Meeting, Highway Research Board.* National Research Council, Washington, DC, 1941, pp. 415–434. Reproduced with permission of the Transportation Research Board.

conditions favor aggregation (and thus larger particles). Dispersive clay (Sec. 2.11) is more sensitive to the environment than silty clay. Explanations on the effect of pore fluid as reflected by pH on gradation curves on dispersive clay are given by Benson (1991).

The effect of ionic treatment on Putnam soil (Heavy Missouri silt-loam) as reflected by liquid limit (Sec. 2.6), field moisture equivalent, plastic limit, and shrinkage limit are presented in Figure 4.12. Of these parameters, the liquid limit shows the most variation.

4.11 Geomorphic process (aging process) of soil

4.11.1 The concept of geomorphic process

A parent material in a given location, after a certain time period and due to various environmental factors, will gradually change its form or properties. The stages or phenomena of such changes are called the *geomorphic process* as introduced by Fang (1986). The rate of change depends on how severe the environmental conditions are. There are five possible processes which could occur in soil and rock as a short-term process (mechanical causes) or long-term process (environmental causes). The long-term processes include those characterized as (a) chemical (b) physicochemical (c) geomicrobiological, and (d) soil-forming. Figure 4.13 shows the concept of geomorphism of soil-rock. In examining Figure 4.13, the mechanical alteration is considered a short-term process while the others are long-term processes.

4.11.2 Short-term and long-term geomorphic process

Short-term and long-term geomorphic processes are illustrated in Figure 4.14. The short-term process mainly is a mechanical process (loading). The long-term geomorphic process includes multimedia energy fields, for example, chemical, physicochemical, and geomicrobiological processes.

An example of long-term geomorphic processes is the gradual conversion of a municipal solid waste landfill into mostly organic soil. This conversion involves

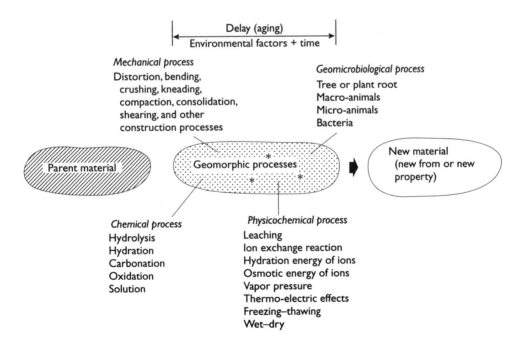

Figure 4.13 The concept of geomorphic (aging) process of soil and rock.

Source: Fang, 1986.

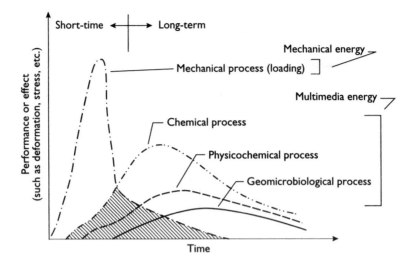

Figure 4.14 Schematic diagram illustrates the effects of short-long-term processes on soil behavior. Source: Fang, 1986.

myriad chemical and biological reactions that occur over varying time periods. Chemical processes can change soil properties and forms as indicated in Figures 4.13 and 4.14 through carbonation, hydration, hydrolysis, and oxidation as noted in the following equations.

a   Carbonation: $Ca(OH)_2 + 2CO_2 = Ca(HCO_3)_2$ (4.9)
b   Hydration: $CaSO_4 + 2H_2O = CaSO_4 \cdot 2H_2O$ (4.10)
c   Carbonation and hydrolysis (solution):
    $2KAlSi_3O_8 + 2H_2O + CO_2 = H_4Al_2Si_2O_9 + 4SiO_2 + K_2CO_3$ (4.11)
d   Oxidation: $2Fe + 3O = Fe_2O_3$ (4.12)

The above equations illustrate how geomorphic processes change soil-rock behavior. In examining Equation (4.11), the parent material of orthoclase is given by the formula $2KAlSi_3O_8$, and the environmental factors are carbonation, CO_2, and hydrolysis, $2H_2O$, after which it becomes kaolinite ($H_4Al_2Si_2O_9$), amorphous silica ($4SiO_2$), and potassium carbonate (K_2CO_3). As time passes and local environmental conditions change, the properties and forms of soil will continue to change. Orthoclase can change into kaolinite and potassium. Likewise, kaolinite and others will change into something else if local environmental conditions change. The rate of change depends on the severity of the environmental conditions. Many "geotechnical surprises" are caused by geomorphic processes during the useful life of the underlying soil. Long-term predictions based on short-term studies without consideration of the local environmental factors are the major reasons why premature or progressive failures frequently occur.

4.11.3   Soil forming process

1   *Soil-forming factors*: Soil forming processes are basically the combination of physicochemical, chemical, and geomicrobiological processes. Soil-forming

Table 4.8 Soil-forming factors

1   Climate
    (a) Temperature (hot–cold; freeze–thaw cycles)
    (b) Rainfall (dry–wet cycle)
2   Biotic
    (a) Flora (tree/vegetation)
    (b) Fauna (bacterial)
3   Edaphic
    (a) Parent soil materials
    (b) Clay minerals
4   Topographic
    (a) Surface relief
    (b) Drainage pattern

factors are the factors concerning the development of soils from accumulated materials that originate from rock weathering can be grouped into four general categories namely climate, biotic, edaphic and topographic, and their subgroups as presented in Table 4.8.

2   *Soil-forming subgroups*: Soil forming is divided into four subgroups namely podzolization, lateritization, carbonation, and alkalization: (a) *podzolization* is a dissolving process which involves true solutions and soil, and hence it is a deteriorative process which takes place in acid media in all humid regions, (b) *lateritization* is characterized by removal of silica and residue of the sesqui-oxides (Sec. 2.9) to the extent that the alumina–silica ratio (Al/Si) becomes 1.0 to less than 2.0. It also is called *desilicification*, (c) *carbonation* is the conversion of soil calcium into carbonates which accumulate in the subsoil at different depths, and (d) *alkalization* consists mainly in accumulation of soluble salts which form in the weathering of soil mineral particles.

4.12   Bacterial attack and corrosion process

4.12.1   *Bacterial attack*

1   *Characteristics of bacteria*: Bacteria are single cell forms. Four major elements, carbon (C), hydrogen (H), oxygen (O), and nitrogen (N) compose about 98.5%, by weight, of the atoms of the cells. Phosphorus (P) and sulfur (S) together make up the additional percent. In general, there are two basic types of bacteria: Heterotrophic and Autotrophic. The heterotrophic type of bacteria requires energy and carbon directly from soil's organic matter, however, the autotrophic type requires energy from oxidation of mineral constituents.

2   *Soil–water–bacterial interaction*: Most bacteria exists at a shallow depth from the ground surface, especially in the A-horizon (Sec. 2.4). Four million bacteria can exist in 1 g of soil (Millar and Turk, 1943). Disturbed soil has more bacteria than undisturbed because the air encourages growth of more (aerobic) bacteria. Bacterial activities can cause acceleration of ion exchange reaction and accession of the decomposition process in soil–water system and consequently will influence soil–water properties such as volume change, compressibility and shear strength.

4.12.2 Corrosion process

1 General discussion: Corrosion is part of the geomorphic processes as discussed in previous section. The corrosion of exposed structural components, particularly those of bridges, pavements, and waterfront structure is a common problem. There are three media in which corrosion occurs: air, water, and ground soil. Bacteria and chemical corrosion are closely related. Bacteria can cause changes in the characteristics of soil–water systems. A typical example of this reaction is the conversion of H_2S into H_2SO_4 through bacterial action, as show in Equation (4.12).

Bacteria

$$H_2S + 2O_2 \longrightarrow H_2 SO_4 \tag{4.12}$$

2 Soil–water–corrosion interaction: In civil engineering, most of the structures erected on the ground surface have extensive underground components and various types of foundations. These components may be subject to underground corrosion over time. Underground corrosion is a more complex phenomena than atmospheric corrosion due to the larger number of variables and unknowns involved. Further discussions on soil–water corrosion are given by Fang (1997).

4.13 Summary

Water is such a common substance that it's not often given a second thought. In fact, the properties of water are quite complex and lead to even greater complexity when mixed with soil. Geotechnical engineering as a field is necessary in part because of the physical, chemical, and biological changes initiated by water. To evaluate the soil–water system, it is necessary to treat the soil element in three-phases: solid, liquid, and gas. The liquid portion can be divided into two regions: gravity water and environmental water. Environmental waters are strongly influenced by the local environmental and are less easily removed as compared to gravity water. Environmental waters are also the most likely place for ion exchange and bacterial activities to occur. Soil–water interaction occurs in various energy fields, and among these energy fields, the thermal-electric energy fields plays the most important role for their interaction. The significance of geomorphic processes (aging processes) in both the short and long-term has been discussed. In particular, ongoing reactions serve to alter the physical and chemical nature of soil and rock.

PROBLEMS

4.1 What is the pH value of a 0.01 mol/l solution of sodium hydroxide (NaOH)?

4.2 Why is soil–water interaction a complex system? Why will soil–water interaction at various energy fields differ?

4.3 Define: saturation, sorption, absorption, and adsorption. What is the mechanism of soil adsorption and its significance in the soil–water system?

4.4 Convert the unit weight of water at 4°C equal to 1 g/cm^3 into SI unit (kN/m^3).

4.5 The dynamic viscosity of water at 5°C is equal to 1.519×10^{-3} Ns/m^2. What are the absolute and kinematic viscosities of the same water at 15°C?

4.6   Define the sensitivity of clay, state how it is usually found, and discuss briefly its significance in engineering problems.

4.7   Explain why some soil is sensitive to moisture as reflected by the shear strength or volume change and others are not? Why is clay less sensitive to liquefaction than sand?

4.8   Explain and compare the reactions and characteristics of the soil–water interaction in the saturated and unsaturated conditions.

4.9   What is the geomorphic process (aging effect) of soil? What are the short- and long-term geomorphic processes? Why are long-term geomorphic processes more important than short-term processes?

4.10   What are the major differences of corrosion processes in the air, water, and ground soil?

4.11   What are the soil-forming processes? Which one is more critical to soil's long-term behavior.

Chapter 5

Hydraulic conduction phenomena

5.1 Introduction

As discussed in Chapter 4, the amount of water existing in the soil mass significantly influences engineering behavior. Beyond its mere presence, the characteristics of water movement and hydraulic conduction in soil are also important and have practical implications. In this chapter, discussions are focused on the hydraulic conduction phenomena near and/or below of the ground surface. The major factors involved in hydraulic conduction include climatological conditions and topographical features such as rainfall, infiltration, percolation, evapotranspiration, sorption, retention, and leachate, as well as the characteristics of soil–water system itself. *Climatic factors* include: wind, rain, snow, ice, frost penetration and wet–dry, hot–cold, and freeze–thaw cycles; while *topographical factors* include: ground features, basin characteristics, channel characteristics, and vegetation/tree patterns. Rainfall or nearby surface water intrude into the soil layer by three major processes: infiltration, percolation, and retention. For example, precipitation that contacts the ground surface may follow several routes: (a) interception by the vegetative canopy (trees, plants etc.), (b) retention in the depressions upon the land surface, (c) direct infiltration, (d) Evaporation, (e) evapotranspiration, and (f) surface runoff. A portion of the rain falls directly into a stream channel and is indicated as channel precipitation. Rain is collected on the ground surface as surface retention. Topographical features are also important factors relative to surface and subsurface drainage systems as discussed in Sec. 5.7.

5.2 Infiltration, percolation, and retention

5.2.1 Infiltration and infiltration capacity

1 *Infiltration*: Infiltration is a part of the hydrologic cycle and was first discussed by Hortin in 1933. He defined infiltration as the passage of water through the ground surface into the subsurface soil layer. Infiltration normally begins at a high rate and then decreases to a minimum. The factors that affect both the amount and the rate of infiltration are primarily those which characterize the soil particle sizes and their relative permanency such as (a) characteristics of clay mineral structure and porosity, (b) time in relation to the amount of infiltration, and (c) type of land usage in relation to the infiltration.

2 *Infiltration capacity*: Infiltration capacity is the maximum rate at which a given soil condition can absorb rain as it falls. This value decreases exponentially in time from a maximum initial value to a constant rate. It also decreases exponentially with time as the soil becomes saturated and soil particles swell.

5.2.2 Infiltration through soil layer and pavement joints

1 *Infiltration through soil layer*: Rainfall or surface drainage water infiltrated into the subsurface soil layer is an important parameter for stability evaluations of soil-foundation systems or earth slope stability problems. The thickness of this saturated zone can be estimated by Equation (5.1). This equation is modified based on the previous work done by Beattie and Chau (1976) to include that soil with surface cracks and contaminated surface water intrusion.

$$h = (\lambda)(\zeta)\frac{kt}{(S_f - S_0)n} \tag{5.1}$$

where h = thickness of the saturated zone, λ = correction for surface cracks (Hair cracks = 1.05, Alligator cracks = 1.10, Galley cracks = 1.15.), ζ = correction for polluted pore fluid (Based on viscosity of pore liquid; Water = 1.0), k = hydraulic conductivity, t = time, S_f = final degree of saturation, S_0 = initial degree of saturation, and n = porosity.

2 *Infiltration through pavement joints*: Infiltration through pavement and base course materials have been discussed by Ridgeway (1976) and Moulton (1991). Two semi-empirical equations have been developed for estimating the amount of water infiltrating through concrete and bituminous pavements, base and subbase courses. These equations are based on water passage through the pavement joints without surface cracks on the pavement surface or presence of contaminated surface water. Ridgeway (1976) and Moulton (1991) provide more detailed discussion, construction layouts, and numerical examples.

5.2.3 Percolation and retention

1 *Percolation*: Percolation is the water movement within the soil mass. Percolation and infiltration are closely related but infiltration cannot continue unless percolation provides sufficient space such as voids in the soil layer for infiltration water. Retention is also a part of the water movement process and is closely related to channel precipitation and overland flow. Water retention for soils of intermediate texture (e.g. silty soils) is partly due to swelling forces and partly to capillary forces. Units for measuring retention include suction, capillary potential or free energy, and pF value (Sec. 5.11). The instruments suitable for measuring suction include tensiometer, pressure membrane apparatus, and vapor adsorption method. Infiltration and percolation processes relating to raindrop, nitrate, soil structure, soil columns and contaminated soil sediment, and water from an agriculture and chemistry viewpoint are given by Conklin (2003).

2 *Retention*: Retention is defined as a capability of a soil to retain water in the soil mass. The rainfall contribution to stream flow from a storm is relatively a constant. That portion of the rain falling directly into the stream channel is indicated as

channel precipitation. At the beginning, almost all of the rainfall is collected on the ground as surface retention. After a sufficiently long time, surface retention approaches a steady state.

5.3 Capillarity phenomena

5.3.1 General discussion

Capillary rise is caused by fluid surface tension. It occurs at the interface between two different materials. For soil, it occurs between surface of water, mineral grains, and air. Basically, surface tension results from differences in forces of attraction between the molecules of the materials at the interface. The general equation for computing the capillary rise in soil derives from the basic physics of a glass tube containing a liquid (water) as shown in Equation (5.2).

$$h_c = \frac{2T_s \cos \alpha}{gr(\rho_1 - \rho_g)} \tag{5.2}$$

where h_c = capillary rise, T_s = surface tension, α = contact angle between liquid and glass tube, g = acceleration due to gravity, ρ_1 = unit weight of liquid, ρ_g = unit weight of liquid of gas (usually neglected for air), and r = radius of glass tube.

5.3.2 Rate of capillary rise

From a practical point of view, the rate of capillary rise is more important than maximum height of capillary rise. In such cases one employs the equation

$$t = \frac{nh}{k}\left(\ln \frac{h}{h-z} - \frac{z}{h}\right) \tag{5.3}$$

where t = time required for the meniscus of the capillary water to rise to the height, z, above the free water level; n = porosity, h = capillary potential height of capillary rise, k = coefficient of permeability of the soil; and z = distance of capillary meniscus from groundwater level.

Capillary rise in soils can be directly measured in the laboratory, under in situ conditions or estimated from an equation such as Equation (5.4). Equation (5.4) is a simplified equation derived from basic capillary rise Equation (5.2). Equation (5.4) is frequently used in geotechnical engineering,

$$h_c = \frac{0.3}{d} \tag{5.4}$$

where h_c = capillary rise (cm) and d = diameter of voids between soil particles (cm). Equation (5.5) is derived from Equation (5.4) as illustrated in Example 5.1.

$$h_c = \frac{0.3}{D_{10}} \tag{5.5}$$

In addition, Terzaghi (1942) also proposed that, h_c can be estimated from the void ratio, e, and D_{10} as

$$h_c = \frac{C}{eD_{10}} \tag{5.6}$$

where C = constant which varies from 0.1 to 0.5, e = void ratio (Sec. 3.2.2), and D_{10} = effective size (see Example 5.1)

EXAMPLE 5.1
Derive the Equation (5.5), the capillary rise in soil

SOLUTION
Surface tension of water, T_s at 20°C = 0.072.8 N/m (72.8 dynes/cm) (see Table 4.1)
    Unit weight of liquid (water), γ_ω = 9.79 kN/m^3 (62.8 pcf). Substituting in Equation (5.2).

$$h_c = \frac{(4)(72.8 \text{ dynes/cm})(10^6 \text{ cm/m}^3)}{(9.79 \text{ kN/m}^3)(10^8 \text{ dynes/kN})(\mathbf{d})} = 0.3\mathbf{d} \text{ (cm)}$$

Since **d**, the diameter of voids between soil particles is difficult to obtain, Terzaghi (1942) suggested that $\mathbf{d} \approx D_{10}$, where D_{10} = effective size of the soil particle (Sec. 3.3), then $h_c = 0.3/D_{10}$ (Eq. 5.5).

    There are several measuring techniques for estimation of the capillary rise in soils. ASTM standards used for measuring capillary-moisture relationships include the pressure membrane apparatus (ASTM D3152) for fine-grained soils and a porous plate device for coarser grained soils (ASTM D2325). A tensiometer may also be used to measure the suction and assess potential capillary rise in a given soil under field conditions.
    In many practical problems, horizontal capillary action is more important in comparison with that in the vertical direction, because it is related to the subsurface moisture migration, potentially leading to water leakage through earth retaining structures. Also, horizontal capillarity can allow leachate migration through landfill clay liners (Ch. 16) and low-permeability soil layers.

5.3.3 Factors affecting capillary rise

Factors affecting capillary rise include (a) effect of time and soil particle size: a curvilinear relationship between time on capillary height for sand, silt, and clay is observed. This relationship is closely related with soil particle size as indicated in Figure 5.1; (b) effect of moisture content and types of pore fluids. If the soil particles are not completely wetted by water (unsaturated soil), the contact angle between the menisci and the soil particles will be greater than zero. An increase in the contact angle will tend to increase the curvature of the meniscus and thereby increase the capillary potential of the soil at a given water content; and (c) effect of temperature, because surface tension of water is an inverse function of temperature. Hence, a

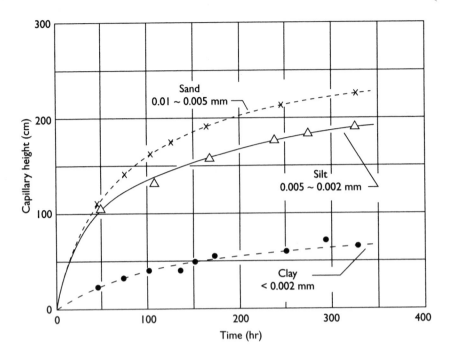

Figure 5.1 Capillary height versus time for sand, silt, and clay.

Source: Tianjing University, 1979 with additional data.

decrease in temperature increases the surface tension and decreases the capillary potential as well as other factors such as dissolved salts, because an increase in the amount of dissolved salt in the soil-water increases its surface tension and thereby lowers the capillary potential of the soil.

5.3.4   Comments and discussion on capillary rise

Two major factors affecting the height of capillary rise are time and soil particle size. Based on laboratory tests, the first 200 h are critical for most soils. After that the capillary rises level off (See Fig. 5.1). Other comments are (a) for fine-grained soil, the value of h_c is much less than would be predicted from Equations (5.5) and (5.6). In most field observations, h_c will not exceed more than several meters; (b) for granular soil, such as clean uniform sand or gravel, the value of h_c is dominated by the effective grain size, D_{10}. Equation (5.5) is valid when D_{10} is greater than 0.002 mm (7.8 × 10^{-5} in.); (c) In general, the value of h_c in fine-grained soils is influenced by the grain size distribution, clay mineral composition, pore fluid, stress history as well as environmental factors, and (d) the rate of the capillary rise depends on the soil particle sizes. For example for soil particle sizes between 0.01 mm (3.9 × 10^{-4} in.) and 0.005 mm (2 × 10^{-4} in.), 7 days are required to reach the capillary height of 2 m (6.56 ft). However, for soil particles smaller than 0.002 mm (7.8 × 10^{-5} in.) much more time is required to reach the same height, perhaps several months.

5.4 Hydraulic conductivity

5.4.1 Darcy's law and flow velocity

1 General discussion: For steady state flow, the quantity of flow, Q, per unit of time across an element of area, A, perpendicular to the direction of flow is a constant value. Flow velocity is the flow through porous media following Darcy's Law as shown in the following equation:

$$Q = vA = kiA \qquad (5.7)$$

where Q = the rate of flow, v = velocity, k = a constant known as *Darcy's coefficient of permeability*, i = hydraulic gradient, and A = cross-sectional area for which flow can pass through. Equation (5.7) can be rewritten as

$$v = ki = Q/A \qquad (5.8)$$

2 Flow velocity and seepage velocity: If the hydraulic gradient, i, is unity, then the velocity, v, in Equation (5.8) is equal to k. Also, v is called *approach velocity* or *superficial velocity* (Taylor, 1948). The *average effective velocity*, v_s, also known as *seepage velocity* of flow through the soil can be computed as

$$v_s = v\, A/A_v = v\, V/V_v = v/n \qquad (5.9)$$

where v_s = seepage velocity, A = total cross-sectional area of the soil specimen, A_v = area of voids in soil specimen, V = total volume of soil specimen, V_v = volume of voids in soil specimen, and n = porosity.

5.4.2 Hydraulic conductivity equations

1 General discussion: All hydraulic conduction equations are based on the fact that in first approximation, the rate of conduction is directly proportional to the gradient of the intensity factor ($\Delta P/\Delta L$) and to a coefficient of conductivity (permeability), k, which is a parameter of the physical system. The conductivity coefficient, k, is the inverse of resistivity, $R = 1/k$.

$$\frac{dV}{dt} = ki = \frac{1}{R}i \qquad (5.10)$$

where dV/dt = volume of flow per unit cross section of system in unit time, k = coefficient of conductivity, R = coefficient of resistivity, i = gradient, $\Delta P/\Delta L$ = ratio of driving potential difference to length over which difference occurs.

2 Poiseuille–Hagen equation: Equation (5.10) becomes Darcy's law (Eq. 5.7) for describing flow of water through soils if k is the coefficient of hydraulic conductivity or permeability. Because of the typically laminar nature of flow in soils, theoretical calculations of k have been attempted by a number of workers on the basis of the fundamental Poiseuille–Hagen equation for flow through capillary tubes.

$$\frac{dV}{dt} = \frac{\pi r^4}{8\eta}i; \tag{5.11}$$

where r = radius of capillary and η = coefficient of viscosity of liquid.

3   *Kozeny–Carman equation*: Best known among the attempts of theoretical derivation of k is perhaps the Kozeny–Carman equation.

$$k = \left(\frac{1}{\eta}\right)\frac{1}{k_\mathrm{p}t^2S^2(1-n)^2}\,\frac{n^3}{} \tag{5.12}$$

where k_p = pore shape factor, t = tortuosity, S = specific surface per unit volume, n = porosity, and η = viscosity.

4   *Winterkorn equation*: Based on Equation (5.12), a simpler equation that gives essentially the same results for noncohesive granular soil is given by Winterkorn (1942) and Winterkorn and Fang (1975).

$$k = \frac{nr^2}{8\eta} \tag{5.13}$$

Since $2/r = S/n$, we obtain

$$k = \frac{n}{2\eta}(n^2/S^2) \tag{5.14}$$

where k = coefficient of permeability, n = porosity, η = viscosity, r = radius of the capillary, and S = specific surface per unit volume. In Equation (5.14), note that n/S can be calculated from the height of capillary rise and n can be determined independently. Table 5.1 gives the results of actual permeability measurements together with coefficients calculated by Equation (5.14).

5   *Comments on hydraulic conductivity equations*: Equations (5.10) to (5.14) are derived for saturated laminar flow and do not hold for other types, for example, flow or creep along pore walls or flow through a porous system containing air bubbles, that is, unsaturated flow. In particular, because of air bubble expansion with increasing temperature, a warm rain may percolate through a soil more slowly than a cold rain even though the viscosity of water decreases with increasing temperature. Moreover, water interacts significantly with mineral surfaces and there is a secondary structure which forms in clay-containing soils. As such, there are no robust ways to theoretically calculate the hydraulic conductivity of natural clay-containing soils and a meaningful hydraulic conductivity must be determined experimentally.

5.4.3   *Laboratory hydraulic conductivity measurements*

1   *General discussion*: The hydraulic conductivity or coefficient of permeability, k, as defined in Equation (5.7) can be measured in the laboratory and/or in the field.

Table 5.1 Interrelationships between grain size, capillary rise, surface area, porosity, and hydraulic conductivity of granular soil

(a) Grain size			(b) Saturated capillary rise (cm)	(c) Surface area per volume (cm²/cm³)	(d) Surface area per pore volume calculated (cm²/cm³) from (b)	(e) Porosity calculated from (c) and (d)	(f) Hydraulic conductivity k (10⁻⁴ cm/s)	
passing sieve No.	Retained sieve No.	Average diameter (cm)					Experimental	Calculated in columns from data (d) and (e)
10	20	0.118	6.4	51	85.4	37.5	50	26
20	30	0.069	9.4	81	125	39.3	23	13
30	40	0.049	13.2	122	176	41.0	13	6.6
40	60	0.031	20.0	194	267	41.0	6.7	2.7
60	80	0.021	29.8	290	397	42.3	5.7	1.3
80	100	0.016	35.6	370	475	43.8	2.7	0.97
100	140	0.012	47.0	488	628	43.8	1.6	0.55
140	200	0.0087	67.0	690	894	43.5	0.7	0.27
200	270	0.0062	90.5	970	1210	44.4	0.3	0.15

Source: From Foundation Engineering Handbook, 2nd Edition, H.Y. Fang, ed., ch. 3, Soil Technology and Engineering Properties of Soils, Winterkorn, H. F. and Fang, H.Y., Copyright (1991) by Van Nostrand Reinhold. With kind permission of Springer Science and Business Media.

There are numerous techniques available directly or indirectly to obtain this value. In the laboratory test, four general approaches are (a) constant-head permeameter, (b) falling-head or variable-head permeameter, (c) measured or computed from consolidation test data, and (d) computed from particle surface area and its porosity. Use of a permeameter for measuring the hydraulic conductivity has been standardized (ASTM D2434, D5084). Use of laboratory consolidation tests to compute the coefficient of permeability will be discussed in Section 9.3.4. Field measurements by pumping tests will be discussed in Section 5.6.

2 *Constant-head permeameter:* The ASTM Standard method for the determination of permeability of granular soils with constant head (ASTM D2434) is based on the following equation:

$$k = \frac{V}{Ait} \tag{5.15}$$

where V = water yield in time t, t = time, A = the cross-sectional area of the specimen, and $i = \Delta P/\Delta L$, that is the difference in pressure at the two ends of the specimen.

3 *Falling-head (variable-head) permeameter:* The falling-head type is suitable for cohesive fine-grained soils where the time required to obtain a measurable flow with the constant-head configuration is too great. Detailed testing procedures are given by standard laboratory testing manuals (Liu and Evett, 2003).

$$k = 2.3 \frac{La}{At} \log_{10} \frac{h_1}{h_2} \tag{5.16}$$

where L = length of soil specimen, a = cross-sectional area of the standpipe, A = cross-sectional area of soil specimen, t = time, and h_1, h_2 = hydraulic head levels. Typical ranges of coefficient of permeability for common construction materials are sand–gravel mixture = 10^{-1}–10^{-3} cm/s (3.3×10^{-3}–3.3×10^{-5} ft/s), sand = 10^{-3}–10^{-5} cm/s (3.3×10^{-5}–3.3×10^{-7} ft/s), silty clay = 10^{-5}–10^{-7} cm/sec (3.3×10^{-7}–3.3×10^{-9} ft/s), clay = less than 10^{-7} cm/sec (3.3×10^{-9} ft/s), asphalt pavement = 10^{-6} 10^{-8} cm/sec (3.3×10^{-8}–3.3×10^{-10} ft/s), concrete pavement = 10^{-5}–10^{-7} cm/s (3.3×10^{-7}–3.3×10^{-9} ft/s), clay liners = less than 10^{-7} cm/s (3.3×10^{-9} ft/s).

EXAMPLE 5.2
A soil specimen 10 cm (3.9 in.) in diameter and 5 cm (2.0 in.) in length was tested in a falling-head permeameter. The head dropped from 45 cm (17.7 in.) to 30 cm (11.8 in.) in 4 min and 32 s. The area of the standpipe was 0.5 cm² (0.078 in.²)

a What was the capillary rise of water in the standpipe?
b Considering the effect of capillary rise in the standpipe, compute the coefficient of permeability of the soil in units of cm/s.
c The soil sample had a specific gravity of solids equal to 2.67 and the sample weighed 10 g (0.22 lb) dry. Compute the average seepage velocity.

SOLUTION

$A = \Pi r^2 = (3.1416)(5.0)^2 = 78.5$ cm ^{2}; $a = 0.5$ cm^2;

$r_{pipe} = (a/\Pi)^{0.5} = (0.5/3.1416)^{0.5} = 0.40$ cm

a   Capillary rise from Equation (5.2)

$$h_c = \frac{2(0.00075 \text{ N/cm}) \cos(0)}{(981 \text{ cm/s}^2)(1 \text{ g/cm}^3)(0.40 \text{ cm})} \times \frac{(1 \text{ kg m/s}^2)}{1 \text{ N}} \frac{100 \text{ cm}}{1 \text{ m}} \frac{1000 \text{ g}}{1 \text{ kg}}$$

$$= 0.38 \text{ cm} \approx 0.40$$

b   Coefficient of permeability computed from Equation (5.7)

$h_0 = 45 - 0.4 = 44.6$ $h_1 = 30 - 0.4 = 29.6$

4 min 32 s. = 272 s.

$$k = \frac{(5 \text{ cm})(0.5 \text{ cm}^2)}{78.5 \text{ cm}^2 (272 \text{ s})} \log_{10}\left(\frac{44.6}{29.6}\right) = 2.1 \times 10^{-5} \text{ cm/s}$$

c   Seepage velocity computed from Equation (5.9)

v = discharge velocity = ki

$k = 2.1 \times 10^{-5}$ cm/s from step (b)

i = hydraulic gradient = h/L (44.6–29.6)/2 = 7.5

V = total volume = Ah = (78.5)(5) = 392.5 cm^3

n = porosity = $V_v/V = (V - V_s)/V = (V - \rho_s M_s)/V$ = [392.5 − (2.67)
×(100)]/392.5 = 0.32

The afore mentioned relationships for hydraulic conductivity in soil correspond to pressure gradients. Other gradients or potentials such as thermal, electric and electromagnetic also lead to the movement of fluid or mass and will be discussed in Chapter 6.

5.4.4 Factors affecting hydraulic conductivity

1   *Particle size and void ratio*: Particle size has a pronounced influence on hydraulic conductivity, k. Figure 5.2 presents the coefficient of permeability versus void ratio for bentonite and kaolinite with various pore fluids including aniline ($C_6H_5NH_2$), nitrobenzene ($C_6H_5NO_2$), ethyl alcohol (C_2H_5OH), and water (H_2O). Figure 5.3 presents coefficient of permeability versus void ratio for a homoionic bentonite clay. The permeability values are computed from laboratory consolidation tests (Sec. 9.3.4). Note the large variations as a function of void ratio.

2   *Type of pore fluid*: The type and amount of pore fluid influences hydraulic conductivity significantly as illustrated in Figure 5.4 by various salts such as ($FeSO_4$), (FeO), ($NaNO_3$), and (Na_2SO_3). The effect of pore fluid on hydraulic conductivity as reflected by dielectric constant will be discussed further in Section 6.10.

3   *Temperature*: Temperature will affect the viscosity of pore fluid as discussed in Section 4.3, and consequently it will affect hydraulic conductivity. The effect of

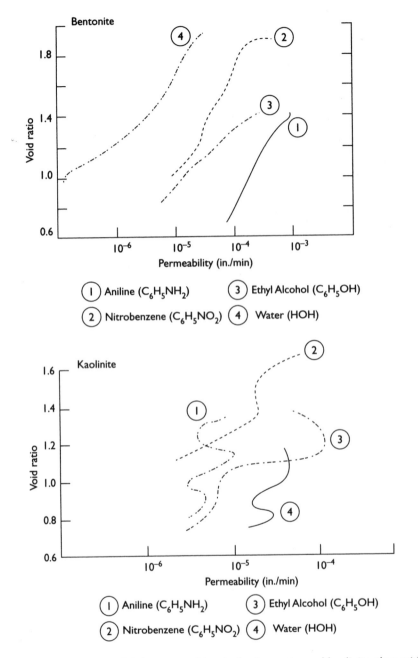

Figure 5.2 Coefficient of permeability versus void ratio for bentonite and kaolinite clays with various pore fluids.

Source: Waidelich, W. C., "Influence of liquid and clay mineral type on consolidation of clay-liquid systems." In *Highway Research Board Special Report* No. 40, Highway research Board. National Research Council, Washington DC, 1958, pp. 24–42. Reproduced with permission of the Transportation Research Board.

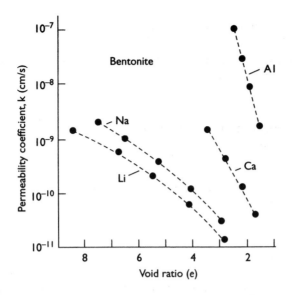

Figure 5.3 Comparison of permeability values homoionic bentonite soil.

Source: Vees, E. and Winterkorn, H. F., "Engineering propertied of several pure clays as functions of mineral type, exchange ions, and phase composition." In *Highway Research Record* No. 209, Highway Research Board. National Research Council, Washington DC, 1967, pp. 55–65. Reproduced with permission of the Transportation Research Board.

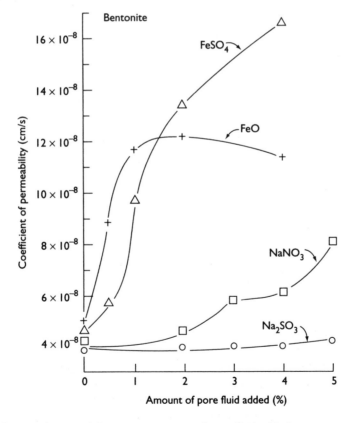

Figure 5.4 Coefficient of permeability versus amount of pore fluid added.

Source: Alther, Evans, Fang and Witmer (1985). Copyright ASTM INTERNATIONAL reprinted with permission.

temperature on the coefficient of permeability of a slightly organic clay is given by Habibagahi (1976) and others.

5.4.5   Hydraulic conductivity of contaminated fine-grained soils

The effect of pore fluid on hydraulic conductivity has become particularly interesting to geotechnical engineers when the encounter municipal, hazardous or nuclear waste sites. Several national and international technical symposia have been organized in recent years.

1   Development of test equipment: The measurement of the hydraulic conductivity of fine-grained soils with hazardous and toxic permeants requires equipment specifically designed and constructed for that purpose. A triaxial type permeameter was developed at Lehigh University as reported by Evans and Fang (1986). It is also referred to as the *Evans and Fang triaxial-permeameter.* The testing system developed for hydraulic conductivity testing with hazardous and toxic permeants has three major components: the control panel, the permeability board and triaxial cell, similar to a modern flexible wall system used in permeability or shear strength testing. Requisite strength and chemical compatibility of the system was achieved through optimizing the use of stainless steel and aluminum for rigidity and Teflon for any location in contact with the permeant. Equipment design and fabrication considerations for testing of hazardous and toxic permeants are discussed in detail by Evans and Fang (1986, 1988). Recently, numerous types of triaxial-permeameters have been developed or modified for conducting studies on contaminated fine-grained soil and many of them are commercially available.

2   Discussion of test results: Typical test results from a triaxial-permeameter are shown in Figure 5.5(a) and (b). Sand-bentonite mixtures are used as a base-case material. Figure 5.5(a) presents the comparison of hydraulic conductivity versus water and aniline ($C_6H_5NH_2$); and Figure 5.5(b) presents the water versus carbon tetrachloride (CCl_4). Significant differences in hydraulic conductivities are found, largely as a result of changes in the electrical double-layer surrounding the bentonite (clay) particles (Ch. 4).

5.5   Stress, pressure, and energy of soil–water system

As discussed in Section 3.2, the voids in a soil element may be filled with water, which carries normal (but not shear) stress. Thus, the total load on the soil element may be carried partially by stress in the grains of solid and partially by stress on the water in the voids between the solid grains. Since the effects of stresses on solid grains and on water are entirely different and the total load or total for testing stress usually represents a known quantity, it is desirable to develop means of separating total stress into its component parts carried by solids and by water. This simple explanation has been made by Terzaghi (1925) and Rutledge (1940) for computing the neutral and effective stresses in soils.

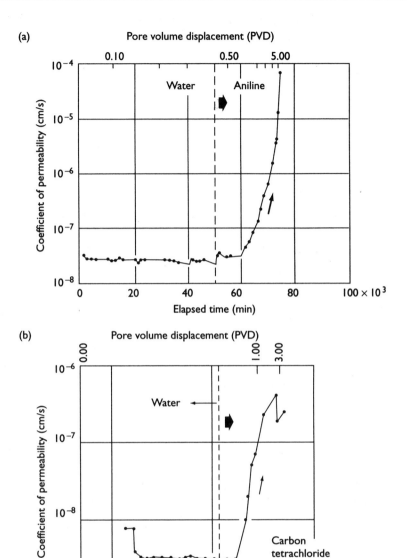

Figure 5.5 Coefficient of permeability versus time for bentonite–sand mixture. (a) Comparison of water vs aniline and (b) Comparison of water vs carbon tetrachloride.

Source: Evans and Fang (1986). Copyright ASTM INTERNATIONL. Reprinted with permission.

5.5.1 *Total, neutral, and effective stresses*

1   *Total stress, σ*: Total stress is the sum of all stresses normal to a plane per unit area of soil element along that plane.
2   *Neutral Stress, u*: Neutral stress is the stress per unit area carried by the water in a soil element. Neutral stress is always normal stress since water cannot transmit

shearing stresses. It is always equal in all directions at any point. Hydrostatic pressure, which increases uniformly with depth, is the most common example of neutral stress.

3 *Effective stress, σ'*: Effective stress is the average normal stress transmitted from grain to grain per unit area of soil element:

$$\sigma' = \sigma - u \tag{5.17}$$

where σ' = effective stress, σ = total stress, and u = neutral stress.

5.5.2 Hydrostatic and excess hydrostatic pressures

1 *Hydrostatic pressure*: Hydrostatic pressure is the pressure in a liquid under static conditions; the product of the unit weight of the liquid and the difference in elevation between the given point and the free water elevation.

2 *Excess hydrostatic pressure*: Excess hydrostatic pressure or hydrostatic excess pressure is the pressure in water greater or less than hydrostatic pressure. When such pressure exists in unconfined bodies of water, the result is water flow. Conversely, if water in the voids of soil is in motion, hydrostatic excess pressures will be found in the water. Equation (5.17) is called the *effective stress equation*. The neutral stress in soil is, thus, the algebraic sum of the hydrostatic pressure and any deviation from hydrostatic pressure that causes flow. If no flow exists, the distribution of neutral stress will always be hydrostatic throughout the depth of the water.

EXAMPLE 5.3
The water table in a deep deposit of very fine sand is 2.3 m (7.5 ft) below the ground surface. Above the water table, the sand is saturated by capillary water. The unit weight of the saturated sand is 2.0 gm/cm³ (124.8 pcf). What is the effective vertical pressure on a horizontal plane at a depth 4 m (13.1 ft) below the ground surface?

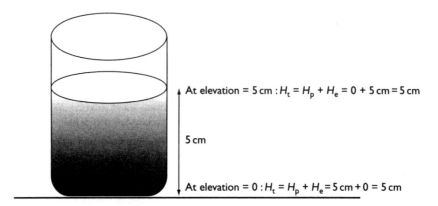

At elevation = 5 cm : $H_t = H_p + H_e = 0 + 5\,cm = 5\,cm$

5 cm

At elevation = 0 : $H_t = H_p + H_e = 5\,cm + 0 = 5\,cm$

Figure 5.6 Schematic diagram illustrating various heads, notice there is no flow in a glass of water since there is no change in total head.

SOLUTION

$\sigma = \gamma H = (2.0 \text{ g/cm}^3)(4.0 \text{ m})(100^3 \text{ cm}^3/\text{m}^3)(1 \text{ kg/1000 g}) = 8000 \text{ kg/m}^2$

$u = (1.0 \text{ g/cm}^3)(1.7 \text{ m})(100^3 \text{ cm}^3/\text{m}^3)(1 \text{ kg/1000 g}) = 1700 \text{ kg/m}^2$

$\sigma' = \sigma - u = 8000 - 1700 = 6300 \text{ kg/m}^2$

5.5.3 Energy or head

There are three types of energies used in the soil-water system namely kinetic energy, elevation energy, and pressure energy. The relationship between energy and head is

$$\text{Head} = \frac{\text{Energy}}{\text{Mass}} = ML/M = L \tag{5.18}$$

where M = mass, and L = length. To express both potential and kinetic energies of fluid, the term *head* is commonly used which is the energy per unit of mass. There are three types of head involving fluid flow in soil; namely, velocity head, pressure head, and elevation head.

1 *Velocity head, h_v*: Velocity head is defined as

$$h_v = v^2 / 2g \tag{5.19}$$

where v = velocity, and g = acceleration of gravity. Flow caused by velocity head is a small amount (0.0003–0.0008 cm) in comparison with causes by other heads. In soil engineering commonly it is negligible, however, for estimation of pollution migration purposes, it may be relevant.

2 *Pressure head, h_p*: The pressure head is the height to which water rises in the piezometer above the point under consideration. In geotechnical engineering applications, the pressure head is needed for computing the porewater pressure and effective stress as shown in Equation (5.20):

$$h_p = \frac{\text{water pressure}}{\text{unit weight of water}} \tag{5.20}$$

3 *Elevation head, h_e*: The elevation head is the vertical distance from the datum.

4 *Total head, h_t*: The velocity head plus pressure head and elevation head is called *total head* or the *piezometer head*. The total head in the water at any point is

$$h_t = h_v + h_p + h_e \tag{5.21}$$

where h_v = velocity head, h_p = pressure head, and h_e = elevation head. Figure 5.6 presents a schematic diagram of elevation, pressure and total heads and *Example 5.7* illustrates a computation procedure for these heads.

EXAMPLE 5.4

Given the following constant-head permeameter, solve for and plot the pressure head, elevation head, and total head as a function of elevation.

SOLUTION

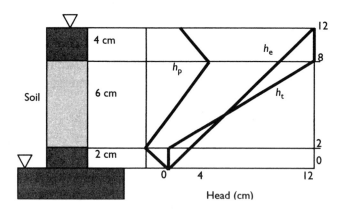

The general approach to these types of problems is to determine elevation (h_e) and total (h_t) heads and then subtract out the pressure head (h_p). First, we know that elevation head increases linearly from the assumed datum to the top of the system, from 0 to 12 cm the above figure. We also know that there is no change in total head from 0 to 2 and then from 12 to 8. Assuming the soil is homogeneous, the change in total head between 2 and 8 is linear and can be fitted with a straight line. Pressure head is then taken as the difference between total and elevation head.

5.5.4   Piezometers and pore pressure measurements

A *Piezometer* is used to measure the hydrostatic or excess porewater pressures in the pores of saturated soils. It is a hollow tube made from metals or plastic. Information obtained from these measurements is important in the study of settlement, bearing capacity, earth pressure, and slope stability. There are four basic types of piezometers available commercially: open standpipe, hydraulic, pneumatic, and electric.

1   *Open standpipe*: An open standpipe, also known as a water level indicator, is used to measure the phreatic surface. The first such device was installed in 1934 by the US Bureau of Reclamation in the embankments at Hyrum Dam in Utah, and at Agency Valley Dam in Oregon. In 1949, Casagrande modified and used it to measure the rate of dissipation of excess porewater pressures at Logan Airport, Boston Massachusetts. It consisted of a tubular porous stone tip embedded in sand and sealed down a borehole. A 3/8-in. (9.5 mm) diameter plastic riser tube connected the tip with the ground surface. The hydraulic head is measured by sounding down the tube to the water surface.

2   *Hydraulic piezometer*: Generally, there are two types of hydraulic piezometers: The foundation type and the embankment type. The foundation type is installed down a borehole and the embankment type is positioned in a trench. The

piezometer tip is a moulded plastic cylinder containing a porous disc. Two tubes connect the tip to a pressure gauge which measures the hydraulic head directly.

3   *Pneumatic piezometer*: A sealed porous tip containing a pressure sensitive valve and diaphragm is connected to a measuring device with two tubes. Air or light oil is pumped into one of the tubes until the pressure in the line equals the pore-water pressure acting on the opposite face of the diaphragm at which time the pore pressure is measured.

4   *Electric piezometer*: In an electric piezometer, the porewater pressure acts against a flexible diaphragm through a porous tip. Electric sensing devices mounted on the opposite face of the diaphragm measure the deflection which is proportional to the porewater pressure.

5.6   Field pumping test

5.6.1   *Principles of pumping test*

A field pumping test is one of the oldest test methods for determining groundwater characteristics, in particular, the hydraulic conductivity, transmissivity, and storativity. The principles of pumping are explained as follows: since water is practically incompressible, two types of equations are generally used to analyze field pumping test data. One is the Thiem equilibrium equation (Thiem, 1906), and the other is the Theis nonequilibrium equation (Theis, 1935). The equilibrium equation can be used to determine the hydraulic conductivity of an aquifer if the rate of discharge of a pumped well is known, and if the drawdown is known in the observation wells at various distances from the pumped well. Drawdown is noted after the cone of depression has been stabilized, that is, it is at equilibrium. The process of reaching equilibrium can be quite time consuming. The nonequilibrium equation permits the determination of aquifer properties when the rate of discharge of a pumped well is known, and when the drawdown as a function of time is determined for one or more observation wells at given distances from the pumped well.

5.6.2   *Steady flow – Thiem equilibrium equation*

The equilibrium equation was developed by Gunter Thiem of Germany in 1906 for the determination of hydraulic conductivity, k. The equation was based on the following assumptions: (a) the aquifer is homogenous and isotropic with respect to hydraulic conductivity, and of infinite areal extent; (b) the hydraulic conductivity is independent of time, and the flow is laminar and steady; (c) the discharging well penetrates and receives water from the entire thickness of the permeable, water-bearing stratum; and (d) the well is pumped continuously at a constant rate until the flow of water to the well is stabilized. Using plane polar coordinates with the well as the origin, the radial flow equation for a well completely penetrating a confined aquifer (Fig. 5.7) is given by

$$Q_w = A\,v = 2\,\pi\,r\,b\,k\,dh/dr \qquad (5.22)$$

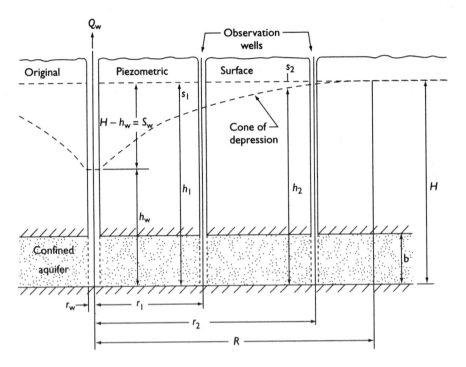

Figure 5.7 Steady flow to a well in a confined aquifer.

where Q_w = well discharge, $[L^3/T]$, A = cross-sectional area perpendicular to flow = $2\pi r b$, $[l^2]$, b = thickness of confined aquifer, $[l]$, k = hydraulic conductivity $[l/t]$, r = radial distance to any point from axis of well $[l]$, v = flow velocity $[l/t]$, h = head at any point in the aquifer at time, t, $[l]$, and $dh/dr = i$ = hydraulic gradient [dimensionless]. Separation of variables gives the following differential equation:

$$dh = \frac{Q_w}{2\pi bk}\frac{dr}{r} \tag{5.23}$$

The boundary conditions for Equation (5.23) are at the well: $h = h_w$ and $r = r_w$, and at the edge of the area of well influence $h = H$, and $r = R$. Integrating Equation (5.23) between limits as indicated.

$$\int dh = \frac{Q_w}{2\pi bk}\int \frac{dr}{r}$$

$$H - h_w = \frac{Q_w}{2\pi bk}\ln\frac{R}{r_w} \tag{5.24}$$

Thus, the hydraulic conductivity, k, can be calculated:

$$k = \frac{Q_w}{2\pi b}\frac{\ln(R/r_w)}{H - h_w} \tag{5.25}$$

Since any two points will define the drawdown curve, Equation (5.25) can be written in terms of drawdowns measured in two observation wells. For this case, the equation for hydraulic conductivity, k, becomes

$$k = \frac{Q_w}{2\pi b} \frac{\ln (r_2 - r_1)}{s_1 - s_2} \qquad (5.26)$$

where $s_1 = H - h_1$, and $s_2 = H - h_2$. The values r_1 and r_2 refer to the radial distance from the axis of the pumped well to observation well #1 and #2, respectively, while s_1 and s_2 refer to the drawdown at observation wells #1 and #2, respectively.

EXAMPLE 5.5   Pumping well confined aquifer
The well in Figure 5.7 has been pumping at 500 l/h for long enough to achieve equilibrium. The two observation wells are positioned at distances of 25 and 50 m away and the observed drawdown was 5 and 1 m, respectively. For an aquifer thickness of 10 m, compute the hydraulic conductivity, k.

SOLUTION
From Equation (5.26):

$$k = \frac{(500 \; l/h)\cdot(1 \; m^3/1000 \; l)\cdot(1 \; h/3600 \; s)}{2\pi(10 \; m)} \cdot \frac{\ln (50 \; m - 25 \; m)}{5 \; m - 1 \; m} = 1.8 \times 10^{-6} \; m/s$$

5.6.3 *Non-steady flow – Theis nonequilibrium equation*

1   General discussion: The partial differential equation in polar coordinates governing non-steady well flow in an incompressible confined aquifer of uniform thickness is

$$\frac{\partial^2 h}{\partial r^2} + \frac{1}{r} \frac{\partial h}{\partial r} = \frac{S}{T} \frac{\partial h}{\partial t} \qquad (5.27)$$

where S = storage coefficient [dimensionless], and t = time since the flow started [L]. The terms h and r have been defined in Section 5.6.2. Transmissivity, T, is defined as the product of the aquifer thickness and hydraulic conductivity, $T = kb$, while Storativity, S, is defined as the volume of water that an aquifer releases or takes into storage per unit surface area of the aquifer per unit change in the component of head normal to that surface. Theis (1935) obtained a solution for the Equation (5.28) based on the analogy between ground water flow and heat conduction. By assuming that the well is replaced by a mathematical sink of constant strength, that after pumping begins, h, approaches, H, and, r, approaches ∞, the solution is:

$$s = H - h = \frac{Q_w}{4\pi T} \int \frac{e^{-u}}{u} du \qquad (5.28)$$

where $u = r^2 \, S/4Tt$. Equation (5.28) is known as the nonequilibrium Theis equation. This equation permits determination of aquifer properties: storativity, S; transmissivity, T, and hydraulic conductivity, k. The exponential integral in Equation (5.28) has

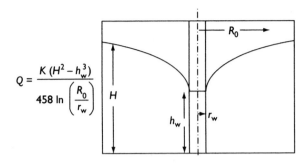

$$Q = \frac{K\,(H^2 - h_w^3)}{458 \ln\left(\dfrac{R_0}{r_w}\right)}$$

Figure 5.8 Steady flow to a well in unconfined aquifer.

been assigned the symbol, $W(u)$, which is called the "well function of u." Several investigators have developed approximate solutions to Equation 5.28, which make use of graphs and charts to solve the integral. Included among these are the Theis method of superposition, the Theis recovery method and Cooper and Jacob method (Cooper and Jacob, 1946). Further discussion of these equations may be found in groundwater (Freeze and Cherry, 1979) or hydrogeology (Fetter, 2001) texts.

5.6.4 Unconfined aquifer

An unconfined aquifer, also known as a water table aquifer, is one that may receive water directly from precipitation events. The aquifer thickness may change according to the rate of recharge or discharge. The analysis of aquifer properties proceeds in a fashion similar to that described for confined aquifers, although the behavior is actually more complicated. Figure 5.8 shows a picture of an unconfined aquifer.

5.7 Drainage and dewatering systems

5.7.1 Functions of drainage and dewatering systems

An excess amount of water trapped in the soil mass will reduce the effective stress as discussed in Section 5.5. This unwanted water must be removed by drainage and dewatering systems. In order to design an adequate drainage system the engineer must have all the information available on the amount of rainfall as discussed in Section 5.1 in the area to be drained. The three main aspects of rainfall for the designer to consider are intensity (the rate of fall), duration (the length of time of intensity), and frequency (the probable period of time for repetitions of combinations of intensity and duration). In other cases, the primary information required is the height of the groundwater level relative to a given construction activity, such as an excavation. Dewatering systems are considered as a temporary measure during the construction stage or for special reasons. Some of typical drainage and dewatering systems are (a) *surface drainage* including open ditch and vertical drain, (b) *subsurface drainage* including horizontal drain and vertical drain, and (c) *dewatering system* including pumping test, sand drain, and wellpoint. The applicability of various

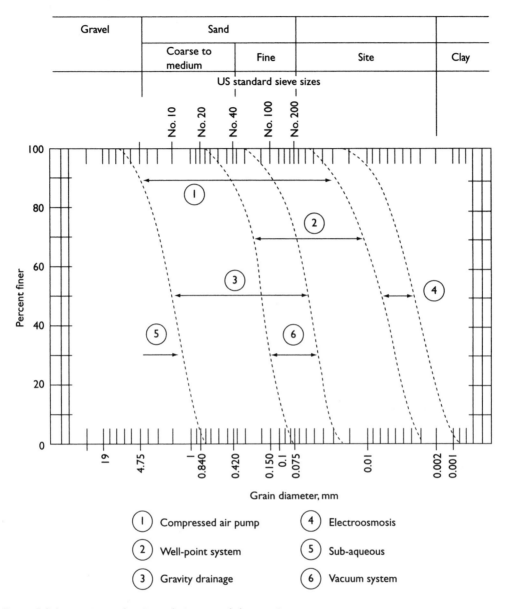

Figure 5.9 Limitations of various drainage and dewatering systems.

drainage and dewatering systems as a function of grain size are presented in Figure 5.9.

5.7.2 Surface drainage system

1 Methods for estimation of discharge: For the design of small surface drainage systems on farm lands or along highways, railroads, and airports, the peak discharges under consideration are mainly those from small rural watersheds. A small watershed is very sensitive both to high-intensity rainfalls of short duration and to land use. Many methods have been developed and used for the determination of design discharges of small watersheds. Some of the methods combine the hydrologic and hydraulic design into one step by which the waterway area of a open ditch or a culvert is estimated directly. At present, there are several methods for small watershed areas including (a) *Method of judgment*: The determination of design discharges or waterway areas is dependent on practical experience and individual judgment. Sometimes, an empirical rule of thumb is developed to replace judgment. (b) *Method of formulas*: Using existing available formulas, tables, and design curves from handbook or specifications. (c) *Method of direct observations*: involves conducting careful surveys of drainage area and stream characteristics and then making a precise hydrologic analysis and hydraulic study.

 2 Typical types of surface drainage systems: (a) Open ditches are commonly used for removing the surface water along highway pavement, shoulder and right-of-way. A drainage culvert is designed as a structure to convey water through or under a roadway, runway, taxi-way, or other obstruction. The choice of arch, box. circular, elliptical, or oval cross-section, and single or multiple installation will depend on capacity, headroom, and economy; and (b) Ponding is a temporary storage basin. If the rate of outflow from a drainage area is limited by the capacity of the drainage facility serving the area, usually a drainpipe is needed. Whenever the rate of runoff at a structure such as an inlet exceeds the drain capacity, temporary storage or ponding occurs. The rate of outflow from a ponding basin is affected by the elevation of the water at the drain inlet, and it will increase as the head on the inlet increases.

5.7.3 Subsurface drainage systems

Subsurface drainage, in general, consists of providing intercepting drains to divert subterranean flows, draining wet masses or areas and controlling moisture in the subsurface soil layers. Subdrains should be designed to function as subsurface drains only and should not operate to remove surface drainage.

 1 Vertical drainage system: The purpose of vertical drains is to remove surface and subsurface waters from highway and airfield pavement surface and pavement components such as base, and subbase layers. Also, it is useful for control of slope failure along the problematic soil and rock such as shale and limestone areas. For vertical drain installations, a drill equipped with a 9 in. (230 mm) hollow, continuous-flight auger is used to drill down through the fill section to the top of the underlying box. A 2.5 in. (64 mm) diameter diamond core bit is then lowered inside the hollow auger to drill through the top of the concrete box. After removing the diamond bit, the wall of the larger hole is thoroughly washed while slowly rotating and moving the auger up and down in the hole. A wire hail screen formed in the shape of a cone is driven upward into the hole in the concrete from inside the box to prevent the drain aggregate from dropping out. The larger hole is backfilled with granular underdrain aggregate from the top of the box up to the ground or pavement surface by carefully pouring the aggregate through the center opening in the hollow auger as it was slowly removed from the hole.

 2 Horizontal drainage system: A horizontal drain is a simple small diameter well that is drilled nearly horizontally into a hillside or fills to remove internal moisture or seepage water. The California Division of Highways was the first to introduce this type of drainage systems about 1939 for control of slope failures along the highway. Many of early installations were made with water driven drills known as *hydraugers*.

Horizontal drains are usually installed as part of the soil stabilization or ground improvement of cut slopes in wet, unstable ground, or old existing highway or railway slopes. They frequently are installed at 2 or 3 levels on benches in cut slopes while excavations are being deepened. For shallow cuts, horizontal drains are usually installed near the ground surface. Collector pipe manifolds or paved ditches are generally provided to take the water away to locations where it can be discharged without the danger of re-entering the slope where it can cause further instability of cuts.

3 Two-layer subsurface drains: Two-layer subsurface drains methods were developed by the California Highway Department in early 1960 because of the inconsistent performance results of single-layer subsurface drainage systems.

5.7.4 Dewatering systems

In general, a dewatering system is a temporary system during the construction stage. There are various types, including wellpoint, sand drain, and pumping tests.

1 Wellpoint dewatering systems: The use of wellpoint systems in construction works started about around 1900. One of the earliest wellpoint systems was installed by a contractor in the Chicago area for a foundation excavation. The concept of a wellpoint dewatering system is similar to the field pumping test as discussed in Section 5.6. A wellpoint is a small diameter tube or pipe fitted with plastic or metal screens which permit water to enter without the loss of adjacent or surrounding soil. They are often equipped with metal points that permit them to be driven or jetted into soil layers. Wellpoints have been most successfully used in silty/sandy soil. If the material is very dense, wellpoints must usually be installed in drilled holes, which greatly increases the cost of an installation. Wellpoints are typically installed in a line or ring surrounding an excavation and are connected through a manifold to a pump that extracts seepage to lower the water table in the area to be excavated. The required spacing, usually between 3 and 12 ft, (1–4 m) depends on the type of soil and the desired amount of groundwater lowering.

2 Sand drains: Sand drains are a small diameter vertical hole in the ground filled with clean filter sand or pea gravel developed by D. E. Moran in 1926. It also is called a *drain well*, sand pile, and sand wick. Holes of 12 in. (30.48 cm) or greater diameter are bored and are filled with clean uniform sand. The top of these drains are interconnected by sand trenches or blanket. This configuration is also used to accelerate the consolidation of a fine-grained deposits by reducing the drainage length and releasing excess pore pressure.

3 Drainage pipe: Drainage pipe is a major part of drainage and dewatering systems. For a particular project use of proper type of pipe is essential. Stevenson (1978) made a survey based on the contractor's opinions about builders of water and sewage treatment plants. For corrosion protection, 41% of contractors prefer cement linings while 33% prefer fusion-type epoxy linings, although these preferences vary with region and of course time as new lining technology is developed.

4 *Root Control:* Root control is also a major part of drainage system mainte-nance, especially for the subsurface. The purpose of root control or removal is (a) prevent root blockages, (b) restore full capacities and self-scouring velocities of water, (c) prevent destruction of pipe, and (d) reduction of septic conditions and hydrogen sulfide generation to increase pipeline life expectancy.

5.8 Seepage flow, flow net, and free water surface

5.8.1 General discussion

Seepage is the movement of gravitational water through the soil. *Seepage force* is the force transmitted to the soil grains by seepage. The *seepage velocity* is the rate of discharge of seepage water through a porous medium per unit area of void space perpendicular to the direction of flow. *Seepage flow* is also called *viscous flow*.

5.8.2 Governing differential equation

Consider the steady state flow of water in and out of an element of saturated soil of volume dx, dy, and dz in Figure 5.10.

For continuity the flow into the element must equal the flow out of the element, therefore

$$dq_x + dq_y + dq_z = 0 \qquad (5.29)$$

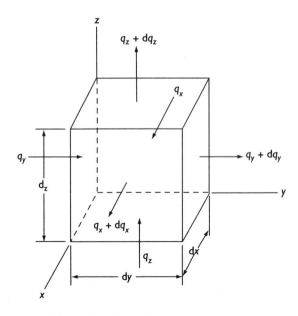

Figure 5.10 Three-dimensional flow through an element.

Utilizing Darcy's Law (Eq. (5.7)) for a differential total flow rate,

$$dq_x = k_x \, di_x \, A = k_x \, di_x \, dydz$$
$$dq_y = k_y \, di_y A = k_y \, di_y \, dzdx \qquad\qquad (5.30)$$
$$dq_z = k_z \, di_z A = k_z \, di_z \, dxdy$$

but

$$i_x = dh/dx; \quad di_x = \partial^2 h/\partial x^2 dx$$
$$i_y = dh/dy; \quad di_y = \partial^2 h/\partial y^2 dy \qquad\qquad (5.31)$$
$$i_z = h/dz; \quad di_x = \partial^2 h/\partial z^2 dz$$

By substituting Equations (5.31) into (5.30) and then the results into Equation (5.29) we have the following:

$$k_x \, \partial^2 h/\partial x^2 + k_y \, \partial^2 h/\partial y^2 + k_z \partial^2 h/\partial z^2 = 0 \qquad\qquad (5.32)$$

If the soil is isotropic then $k_x = k_y = k_z = k$, and Equation (5.32) reduces to

$$\frac{\partial^2 h}{\partial x^2} + \frac{\partial^2 h}{\partial y^2} + \frac{\partial^2 h}{\partial z^2} = 0 \qquad\qquad (5.33)$$

Equation (5.33) is Laplace's equation. The solution to Equation (5.33) for a specific problem can be obtained in a number of ways. Such as (a) close form solution: a rigid analytic solution for Laplace's equation can be obtained for some seepage problems, (b) open form solution: by approximating the partial derivatives of Laplace's equation by their finite difference equivalent a numeric solution can be obtained using the computer, and (c) graphical solution: an approximate solution which is quite convenient and relatively simple to use, This type of solution is commonly used in geotechnical engineering.

5.8.3 Flow through idealized sections

Consider the steady state seepage of porewater through a homogeneous, isotropic soil mass as shown in the Figure 5.11.

Water flow will obey Laplace's equation, which is a solution of two families of intersecting (conjugate) curves. The family of flow lines represents trajectories of water particle movement. The existence of these flow lines can be clearly demonstrated with laboratory models. The intersecting family of lines defines the equipotential lines along which the total head is constant. These curves can also be shown using an electrical analogy model whereby conducting and resisting elements are used to simulate flow and head loss. In order to physically sketch a flow net, it is necessary to keep three facts in mind:

1 Observe all boundary conditions;
2 Have flow line and equipotential lines intersect at right angles.
3 Have the resulting areas created by adjacent flow lines and equipotential lines form curvilinear "squares." A good check is to see if a circle can be inserted that

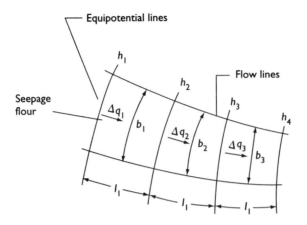

Figure 5.11 General seepage flow net.

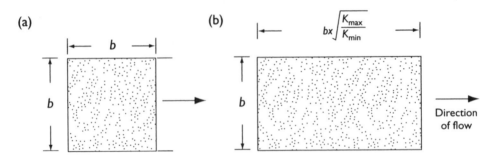

Figure 5.12 Characteristics of flow nets: (a) Natural scale of flow net (homogeneous) and (b) Transformed scale of flow net (heterogeneous).

is tangent to all sides of the "square." The art of sketching flow nets is initially quite difficult, but a reasonable amount of proficiency is quickly developed. It is best to build flow and equipotentials together from one side (preferably upstream) toward the other. If the end interval does not create a "square," a fractional portion is allowed. Seepage software exists to simplify this process.

5.8.4   Characteristics of seepage flow in soil layer

1   *Flow line and equipotential line*: As indicated in Figure 5.11, the flow line is the path that a particle of water follows in its course of seepage under laminar flow conditions. Equipotential lines are perpendicular to flow lines and indicate the extent to which water will rise in piezometer tubes (Sec. 5.5.4).

2   *Flow net for homogeneous soil layer*: Graphical techniques for developing a flow net are useful to estimate the quantities of flow through natural soil or earth structures. A flow net consists of two groups or families of curves which bear a fixed relationship to each other. The flow channel is the portion of a flow net bounded by two adjacent flow lines as shown in Figure 5.12(a).

3   *Flow net for heterogeneous soil layer*: Figure 5.12(b) shows how the scale of an idealized flow cell is transformed to reflect variable hydraulic conductivity. An effective value of permeability, designated by k', is determined for the modified section.

Figure 5.13 presents various types of flow nets currently used in geotechnical engineering, including retaining walls, earth dams, wells, and excavations. Cedergren (1974) provides more details on the construction procedures for various types of flow nets.

5.9   Protective filters

5.9.1   Characteristics of filter

A *protective filter* is a layer or combination of layers of pervious materials designed and installed in such a manner as to provide drainage, yet prevent the movement of soil particles mobilized by flowing water. The main objectives are to relieve excessive seepage pressures and prevent the subsurface erosion of fine-grained soils (Sec. 16.5). Indications are that their first use may have been in connection with the dissipation of uplift pressure that had caused damage to the concrete aprons of some overflow dams. A filter material, regardless of how it is used, must satisfy two requirements: (a) it must be stable within itself, that is, its fine particles must not be susceptible to disturbance by seepage force; and (b) it must prevent intrusion of the adjacent soil into its pores to prevent clogging.

The first requirement is met by having the filter considerably more permeable than the adjacent base material. An increase in permeability will cause a corresponding decrease in the hydraulic gradient, which will result in reduced seepage force. In this way, a protective filter is able to prevent the occurrence of problems that might have arisen in connection with uplift pressure, boiling or quick condition, and erosion. The second requirement is met by controlling the pore size if the filter material adjacent to the soil being protected. If the movement of base material into the pores of the filter cannot be prevented, the filter will quickly become as poor as the soil it has replaced.

5.9.2   Function of protective filter

Bertram (1940) reported the results of an experimental investigation of the filter characteristics of two selected uniformly graded sands by subjecting them, in various filter-base combinations, to permeability tests under hydraulic gradients. The materials used included a uniformly graded round grained quartz sand (Ottawa sand) and a uniformly graded angular crushed quartz. The possibility of two types of failure was recognized, clogging of the pores of the filter with base material and complete passage of the base through the filter. Stability of the system (i.e. prevention of base movement into the filter) was indicated by the attainment of a constant permeability and conclusions were drawn from comparison of weights and grain size distributions of each type (filter and base) as placed and after a state of equilibrium had been reached. In order to fulfill the objectives of a filter material, an

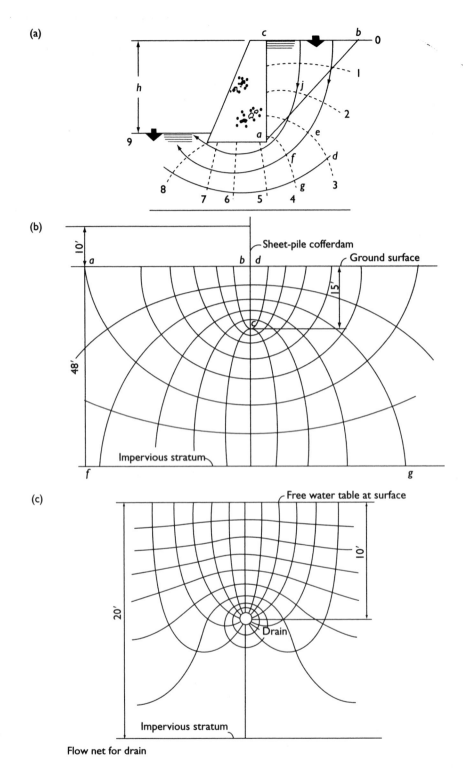

(a)

(b)

Sheet-pile cofferdam

Ground surface

Impervious stratum

(c)

Free water table at surface

Drain

Impervious stratum

Flow net for drain

Figure 5.13 Typical flow net examples. (a) Flow net for retaining wall; (b) flow net for sheet-pile cofferdam; (c) flow net for drain; (d) flow net for masonry dam; (e) flow net for earth dam with rock-fill toe; (f) flow net for earth dam with down stream drain and (g) flow net for earth dam clay core wall.

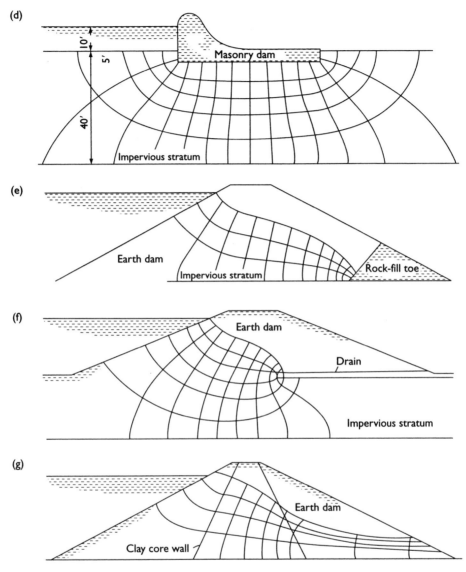

Figure 5.13 Continued.

empirical design based upon a modification of the original method assessing the grain size distribution developed by Terzaghi (1942) had been established. The approach was later substantiated by tests conducted on protective filters used for the protection of soils in the construction of earth dams and other related projects as follows:

1   *Clogging*: To prevent clogging the pipe with small particles infiltrating through the openings, the following requirements must be satisfied:

$$\frac{d_{85} \text{ of filter material}}{\text{Size of opening in pipe}} = 2 \tag{5.34}$$

where d_{85} = the diameter at which 85% of the material passes.

2   *Movement of particles*: To prevent the movement of particles in the protected soil, the following conditions must be satisfied:

$$\frac{d_{15} \text{ of filter material}}{d_{85} \text{ of protected soil}} = 5 \tag{5.35}$$

When the protected soil is plastic and without sand or silt particles, the d_{15} (diameter at which 15% passes) size of the filter material need not be less than 1 mm (0.0394 in.).

3   *Free water movement*: To permit free water to reach the pipe, the filter material must be many times more pervious than the protected soil. It has been found that this condition is fulfilled when the following requirements are satisfied:

$$\frac{d_{15} \text{ of filter material}}{d_{85} \text{ of protected soil}} = 5 \tag{5.36}$$

Thus, a rigid (concrete) pavement base course or subbase might be designed as a filter to achieve any one or all of three purposes: (a) prevention of intrusion of subgrade material into the base course (as a result of pumping action from dynamic loading by passing vehicles); (b) prevention of the loss or redistribution of the fines on the base course or subbase as the result of a pumping action; and (c) prevention of intermixing of subbase and base course at their surfaces.

5.9.3   *Selection of filter materials*

Example aspects of filter design criteria for rock slope protection and highway roadbed construction are presented in Figure 5.14. Geotextitles are commonly used as a protective filter. Their advantages and limitations will be discussed in Chapter 15. Industrial by-products such as steel slag or waste rubber tire may also be used.

5.10   Creeping flow and mass transport phenomena

As discussed in Sections 3.6 and 6.8, soil is composed of electrically negative mineral surfaces while water is composed of electric water dipoles and predominantly positively charged ions. It then follows that the soil–water system possess a highly electric character and hence, will respond to the application of an electric potential. The interaction of these electrically charged components of the system is a function of temperature. Therefore, the thermal-electric properties are closely related and exist in the soil–water system in the natural environment.

5.10.1   *Creeping flow*

The nature of soil–water interaction in the environment is a function of both the liquid and solid phases. To describe a flow that has a low velocity, is unsteady,

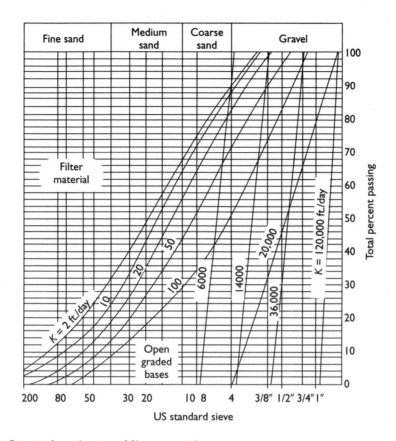

Figure 5.14 Criteria for selection of filter material.

Source: Cedergren, H.R. *Seepage, Drainage, and Flow Nets*, 3rd Edition © 1974 Wiley. This material is used by permission of John Wiley & Sons, Inc.

nonuniform, or sometimes discontinuous, and may contain solid or gaseous forms of toxic or hazardous substances, the term *creeping flow* has been introduced (Fang, 1986). Soil is composed of electrically negative mineral surfaces while water is composed of electric water dipoles and predominantly positively charged ions. Also, due to the complex nature of soil pollution the term "contamination" is used loosely in the text. These terms will be reflected in the context of the temperature or ion characteristics of the solution.

5.10.2 Mass transport phenomena

Water always contains some gases because natural water contains variable amounts of dissolved oxygen (O), nitrogen (N), and carbon dioxide (CO_2). Water in a soil system may simultaneously be present as vapor, as a liquid of varying viscosity, and as a solid of varying plasticity. Except in the vapor phase, water possesses a well-developed structure of highly electric character due to its own polarized molecular nature, as well as being under the influence of the electrically charged surfaces and of ions in solution. Since the thermal energy of the molecules and

Table 5.2 Comparison of general characteristics between hydrualic conductivity and mass transport phenomenon

Mass transport phenomenon	
Hydraulic conductivity Due to mechanical (hydrostatic) potential	Mechanical energy field

Energy conductivity (environmental)
Hydration energy
    Due to the hydration energy of ions,
        related to the heat of wetting
Osmotic energy
    Due to the osmotic energy of ions
        either held in a kind of Donnan
        equilibrium on the solid particle
        surfaces or free in the aqueous
        solution
Capillary potential
    Due to the surface tension of water
        and the size and geometry of the
        soilpores
Electric potential
    Electro-osmosis: eletrokinetic
        phenomena
Thermal potential
    Thermo-osmosis: thermal electric effect
Magnetic potential
    Electromagnetic force
Vapor pressure potential
	↑
	
	Multimedia energy fields
	
	↓

Source: Fang (1997) with modifications.

ions of the aqueous solutions act counter to the structure-forming factors, water structures and their physical and physicochemical consequences are thermal-sensitive which leads to a tremendous complexity and to interrelationships between all physical and physicochemical factors in soil–water systems. In view of the manifold properties of the water substance and modifications in the soil environment, water in soil responds to the imposition of any energy gradient be it mechanical, thermal, electric, magnetic as discussed in Section 1.3 or other and responds to or exhibits coupling effects indicative of the disturbance of the other energy fields. Such response usually results in *mass transport phenomena;* that is, flow through a porous medium that may not be described simply by hydraulic conductivity and Darcy's Law, as is commonly used in geotechnical engineering (Tables 5.2 and 5.3).

5.11   Soil–water suction and diffusivity

The term known as soil–water suction or tension was started by Buckingham in 1907 based on the work by Briggs in 1897, and introduced into geotechnical engineering

Table 5.3 Porewater produced during mass transport phenomena

Mechanical (hydrostatic) pressure[a]	Mechanical energy field
Environmental pressure	
Pressure produced from heat of wetting	
Pressure produced from the kinetic dispertive force	Multimedia energy field + Mechanical energy field
Electric-motive force (emf)	
emf produced from thermal–electric effect	
emf produced from electro–electric force	
Water–vapor force	

Source: Fang (1997) with modifications.

Notes

a Pore pressure (u) commonly used in the geotechnical engineering for computing the effective stress (σ). The pressure is due to hydrostatic head only is given by $u = h\gamma_w$, where u = pore pressure, h = hydrostatic head, and γ_w = unit weight of water.

by Schofield in 1935 and Schofield and Da Costa in 1938. The significance of suction and the measurement techniques were established by Coleman in 1949. After an International Conference on Pore Pressure and Suction in Soil in 1961, this concept has been well accepted by geotechnical and highway engineers.

5.11.1 Suction and negative porewater pressure

The porewater in natural and compacted soils is often in a state of tension; that is, its pressure is negative relative to atmospheric conditions. This negative pressure is called *suction*. Numerous devices capable of measuring the suction include ceramic resistance blocks, tensiometers, pressure plates, osmotic tensiometer, and thermocouple psychrometers. These measuring devices have been examined and reviewed by Croney and Coleman (1961). The unit for soil–water suction is called the *pF* unit. On the pF scale of measuring suction, the logarithm to base 10 of the suction expressed in centimeters of water is equivalent to the pF value. Thus, 10 cm of water = 1 pF, and 1000 cm of water = 3 pF (approximately one atmosphere). Soil in equilibrium with free water has a pF of zero, and oven dried, almost pF = 7. Distinguishing between soil suction and negative porewater pressure can be explained by the following equation:

$$u = s + \alpha\,p \tag{5.37}$$

where u = negative porewater pressure, s = soil–water suction, α = change of negative porewater pressure with applied pressure at constant moisture content, and p = applied pressure. The value of (α) can be measured directly by a loading test on a sample of known suction. Equation (5.37) can be used to estimate the ultimate moisture distribution beneath pavements and structures on saturated or unsaturated ground soils.

5.11.2 Soil–water diffusivity

The soil–water diffusivity concept was introduced into geotechnical engineering by Childs and Collis-George in 1950. Their theory for the flow of water in unsaturated soil assumes that Darcy's law can be written as a diffusion-type water-flow equation in homogeneous soils where gradients of water content rather than gradients of total potential are expressed as

$$q = D\, i_{\omega} - k \tag{5.38}$$

where q = water flux, ω = soil–water content on a volume basis, k = hydraulic conductivity of soil, D = soil–water diffusivity, and i_{ω} = water content gradient. In Equation (5.38) both k and D are the function of the soil–water content. The diffusivity, D, is also equal to

$$D = k/C \tag{5.39}$$

where C = the specific soil–water storage capacity = $\partial\omega / \partial h$, where, h = soil–water pressure head (suction). The parameters of k, ω, and h can be determined experimentally as described by Elzeftway and Dempsey (1976). In the field condition, the water table is established at the bottom of a compacted subgrade soil column, and the movement of water through the soil can be measured under isothermal conditions. A gamma-ray method (ASTM C1402) can be used for nondestructive measurement of the water content and a tensiometer-pressure transducer arrangement to measure the soil–water pressure (suction). The unsaturated hydraulic conductivity, k, diffusivity, D, and soil–water characteristics parameters can be evaluated from these data to predict the movement of moisture through unsaturated ground soils. A typical relationship of soil–water content to soil–water suction and soil–water diffusivity for a silty clay soil is presented in Figure 5.15. A thorough overview of unsaturated soil mechanics is given by Lu and Likos (2004).

5.12 Diffusion and migration

5.12.1 Diffusion

Diffusion is the process whereby ionic or molecular particles move in the direction of their concentration gradient under the influence of their kinetic energy. The process of diffusion is often referred to as *self-diffusion, molecular diffusion,* or *ionic diffusion*. It is a type of chemical transport phenomena. According to Fick's First Law, the mass of diffusing substance passing through a given cross-section per unit time is proportional to the concentration gradient as

$$F = -D\, dC/dx \tag{5.40}$$

where F = mass flux, D = diffusion coefficient, C = solute concentration, and dC/dx = concentration gradient, which is a negative quantity in the direction of diffusion. The diffusion coefficients are temperature dependent. In soil systems, the range of diffusion coefficients is typically on the order of 10^{-9} to 10^{-10} m²/s at 25°C.

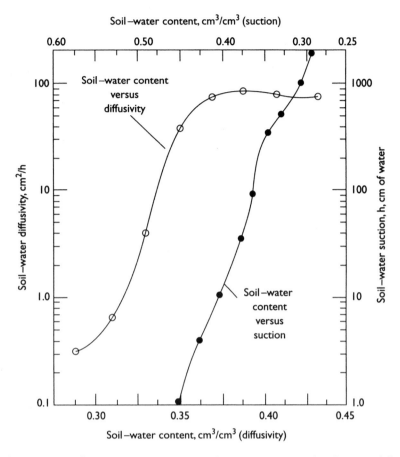

Figure 5.15 Relationship of soil–water content to soil–water suction and soil–water diffusivity for a silty clay soil.

Source: Elzeftawy, A. and Dempsey, B. J., "Unsaturated transient and steady state flow of moisture in subgrade soil." In *Transportation Research Record* No. 612, Transportation Research Board. National Research Council, Washington DC, 1976, pp. 56–61. Reproduced with permission of the Transportation Research Board.

If the soil–water system is polluted, the ranges of coefficient change dramatically. An important consideration in landfill design is the fact that chemicals may migrate under diffusion, advection (bulk fluid movement), or more commonly through combinations thereof. It is even possible for a chemical to move in the opposite direction of fluid flow.

5.12.2 Migration

Migration phenomena of soil results from the dynamic behavior of soil moisture movement. Dynamic equilibrium between water in the various states, within individual water molecules will pass back and forth at stabilized rates, between the vapor and the liquid states, between capillary and free water. Furthermore, within a volume of soil at uniform temperature, water molecules in the various states will diffuse back and forth from one place to another. If there is a temperature difference in two locations, water will move, but then, it will be called *moisture migration* under thermal gradients. In certain cases, it is called *thermoosmosis* (Sec. 6.3.3). Migration is a transient redistribution of water in the soil–water system rather than a continuing flow, and water exists in the soil and may migrate to some place else, if the local environmental conditions change. There are many possible factors such as capillarity, vapor diffusion, moisture film transfer, electromotive force (Sec. 6.11), and electrokinetic force (Sec. 6.11).

5.13 Summary

Water has a pronounced influence on the engineering behavior of soil, as revealed by the concept of effective stress. The effective stress refers to the grain to grain contact force per unit area, which may change depending on the extent to which water is present and at positive or negative pressure. Water below the groundwater table is under positive pressure, which serves to reduce the effective stress while above the water table capillary action retains water below atmospheric (negative) pressure. Negative porewater pressure increases the effective stress. The degree of capillary rise may be computed through various equations which generally correlate with the size of particles or voids.

The rate of fluid movement in soils is described by Darcy's law, which states that the velocity of flow is related to the hydraulic gradient through a constant of proportionality, given as the coefficient of permeability or hydraulic conductivity. The hydraulic conductivity may be determined theoretically or experimentally, although preference is given to experimental measurements.

The value of hydraulic conductivity is important in many areas, including the design of dams, dewatering systems, water supply wells, landfill liners, and drainage systems. Filters are often used to ensure that water may flow through soils without mobilizing the individual particles downstream. Design criteria involve an assessment of the grain size distribution of both the filter and the material to be protected.

Creeping flow and mass transport phenomena were introduced in this chapter. Creeping flow is a flow which has low velocity, is unsteady, non-uniform, or sometimes discontinuous and may contain solid or gaseous forms of toxic or hazardous substances. Mass transport by diffusion was described as given by Fick's law. It is important to recognize that mass (e.g. chemicals, pollutants, etc.) may be transported by flowing water (bulk fluid movement, advection) or by virtue of the concentration gradient alone. Indeed, pollutants may travel against the prevailing hydraulic gradient if the concentration gradient is in the opposing direction.

PROBLEMS

5.1 Discuss hydraulic conductivity phenomena near the ground surface. Why are weather conditions important?

5.2 Derive the capillary rise, h_c, of Equation (5.2).

5.3 What is the maximum height of capillary rise of water in a material having a uniform pore diameter of 0.0001 cm ($3.9 - 10^{-5}$ in.)? Why is the rate of capillary action more important than just the height of capillary rise? Explain.

5.4   Distinguish between positive and negative porewater pressures in the soil–water system. What is the unit for negative porewater pressure?

5.5   At a certain location, the ground surface is level, and the first 3.4 m (11.2 ft) of soil is a saturated silty clay formation. Beneath the clay is a layer of clean sand containing water under artesian pressure. The pressure in the water just under the clay layer is 0.2 kg/cm² (2.8 psi). The unit weight of the clay is 2.10 g/cm³ (131 pcf). The submerged unit weight of the sand is 0.95 gm/cm³ (59.3 pcf). Find the effective vertical pressure in kg/cm² at a point in the sand layer 5 m (16.4 ft) below the ground surface.

5.6   At a certain location in a large lake, the depth of water is 20 m (65.6 ft). The lake bottom consists of loosely deposited sediments with a thickness of 50 m (164.0 ft). The void ratio of the bottom material is 2.0, the specific gravity, G_s = 2.72. Calculate the vertical component of the effective stress at a depth of 5 m (16.4 ft) in the sediment (25 m below the surface of lake).

5.7   A pumping test was made in a pervious strata extending to a depth of 40 ft (12.2 m), where an impervious stratum was encountered. The original water table was at the ground surface. The test well has a diameter of 2 ft (0.61 m). A yield of 120 gal/min (454 l/min) was established by a steady pumping that produced a maximum drawdown of 25 ft (7.62 m) in the test well. Assuming a radial distance of 500 ft (152.5 m) to where the drawdown is zero, calculate the coefficient of permeability, k, in both units.

5.8   A large, open excavation was made in a layer of clay with a saturated unit weight of 112 pcf (1.80 g/cm³). Exploration of the site before excavation showed a stratum of dense sand at a depth of 40 ft (2.2 m) below the ground surface. It was observed that the water had risen to an elevation of 15 ft (4.6 m) below ground surface in one of the exploration drill holes. Calculate the critical depth of the excavation after which the bottom would be cracked and a boiling condition would exist.

5.9   A soil has a unit weight of 17.1 kN/m³ (108.9 pcf) and a water content of 7%. How much water, in gallons, should be added to each cubic yard of soil to raise the water content to 12% (assume that the void ratio remains constant).

5.10 Discuss how and why contaminated pore fluid will affect the permeability of clay-like soils. Also, explain why contamination of a fine-grained soil is more problematic than a coarse-grained soil.

Chapter 6

Thermal and electrical properties of soils

6.1 Introduction

As discussed in Chapters 4 and 5, thermal–electrical and magnetic properties play an important role in soil behavior, because soil itself is a thermal-sensitive, electrical-magnetic material. Moreover, soil is very sensitive to local environmental factors, especially climatic variables such as: precipitation, wind, evaporation, groundwater table fluctuation, and frost penetration. These factors directly or indirectly relate to a thermal, electric, and magnetic influence on the engineering behavior of soil.

Heat is a form of energy and has been conceived as a form of motion within the particulate components of matter. Rumford and Davy furnished scientific proof of this concept in 1798–1799. Types of heat sources penetrating into the ground soil layer can be grouped into two major categories: natural and man-made sources. Natural heat sources include solar energy, geothermal energy, natural decay of radioactive elements, and wild fire caused by drought or thunderstorms. Anthropogenic sources cover the heat released from steam pipes, electrical cable lines, heat pump systems, landfill wastes, nuclear wastes, and underground mine fires.

6.1.1 Thermal regime and climatic behavior near the ground surface

Climatic conditions such as hot-cold, wet-dry and freeze–thaw cycles near the ground surface will directly and/or indirectly affect soil behavior. Therefore it will affect the stability of various geotechnical structures. As indicated in Figure 1.4, climatic conditions will change the state of matter under freeze-thaw and wet-dry conditions. Among these conditions, precipitation (Sec. 5.1) is particularly critical. Precipitation is any form of moisture that falls to the earth from a cloud, or that condenses on the Earth's surface. Precipitation can be liquid or solid and includes drizzle, dew, rain, snow, sleet, hail and frost. These elements are directly related to the in situ moisture content of ground soil, characteristics of groundwater table, and depth of frost penetration. Consequently, it will affect the bearing capacity, earth pressure, settlement as well as stability of earth slopes.

6.1.2 Groundwater level

Groundwater level fluctuation is also a major phenomenon in the thermal energy field. Observations in both laboratory and field investigations suggest that groundwater levels in temperate climates fall during the winter as the air temperature reduces and rise approximately the same amount in the spring as the air temperature increases, when corrected for recharge. Some relatively important observations of groundwater level are reported as follows: (a) the winter water table declines as there is upward movement of capillary moisture towards frost layers near the surface; (b) the water table lowering is associated with depth of frost and accompanied by increased soil moisture in the frost zone above the water table; and (c) the change in soil moisture content and water table elevation is associated with soil temperature. As the soil temperature decreases during the winter, the water table falls and the soil moisture content increases in the surface horizons by migration of water from the subsurface soil. Recall that the groundwater table is the point at which water is present at atmospheric conditions, and water above this point are held at less than atmospheric, or negative gauge pressure. The significance of groundwater table fluctuation has been studied at the AASHO Road Test. Three different cover conditions were considered: Natural ground surface under vegetation cover; under asphalt pavement, and under concrete pavement. In addition, a comparison between non-traffic and traffic areas was undertaken. A total 26 groundwater table measurement installations were made on the centerline of pavements for all the test sites. Typical test results and discussion are reported elsewhere (HRB, 1962).

6.2 Measurable parameters of heat

6.2.1 Measurable thermal parameters

Once heat is transmitted into the ground, the ability of the soil to retain or dissipate heat is dependent on its heat capacity and thermal conductivity. There are five basic parameters to measure the characteristics of heat: (a) *mass heat capacity*: the mass heat capacity is defined as the quantity of heat, Q, required to raise a unit weight of material one degree; (b) *volumetric heat capacity*: the volumetric heat capacity is defined as the quantity of heat required to raise a unit volume of material one degree; (c) *specific heat*: specific heat is the ratio of mass heat capacity of the material, divided by the mass heat capacity of water. It is a dimensionless quantity, the typical values of specific heat for air = 0.237, ice = 0.463, water = 1.000, wood = 0.327, limestone = 0.216, quartz = 0.188; (d) *thermal conductivity* or *thermal resistivity*: for practical application, the values for thermal conductivity and resistivity are important. Detailed discussion of these two parameters will be presented in Section 6.4; and (e) *thermal diffusivity*: the thermal diffusivity is the quotient of the thermal conductivity and the heat capacity per unit volume.

6.2.2 Temperature and heat

1 *Temperature*: Temperature is a measure of the internal motion of an object's constituent molecules. The greater the motion, the greater the internal energy and the higher the temperature. There are three commonly used temperature scales, namely Fahrenheit (°F), Centigrade (°C), and Absolute or Kelvin (°K). On the Fahrenheit scale, the freezing point of pure water is 32°F, and boiling point is 212°F. The Centigrade scale is fixed at 0°C for the freezing point of pure water, and 100°C for

the boiling point. The Kelvin or Absolute scale is similar to the Centigrade scale in that the divisions on the scale are the same size, but the zero on the Kelvin or Absolute scale is $-273°C$, and the boiling point of water is $373°K$.

2 *Heat*: Heat is thermal energy in transit. If a hot object is placed in contact with a cold object, heat will flow from the hot to the cold object; some of the molecular motion of the hot object will be transferred to the cold object. The unit of heat is the calorie (cal). One calorie is defined to be the amount of heat required to raise the temperature of one gram of water by $1°C$.

6.3 Heat transfer process and soil–heat interaction

6.3.1 Heat transfer process in ground soil

Heat transfer in soil occurs through three basic processes: conduction, convection, and radiation. (a) *conduction*: conduction is a process in which heat energy is transferred among molecules within a substance, or between two substances in physical contact, or by direct molecular interaction. The rate of heat flow or thermal conductivity of a substance is dependent on the capacity of its molecules to give or absorb heat; (b) *convection*: convection is defined as the heat transfer between a surface and a moving liquid or gas or the transfer of heat by the movement of the molecules from one point to another. In convection processes, heat always moves from warm to cool. The larger the temperature difference between two substances, the more heat will transfer; and (c) *radiation*: radiation is characteristic of a material. All materials are constantly radiating thermal energy in all directions because of the continual vibrational movement of molecules as measured by temperature at their surface.

In most soil, the heat transfer process within the soil mass is conduction or convection or both as illustrated in Figures 6.1(a) and(b). In examining Figure 6.1(a), the thermal conductivity for unsaturated soil with air is smaller than fully saturated soil (no air) as shown in Figure 6.1(b), because the air is a poor heat conductor.

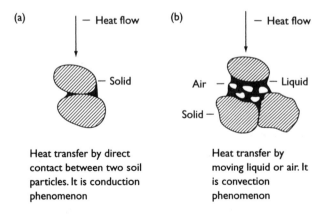

(a) — Heat flow — Solid

Heat transfer by direct contact between two soil particles. It is conduction phenomenon

(b) — Heat flow Air — — Liquid Solid —

Heat transfer by moving liquid or air. It is convection phenomenon

Figure 6.1 Comparison of characteristics of heat flow in soil. (a) Saturated soil; and (b) Unsaturated soil.
Source: Fang (1997).

6.3.2   Phenomena and reaction during soil–heat interaction

Because of water movement in the vapor phase and the lightness of the water molecule, the thermal conductivity in the vertical direction of a moist soil system may be considerably larger than the vertically downward component. Since heat and moisture move together from higher to lower temperatures and the movement and concentration of either or both may greatly affect the service quality and durability of a road, or a foundation structure. As such, these factors should be considered at the design stage. It is particularly important to avoid great differences in thermal and moisture transmission properties of the different constituent layers of a soil-structure system down to the depth of influence. This depth of influence has two components, (a) the depth to which the loading condition influences the surrounding soil, for example, Boussinesq equations (Ch. 9) and (b) the extent to which there is still a marked amplitude of the daily and seasonal temperature waves. Resulting phenomena from the above soil–heat interactions include: thermoosmosis, thermomigration and thermal storage capacity.

6.3.3   Thermoosmosis

If one applies a hot plate at one end of an otherwise well insulated cylindrical or prismatic specimen of a moist or water saturated soil and a cold plate at the other end, a number of thermal, electrical and mechanical phenomena can be observed. Among these effects is the movement or shift of water within the specimen from the hot-side to the cold. This phenomenon can be called *thermoosmosis* if we take the term osmosis in its original Greek meaning of "pushing," without implying a specific theoretical picture for the mechanism(s) involved in this pushing. In other words, thermoosmosis is moisture migration (Sec. 5.12) in response to a thermal gradient. It is also called *thermomigration*. Several points may be made relative to thermoosmosis:

1   For the condition in which a cohesive material yields a maximum amount of water under a thermal gradient, it contains significant amounts of entrapped air that expands on the warm-side and contracts on the cool-side and thus produces a hydrostatic pressure gradient.
2   Micro-convection phenomena may occur in the entrapped and continuous air phases within the system and add their share to the total water transmission.
3   Application of the temperature gradient causes gradients in the surface tension of the water films, in the thickness of the exchangeable double layer near the particle surface, in the hydration of the exchangeable ions, in the solubility of water in the solid surface and in the geometrical structure as well as in the dissociation and association constants of the water substance itself.

6.3.4   Thermal storage capacity in soils

The heat balance of soil includes heat exchange, phase changes, heat flow, and radiation. Once heat is transmitted into the ground, the ability of the soil to retain or dissipate heat is dependent on its heat capacity and thermal conductivity. In order for us to visualize the thermal storage capacity and other related parameters, an idealized

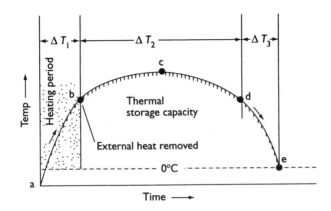

Figure 6.2 Temperature–time relationship and thermal storage capacity.

Source: Chaney *et al.* (1983). Copyright ASTM INTERNATIONAL. Reprinted with permission.

schematic diagram illustrating temperature–time relationship of ground soil is shown in Figure 6.2. In examining Figure 6.2, point **a** to point **b** represents the heating period for which heat energy is absorbed in the soil at the depth being monitored. The time for this heating period is denoted ΔT_1. The period of time ΔT_2 from point **b** to point **d** corresponds to the period where external heat is no longer applied to the ground surface. Although the external source of heat has stopped, energy is still being received at depth from point **d** to point **e** and denoted by ΔT_3.

From Figure 6.2, the thermal storage capacity of the soil is defined as the area under the curve described by points **b**, **c**, **d**, and **e**. In order to have a higher thermal storage capacity, we must have larger values for ΔT_2 and ΔT_3.

6.4 Thermal conductivity and resistivity

6.4.1 Characteristics of thermal conductivity

1 *Characteristics of thermal conductivity*: Thermal conductivity is a measure of a material's ability to transmit heat and is defined as the quantity of heat which flows normally across a surface of unit area of the material per unit time per unit temperature gradient normal to the material's surface. It is the proportionality factor, K, in Equation (6.1):

$$\frac{dQ}{dt} = KA\frac{dT}{dx} \qquad (6.1)$$

where dQ/dt = rate of heat flow, K = constant, A = cross-sectional area normal to the direction heat flow, and dT/dx = temperature gradient normal to the surface. The unit of thermal conductivity is most commonly expressed in the units cal/cm^2-s-°C/cm (cal/cm-s-°C) or Btu/ft-h-°F.

EXAMPLE 6.1   Thermal conductivity conversion
A sample of soil has a thermal conductivity of 1.73 watt/m°C. Calculate the corresponding value in Btu/ft-h°F.

SOLUTION
Recall that a watt = 1 joule/s

$$1.73 \frac{\text{watt}}{\text{m}°\text{C}} \cdot \frac{\text{joule}}{\text{watt} \cdot \text{s}} \cdot \frac{3600 \text{ s}}{\text{hr}} \cdot \frac{\text{Btu}}{1055 \text{ joule}} \cdot \frac{\text{m}}{3.28 \text{ft}} \frac{(5/9)°\text{C}}{°\text{F}} = 1 \frac{\text{Btu}}{\text{ft} \cdot \text{hr} \cdot °\text{F}}$$

   2   Thermal conductivity relating to soil texture and mineral structures: The thermal conductivity of soil is dependent upon many parameters, the most important of which are the mineral composition, soil texture, density, moisture content, degree of saturation, organic content, and temperature. Based on research findings from various sources, Kersten (1949), Van Rooyen and Winterkorn (1959), the following results are summarized: (a) the thermal conductivity of soil existing at temperature above the freezing point increases with an increase in mean temperature. In most cases, the value does not vary appreciably in the temperature range 25°F to −20°F (−4°C to −29°C); (b) at the optimum moisture content, (Sec. 7.2) the thermal conductivity of a frozen, compacted specimen averages close to 20% greater than that of an unfrozen specimen; (c) at constant moisture content, an increase in density results in an increase in conductivity. The rate of increase is about the same at all moisture contents; and (d) for saturated, unfrozen soils, the conductivity decreases with decrease in density. For saturated, frozen soils, no well-defined relationship between density and conductivity was found.
   The thermal conductivity varies, in general, with the texture of soils. At a given density and moisture content, the conductivity is relatively high in coarse textured soils such as gravel or sand, somewhat lower in sandy loam soils, and lowered in fine textures soils such as silt loam or clay. The thermal conductivity of a soil is dependent upon its mineral composition. Sands with a high quartz content have greater conductivities than sands with high contents of such minerals as plagioclase and feldspar. Soils with relatively high contents of kaolinite have relatively low conductivities.
   3   Prediction of thermal conductivity from laboratory results: To predict thermal conductivity the following empirical formulas were suggested by Kerstern (1949):

  a *Silt and clay soils*:
     Unfrozen: $k = [0.9 \log \omega - 0.2] \, 10^{0.01 \gamma}$ (6.2)

     Frozen: $k = 0.01(10)^{0.022 \gamma} + 0.085 \, (10)^{0.008 \gamma} (\omega)$ (6.3)

  b *Sandy soils*:
     Unfrozen: $k = [0.7 \log \omega + 0.4] \, 10^{0.01 \gamma}$ (6.4)

     Frozen: $k = 0.076 \, (10)^{0.013 \gamma} + 0.032 \, (10)^{0.0146 \gamma} (\omega)$ (6.5)

where k = the thermal conductivity, unit is in British thermal units per square foot per inch per hour per degree Fahrenheit, (see Example 6.1 for conversion into SI units), ω = the moisture content (%), and γ = dry density of soil, pcf. Figure 6.3 is plotted with logarithm of thermal conductivity versus porosity of kaolinite clay at

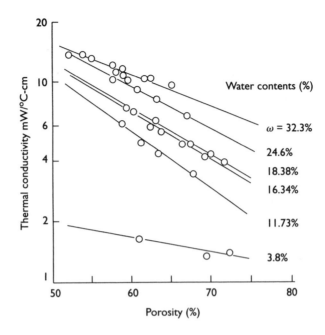

Figure 6.3 Thermal conductivity versus porosity.

Source: Reno, W. H. and Winterkorn, H. F. thermal conductivity of kaolinite clay as a function of type of exchange ion, density and moisture content. In *Highway Research Record* No. 209, Highway Research Board. National Research Council, Washington DC, 1967, pp. 79–85. Reproduced with permission of the Transportation Research Board.

various water content, In all cases, decreasing the water content and increasing the porosity will decrease the thermal conductivity.

6.4.2 *Thermal resistivity*

The reciprocal of thermal conductivity is referred to as *thermal resistivity* and is a measure of a material's ability to resist heat flow. It's units are cm-s-°C/cal or cm-°C/watt also referred to as a *thermal ohm*. The usual method of measuring thermal resistivity of soil in the field is to insert a 'thermal needle' into the ground to observe the temperature-time characteristic resulting from a given heat input. This needle consists of a small diameter metal tube, a heating element and thermocouples. Thermal needles for field use have been developed since 1952. A detailed description of various types of thermal needles with theoretical background are given by Van Rooyen and Winterkorn (1959), Mitchell and Kao (1977), and Chaney *et al.* (1983). Small thermal needles of the order of one foot in length are available commercially or may be made on the same general principles as the field needles. These small needles are useful for measurements on laboratory samples of soils. Automotive-type storage batteries are usually used as sources of power for heaters. The thermocouple or thermistor may be of any good commercial type, adapted to the specific metals used in the needle and suitably designed for field use. A Princeton type thermal

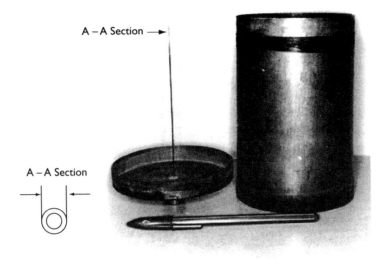

A – A Section

A – A Section

Figure 6.4 Princeton University type of thermal needle for measuring thermal resistivity of compacted fine-grained soil in laboratory. (a) Thermal needle 5–7/8 in.long, 0.015 in. O.D; and (b) Compaction mold 6-in.height, 3–7/8 in. I.D.

needle developed at Princeton University for laboratory use is shown in Figure 6.4. Detailed descriptions and test procedures of thermal needle are given by Winterkorn (1970b).

6.4.3 Calculation and data presentation of thermal resistivity

1 Computation procedure: The working formula for calculation of thermal resistivity based on the thermal needle probe is given as (Winterkorn, 1970)

$$\rho = \frac{(4\pi L)}{W}\Delta\theta \tag{6.6}$$

where $\rho = 1/k$ = the thermal resistivity of the soil in °C (centimeter per watt or thermal ohm-centimeter), $k = 1/\rho$ = the thermal conductivity of the material, L = length of heater filament (in in.), W = constant power to the heater (watts), I = current (amps), E = average of voltage measured occasionally during the test period, $\Delta\theta$ = rise in temperature °(C), during one logarithmic decade of time (based on the valid portion of the curve which shows a steady rate of temperature increase). Some assumptions for using Equation (6.6) are (a) the probe or heat source must be of infinitesimally small thickness, (b) the probe must be of infinite length, (c) the probe must be of material having the same diffusivity (Sec. 6.4.4) and conductivity as the material under test, (d) the temperature must be measured at the surface of the probe, and (e) the material under test must have uniform temperature throughout, be homogeneous throughout, be of infinite dimension radically from the source, and be bounded by a perfect insulating plane perpendicular to the axis of the line source.

While these assumptions are not met explicitly, testing conditions often approximate them sufficiently. Typical thermal resistivity, °C-cm/watt results are ice = 44.6, water = 165, mica = 170, air = 4000, moist clay = 80–90, loose dry sand = 175.

   2   *Graphical interpretation of thermal resistivity data:* Figure 6.5 presents a three-dimensional plot on a surface depicting thermal resistivity as a function of solid, air, and water phases (Van Rooyen and Winterkorn, 1959). The density of soil is expressed as the volume fraction of solids per unit volume of material on the X-axis, the moisture as the volume fraction of water per unit volume of material on the Y-axis, and the thermal resistivity along the Z-axis. The intersection *c* of the solids and moisture axes corresponds to 100% air. In Figure 6.5 the maximum point *a* on

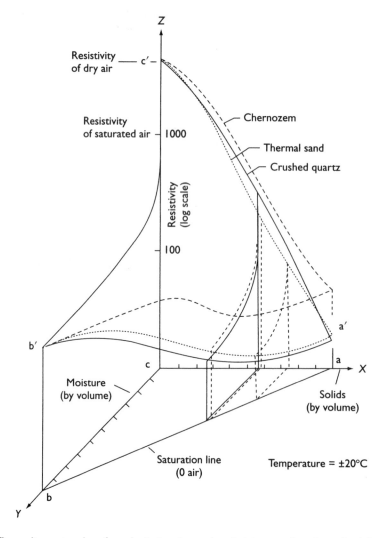

Figure 6.5 Three-dimensional surface depicting thermal resistivity as a function of solid, air, and water phases.

Source: Van Rooyen, M. and Winterkorn, H. F. Structural and textural influences on the thermal conductivity of soils. In *Proceedings of the* 38th *Annual Meeting,* Highway Research Board. National Research Council, Washington DC, 1959, pp. 576–621. Reproduced with permission of The Transportation Research Board.

the X-axis represents 100% solid; **aa'** is then the thermal resistivity of the solids alone. The point **b** on the Y-axis represents 100% water and **bb'** the thermal resistivity of water. The length cc' along the Z-axis constitutes the thermal resistivity of air.

For different soils, the only limiting value in the figure that will show a considerable change will be the effective thermal resistivity of a solid alone, which is represented by the value aa'. The value of bb' may be altered by the amount and type of pollutant in solution, temperature, or presence of other liquids. Similarly, cc' the thermal resistivity of air may be changed by changes in its composition and temperature. Curve ab represents the variation in thermal resistivity for saturated systems in which the solid content varies from 0% to 100%. As an illustration of this variation, the dotted curves were drawn, which are representative of a clay system with lower thermal resistivity in the dry state and somewhat higher resistivities in the saturated state.

Three types of materials are presented in the Figure 6.5 include chernozem soil, thermal sand, and crushed quartz. Chernozem soil is a zonal group of soil (Sec. 2.6) having a deep, dark-colored to near black surface horizon, rich in organic matter, which grades below into lighter-colored soil and finally into a layer of lime accumulation; developed under tall and mixed grasses in a temperate to cool subhumid climate.

6.4.3.1 Comments on equations for calculating the soil thermal resistivity

With all these complex independent, and interdependent variables present, it is evident that the problem of soil thermal resistivity is a complicated one; and that the development of an equation for this quantity is a formidable task. A number of investigators have, however, developed such equations; and a summary of the available results is given below. In general, equations may be classified under two groups: (a) empirical equations based on data obtained by measurement and analyses by graphical or numerical methods; and (b) theoretical equations which are based on imaginary models in which the actual soil structure is simplified in such a way as to permit a mathematical analysis. For dry soils, equations have been presented by several investigators. The major difference between these equations lies in the difference in the models on which they are based.

From the geotechnical engineer's perspective, the equations for dry soils are largely of academic interest. Dry soil is rarely encountered outside of the laboratory, since all soils have a considerable water affinity and have a strong tendency to absorb water from the atmosphere, at least to the extent of their hygroscopic requirements. Frequently, some capillary water diffuses in the soil from groundwater level. In considering the formulas for moist soils, it should be noted that some formulas are theoretical and some are empirical.

EXAMPLE 6.2 Thermal conductivity and resistivity

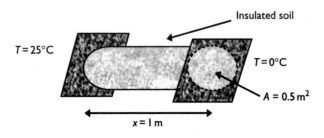

For the above figure, calculate the thermal conductivity of the soil contained in the tube if the heat per unit area $Q = 21.3$ watts. Recall Equation (6.1), rearranged in terms of thermal conductivity, K:

SOLUTION

$$K = \frac{dQ}{dt} \cdot \frac{dx}{dT} \cdot \frac{1}{A} = 21.3 \text{ watts} \cdot \frac{1 \text{ m}}{(25°C - 0°C)} \cdot \frac{1}{0.5 \text{ m}^2} = 1.7 \frac{\text{watts}}{\text{m}°C}$$

The above figure is a conceptual design for experimentally determining the thermal conductivity. This is a useful property, particularly when assessing the extent to which frost or thaw penetration (Sec. 6.7.3) may affect foundation footings, pavements or landfill covers.

6.4.4   Thermal diffusivity

The *diffusivity* is the quotient of the thermal conductivity and the heat capacity per unit volume. It is used in calculations where non-steady state conditions prevail, unlike Example 6.2, which assumed steady state conditions, that is, the temperature gradient from 25°C to 0°C had existed long enough for a constant flow of heat to occur. An example of unsteady state conditions is the calculation of the time it takes to freeze soil that is initially unfrozen. The diffusivity value may be determined by calculation if the thermal conductivity, specific heat and density of a soil are known. Its unit in the SI system is cm^2/s.

6.5   Effect of heat on engineering properties of soils

Heating soils can lead to changes in the chemical structure of the constituent minerals. An extreme example is the manufacture of bricks or pottery from clay. While the temperature effect is quite straight forward in relatively pure clay-water systems within a limit range of temperature, it may cause marked dispersion or flocculation effects depending on clay mineral and type of exchange ion. A comprehensive study on temperature and heat effects on soil is given by Chandrasekharan *et al.* (1969) and some of these results are summarized as follows:

6.5.1   Effect of temperature on soil constants

1   Moisture content of soil: The moisture content has a strong influence on the engineering behavior of soil. Of course, to determine the moisture content in a reasonable amount of time it is generally necessary to heat the soil. There are three basic procedures for determination of moisture content namely (a) oven dry (ASTM D2216), (b) air-dry at room temperature, and (c) microwave oven dry. The air-dry method is time consuming and does not remove all soil moisture. As such, conventional and microwave ovens are typically used. Different oven temperatures yield different dry weights for the soil, particularly for organic soils. Increasing the oven temperature for some organic soils causes organic matter to burn. This results in a lower dry weight and a correspondingly high moisture content. For granular soils such as Ottawa sand, the temperature has little effect over a range of 45–250°C (113–482°F). Often a value of 105°C is specified for use, although this too varies depending on the anticipated mineralogy.

2   *Atterberg limits and unit weight*: Temperature changes influence the liquid and plastic limits proportional to the change with temperature in the viscosity of water. Laguros (1969) tested kaolinitic, illitic, and montmorillonitic clays and found their liquid and plastic limits decreased with an increase in temperature, with the greatest effect observed for montmorillonite. Soil temperature has an effect on the moisture–unit weight relationship, particularly for soils high in clay content.

3   *Compressibility and volume change*: In general, the effects of temperature on compressibility and volume change include (a) an increase in compressibility with increase in temperature, the greatest effects being observed during secondary consolidation (Sec. 9.2). The coefficient of consolidation, c_c, increased radically between 40°F and 70°F (4.4–21.1°C); (b) decrease in soil volume with increase in temperature at constant pressure with the temperature being the only external variable; (c) temperature increases caused immediate volume changes with magnitudes dependent upon the magnitude of the temperature change; and (d) magnitude of the initial stress (excess-pore-pressure) has a secondary effect on the magnitude of the volume change.

4   *Comments on effect of temperature or heat on soil*: It must be kept in mind that with the exception of the Atterberg limits and compaction tests, the effect of temperature on soil–water systems was determined at constant water content or with free accessibility to a water reservoir. If the change in temperature is associated with a change in moisture content then the total effect is the sum, or the difference, as the case may be, of both temperature and moisture change effects. This, of course, is the reason why field CBR (Sec. 12.9.1) and plate-load test results (Sec. 12.9.2), various shear results, etc. yield higher values during the warmer summer months. The general effect of temperature on the engineering properties of soil is summarized in Table 6.1. However, the information contained therein is to be used with caution because of the possible modifying or counteracting effects due to the previously mentioned geometric, granulometric and structure factors.

Table 6.1 General effect of temperature on the engineering properties of non-swelling soils

Soil properties	Temperature	
	Low	High
Atterberg limits		
Plastic limit, ω_p	High	Low
Liquid limit, ω_L	High	Low
Moisture–unit weight relationship		
Maximum unit weight, γ_o	Decrease	Increase
Optimum moisture content, OMC	Increase	Decrease
Hydraulic conductivity, k	Decrease	Increase

6.5.2 Factors affecting heat on soils

The effect of heat on soils is dependent in part on the particle size composition as well as the presence or absence of: (a) a granular bearing skeleton, (b) secondary structure of silt–clay aggregates, and (c) the mechanical, thermal, and moisture history of the soil system.

1 Granular soils: Granular soils such as sand, gravel, cobble, and their mixtures derive their mechanical resistance properties from friction and interlocking which are affected to only a small degree by the presence or temperature of water. They are members of the large class of macromeritic systems (Sec. 3.5.4), and their mechanical properties are in full accordance with the laws established for such system.

2 Cohesive soils: In the case of cohesive soils a distinction must be made between those that, in the compacted state, possess a granular skeleton and those without. Cohesive soils with a granular bearing skeleton are members of the large class of construction materials that have been designated as *collameritic systems*. The soil with granular skeleton is defined by a combined volume of silt–clay plus associated maximum in situ water content that is smaller than the intergranular spaces left free by the skeleton. The maximum water content is geometrically defined. As for soils without a granular bearing skeleton, the least desirable are the silts that change from a semisolid to a liquid state at a small increase in moisture content. The greater the clay content and its water affinity of a soil, the smaller the change of its consistency with an increase in water content.

6.5.3 Soil properties at high temperature range

1 Heat treatment or soils stabilization by heat process: Heat treatment or thermal stabilization is defined as an irreversible and effective increase of the shear strength of soil or rock. Based on experimental work (Temperature range from 200–600°C) reported by Wohlbier and Henning (1969), the heat treatment of kaolinite clay results in a permanent increase of shear strength. In the case of a material in which the increased strength is only due to a change in capillary tension and not structure, the effect of stabilization when reducing this capillarity by adding water is somewhat reversible. Also, Chandrasekharan *et al.* (1969), studied the effect of heat treatment on Indian black soil (Expansive soil), and found a moderate reduction of water affinity in the initial ranges of heat treatment between 25°C and 250°C.

2 Soil treatment by fusion process: Fusion is a heat process that uses high temperatures to melt a solid into liquid. It also refers to reactions between small atomic nuclei to form a larger one. The solid phase of a soil particle may consist of many different minerals, some of which contain water as an integral part of their crystal lattice structure. Most solid particles also hold surface water layers by means of molecular attraction (adsorption). The fusion process has been used for soil stabilization and ground improvement. Detailed discussions on these aspects are given by Post and Paduana (1969).

6.5.4 Albedo

Albedo, also known as the *reflection coefficient*, is defined as the ratio of solar radiation reflected to that which is received, per unit of surface and time in percent.

Albedo data will reflect the characteristics of structural surface with respect to solar energy. The typical results and ranges for various construction materials are asphalt pavement <10, ocean water = 3–10, grass, 15–25%, woods = 5–20, sandy soil = 15–40, snow, compacted = 40–70, snow, fresh = 75–95.

6.6 Effect of heat on performance of soil-foundation system

6.6.1 General discussion

1 Temperature variations: The importance of climatological factors relating to the performance of highway pavements has been pointed out by Eno in 1929. These factors include temperature, frost, sunshine, wind, humidity, precipitation, runoff, and evaporation. In 1944 Winterkorn used physicochemical concepts to explain how these factors affected the performance highway components. Since then, many investigators have attempted to develop measuring techniques and analysis methods that isolate each of these factors. Geotechnical engineers believed that moisture content and temperature of pavement components, as well as depth of frost penetration and groundwater fluctuation had the most significant effects on pavement performance. Ultimately, all these factors derive from the thermal regime on the earth surface. Soil temperature has been shown to vary in a somewhat regular pattern, reflecting both the annual and diurnal cycles of solar radiation. Superimposed on these regular cycles are fluctuations of variable duration and amplitude created by changing climatic conditions.

2 24-hr temperature study: The results of a 24-hr temperature study conducted at AASHO (AASHO is the predecessor to AASHTO (American Association of State Highway Transport official)) Road Test indicate that temperatures within both asphalt and concrete pavement sections exhibited similar trends in response to air temperature fluctuations. Results also indicate that, in the early morning when the air temperature was low, the temperature of the pavement bottom was higher than that of the pavement surface. When the air temperature rose, the temperature at the top of the pavement surface rose sharply as compared with that at the middle and bottom part of the pavement. This phenomenon is a function of the thermal conductivity and heat capacity.

3 Isotherms of pavement sections: Figure 6.6 shows isotherms for the soil-pavement system of non-traffic pavement section. The upper part of the figure was obtained when the air temperature was a maximum (80°F). The lowest part was obtained the same day. Figure 6.7 presents temperature variations in soil-pavement system of concrete pavement including shoulders. In examining Figure 6.7, the temperature variations happened to be on shallow depth of pavement surface.

6.6.2 Seasonal affects on the performance of soil-pavement systems

The influence of seasonal factors on the performance of soil-pavement systems is discussed as follows:

1 Subsurface soil strength: The variation of moisture content, California Bearing Ratio (CBR Sec. 12.9.1), and elastic modulus (Sec. 12.9.4) over a period of two years

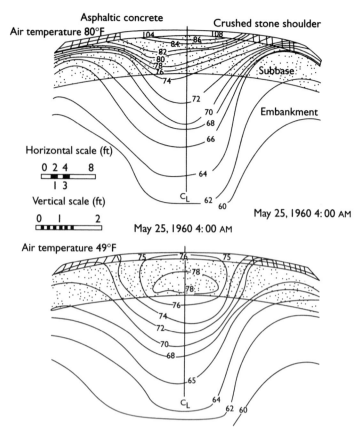

Figure 6.6 Soil-pavement system isotherms (Asphalt pavement).
Source: Data from AASHO Road Test, HRB, 1962b.

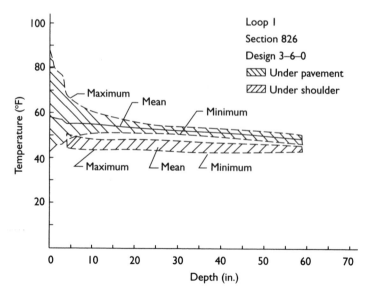

Figure 6.7 Temperature variations in soil-pavement system (Concrete pavement).
Source: Data from AASHO Road Test, HRB, 1962b.

is shown in Figure 6.8. All data were obtained from non-traffic test sites. It may be seen that, as may have been predicted, there was a tendency for the strength of the pavement components (base, subbase, and embankment soil) to increase as the moisture content decreased. It should be noted that tests on the traffic sites yielded moisture content, CBR, elastic modulus and density values that were different in magnitude from those obtained in non-traffic test sites. Normally, moisture contents were less, and CBR and elastic modulus values were greater on the traffic sites than on the non-traffic sites. This was particularly evident in the spring time.

2 *Embankment soil pressures:* Variations in embankment pressure with seasons of the year are shown in Figure 6.9. The pressure cell used on the embankment soil surface utilizes strain gages to record the pressure transmitted to the face of the unit. The gages were cemented to a flexible diaphragm mounted in the interior of the cell. The pressures were measured with the loaded wheel (single axle vehicle) stopped at 6-in. intervals from points 2 ft (0.61 m) ahead and 2 ft (0.61 m) behind the location of the cells. The pressure reached a maximum value during the spring and early summer and decreased subsequently. The pattern of pressure variation with seasons was similar to that observed for creep speed deflections. The effect of design (total thickness) of the pavement on transmitted pressure is also studied. A fairly orderly effect is evident for the spring and summer periods.

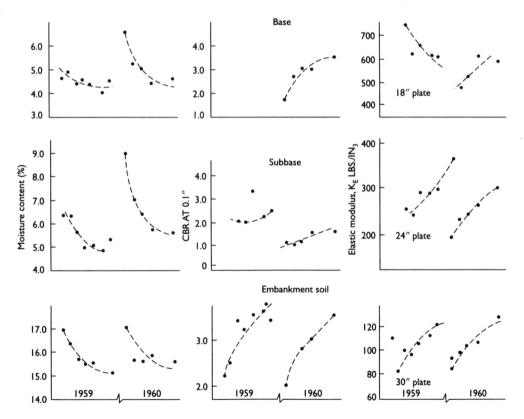

Figure 6.8 Seasonal variation in strength characteristics of pavement components.

Source: Data from AASHO Road Test, HRB, 1962b.

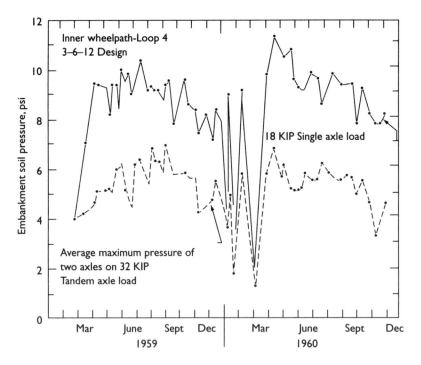

Figure 6.9 Seasonal variation of embankment soil pressure.

Source: Data from AASHO Road Test, HRB, 1962b.

6.7   Freezing–thawing behavior of soil

6.7.1   Characteristics of frozen soil

In general, the engineering properties of frozen soils are temperature dependent in the range from 0°C to−10°C, where the phase composition of water changes appreciably. Its important to note that while the temperature of a soil–water system may be lower than 32°F (0°C) there is almost always some unfrozen water. The temperature dependence is related to changes in the amount of this unfrozen water. The effect of temperature is greater in soil types that contain large quantities of unfrozen water. Soil types, particle size, and water content of the soil play an important role relating to the characteristics of frozen soil. The amount of unfrozen water increases in the direction of sand < silt < clay. The engineering problems for frozen soil include: Frost action, frost depth (penetration), and frost heave, as discussed in the following sections.

6.7.2   Frost action and freezing index

1   Frost action: Frost action involves the freezing and thawing of moisture in materials and the resultant effects on these materials or the structures with which they are associated. When water freezes to form ice, it increases in volume by approximately 9%. However, frost action may lead to the creation of ice lenses which result in surface heaves that are many times larger than 9%. The additional movement is the result of suction and capillary forces which serve to draw in more water to the freezing front. This action is most prominent in frost-susceptible soils, which tend to have a balance between capillarity and permeability. For example, clay soils have great capillarity but low-permeability, and hence it is difficult for significant quanti-ties of moisture to be transmitted to the freezing zone. Coarse sands and gravels have great permeability, but negligible capillarity with which to draw in moisture. As such, silts tend to be most frost-susceptible.

2 *Freezing index*: The freezing index, F, is used as a measure of the combined duration and magnitude of below-freezing temperature occurring during any given freezing season. It is the number of degree-days between the highest and lowest points on the cumulative degree-day-time curve for one freezing season. There are two types of freezing index. The index determined for air temperatures at 4.5 ft (1.4 m) above the ground is commonly designated as the *air freezing index*; while that determined for temperatures immediately below a surface is known as the *surface freezing index*.

6.7.3 Frost penetration and heave in soil

When the ground surface temperature falls below the freezing point (0°C) freezing of subgrade soil may occur. When water changes from a liquid to a solid state, its volume expands. Damage to shallow foundations (footings), walls, pavements may occur as a result of the frost penetration. Frost penetration or frost depth can be determined both theoretically and experimentally. The theoretical approaches for the estimation of frost penetration in this section include the simplified Stefan method (Berggren, 1943) and the modified Berggren method (Aldrich and Paynter, 1956).

1 *Simplified Stefan method*: The assumptions involved in the Stefan method include (a) homogeneous soil, (b) the temperature gradient in the frozen zone is linear, (c) the porewater existing in the soil is not moving, and (d) considers only the volumetric heat of latent fusion. Then the depth of frost can be estimated from Equation (6.7).

$$z_p = \sqrt{\frac{48k_f F}{L}} \tag{6.7}$$

where z_p = depth of frost penetration, k_f = thermal conductivity of frozen soil, F = freezing index, and L = latent heat of fusion. L is given by the product of the latent heat of fusion of water, water content and dry density, that is, $L_w \omega \gamma_d$, where L_w = 143.4 Btu/lb (333 kJ/kg). Equation (6.7) can be further modified for a multi-layered soil or pavement system (Jumikis, 1977). Miller and Lee (1997) used Jumikis' modified equation for computing depth of frost penetration in landfill cover systems. Frost penetration in the topmost layer is calculated in the same manner as the single-layer method. Discussion of the limitations of simplified equation is given by Miller and Lee (1997).

2 *Modified Berggren method*: This method was developed by Berggren in 1943 and modified by Aldrich and Paynter (1956) for the estimation of frost penetration below highway and airfield pavements by the following formula:

$$z_p = \lambda \sqrt{\frac{48k_f F}{L}} \tag{6.8}$$

where k_f, F, L are defined in Equation (6.8) and λ is a dimensionless correction factor. The correction factor, λ is always less than unity and depends on the thermal properties of the unfrozen and frozen soil and is a function of two parameters, the thermal ratio, α, and the fusion parameter, μ.

$$\alpha = \frac{T_u}{T_f} = \frac{T_u t}{F} \tag{6.9}$$

$$\mu = \frac{C_f T_f}{L} \cong \frac{C_f F}{Lt} \tag{6.10}$$

where α = thermal ratio, μ = fusion parameter, T_u = temperature above freezing, T_f = temperature below freezing, t = the duration of the freezing period, and

$$C_f = \gamma_d c_s + \frac{\gamma_d \omega c_i}{100} = \left(0.17 + \frac{\omega}{200}\right)\gamma_d \tag{6.11}$$

in which C_f = volumetric heat of frozen soil, γ_d = dry density of material, ω = water content (%), c_s, c_i are the specific heats of solids, ice, respectively.

3 *In situ frost depth (penetration) measurements*: There are several in situ frost depth measurement devices available. A device was developed at the AASHO Road Test (Andersland and Carey, 1957) by which determination of depth of frost could be made without disturbing the soil-pavement system. The system was based upon the knowledge that electrical resistance of a soil–water system changes rapidly upon freezing. Pairs of electrodes buried in the soil later were connected to leads that extended to the ground surface. Measurements of the resistance across these electrodes indicated the depth to which the soil–water system had frozen. Results showed the mean frost penetration extended below the pavement surface for four cover conditions studied. It may be seen from results that, in general, frost penetration was greater under concrete pavements than under bituminous pavements, due to the greater heat conductivity of portland cement concrete. However, it should be noted that there was no subbase material in the case of rigid pavements. In fact, some rigid pavement sections had neither subbase nor base material. Figure 6.10 presents the approximate depth of frost penetration in United States. Local variations may be large, especially in mountainous areas. These data are useful for the preliminary analysis and design of foundation structures and ground improvement systems.

4 *Frost heave*: Frost heave is the raising of the ground surface due to the accumulation of ice in the underlying soil. It occurs as a result of suction associated with the freezing process, causing water to migrate into the frozen soil zone and form ice lenses. This process will ultimately produce heavy pressures and large displacements. The Cold Regions Research and Engineering Laboratory (CRREL) of the US Army Corps of Engineers has developed a scale that relates frost susceptibility to the average rate of heave in mm/day, namely: 0–0.5 – negligible, 0.5–1.0 – very low, 1.0–2.0 – low, 2.0–4.0 – medium, 4.0–8.0 – high, greater than 8.0 – very high. Similarly, CRREL has determined that frost susceptibility increases according to soil grouping, from F1 (least susceptible) to F4 (most susceptible) where F1 includes gravelly soils with 3–20% material finer than 0.02 mm, F2 = Sands with 3–15% finer

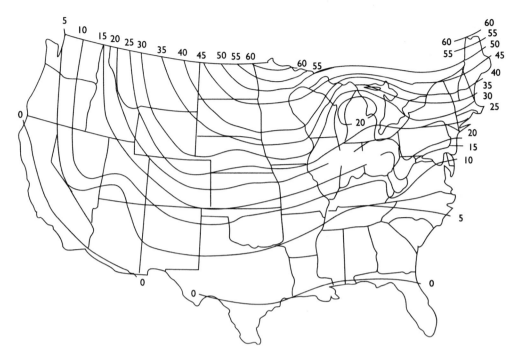

Figure 6.10 Approximate depth of frost penetration in the United States.

Source: Aldrich, H. P., Jr. Frost penetration below highway and airfield pavements. In *Highway Research Board Bulletin* No. 135, Highway Research Board. National Research Council, Washington DC, 1956, pp. 124–149. Reproduced with permission of the Transportation Research Board.

than 0.02 mm, F3 = includes gravelly soils with more than 20% finer than 0.02 mm and sands, except in the case of fine silty sands with more than 15% finer than 0.02 mm, as well as clays with a PI greater than 12%, with the exception of varved clays, and the final category F4 = includes silts and sandy silts, fine silty sands with more than 15% finer than 0.02 mm, lean clays with a PI less than 12% and varved clays.

5 *Permafrost*: Permafrost is defined as permanently frozen ground. Frozen ground poses few engineering problems if it is not disturbed. But changes in the surface environment lead to thawing of the permafrost, which in turn produces unstable ground susceptible to soil creep and landslide, slumping, and subsidence. In Arctic and Antarctic regions, the soil remains frozen throughout the year to great depths up to 1000 ft (305 m) and more. A special characteristic of permafrost is the thaw in warm months which converts the top layer of ground into wet and unstable material.

6.7.4   Behavior of thawing soils

When ground temperature increases after beyond the prevailing freezing point, soil starts thawing. At that time, the strength of soil decreases. Experience shows that shortly after the thawing of the soil beneath the pavement, the soil loses its bearing capacity for a certain period of time. Frost boil is an important phenomenon during soil thawing process.

1   *Frost boil*: Frost boil refers to the softening of soil that occurs during a thawing period due to liberation of water from ice lenses or layers. The spring breakup takes place during the frost-melting period. The moisture accumulates in the previous frozen zone of soil, after thawing of the ice lenses, resulting in a lowering of the bearing capacity (Sec. 12.12.2) of the soil system for a period of time.

2   *Thawing index*: Thawing index (T) is used as a measure of the combined duration and magnitude of above-freezing temperatures occurring during any given thawing season. It is the number of degree-days between the lowest and highest points on the curve for cumulative degree-days versus time for one thawing season. The index determined for air temperature at 1.5 m (4.92 ft) above the ground is designated as the *air thawing index* (T_a), while that determined for temperatures immediately below a surface is known as the *surface thawing index* (T_s), The *mean thawing index* (T_m) is the index determined on the basis of mean temperatures.

EXAMPLE 6.3
Determination of frost depth
Determine the maximum frost penetration into a homogeneous deposit for the following winter season and conditions:

Mean annual air temperature $= 2.5°C$
Freezing index $= 1350$ degree-days
Length of freezing season $= 150$ days
Soil conditions: $\gamma_b = 15.7$ kN/m³, $w = 30\%$, $k = 1.7$ W/m/°C

SOLUTION

$\gamma_d = 15.7/(1 + 0.3) = 12.1$ kN/m³

$L = 3.40 \times 10^4(0.30)(12.1) = 123184$ kJ/m³ $= 123.2$ MJ/m³

$T_u = 2.5 - 0 = 2.5°C$ (i.e., temperature above freezing)

$T_f = F/t = 1350$ degree-days/150 days $= 9°C$ (temperature below freezing)

$\alpha = T_u/T_f = 2.5/9 = 0.28$

$C = 12.1(72.4 + 427(0.30)) = 2426$ kJ/m³ $= 2.43$ MJ/m³

$\mu = (C/L)T_f = (2.43/123.2)(9) = 0.177$

$\lambda =$ From Table (e.g. figure 12.52, p. 288, Mitchell, 1993) $= 0.91$

$$Z = 0.91 \cdot \left(\frac{2 \cdot (1.7 \text{ W/m°C}) \cdot (1350°C \cdot \text{days}) \cdot (86400 \text{ s/day}) \cdot (1 \text{ MJ}/10^6 \text{J})}{123.2 \text{ MJ/m}^3} \right)^{1/2}$$

$Z = 1.63$ m

6.8 Electrical properties of soil

6.8.1 *Characteristics of particles and electricity*

As discussed in Section 1.6.2, ordinary atomic particles are neutral, that is, they do not carry an electrical charge. However, under certain circumstances atoms becomes electrically charged. Such charged atoms are called *ions*. Since atoms are able to form ions, atoms also contain basic units of a negative charge which can be lost to form positive ions or gained to form negative ions. These units of charge are the same units of charge which characterize electricity. Such units of negative charges are known as *electrons*. The positive charge is necessary to neutralize the negative electrons in a neutral atom. Such positive subatomic particles present in the atom is called *proton* (+). If a subatomic particle carries no electrical charge (neutral), it is known as a neutron (0). Basic types of particles which serve as building units of matter are presented in Table 1.1. There are two kinds of electrical charges, positive and negative charges. The positive charge in ordinary matter is carried by protons, and the negative charge by electrons. Charges of the same sign repel each other, charges of opposite signs attract each other. The net charge is equal to the total positive charge minus the total negative charge. According to the *Principle of Conservation of Charge*, the net electric charge in an isolated system always remains constant.

6.8.2 *Electrical energy field*

Electrical energy existing in the soil–water system plays an important role because all other energy such as mechanical, thermal or magnetic hinges on the characteristics of electric energy as illustrated in Figure 1.3. Electrical characteristics in the soil–water systems have multiple phases due to the following reasons: (a) Both soil and water have inherent electrical characteristics as discussed in Sections 3.7 and 4.3 (b) Electrical energy is closely related to thermal and magnetic energy and difficult to separate in the natural environment; and (c) Electrical–chemical interaction in the soil–water system is sensitive to local environments and in many cases, these interactions are not clearly understood. The magnitude and behavior in the soil–water system cannot be measured effectively; and some soil–electricity interaction in the environment is also not clearly understood and a theoretical approach sometimes is over simplified. Differences between theory and experimental observations are often significant.

Soil surface electrochemistry is important for studying the electrical properties of soil. The basic concept of electrochemistry was established by Michael Faraday in 1834, who discovered the fundamental law of electrolysis. However, the application of his concept in engineering is a relatively new interdisciplinary subject. It is a combination of chemistry, physical-chemistry, engineering, and their interactions with electricity and environment. Progress for development of electrochemistry also hinges on the development of other subjects such as clay mineralogy and ion exchange reactions in the soil–water system. Clay mineralogy has been discussed in Section 3.9 and ion exchange reactions have been discussed in Section 4.7. In this chapter only the electrical properties of soil-water system are emphasized.

6.8.3   Mechanisms of soil–electric interaction

The interaction between soil and electricity depends on soil particle size, mineral structure, mineral surface conditions and characteristics of pore fluid, as well as ion exchange capacity and properties of electrolytes. The ion movement direction in soil-water system follows the direction of electric current, and the influence area (electric energy field) is related to the magnitude of electrical charge and characteristics of soil-water system. Also, there are two distinct cases in the soil–electricity interaction, given as:

1   *Dry soil condition*: When soil is dry, the electric resistivity in general is very high, because there is very little interaction between electric charge (input energy) and ions existing in the soil;
2   *Moist soil condition*: For moist soil, the electric conductivity increases and resistivity decreases significantly, due to the moist film around soil particles which serves as a bridge linking the electric charge and the ions existing in the soil.

6.8.4   Soil–electricity reactions and phenomena

1   *Electrode and electrode reaction*: Electrodes consists of two metal elements or rods (Fig. 6.11) connected to an electron "pump" (e.g. battery) immersed in a solution containing ions. Cations are attracted toward the negative electrode and anions toward the positive electrode. The electrode that attracts the anions is called the

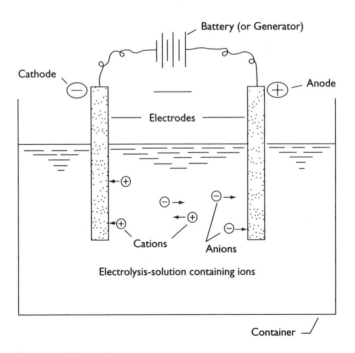

Figure 6.11 Typical setup of electrolysis apparatus.
Source: Fang (1997).

anode (+) and the electrode that attracts the cations is called the *cathode* (−). The arrangement of such system is referred to as a *cell*.

A reaction in which a species loses or gains electrons at an electrode and is converted to a new species is called an *electrode reaction*. An electrode reaction only occurs when a simultaneous reaction involving a corresponding gain/loss of electrons occurs at the other electrode in the system. In another words, electrode reactions are chemical reactions involving electron transfer.

2 *Electrolysis*: Electrolysis is the process of subjecting a solution to a condition which will produce electrode reactions. Electrolysis characteristics control the relationship between the quantity of electricity passed through the cell and the amounts of substances produced by oxidation and reduction at the electrodes. The quantity of electrical charge associated with one mole of electrons is called the *farad*. The relationship between farad and other units commonly used are 1 farad = 1 mole of electrons = 96,500 coulombs.

3 *Electrolytes*: Substances that form ions in aqueous solution are called *electrolytes*. Some substances that ionize completely are called *strong electrolytes*. Others ionized only slightly are called *weak electrolytes*. A substance that forms a solution that does not conduct electricity is called a *nonelectrolyte*. Many molecular substances such as alcohol and sugar are nonelectrolytes.

4 *Oxidation–reduction reaction (redox reaction)*: All chemical elements can accept or donate electrons. If an element loses electrons by a substance, it is called *oxidation* or if it gain electrons, it is called *reduction*. Oxidation and reduction always occur together because a substance can only donate electrons if another substance can accept them, and vice-versa. A reaction involving oxidation and reduction is called *oxidation–reduction reaction* also referred to as *redox reaction*.

5 *Polarization and proton migration*: As discussed in Section 1.4.2, indications are that all atoms and hence the molecules formed thereof contain positive and negative electric charges, (a) contained positive charges being associated with the atomic nucleus; and (b) contain negative charges are embodied in the electrons that surround the nuclei. In the electric energy fields, the volume occupied by the electrons can be distorted and shifted with respect to the position of the associated nuclei. This phenomenon is called *polarization*. In the electric or polar system composed in soil minerals, dissolved ions and water dipoles and the various electric fields associated with these constituents tend to readjust themselves in the direction of the lowest free energy or highest entropy of the system. A primary role in this readjustment is caused by proton migration and proton exchange as claimed by Eyraud *et al.* (1965).

6.9 Electrical behavior of soil–water system

6.9.1 *General discussion*

From basic physics, we learn that if 1 joule (J) of work is required to move a coulomb (C) of positive charge from one point to another, the potential differential between the two points is a volt (V). Figure 6.12 shows the schematic diagram of electric current and resistivity. The measure of electric current is the amount of charge that passes a given point per unit of time. If 1 (C) of charge passes a point in 1 second(s), the current is defined to be 1 ampere (A).

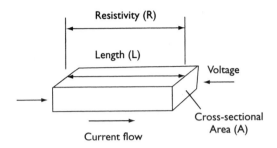

Figure 6.12 Schematic diagram of electric current and resistivity.

6.9.2 *Electric current, conductor, insulator, and voltage*

1 *Electric current*: As indicated in Figure 6.12, electric currents in metal wires always consist of flow of electrons (e); such currents are assumed to occur in the direction opposite to that in which the electrons move. Since a positive charge going one way is for most purposes equivalent to a negative charge going the other way, this assumption makes no practical difference. Both positive and negative charges move when a electric current is present in a liquid or gaseous conductor.

2 *Conductor and insulator*: A conductor is a substance through which electric charge can flow easily and an insulator is one through which electric charge can flow only with great difficulty. Metals, many liquids and plasmas are conductors. Plasma is the gas whose molecules are charged. Nonmetallic solids, certain liquids, and gases whose molecules are electrically neutral are insulators. A number of substances, called semiconductors are intermediate in their ability to conduct electric charge.

3 *Voltage*: Voltage, V, is defined as the ratio of work, W, to the electric charge, Q. 1 (V) = 1 (J/C), where J = joule and C = coulomb.

4 *Conductance*: Conductance, g, is reciprocal of resistance, R , and may be defined as being that property of a circuit or of a material which tends to permit the flow of an electric current, I. The unit of conductance is the reciprocal of ohm (1/ohm or mho).

5 *Capacitor*: A capacitor is a system that stores energy in the form of an electric field. A capacitor consists of a pair of parallel metal plates separated by air or other insulating material. The potential difference (V), between the plates of a capacitor is directly proportional to the charge (Q), on either of them, so the ratio (Q/V), is always the same for a particular capacitor.

6 *Capacitance*: Capacitance, C, is the ratio of (Q/V). The unit of capacitance is the farad, F, where 1 F = 1 coulomb/volt.

7 *Electromotive force*: An electromotive force (emf) is defined as any force that can cause an electric charge to move and thereby give rise to an electric current. A battery is a typical example of a source of emf. The emf also can be generated from thermal or electromagnetic energy. The unit of emf in general is cm-volt. Also, it must be noted that "electromotive force" is a misleading term, since emf

refers to a quantity related to energy and not to force. It is for this reason, as a common practice, reference is made to (emf) or (EMF), or represented by the symbol (E), and not to the misnomer "electromotive force".

6.10 Dielectric constant (D, ϵ)

6.10.1 Definition and characteristics of dielectric constant

The concept of *dielectric constant* was proposed by Quinke in 1859–1861. It is a measure of a material's ability to perform as an insulator. That is, it is a measure of the capacity of a material to reduce the strength of an electric energy field. The higher the dielectric constant of the material, the more the material behaves as an insulator. The dielectric constant of water is very high in comparison with other liquids as water = 80. Dielectric constant is not measured directly. It can be obtained through measuring and computing from the following equation as

$$D = C_s \, d/A \qquad\qquad (6.12)$$

where D = dielectric constant, C_s = capacitance, F (farad), d = length of a specimen, and A = cross-sectional area of specimen. For a typical soil specimen, where $d = 0.7$ cm, and $A = 2.85$ cm^2, (Fang, 1997) then

$$D = 2.775 \, C_s \qquad\qquad (6.13)$$

6.10.2 Relationship between dielectric constant and soil properties

1 *Dielectric constant versus Atterberg Limits*: The effect of dielectric constant on soil behavior is significant, because the dielectric constant is influenced by both ion concentration and types of ions in the soil–water systems. Acar and Olivien (1989) show the effect of organic fluids as reflected by dielectric constant on the Atterberg limit of Georgia kaolinite clays in that the dielectric constant of the solvent decreases, the liquid limit increases, while plasticity index remains constant. Kaya and Fang (1997), using the Atterberg limit of solvents as a function of dielectric constant as shown in Figure 6.13, found these observations are in agreement with the observation of Fernandez and Quigley (1985), in that the dielectric constant affects the flocculation and channelization in the soil structure. A decrease in dielectric constant causes aggregation and channelization within a soil mass, thus causing an increase in the flow area.

2 *Dielectric constant versus volumetric changes*: Kaya and Fang (1995, 1997) have reported on the swelling of kaolinite, illite, and smectite. They found that there is a significant relationship between the dielectric constant of the solvent and swelling. An increase in swelling with a decrease in the dielectric constant was attributed to an increase in the flocculation of the clay particles with a decrease in the dielectric constant. Acar and Olivieri (1989) reported the effect of pore fluids, as reflected by the dielectric constant, on free swell of montmorillonite clay as shown in Figure 6.14. Also, significant relationships among various organic fluids have been found.

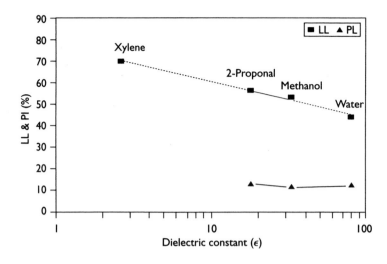

Figure 6.13 Relationship between dielectric constant and liquid limit.

Source: Kaya, A. and Fang, H. Y., Identification of contaminated soils by dielectric constant and electrical conductivity, *Journal of Environmental Engineering ASCE*, v. 123, No. 2, pp. 169–177. © 1997 ASCE. Reproduced by permission of the American Society of Civil Engineers.

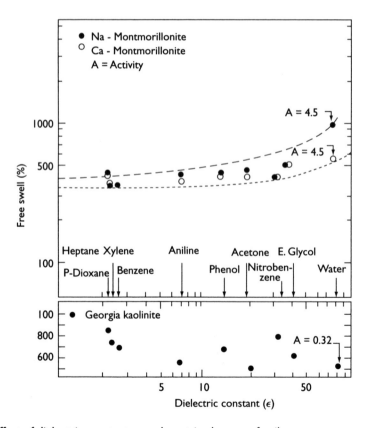

Figure 6.14 Effect of dielectric constant on volumetric changes of soil.

Source: Acar, Y. B. and Olivieri, I Pore fluid effects on the fabric and hydraulic conductivity of laboratory compacted clay. In *Transportation Research Record, No. 1219*, Transportation Research Board. National Research Council, Washington DC, 1989, pp. 144–159. Reproduced with permission of the Transportation Research Board.

3   *Dielectric constant relating to CEC and zeta potential*: The cation exchange capacity (CEC) of kaolinite clay is determined by sodium extraction. The result of the CEC of kaolinite is presented in Figure 6.15(a). In examining Figure 6.15(a), it is indicated that CEC linearly increases with an increase in the dielectric constant of the pore fluid. Zeta (ζ) potential of soil is an important parameter in the electric energy field. Detailed discussion on this aspect will be presented in Section 6.13.3. Zeta potential versus dielectric constant is presented in Figure 6.15(b). In examining Figure 6.15(b), it can be seen that the dielectric constant of the pore fluid decreases and reaches zero value within the experimental error range which can be attributed to the fact that surface charge density of the soil particles decreases as proton surface charge density gets smaller and smaller.

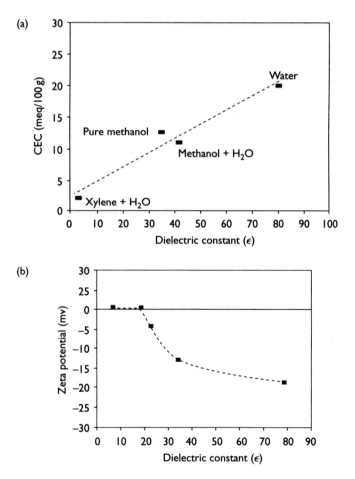

Figure 6.15 Relationship of dielectric constant with CEC and zeta potential. (a) Relationship with CEC; and (b) Relationship with zeta potential.

Source: Kaya, A. and Fang, H. Y., Identification of contaminated soils by dielectric constant and electrical conductivity, *Journal of Environmental ASCE*, v. 123, No. 2, pp. 169–177. © 1997 ASCE. Reproduced by permission of the American Society of Civil Engineers.

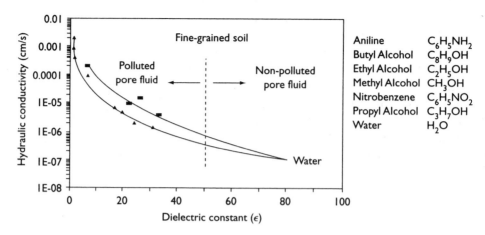

Figure 6.16 Hydraulic conductivity versus dielectric constant.

6.10.3 Identification and characterization of contaminated soil by dielectric constant

Figure 6.16 presents the relationship of hydraulic conductivity versus dielectric constant for various pore fluids including both contaminated pore fluids such as: aniline ($C_6H_5NH_2$), ethyl alcohol (C_2H_5OH), nitrobenzene ($C_6H_5NO_2$), and water. Figure 6.16 can be used for identification and characterization of contaminated soils as relating to hazardous/toxic waste sites.

6.11 Electrical conductivity and resistivity of soil

6.11.1 Electrical conductivity

The ease with which electricity may flow through soil is governed by its *electrical conductivity*. Electric potentials may also give rise to the movement of water. The behavior of such movement of fluid under electric potential is called *electrokinetic phenomena*. The phenomena can be divided into two major groups as (a) electrokinetic phenomena between soil particles; and (b) electrokinetic phenomena in clay suspension. Further discussions on this aspect will be presented in Section 6.12.4. Figure 6.17 presents the electrical conductivity versus porosity for illite, kaolinite, and bentonite clays with three temperatures. Theoretical and experimental explanations on electroosmosis based on surface-chemical properties of clay mineral and soil are also given by Winterkorn (1942, 1958).

6.11.2 Characteristics of electric resistivity

The property of an electric circuit tending to prevent the flow of current and at the same time causing electric energy to be converted into heat energy is called *electrical*

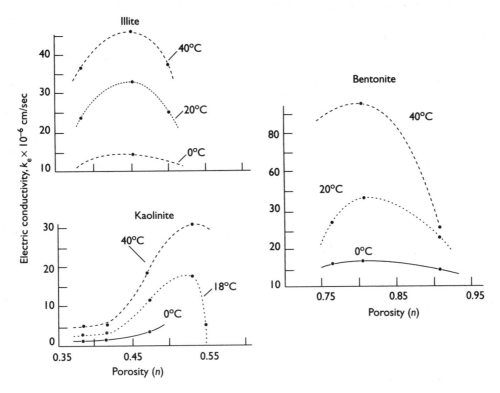

Figure 6.17 Electric conductivity versus porosity for illite, kaolinite, and bentonite clays with three temperatures.

Source: Winterkom H. F. Mass transport phenomena in moist porous systems as viewed from the thermodynamics of irreversible processes. In *Highway Research Board Special Report* 40, Highway Research Board. National Research Council, Washington DC, 1958, pp. 324–338. Reproduced with permission of the Transportation Research Board.

resistance. The ohm is the practical unit of resistance and defined as that resistance which will allow 1 ampere, A, to flow if 1 volt, V, is impressed across its terminals. An ohm has such a value that 1 ampere flowing through it for 1 produces as heat 1 joule of energy. A schematic diagram of electric current and resistivity is presented in Figure 6.12.

Using electric resistivity is the most common method for measuring the characteristics of soil–water systems as well as engineering properties of soil-rock. The applications are listed as follows:(a) environmental effects on soil or water characteristics, such as moisture content, temperature, and quality of water; (b) mechanical properties of soil, such as unit weight, porosity, Young's modulus, and strength; (c) frost depth (see Sec. 6.7); (d) locating subsurface materials (such as gravel, sand and soil deposits); (e) locating buried geostructural members such as underground pipes, storage tanks, as well as hazardous/toxic wastes; and (f) locating leak location for geomembrane liners. Resistivity is related to the moisture content of material, because electricity is a function of ion exchange capacity. Resistivity is also affected by temperature. When temperature increases the resistivity decreases.

6.12 Electrokinetic phenomena in soil–water system

6.12.1 General discussion

The *electroosmosis* phenomenon was discovered by Reuss in 1807, who found if an electrical current is applied to a porous membrane or to a capillary tube, movement of water will occur from the anode to the cathode and that as soon as the electrical current is stopped, the water movement will also stop. The classical theory to explain this phenomenon is due to Helmholtz in 1879 and refined by Smoluchowski in 1914. It was established originally for the simple condition of flow in a glass capillary and then developed in a way analogous to a Darcy type equation. The phenomena can be divided into two major groups:

1   *Mechanical phenomenon of electroosmosis*: (a) *Electroomosis*: This refers to fluid flow through soil particles where only the fluid moves and the soil particles remain stationary. (b) *Electrokinetic Phenomena*: Soil particles in suspension moving under an electrical gradient is called electrokinetic phenomena.
2   *Physicochemical phenomenon of electroosmosis*: Electrophoresis is the physicochemical behavior of a particle in clay suspension. Charged particles in the clay suspension are attracted or repelled from each other and negatively charged particles move towards the anode.

6.12.2 Helmholtz–Smoluchowski classical theory

The Helmholtz–Smoluchowski theory is one of the most widely used in describing electroosmotic phenomena. Assumptions for the Helmholtz–Smoluchowski theory include (a) the liquid carries a charge opposite to that of the rigid wall and the two charge systems for an electric double layer along the wall; (b) the thickness of the double layer is extremely, but not infinitesimally small; (c) the layer of liquid molecules in contact with the wall is immovable while the rest of the molecules in the liquid within the electric double layer are movable; (d) within the double layer, laminar flow occurs such that the velocity rises from zero next to the solid wall to a maximum at the center of the counter-charge layer in the liquid. Hence, the velocity remains constant for the remainder of the cross-section; (e) the externally applied potential acts on the charges of the double layer; and (f) the wall is an insulator and the contained liquid possesses electrolytic conductivity. The general equations for this theory are given as follows.

a   *Bulk velocity of flow*: Analysis of such a system gives for steady flow conditions under potential gradient E (volt/cm) the bulk velocity, v, as:

$$v = \frac{D\varphi E}{4\pi\eta} \qquad (6.14)$$

where v = bulk velocity, D = dielectric constant of the liquid, φ = electric potential of the double layer, E = the potential drop, η = viscosity of the liquid;

b   *The volume of flow per unit time*: Multiplying both sides of Equation (6.14) with $r^2\pi$, the cross-section of the capillary, we obtain V'/t the volume of flow per unit time.

$$V'/t = \frac{r^2\pi D\varphi E}{4\pi\eta} \qquad (6.15)$$

where V'/t = volume of flow per unit time;

c   *The volume of water in unit time through a unit cross-section*: Multiplying both sides of Equation (6.15) with the number of capillaries per unit cross-section and the numerator and denominator of the right side with the unit length, L, we obtain:

$$V/t = \frac{D\varphi(n)E}{4\pi\eta L} \qquad (6.16)$$

and

$$V/t\,E = \frac{D\varphi n}{4\pi\eta L} \qquad (6.17)$$

Electroosmotic transmission coefficient: (a) non-polluted pore fluid (Water): Equation (6.16) is the volume of water delivered in unit time through a unit cross-section under an electric potential gradient of 1 V/cm and represents an electroosmotic transmission coefficient k_e which is analogous to the coefficient k in Darcy's equation. Because of the expected deviation of the values for D, and in Equation (6.17) in the strong electric fields of the double layer from those for the bulk liquid and the resulting uncertainty, one usually combines the term $[D\phi v; /4\pi\eta]$ into a constant k_{eo} characteristic for the specific system and writes

$$k_e = k_{eo}n \qquad (6.18)$$

where k_e = electroosmotic transmission coefficient, k_{eo} = constant characteristic for the specific system, and n = porosity. Equation (6.18) holds well for natural undisturbed silty soils, but not for systems with extremely fine pores such as well disturbed clay soils and membranes of animal and vegetable origin; (b) polluted pore fluid: for polluted pore fluid, Winterkorn (1958) suggested, suitable equations can be developed on the assumption that the counter ions are located throughout the pore liquid and do not form an electric double-layer. The electroosmotic transmission coefficient may also be given as

$$k_e = C_e(1 - n)^{2/3}\,n \qquad (6.19)$$

where C_e = constant expressing geometrical and solid–liquid interaction factors and n as given previously.

Limitations of Helmholtz–Smoluchowski classical theory: If the Helmholtz–Smoluchowski classical theory is used for evaluation of electrokinetic phenomena, the following limitations must be noted. (a) The theory is based on the electric double-layer thickness of colloid chemistry and holds quite well for pores that are not too small and for walls whose electric-double-layer is not too diffuse; and (b) The theory can be applied only to a narrow range of soil types as pointed out by L. Casagrande (1952) such as undisturbed or remolded silty clays. It cannot be applied to well mixed clay soil commonly used for the landfill clay liners in controlling hazardous and toxic soils or with animal and vegetable origin, such as the diverse material found in landfill areas (Winterkorn, 1958).

6.12.3 Electric-double-layer and Zeta potentials

1 *Electric-double-layer*: For an idealized soil particle and its associated ions, if an external difference in potential is applied, it can be observed that soil and a thin layer of attracted cations and water molecules will move with the clay particles toward the anode, while the diffuse system of counterions and the water associated with it will move toward the cathode. This system and the phenomenon are referred to as the *diffuse-double-layer* or called *diffuse-electric-double layer*. In simpler terms, the double-layer consists of the clay particles, adsorbed cations, and water molecules in one layer, while the other layer is the diffuse swarm of counterions. The theory explaining this system is called *diffuse-double-layer theory* (Gouy, 1910; Chapman, 1913), or sometimes referred to as *Gouy–Chapman double-layer theory*.

Based on the Gouy–Chapman model, the effect of pore fluid properties upon the double-layer thickness has been reviewed (Evans, 1991). Further, it has been shown that a reduction in thickness reduces the inter-particle repulsion forces and thus increases the tendency for a flocculated or aggregated soil structure. The equation for the double-layer thickness as predicted by the Gouy–Chapman theory is given as presented by Van Olphen (1977):

$$t = \sqrt{\frac{\epsilon K T}{8\pi n e^2 v^2}} \tag{6.20}$$

where t = double-layer thickness, ϵ = dielectric constant, K = Boltzmann's constant, T = temperature, n = electrolyte concentration, e = elementary charge, and v = valence of cations in pore fluid.

2 *Zeta potential*: Zeta potential (ζ) is the electric potential developed at solid–liquid interface in response to movement of colloidal particles. Under the influence of an applied potential, the particle and a fixed film containing the ions between the particle surface and the boundary line move toward the negative electrode. The thickness of the double-layer affects the magnitude of the zeta potential, ζ. This potential is also influenced by ion exchange capacity and size of ion radius as cited by Fang (1997).

6.12.4 Electrokinetic dewatering and decontamination processes

1 *Dewatering*: The first practical application of electrokinetic dewatering was made in 1939 on a long railroad cut in Salzgitter, Germany (L. Casagrande, 1952).

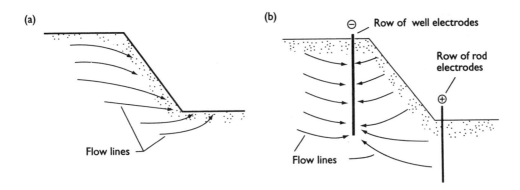

Figure 6.18 Comparison of flow characteristics. (a) Hydraulic flow; and (b) Electrokinetic flow.

Source: Casagrande, L., Electro-osmotic stabilization of soil, *Journal of the Boston Society of Civil Engineers*, v. 39, Jan. pp. 51–83. © 1952 ASCE. Reproduced by permission of the American Society of Civil Engineers.

A test well had been drilled and it was found that because of the fine texture of silty clay, traditional pumping wells (Sec. 5.7.4) would not be very effective in stabilizing the slopes. Comparison between hydraulic flow and electroosmotic flow for the flow of water toward the face of a bank is shown in Figure 6.18. Figure 6.18 shows how, through careful installation of electrodes, the flow of water toward the face of the bank or the bottom of an excavation can be reversed. The economical use of electroosmosis is confined to a narrow range of fine-grained soils, where other methods of drainage or consolidation would be too expensive. Figure 6.19 shows the comparison of mechanisms between dewatering and decontamination by electrokinetic process.

2 Soil decontamination process: As illustrated in Figure 6.19(b), the mechanism between dewatering and decontamination by electrokinetic process is different. To remove unwanted contaminated water or pollutants trapped between soil particles, the spaces between soil particles must remain open to flow. In many cases, due to swelling, shrinkage air bubbles, soil chemistry, soil bacteria, and other soil-environmental interaction, the characteristics between soil particles are unpredictable. A recharge of a non-polluted water system as shown in Figure 6.19(b) should be installed during the electrokinetic soil decontamination process.

3 Electrochemical process: The principal of the electrical-chemical process (L. Casagrande, 1952; Karpoff, 1953) is similar to the electrokinetics process. During the process, an introduction of new ions to the soil for the positive electrodes (aluminum) or an introduction of new ions such as Al^{3+}, Ca^{2+}, Mg^{2+}, etc. to the soil through perforated iron pipe anodes is made. This method of ground improvement produces ionization of the electrolytes in the porewater system. Electrochemical treatment of the saturated unstable soils brings more improvement in the physical properties than that of electroosmotic alone. The principal changes effected by this treatment are (a) decrease in moisture content, (b) decrease in hydraulic conductivity, (c) increase in density, (d) increase in cohesion and shear strength, and (e) increase in bearing capacity.

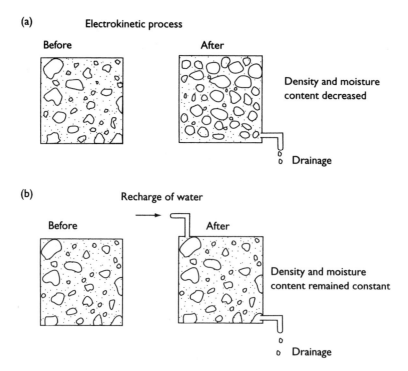

Figure 6.19 Mechanisms of dewatering and decontamination by electrolytic process. (a) mechanism of dewatering process, and (b) mechanism of soil decontamination process.

Source: Fang (1997).

6.13 Thermo-electromagnetic phenomena

6.13.1 Electroviscous effect

When water flows under a hydraulic gradient into soil voids between negatively charged mineral surfaces, the exchangeable cations are swept downstream and a streaming potential develops. This is a phenomenon for which Elton (1948) postulated that the streaming potential exerts an electrical retarding force on the exchangeable cations. This retarding force is transmitted to the surrounding water by viscous drag and supplements the normal viscous drag and retarding force. This phenomenon is the electroosmotic flow in the reverse direction. The result is equal to the hydrodynamic flow minus the counter-electroosmotic flow.

6.13.2 Thermoelectric effect

As discussed in Chapter 1, the thermoelectric effect was discovered by J. T. Seebeck in 1822. His discovery of a novel method for the direct transfer of heat into electric energy is now known as the Seebeck or thermoelectric effect. The device for measuring such effects is called a *thermocouple*. In early 1940, Winterkorn and Associates (Winterkorn, 1942) conducted simple laboratory experiments on clays as discussed in Section 6.3.3. The phenomena of thermoosmosis and thermoelectric effect are very closely related, and it is difficult to separate them in a soil–water system. This statement was made by Winterkorn (1958) as

> almost immediately upon the application of a hot face at one end of a soil specimen and of a cold face on the other, long before the temperature gradient within the specimen has become uniform, an electrical potential gradient of quite uniform character is established in the soil system.

6.13.3 Electromagnetic effect

As discussed in previous sections, thermoelectric energies are closely related in the natural soil–water system. Their distribution in soil is in a random pattern due to complex soil–water interactions in the environment. The ion movement in the soil–water system is also in a random motion. This is due to the bombardment of the dispersed particles by molecules of the medium and is called *Brownian movement*. When additional electric current is applied into the soil–water system, the particles remain in random motion, but the energy field boundary will change. When two or more moving electric charges interact in the soil–water system, then the thermoelectric energies change into thermoelectric-magnetic energies.

Magnetism produced by electricity is called *electromagnetism*. In 1989, a new phenomenon has been observed that the fluid flowing through a polluted, unsaturated fine-grained soil can be controlled by thermal-electromagnetic process (Fang, 1997). A laboratory measuring apparatus has been developed to demonstrate such behavior. A comparison between existing methods and this proposed process for soil decontamination indicates that this new process has significantly greater electrical capacity.

6.14 Summary

The transfer of thermal and electrical energy in soil has many implications in geotechnical engineering. The effects of freezing and thawing of foundation soils or pavement systems can be predicted through an assessment of thermal properties. In general, temperature changes will influence soil parameters including consistency, conductivity, compressibility, and shear modulus. Electrical phenomena also have bearing on soil behavior. In particular, changes in electrical resistivity can be measured and correlated with soil types and properties. An electric potential may be used to move water and ions through soil, as may be required in dewatering for excavations or decontamination.

PROBLEMS
6.1 Discuss the heat sources and heat characteristics. Define the difference between temperature and heat.
6.2 What is the thermal storage capacity? Can clay store more heat than sand, why?
6.3 Convert the thermal conductivity of soil from cal/cm-s-°C into BTU/in-hr-°F.
6.4 Solve Example 6.3 if the thermal conductivity was 2.7 W/m/°C.

6.5   Why is the dielectric constant an indication of a polluted soil–water system? Explain.

6.6   What are the mechanism(s) of dewatering and decontamination by electro-kinetic process.

6.7   Why is the geometry of a given electrode/electrode spacing important for measuring electrokinetic phenomena in a soil–water system? Why would electrodes made of metal not be suitable for measuring electrokinetic phenomena in contaminated soil–water system?

Chapter 7

Soil compaction (densification)

7.1 Introduction

Compaction or densification is a simple and low-cost mechanical process for ground improvement. The main purpose of this process is to change ground soil voids by means of mechanical force in order to reduce the hydraulic conductivity (Sec. 5.4), settlement (Sec. 9.7), and to increase the shear resistance (Sec. 10.5) and bearing capacity (Sec. 12.5). The engineering concept of this process was proposed by Proctor in 1933 to estimate field compaction efficiency.

The *field compaction efficiency* (F_E) is defined as the ratio of field and laboratory compaction performances by using the following relationship:

$$F_E = \frac{F_p}{L_p} \tag{7.1}$$

where F_E = field compaction efficiency, F_p = field compaction performance, and L_p = laboratory compaction performance. The value of compaction performance stated in Equation (7.1) can be measured by various types of tests commonly used in geotechnical engineering including, determination of compacted unit weight, moisture content, shear strength, etc. of soil. However, the measurement of unit weight in both laboratory and field is the most simple and effective approach, generally accepted and used method. The procedures for measuring the field unit weight of soil include sand-cone method (ASTM D1556), nuclear method (ASTM D2166), rubber-balloon method (ASTM D2167), drive-cylinder method (ASTM D2937), and others. Details of both field and laboratory compaction performance with emphasis on mechanisms, processes, factors affecting test results, and data interpretations will be discussed in the following sections.

7.2 Unit weight and moisture content relationship

7.2.1 General discussion

The simplest way to illustrate the laboratory compaction performance and process is through the unit weight (density)-moisture content relationship of soil from compaction tests in the laboratory. The laboratory compaction test procedures have

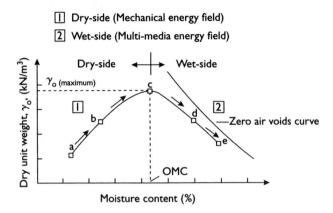

Figure 7.1 Typical dry unit weight versus moisture content curve for a silty clay by the laboratory compaction.

been standardized by both the American Society of Testing and Materials (ASTM) and American Association of State Highway and Transportation Officials (AASHTO). There are two major types of test procedures: the Standard (ASTM D698; AASHTO T991) and Modified (ASTM D1557; AASHTO T180). The major difference between these two methods is that the modified method gives a higher compactive effort (CE) than the standard procedure. A typical relationship between dry unit weight versus moisture content for silty clay in the laboratory test is shown in Figure 7.1.

7.2.2 Dry unit weight and optimum moisture content

The dry unit weight (dry density) can be computed from the wet unit weight (wet density) and the corresponding moisture content. The wet unit weight is defined as indicated in Equation (7.2) and dry unit weight can be computed from the wet unit weight as indicated in Equation (7.3). The point, c, shown in Figure 7.1 is the peak value of dry unit weight is called the *maximum dry unit weight* (maximum dry density) and the corresponding moisture content is called the *optimum moisture content* (OMC). The moisture content below the OMC is referred to as *dry-side* and above the OMC is referred to as *wet-side*. The mechanisms of dry-side and wet-side will be further discussed in Section 7.3 under the title of compaction theories and mechanisms. The right side of Figure 7.1 shows the *zero-air-voids curve* (ZAV curve), which will be explained in Section 7.2.3.

$$\gamma_m = \frac{W}{V} \tag{7.2}$$

where γ_m = wet unit weight of soil, W = net weight of wet soil in container (compaction mold), and V = volume of container. Moisture content samples are

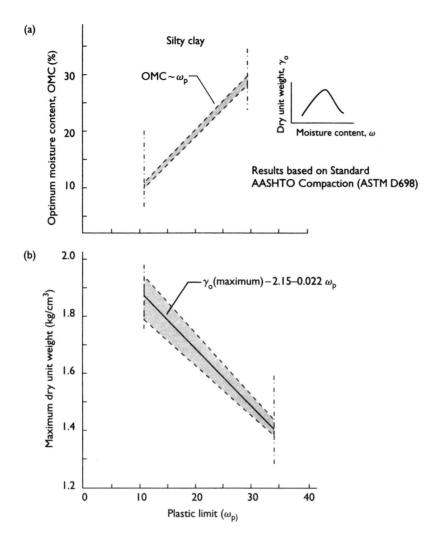

Figure 7.2 Estimation of the optimum moisture content (OMC) and the maximum dry unit weight γ_d of fine grained soil from plastic limit ω_p. (a) optimum moisture content versus plastic limit, and (b) maximum dry unit weight versus plastic limit.

taken from compacted soil specimens and the dry unit weight of soil can be computed from Equation (7.3) as,

$$\gamma_o = \frac{\gamma_m}{1 + \omega} \tag{7.3}$$

where γ_o = dry unit weight of soil, γ_m = wet unit weight of soil, and ω = moisture content of soil (%). As reported by Shook and Fang (1961), several types of unit weight–moisture relationship curves of soils are obtained depending on soil types. Figure 7.2(a) is typical type of silty clay commonly used for highway embankment construction. For this type of soil and the indicated range, the OMC and plastic limit are closely related as:

$$OMC \cong \omega_P \tag{7.4}$$

Figure 7.2(b) shows the relationship of maximum dry unit weight versus plastic limit. The approximate relationship for silty clay is:

$$\gamma_o(\text{maximum}) = 2.15 - 0.021\,\omega_P \tag{7.5}$$

Where OMC = optimum moisture content (%), ω_P = plastic limit, and γ_o (maximum) = maximum dry unit weight kg/cm³.

EXAMPLE 7.1
Prove the relationship between wet and dry unit weights of soil as indicated in Equation (7.3).

SOLUTION
By definition, $\gamma_o = W_s/V$ and $\gamma_m = W/V$
Using a phase diagram (e.g. Fig. 3.2), $W = W_s(1 + \omega)$
Then $W/V = W_s(1 + \omega)/V = (1 + \omega)$
and $\gamma_o = \gamma_m /1 + \omega$

EXAMPLE 7.2
In a freshly compacted highway fill, a sand-volume apparatus has determined the volume of an excavation to be 0.50 cu. ft. (0.01415 m³). If the weight of material excavated is found to be 58.0 lbs (258.1 N) with a moisture content of 10.4%, what is the dry unit weight of the soil (γ_o), as compacted?

SOLUTION
Given: V = 0.5 cu. ft, W = 58.0 lbs, ω = 10.4%

From Equation (7.1), $\gamma_0 = \dfrac{W}{V(1 + \omega)} = \dfrac{58.0}{0.5(1 + 0.104)} = 105$ pcf (10.7 kN/m³).

7.2.3 Zero-air-void curve

The ZAV curve is defined as the weight of solids per unit volume of a saturated soil mass. Figure 7.1 shows the ZAVs unit weight as a function of water content. The ZAV curve represents an upper bound for compaction, that is, a density greater than given by Equation (7.6) cannot be achieved.

$$\gamma_d = \frac{G_s}{1 + (\omega G_s/S)} \tag{7.6}$$

where γ_d = dry unit weight of soil, G_s = specific gravity of solid, γ_ω = unit weight of water, and S = degree of saturation.

7.2.4 Compactive effort

Compactive effort (CE) is the ratio between compactive energy to the volume of soil. Compactive energy is mechanical energy produced during the compaction test. If the laboratory standard compaction procedure (ASTM D698) is used, then the CE value can be computed by the following relationship:

$$CE = \frac{(\text{Wt. of hammer}) (\text{Ht. of drop}) (\text{No. of layer}) (\text{Blow/layer})}{\text{Volume of compaction mold}} \quad (7.7)$$

Table 7.1 presents the laboratory compaction test and CE, based on Equation (7.7).

EXAMPLE 7.3 illustrates such computations.

EXAMPLE 7.3
Compute the compactive effort (CE) for a Standard Proctor Compaction Test (ASTM D698).

SOLUTION
From Equation (7.3) and based on data from Table 7.1, the compactive effort (CE) for a standard Proctor compaction test can be obtained as:

$$CE = \frac{(24.5 \text{ N})(0.305 \text{ m})(3 \text{ layers})(25 \text{ blow/layer})}{(0.000942 \text{ m}^3) (1000)} = 594.8 \text{ kJ/m}^3$$

In SI units, the compactive energy equals to kilo-joules per cubic meter (kJ/m^3). In fps unit, 1 ft lb/ft^3 = 0.04796 kJ/m^3, then

$$CE = 594.8 / 0.04796 = 12{,}400 \text{ ft lb/ft}^3$$

The compactive effort of compaction in the field, is often related to the number of passes of rolling equipment (e.g. a steel drum or sheepsfoot roller). Figure 7.3

Table 7.1 Standard and modified laboratory compaction test procedure and their corresponding compactive efforts

Variables	Standard compaction (ASTM D698; AASHTO T99)	Modified compaction (ASTM D1557; AASHTO T188)
Mold size	4 in. (10.16 cm)	4 in. (10.16 cm) or 6 in. (15.24 cm)
Volume of mold	1/30 ft³ (0.000942 m³)	1/30 or 1/13.33 ft³ (0.000942 or 0.00212 m²)
Hammer weight	5 lb (24.5 N)	10 lb (44.5 N)
Height of drop	12 in. (0.305 m)	18 in. (45.7 cm)
Layers	3	5
No. of blows	25	55
Soil	> #4 Sieve	> #4 Sieve
Compactive effort	595 kJ/m³ (12,400 lb ft/ft³)	2698 kJ/m³ (55,250 lb ft/ft³)

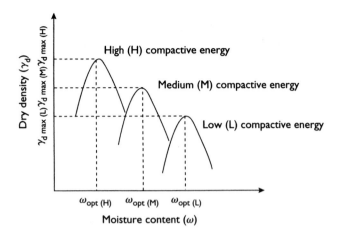

Figure 7.3 Maximum dry unit weight (γ_{dmax}) and optimum moisture content (ω_{opt}) versus compactive energy.

presents the interrelationship of dry unit weight, optimum moisture content, and compactive effort.

7.3   Soil compaction theories and mechanisms

There are several concepts and theories to explain the compaction process of soil and unit weight–moisture content relationships. There are two general types of approaches: (a) to explain the relationship between unit weight versus moisture content of soil, as illustrated in Figure 7.1; and (b) to explain the mechanism of compaction processes.

7.3.1   Soil compaction theories

1   Proctor theory: proctor (1933) assumes that the soil mass is composed of gravel, sands, silts, and clays and that compaction is the act of forcing finer grains into the voids between larger grains. The mechanism of the Proctor theory is explained as follows. (a) It is contended that the water coats the surface of the soil grains and serves as a lubricant which reduces the frictional resistance between the soil particles and permits the compacting force to become more efficient in arranging the fine soils into the voids between the larger particles and in bringing particles generally closer together; and (b) If the moisture content is not sufficient to produce adequate lubrication, the unit weight of the compacted soil will be relatively low because the compacting force is not enough to overcome the frictional resistance between the soil grains.

  2   Hogentogler theory: Hogentogler (1937b) proposed that compacted soils undergo stages of wetting. When the moisture contents are expressed as percentages of the combined volumes of soil solids and moisture, the moisture–unit weight relation becomes a series of straight lines with different slopes. From Figure 7.1, four stages are proposed as: (a) hydration (Points **a** to **b**), (b) lubrication (Points **b** to **c**), (c) swell (Points **c** to **d**), and (d) saturation (Points **d** to **e**).

3   *Lambe theory*: Lambe (1958) used a physicochemical concept to explain the unit weight (density)-moisture content relationship of soil (particularly fine-grained soils). (a) Low density, as shown in point, **a**, of Figure 7.1, is due to insufficient water for the diffuse-double-layer, (Sec. 6.8) which gives a higher concentration of electrolytes and reduces the inter-particle repulsion causing a tendency toward flocculation of the colloids. The flocculation structures (Fig. 3.9) means a low-degree of particle orientation and low density; (b) Increase the moisture content from point **a** to point **b** in Figure 7.1, the double-layer expands, electrolyte concentration reduces and permits a more orderly arrangement of particles which gives higher density; (c) Further increase of moisture content from point **b** to point c indicates further expansion of the double-layer, further reducing the electrolyte concentration and continued reduction in the net attractive forces between particles; and (d) For the higher compactive effort, which gives greater input of work, the more parallel clay particles become.

7.3.2   Mechanism of compaction process based on law of conservation of energy

The *Law of Conservation of Energy* is applied to explain the mechanism of a compaction process as proposed by Li (1956), which is based on the following assumptions: (a) the soil mass in its loose state is structurally homogeneous; and (b) the compaction process is defined as the process of densification of soil by dynamic load application, causing a decrease in air voids due to change in relative position of soil grains. The law of Conservation of Energy is employed to interpret how energy is being spent during the compaction process. There are three basic steps in the compaction process as: work done, energy change, and heat transfer.

1   *Work done*: Work is defined as the product of force and the distance through which the force acts. There are two general types of loading. The two types are static and dynamic loading (Falling hammer) as:

a   *Static loading*: In static loading, the work done on the soil is:

$$W = p \, ds \tag{7.8}$$

b   *Dynamic loading*: In dynamic loading, the work done is:

$$W = \tfrac{1}{2} M \, (v_1 - v_2)^2 \tag{7.9}$$

where W = work done, p = load, s = settlement, M = mass of the applied load, v_1 = initial velocity of the applied load, and v_2 = final velocity of the applied load. In laboratory compaction, the product of the hammer weight and the distance through which the hammer drops is the amount of work done, and is also the energy transferred to the soil mass.

2   *Change in total energy*: In soils containing particles of clay size, the increase in density by compaction changes the total internal energy of the soil. It is a process involving reorientation of soil particles which processes forces of repulsion or attraction due to adsorbed ions and adsorbed water molecules. The physical changes during

compaction, such as the compression of gases in the voids, the increase of the amount of dissolved gases in porewater, the elastic strain of solid particles, and the other complex colloidal phenomena are the results of change in the amount of internal energy.

3 *Heat transfer from the compaction process*: In the process of compaction, there must be relative motion of soil particles of all sizes. Force is required to overcome the frictional resistance developed between particles during the motion. The energy to overcome the frictional resistance may be converted into heat. There are three types of frictional resistance that characterize the relative motion between soil particles during compaction, namely (a) Dry surface friction, (b) Hydrodynamic surface friction, and (c) Boundary surface friction. The significance of the law of Conservation of Energy upon compaction of soil lies in understanding how the energy is being spent during the compaction process and how it may be consumed in many other ways than the densification of soil.

7.3.3 Soil compaction mechanism explained by the particle-energy-field theory

The particle-energy-field theory has been applied to explain the unit weight–moisture relationship of compacted clay by Fang (1997). From Figure 7.1, there are two distinct characteristics of soil existing in the moisture-unit weight relationship, the dry-side and the wet-side. Since the behavior of soil in these two stages is different, particle-energy-field theory discussed in Section 1.6 may help to clarify moisture-unit weight relationship as follows:

1 *Dry-side*: On the dry-side (Points **a** to **b**, in Fig. 7.1), the soil particle arrangement is controlled by mechanical energy as described by the Proctor theory in Section 7.3.2, forcing the smaller soil particles into the voids between the larger particles by compaction effort. There is no major physicochemical interaction between soil–water interactions on the dry-side condition.
2 *Wet-side*: On the wet-side (Points **b** to **c**, in Fig. 7.1), the soil–water interaction between particles will be influenced by physicochemical behavior of soil as described by the Lambe theory in Section 7.3.3.
3 *Contaminated soil*: In the case of contaminated soil, there are physicochemical–biological effects on the soil–water system as discussed in Sections 3.0 and 4.0. In both cases (1) and (2), it is evident that both wet- and dry-sides are influenced by multimedia energy fields such as thermal, electrical, and electromagnetic energies.

7.4 Characteristics of compacted soil

7.4.1 Compressibility and conductivity of compacted fine-grained soil

The general engineering properties of compacted fine-grained cohesive soils will depend greatly on the method of compaction, the compaction effort (Sec. 7.2.4), and the water content at compaction. Previous experience suggests (a) that swelling is greater and shrinkage (Sec. 4.4) is less for clay compacted on the dry-side of optimum (Fig. 7.1); (b) Compressibility (Chapter 9) of compacted clays are a function of the method of compaction and conductivity characteristics of compacted soil (Chapter 5).

7.4.2 Strength characteristics of compacted fine-grained soil

Strength characteristics of compacted fine-grained soil are commonly determined by laboratory unconfined compression, direct shear, triaxial shear, and tensile tests. The general tensile and fracture characteristics of compacted fine-grained soil have been discussed in Chapter 8. The particular influence of moisture content at compaction on the strength of an embankment soil is given in Figures 7.4 and 7.5. The soil was a C-horizon material available on the AASHO Road Test site. The method used in

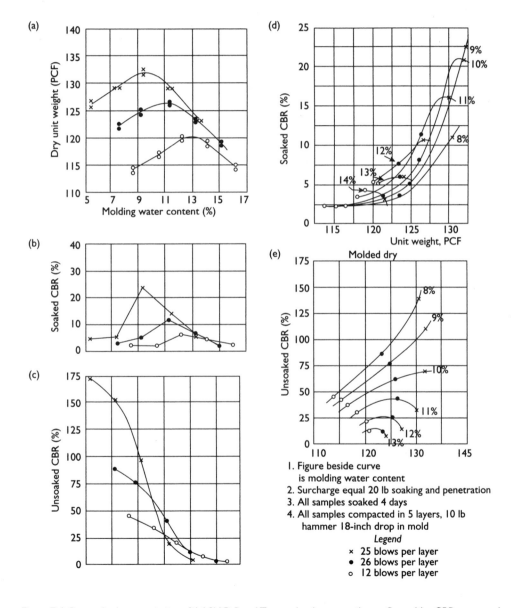

Figure 7.4 Strength characteristics of AASHO Road Test embankment soil as reflected by CBR test results.

Source: Shook, J. F. and Fang, H. Y., "Cooperative materials testing program at the AASHO Road Test." In *Highway Research Board Special Report 66*, Highway Research Board. National Research Council, Washington DC, 1961, 173 p. Reproduced with permission of the Transportation Research Board.

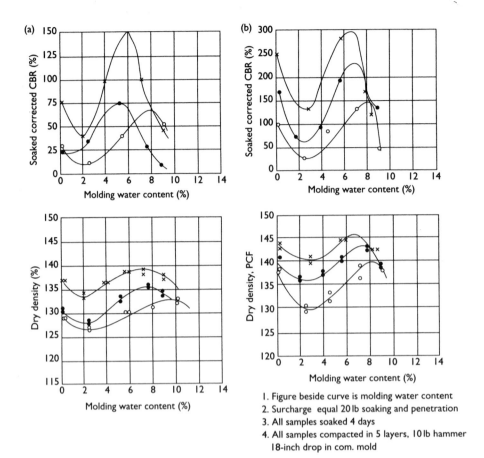

Figure 7.5 Strength characteristics of AASHO Road Test base and subbase materials as reflected by CBR test results. (a) Gravel subbase material and (b) Crushed stone base material.

Source: Shook, J. F. and Fang, H. Y., "Cooperative materials testing program at the AASHO Road Test" In *Highway Research Board Special Report 66*, Highway Research Board. National Research Council, Washington DC, 1961, 173 p. Reproduced with permission of the Transportation Research Board.

preparing the embankment soil samples for testing and the results of general geotechnical properties are discussed by Shook and Fang (1961). Figure 7.4 shows data for a complete set of California Bearing Ratio (CBR) curves for different molding densities, water contents, and number of blows of the drop hammer. Both soaked and unsoaked CBR values are included. Detailed test procedures for the CBR test are presented in Section 12.9.1.

In addition to a completed CBR test, other strength measurements on the AASHO Road Test silty clay include direct shear, triaxial shear, and unconfined compression will be discussed in Section 10.5.

7.4.3 Characteristics of compacted coarse-grained soil

California bearing ratio (CBR) test: The subbase and base materials used on the AASHO Road Test are used to illustrate the strength characteristics of coarse-grained soil. The subbase material was a natural sand-gravel material modified by washing and the addition of fine silica sand in the minus #40 US standard sieve range and a small amount of binder soil. Results of CBR test data for both subbase and base are presented in Figures 7.5(a) and 7.5(b). The behavior is similar to that observed for the embankment soil. Laboratory shear tests on compacted coarse-grained soil will be discussed in Chapter 10.

7.5 Factors affecting compacted soil

7.5.1 General discussion

Factors influencing compaction test results have been discussed by numerous investigators (Johnson and Sallberg, 1962; Hilf, 1991). In addition to the type of soil, the method of compaction and nature of the pore fluid may significantly influence compaction results. Because the viscosity of pore fluid (water) and the degree and intensity of the interaction of mineral surfaces with water are a function of temperature, the unit weight–moisture relationships are also temperature dependent. In general, the temperature effect increases with increasing specific surface of soil particles, and also is a function of the clay mineral and exchangeable ions and the electrolytes in the aqueous phase. Factors affecting compacted soils may be categorized according to whether particles are fine or coarse: (a) Fine-grained soils are sensitive to grain size distribution, maximum particle size, curing duration, recompaction, wetting and rewetting process, temperature; while (b) Coarse-grained soils are influenced most by, grain size distribution and maximum particle size as indicated in Figure 7.7.

7.5.2 Curing duration and compaction process

1   *Curing duration*: The curing period affects the shape of unit weight versus the moisture relationship. To achieve good compaction in highly cohesive soils, moisture must be evenly distributed, which takes time. Also, the direction of moisture transfer (i.e. from wet to dry or dry to wet) on standard laboratory compaction test results has been reported by Tamez (1957). Specifically, the flow path from the dry-side and wet-side are in different energy fields. From wet to dry, the soil-water is in the thermal energy field, however, from dry to wet, the process is in the multimedia energy field as explained in Sections 1.6 and 6.2.

2   *Recompaction and moisture distribution*: Compaction procedures significantly affect maximum unit weight and optimum moisture content. Figure 7.6 illustrates the effects of compaction test procedures (i.e. compaction, recompaction, wetting/drying) for a given soil.

3   *Amount of gravel content*: *The Bureau of Reclamation Earth Manual* (1973) provides maximum and minimum densities of typical sand–gravel soils. Figure 7.7 shows the influence of gravel content on the dry unit weight of soil. Two types of compaction energy were used to achieve maximum dry density, namely, use of a shaking table and drop hammer.

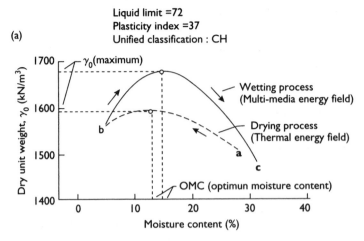

Liquid limit = 72
Plasticity index = 37
Unified classification : CH

(a)

Figure 7.6 Effect of moisture distribution and recompaction on unit weight versus moisture content relationships: (a) Effect of moisture distribution and (b) Effect of recompaction.

Source: Tamez (1957). Copyright ASTM INTERNATIONAL. Reprinted with permission.

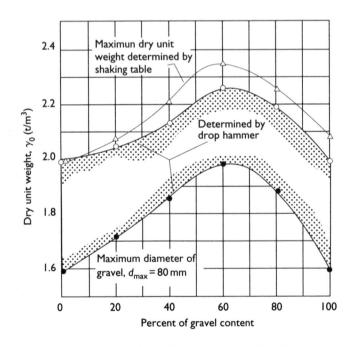

Figure 7.7 Percent of gravel content on unit weight–moisture relationship.

Source: Shi (1981). Used with permission of Taylor & Francis.

7.5.3 Temperature effects on compaction results

1   *Temperature above freezing*: The temperature effect on soil is quite straight forward in relatively pure clay–water systems within a limited range of temperature as discussed in Section 6.5.

2   *Temperature below freezing*: Figure 7.8 shows the effect of freezing temperatures on compaction of fine sand with a trace of silt. In examining Figure 7.8, both standard and modified AASHTO compaction efforts are used. Significant effects

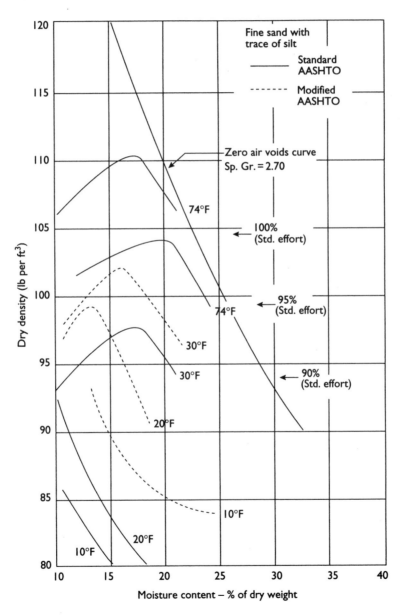

Figure 7.8 Effect of temperature (below 32°F) on unit weight–moisture content relationship.

Source: Fang 1997, Used with permission of CRC Press/Taylor & Francis.

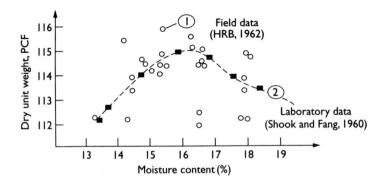

Figure 7.9 Sensitivity of weathering to standard compaction test results on AASHO Road Test silty clay: (a) Field data and (b) Laboratory data.

Source: HRB, "The AASHO Road Test Report 2, Materials and Construction." In *Highway Research Board Special Report* 61B, Highway Research Board. National Research Council, Washington DC, 1962, 173p. Reproduced with permission of the Transportation Research Board.

are observed when the temperature goes below freezing. This effect is mainly due to ice-lense formation. Because freezing affects the unit weight–moisture content relationship, a careful examination of in situ ice content should be made when compacting frozen or partially frozen soils. Alternatively, field compaction during the frozen condition should be avoided.

3   *Weathering effect*: The sensitivity of local weathering to standard compaction tests on AASHO Road Test silty clay is presented in Figure 7.9(a) and (b). In examining Figure 7.9(a), the field unit weight versus moisture content in field conditions exhibits large variations, however, for the laboratory compaction test (Fig. 7.9(b)), a typical relationship between unit weight versus moisture content similar to Figure 7.1 was found.

7.5.4   Effects of exchangeable ions and pollution substances

1   *Effects of exchangeable ions*: The effects of exchangeable ions on the characteristics of compacted Putnam soil (silt loam) as reflected by penetration cone results are presented in Figure 7.10. The process of cone penetration will be discussed further in Section 10.10.2. Encircled points represent data at OMC. The experimental data indicated that when potassium, K, was the exchangeable ion, a higher maximum dry unit weight and lower optimum moisture content was achieved.

2   *Effects on pore fluid*: Table 7.2 presents laboratory compaction test results for three clay minerals with water, H_2O and dimethyl sulfoxide $(CH_3)_2SO$, (DMSO) (see Table 4.5).

Figure 7.11 presents compaction test results on clay liner material with water and leachate from Central Pennsylvania (Fang and Evans, 1988). In examining Figure 7.11, the maximum dry unit weight and optimum moisture content for both water and leachate indicates that, for the particular soil and leachate tested, no significant differences are found.

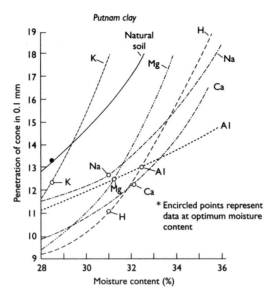

Figure 7.10 Effect of exchangeable ions on optimum moisture content.

Source: Winterkorn, H. F. and Moorman, R. B. B., "A study of changes in physical properties of Putnam soil induced by ionic substitution." In *Proceeding of the 21st Annual Meeting, Highway Research Board*, Highway Research Board. National Research Council, Washington DC, 1941, pp. 415–434. Reproduced with permission of the Transportation Research Board.

Table 7.2 Effect of pore fluid on compaction test results

Clay mineral	Liquid type	Maximum dry unit weight pcf (kN/m³)	Optimum liquid Content (OLC) (%)
Attapulgite[a]	H_2O	39 (6.1)	49
	DMSO	36 (5.7)	36
Kaolinite	H_2O	85 (13.4)	28
	DMSO	71 (11.2)	40
Na-Bentonite	H_2O	65 (10.2)	50
	DMSO	48 (7.5)	76

Source: Andrews, R. E., Gawarkiewicz, J. J. and Winterkorn, H. F., "Comparison of the interaction of three clay minerals with water, dimethyl sulfoxide and dimethyl formamide." In *Highway Research Record* No. 209, Highway Research Board. National Research Council, Washington DC, 1967, pp. 66–78. Reproduced with permission of the Transportation Research Board.

Notes
H_2O – water; DMSO – Dimethyl sulfoxide $(CH_3)_2SO$.
a See Section 4.5.2.

7.6 Field compaction

7.6.1 General discussion

Field compaction (in situ), in general can be grouped into two categories (a) Shallow surface compaction, and (b) Deep compaction. It also can be divided into the

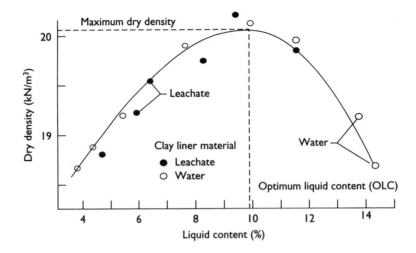

Figure 7.11 Effect of pore fluid on unit weight–moisture content relationship.

Source: Fang and Evans (1988). Copyright ASTM INTERNATIONAL. Reprinted with permission.

compaction of cohesive soils (clay) and cohesionless soils (granular materials). During the compaction process on clay and granular soils different types of equipment and procedures are used. For deep compaction and special cases, the following equipment is used: vibrator beam, dynamic consolidation, blasting, and electro-potential process and will be discussed in Section 7.9.

There are various types of compaction equipment available. Pneumatic-tired rollers, as a type, are suited to compacting any type of soil, provided the values of contact pressure and wheel load are sufficient for the soil being compacted. In compaction operations on shallow depths such as highway embankment, two principles of action are involved:

1   First, it is necessary to place the soil in layers sufficiently thin to permit air and water to be expelled efficiently and easily. Clay-like soil must be placed in thin layers, whereas a sandy soil could be rolled in thick layers;

2   Second, the compression of soil particles requires movement of the individual particles in order to fit them together and fill in the voids as indicated by the Proctor theory in Section 7.3. Before movement can take place, friction must be reduced. Lubrication of the soil particles by means of moisture will help to overcome friction. Too little moisture will not materially reduce friction; too much moisture reduces the effective density and means that the excess porewater must be expelled. There is, then, an optimum or ideal moisture content.

7.6.2   Field compaction on soils at shallow depth

1   *Field compaction on cohesive soil:* Relevant shallow field compaction equipment includes rubber-tired equipment and sheepsfoot rollers. The term sheepsfoot refers to the protrusions that extend out from an otherwise smooth drum roller. These protruding elements serve to knead the cohesive material into a denser state.

2 *Field compaction on coarse-grained soil*: Compaction for coarse-grained soils at shallow depths may involve different equipment. Field compaction equipment includes crawler-type tractor, rubber-tired equipment and steel drum rollers. Unlike fine-grained soils, vibration initiated in the wheels of compaction equipment is often particularly useful in coarse-grained soils. Compaction at greater depths will be discussed in Section 7.8.

7.7 Field compaction controlling methods

7.7.1 General discussion

The performance of field compaction at shallow depth in cohesive soil is often gauged in terms of percent compaction (relative compaction), and moisture content while cohesionless soils are assessed in terms of relative density, compactibility, and in situ moisture content. For deep depths, the controlling methods includes relative density, in situ cone penetration test, and standard penetration test (SPT). Detailed discussions of each method are presented as follows:

7.7.2 Compaction controlling parameters

1 *Percent of compaction* (Relative Compaction): *Percent of compaction* (PC), or *degree of compaction* or *relative compaction* is defined as the ratio of field dry unit weight and maximum dry unit weight determined from laboratory tests as discussed in Section 7.2.

$$PC = \frac{\gamma_o(\text{field})}{\gamma_o(\text{lab})} \times 100 \qquad (7.10)$$

where PC = percent of compaction (%), $\gamma_o(\text{field})$ = field dry unit weight, and $\gamma_o(\text{lab})$ = laboratory determined maximum dry unit weight (Fig. 7.1). Table 7.3 shows typical compaction requirements for various projects.

2 *Relative density*: Relative density (D_r) (Terzaghi, 1925) is a measure of a soil's density relative to its maximum density, expressed in terms of void ratio,

Table 7.3 Typical requirement of percentage of compaction

% Compaction	Type of project
100–105	Airports Interstate highways.
95–100	Major state highways.
85–95	State highways Major county highways
>80	Minor county highways Service roads
>70	Frontage roads, Detours

porosity dry density, permeability, grain size or other metrics. It is given as follows:

a   Relative density based on void ratio, e,

$$D_r = \frac{e_{max} - e}{e_{max} - e_{min}} \tag{7.11}$$

where D_r = relative density (%), e = void ratio of the material being tested, e_{max} = void ratio of the material in its loose state, and e_{min} = void ratio of the material in its densest state;

b   Relative density based on porosity, n,

$$D_r = \frac{(n_{max} - n)(1 - n_{min})}{n_{max} - n_{min}} \tag{7.12}$$

where n_{max} = maximum porosity (looser state), n_{min} = minimum porosity (densest state), and n = porosity of the soil being tested. Alternatively,

$$D_r = \frac{n_{max} - n_{min}}{n_{min}(1 - n_{max})} \tag{7.13}$$

Ranges of relative density for sand are as follows: *Loose sand*: $0 < D_r < 1/3$, *Medium sand*: $1/3 < D_r < 2/3$, *Dense sand*: $2/3 < D_r < 1.0$ (Hilf, 1991);

c   Relative density based on dry unit weight:

$$D_r = \frac{\gamma_{max}(\gamma - \gamma_{min})}{\gamma(\gamma_{max} - \gamma_{min})} \tag{7.14}$$

where γ_{max} = dry unit weight at dense state, γ_{min} = dry unit weight at loose state, and γ = dry unit weight at in situ condition;

d   Relative density based on permeability and grain size: Hilf (1991) notes the following relationship for relative density as

$$D_r = \frac{k_{max} - k}{k_{max} - k_{min}} \tag{7.15}$$

where k_{max} = maximum permeability, k_{min} = minimum permeability, and k = permeability at in situ condition; and

e   Relative density determination based on SPT and grain size: Marcuson and Bieganousky (1977) proposed an experimental relationship of normally consolidated sand. They stated that the relative density of sand can be estimated from SPT test resistance and grain size as reflected by the coefficient of uniformity:

$$D_r = 11.7 + 0.76\{[222\,(N) + 1600 - 53\,(p'_{vo}) - 50\,(C_u)^2]\}^{1/2} \tag{7.16}$$

where N = measured SPT resistance (Sec. 12), p'_{vo} = effective overburden pressure, and C_u = coefficient of uniformity (Sec. 3.3).

3 *Compactibility*: Compactibility (*F*) is defined as the ratio

$$F = [e_{max} - e_{min}]/e_{min} \qquad (7.17)$$

Where F = compactibility (Range of F is about 0.5 to 2.4), e_{max} = void ratio of the soil in its loosest state, and e_{min} = void ratio of the soil in its densest state. In well-graded cohesionless soils, $e_{max} - e_{min}$ is large and e_{min} is small; hence, F, is large and these soils are easily compacted.

4 *Suitability number*: Suitability number (S_N) proposed by Brown (1977) is based on the grain size distribution of the backfill material that controls the rate of compaction. It is defined as

$$S_N = 1.7 \sqrt{\frac{3}{(D_{50})^2} + \frac{1}{(D_{20})^2} + \frac{1}{(D_{10})^2}} \qquad (7.18)$$

where D_{50}, D_{20}, and D_{10} are the diameter in mm through which, respectively, 50%, 20%, and 10% of soil particles will pass (Sec. 3.3). The smaller the S_N, the more desirable the backfill material is. The following is a rating system: S_N = 0–10, Excellent; 10–20, Good; 20–30, Fair; 30–50, Poor; >50, Unsuitable.

5 *Field moisture content*: Coupled with PC, the in situ moisture content is most important. In general, ±2–3% from OMC is suggested. This expectation is always stated in a construction manual or contracted for a specified project.

7.7.3 One-point method for determining maximum dry unit weight

1 *General discussion*: The one-point method is a short-cut, time-saving procedure for determination of maximum dry unit weight during the construction period. There are various types of one-point methods available. For example, a highway embankment was constructed in block-lifts approximately 4 in. (10 cm) thick when compacted. For each construction block, it was required that each lift be tested and accepted before commencing construction of the succeeding lift. To prevent undue delay between the completion of a block-lift and the starting of the next, the testing was accelerated by the development of a one-point procedure for determining maximum dry unit weight of the embankment soil. There are several one-point short-cut methods for determination of maximum dry units available, however, the AASHO Road Test one-point method is most comprehensive and are detailed and discussed in this section.

2 *AASHO Road Test one-point method*: The AASHO Road Test one-point method (HRB, 1962a) is based on the Wyoming and Ohio State Highway Departments. It made use of the penetration resistance and a family of moisture-density curves.

a *Preparation of family curves of unit weight versus moisture content*
The family of curves used at the AASHO Road Test was developed from standard compaction tests (Table 7.1). In the portion of the family curves shown in Figure 7.12, the curves are plots of wet unit weight, moisture content, and penetration resistance reading. The optimum conditions line (heavy

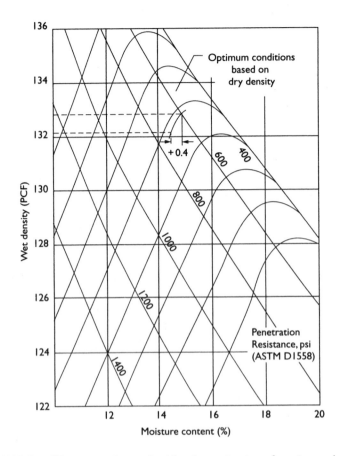

Figure 7.12 AASHO Road Test one-point method for determination of maximum dry unit weight and optimum moisture content of fine-grained soil.

Source: HRB, "The AASHO Road Test Report 2, Materials and Construction." In *Highway Research Board Special Report 61B*, Highway Research Board. National Research Council, Washington DC, 1962, 173p. Reproduced with permission of the Transportation Research Board.

line in Fig. 7.12), is based on maximum dry unit weight and coincides with the 600 psi (4134 kN/m²) penetration resistance needle reading line for that portion. However, it extends below the 600 psi line on the upper part of the complete family of curves and above the 600 psi line on the lower part.

b *Preparation of laboratory soil specimen*
The preparation of a specimen involves pulverizing a soil sample, removing the plus 1/4 in. material, and adjusting the moisture content of the sample to within a few percentage points of optimum. The soil is compacted in three layers in a 4 in. ((10.16 cm) diameter mold, using 5.5-lb (2.49 kg) hammer, and a 12-in. (30.5 cm) drop standard compaction test (ASTM D698)).

c *Computation procedures for optimum condition*
Optimum conditions include both maximum dry unit weight (γ_o) and OMC. The OMC can be obtained from the following procedures: (a) First

determine the wet unit weight, moisture content and penetration resistance for the specimen; (b) Next obtain the wet unit weight corresponding to the maximum dry unit weight; (c) Adjust the moisture content of the specimen to optimum from the family of curves (Fig. 7.7); and (d) Finally compute the corresponding maximum dry unit weight (Example 7.2).

3 *New York DOT one-point methods*: Another rapid test method for earthwork compaction control was developed by the New York State Department of Transportation (NY DOT). This method is also based on a one-point compaction test and a family of moisture-unit weight curves. Time saving is accomplished by a combination of compaction equipment and computation improvements. Through use of a special slide rule, in-place density is obtained from the volume of a hole as measured from the moisture-unit weight curves and compiled in compaction control tables. It can be compared with on-place densities without calculations or interpolations. Moisture determinations are usually not required and conversions from wet to dry density are not necessary.

7.7.4 *Estimating volume of compaction*

Estimating the volume of compaction obtainable with rolling equipment with reasonable accuracy is of importance in determining the number of units required for any particular project, as well as arriving at unit costs. An estimate can be made with the following formula proposed by Highway Research Board (HRB, 1952).

$$V_C = \frac{5280(v)}{27(P)} \frac{(W)(D)(F)(E)}{12} \tag{7.19}$$

where V_C = volume of compaction in cubic yards per hour, v = speed of towing tractor, miles per hour, P = number of passes required, W = width of roller or combination of rollers in feet, D = depth of lift in inches, F = percentage factor of pay yards to loose yards (sand 90%, common earth 80%, clay 70%, rock 50%, all expressed as a percent of loose quantity), E = job efficiency factor allowing for end turns, time losses, (Excellent – 90%, Average – 83%, and Poor – 75%). Data for F and E may be obtained from the US Government Technical Manual TM-5–9500 entitled Principles of Modern Excavation and Equipment.

Two charts have been prepared as shown in Figure 7.13(a) and (b). The charts show graphically, the maximum possible productive capacity of given sheepsfoot and pneumatic-tire rollers for various number of passes and different operating speeds when compacting a 6 in. (152 mm) compacted lift. It may be seen that the productive capacity is directly proportional to the operating speed. These charts together with Equation (7.7), may be used as a guide for estimating roller capacities of a given size and the weight of pneumatic rollers or a sheepsfoot roller with a dual-drum type with 4 in. (10 cm) drums when compacting a 6 in. (152 mm) lift. A numerical example to illustrate the computation procedures is presented in Example 7.4. It should be noted, that each project requires an individual complete analysis of field conditions including soil types, construction equipment characteristics, and labor efficiency.

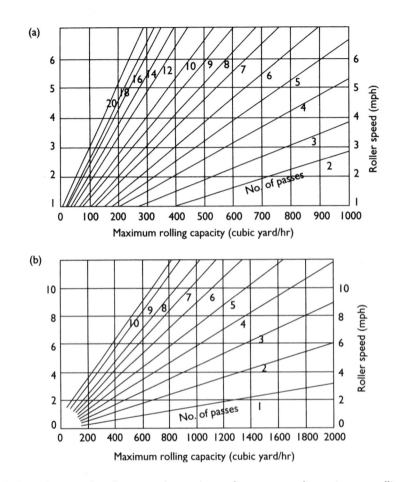

Figure 7.13 Correlation of roller speed, number of passes, and maximum rolling capacity. (a) Sheepsfoot and (b) Pneumatic-tire roller.

Source: HRB, "Construction Equipment." In *Highway Research Board Bulletin* 58, Highway Research Board. National Research Council, Washington DC, 1952, Reproduced with permission of the Transportation Research Board.

EXAMPLE 7.4 (After HRB, 1952)

Given: C-50 Multiple Box Roller and DW-21 Tractor, Speed of Towing Tractor = 4.5 mph, P = 4 passes, W = 9.83 ft. C-50 Rolling Width, D = 12 in. depth of lift, F = 80% factor for common earth, E = 83% average efficiency, Estimating volume of compaction, V_C.

SOLUTION

$$V_C = \frac{4.5 \times 5280}{4 \times 27} \times \frac{9.83 \times 12 \times 0.80 \times 0.83}{12} = 1435 \text{ yd}^3/\text{hr}(1096 \text{ m}^3/\text{hr})$$

Note: Each project requires an individual complete analysis of conditions and characteristics, and the above is for illustration only.

7.8   Field deep compaction and mass compaction

In many cases a thick layer of loose materials overlies the site where it is not practical to excavate and replace it with other material. In such case, in situ deep or mass compaction or other "soil improvement" techniques can be used. In this section, several commonly used deep compaction techniques such as vibrocompaction, dynamic compaction, and blasting are discussed as follows: (a) vibrocompaction, (b) vibrating beam, (c) dynamic consolidation, (d) Blasting (Sec. 7.9), and (e) Electric potential process (Sec. 7.10).

7.8.1   Vibrocompaction

The vibrocompaction method was developed in 1936 in Germany and introduced to the United States by Steurman in 1939. This technique uses vibration for the densification of granular soils, sandy silt and some clays. A torpedo-shaped vibration generator was specifically designed for this. There are several configurations, and in one case it consists of a cylindrical shell about 2 m (6 ft) long and 40 cm (16 in.) in diameter that weighs 1522 kg (3500 lb). An eccentric weight mounted inside the cylinder rotates at 1800 rpm to develop large, horizontal centrifugal forces. Water jets flow from the bottom to the top of the element under pressures of 60 psi (414 kN/m^2). The water jet saturates the loose material and the vibration energy densifies the material. The SPT (Sec. 12.9.3) or cone penetrometer tests (Sec. 12.9.5) have been used to determine the densification of loose sands which will give some indication of the relative density.

7.8.2   Vibrating beam

A *Vibrating beam* is used commonly for densification of slurry wall and concrete mixtures at various depths at in situ conditions. It also has been used for densification of uniform deep sand layers. This technique uses a vibratory-type pile driver to cause the penetration of a beam of specified dimensions to the design depth.

7.8.3   Dynamic consolidation (compaction)

1   General discussion: *Dynamic consolidation*, also called *dynamic densification*, is a mechanical process used to densify loose soil deposits at great depths. It basically involves dropping a large weight (20–40 tons) from heights of 30 m or more. The process used at the present time is not new. The largest construction project using this technique was during the Second World War in early 1940, when an airfield was built in Kunming, southwest China, for the US Flying-tiger B-29 bomber landings. The method is frequently used around the world, especially in China, yet little publicity has encouraged the scientific study of the process. However, in the 1970s, the Menard Group (Menard and Broise, 1975) presented a scientific approach for the analysis of the dynamic densification process in which they included vibration during the in situ consolidation process in correlation with basic geotechnical parameters and field sub-surface investigations. Since then, this method has been widely used in many large scale construction projects for densification of deep granular soils and more recently for the densification of loose sands (Ch. 16), landfills (Ch. 16), as well as some clay deposits. Fang (1997) describes a laboratory study of dynamic consolidation with three distinct types of fill materials used: clay, silt (coal combustion fly ash), and sand at two moisture conditions, dry and saturated. For their work, both the Westergaard and Bousinesq equations (Ch. 9) were successfully used to predict the vertical stress as a function of depth. In general, when comparing theoretical predictions with experimental data, the Westergaard solution under-predicts the stress imposed by dynamic compaction while the Bousinesq solution predicts slightly larger than observed values. As such, these two solutions are useful in defining the upper- and lower-bounds of expected influence.

   2 Dynamic consolidation mechanism and process: To understand the dynamic densification process, it is necessary to examine the mechanism and mechanics of soil-pounder (weight) interaction. As shown in Figure 7.14, when the pounder is applied to the soil mass, deformation will result from immediate elastic and inelastic deformation of the soil structure. In granular soils, porewater drains from the soil resulting in a reorientation of soil particles.

   3 Effective depth due to dynamic consolidation: The greatest concern in this process is the depth of influence (effective depth) due to the dynamic densification. The depth of influence is how deep the pounder, dropped freely from a certain height, will affect the fill material below the ground surface as shown in Figure 7.14. There are several proposed formulas available as summarized in Table 7.4. These equations, 7.20 to 7.24, have been used to describe the effectiveness of the dynamic consolidation process.

   The parameters in Table 7.4 are given as: D_e = effective depth of influence (m), W_x = weight of pounder (weight) (metric tons), h_x = height of free drop (m), d = diameter of pounder (tamper), E = applied impact energy, A = area of impact, s = undrained shear strength of soil, and ψ = coefficient of effective depth. The value is a function of the fill material type, pounder size, and degree of saturation of the fill material.

EXAMPLE 7.5   Dynamic compaction
To prepare the subgrade for a section of Interstate 65 in Alabama, dynamic deep compaction was selected to improve cone penetration values, q_c from values as low

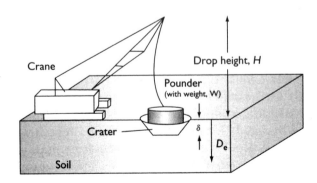

Figure 7.14 Schematic diagram illustrating the load–soil interactions in dynamic consolidation test.

Table 7.4 Summary of effective depth equations

a *Menard and Broise Method (1975)*	
$D_e = [W_x \, h_x]^{0.5}$	(7.20)
b *Leonards–Cutter–Holtz Method (1980)*	
$D_e = 0.5 \, [W_x \, h_x]^{0.5}$	(7.21)
c *Lukas Method (1980)*	
$D_e = (0.65–0.80) \, [W_x \, h_x]^{0.5}$	(7.22)
d *Charles–Burford–Watts Method (1981)*	
$D_e = 0.4d \, [(E/A) \, (1/d) \, (1/s)]^{0.5}$	(7.23)
e *Fang and Ellis Method (1995)*	
$D_e = \psi \, [W_x \, h_x]^{0.5}$	(7.24)

as 25 kg/m^2 to greater than 100 kg/m^2. A conventional crawler crane was used to drop a 20 ton circular weight (diameter = 3 m) from a height of 18.3 m. Estimate the depth of influence, assuming the subsurface is granular and free-draining.

Solution
Using Equation (7.21) (see Table 7.4), we have:

$$D_e = 0.5\sqrt{W_x h_x} = 0.5\sqrt{(20\ \text{ton})(18.3\ \text{m})} = 9.6\ \text{m}$$

7.9 Compaction by blasting techniques

The compaction of loose sand, or silty soil, can sometimes be achieved by the use of blasting techniques. In the case of blasting, the soil at a depth below the explosion is compacted. The soil in the immediate vicinity of the blast may be rather loose and additional compaction by other means will be needed. For further discussion on blasting dynamics, see Ch. 11. The basic procedure is outlined as follows.

7.9.1 Blasting operation procedure

1   The depth of expand hole should be ≤5 m.
2   Repeated blasts are more effective than a single large blast.
3   Selection of Explosion can use Equation (7.25) for estimation of amount of explosive.

$$C = k_q D^3 \tag{7.25}$$

where C = amount of explosive (kg) k_q = coefficient of explosion, from 0.95 to 1.20, and, D = diameter of expanded hole, (m). The diameter of expanded hole, D, can be computed from Equation (7.25) or it can be determined from Equation (7.26) if the amount of explosive is given

$$D = 0.84\ (C) + 0.2852 \tag{7.26}$$

Equation (7.26) is an empirical relationship derived from testing and research at Tianjing University, the corresponding correlation coefficient (r^2) is 0.996.

7.9.2 Check bearing capacity of exploded pile

Bearing capacity of exploded pile can be estimated by Equation (7.27), given as

$$P = R_d A_d \tag{7.27}$$

where P = allowable vertical bearing value of single exploded pile (tons), R_d = allowable bearing value, t/m^2, and A_d = area of exploded pile (m^2). Equation (7.27) is suitable for pile lengths less than 5 m. The value of R_d is a function of soil type. In the case of clay soils, R_d ranges according to the liquidity index, LI (see Ch. 2). If the LI is between 0 and 0.25, then R_d ranges between 50 and 80, while for LI values between 0.25 and 0.60, R_d ranges from 35 to 50. R_d values for clays with LI values greater than 0.60 have not been determined, although this condition is not common. In the case of sands, R_d values range from 40 to 140, with lower values applying to fine sands and higher values applying to more medium and coarse sands. For rock, R_d values range from 150 to 300, depending on the extent of weathering – the greater the weathering, the smaller the value of R_d.

7.10 Soil densification by an electrical process

Soil densification by an electrical process is closely tied with the dewatering process as discussed in Section 6.12. In general, there are three basic processes as: (a) Electrokinetic process, (b) Electrochemical process, and (c) Electromagnetic process.

7.10.1 Densification by electrokinetic process

Densification by the electrokinetic process (Sec. 6.12) has been applied to pile foundations to improve friction pile capacity. The test site was located at the Big River bridge on about 100 m (328.0 ft) of soft varved clay (Sec. 2.11) and loess silt deposits. It is the route of the Trans-Canada Highway, which passes the north shore of Lake Superior. Based on in situ pile load tests, the overall effect of the electroosmosis markedly increased the pile capacity. More than thirty years later, further studies indicated that there was no reduction in the load bearing capacity over this period and the recorded settlement of the bridge foundations was minimal.

7.10.2 Densification by electrochemical process

The electrochemical process (Sec. 6.12.4) is similar to the electrokinetic process, except that during the process, an introduction of new ions to the soil for the positive electrodes (aluminum) or an introduction of new ions such as Al^{3+}, Ca^{2+}, Mg^{2+}, etc. through a perforated iron pipe anodes is made Karpoff (1953). Densification by the electromagnetic process (Sec. 6.13.3), has only been conducted at the laboratory scale.

7.11 Summary

Compaction techniques are perhaps the oldest method of improving soils, that is, increasing their strength, reducing compressibility or reducing permeability. Surface compaction is initiated in the field through the use of heavy machinery such as steel-drum rollers. These rollers may have protruding elements (sheepsfoot roller) which are useful for clay soils or vibrators which are more useful for cohesionless soils. The extent of field compaction is often judged relative to the amount of compaction achieved under laboratory conditions, using some version of the Proctor test. The key variables include maximum dry unit weight and optimum moisture content, which are often specified as part of construction contract documents for a given project. For compaction at greater depths, deep dynamic compaction, vibrocompaction, blasting are among the many techniques that may be implemented. These techniques are particularly useful when site soils are loose and selection of an alternative site or replacement of existing soils is not an option.

PROBLEMS

7.1 Define maximum dry density, optimum moisture content and, the ZAV curve.

7.2 In a dynamic compaction test, what is the effect of increasing the compactive effort on: (a) the optimum moisture content and (b) maximum dry density?

7.3 The undisturbed soil at a given borrow pit is found to have the following properties: moisture content (ω) = 15%, void ratio (e) = 0.60 and, specific gravity of solid (G_s) = 2.70. The soil is to be used to construct a rolled fill having a finished volume of 50,000 yd^3. The soil is excavated with trucks having a capacity of 5 yd^3 each. When loaded to capacity, these trucks are found to contain, on the average, a new weight of soil equal to 13,000 lbs. (a) Determine: Degree of saturation (S); Wet unit weight and dry unit weight of the undisturbed borrow pit material. (b) Assuming that each load is a capacity load, how many truck loads are required to construct the fill?

7.4 From Problem 7.3, in the construction process, the trucks dump their load on the fill. The material is spread and broken up after which a sprinkler adds water until the water content is to equal 18%. The soil and water are thoroughly mixed by means of discers and then compacted until the dry unit weight equals to 110 pcf. (a) How many gallons of water will have to be added (per truckload) assuming that the moisture lost by evaporation during excavation hauling and handling is negligible? (b) If the fill should become saturated at some time subsequent to construction and it does not significantly change in volume, what will its saturation moisture content be? (c) What will the saturation moisture content be if the soil swells to increase its original volume by 10%?

7.5 In a borrow pit, a natural soil $(G_s$ = 2.70) was found to have a moist unit weight of 112 pcf and a water content of 12.0%. It is decided to compact this soil in a highway embankment to a dry unit weight of 115 pcf at a water content of 15%. If 10,000 cubic yards of borrow material are required,

(a) How many cubic yards of borrow material are required?
(b) How many gallons of water (total) must be added to the borrow before compaction (1 ft^3 = 7.48 gallon)?
(c) If the compacted fill later becomes saturated without change in the void ratio, what will its water content and unit weight be?

7.6 Based on Figure 7.2 (AASHO Road Test method) the data for soil that is to be used to construct a highway embankment 35 ft high, showed that 90%

(by weight) was found to pass the #10 US Standard sieve and 60% passed the #40. The liquid limit and plastic limit were found to be 34 and 18 respectively. What dry unit weight would you specify, and why?

7.7 From Problem 7.5, if in situ unit weight and moisture content determinations on the first few layers of highway fill averaged as follows: mass unit weight = 144.0 pcf and moisture content = 8.2%, what specific recommendation would you make to the contractor to permit him to meet specifications at a minimum cost to him?

7.8 From the standpoint of its engineering properties (shear strength, volume change, and permeability), explain why the specification of dry unit weight alone is not satisfactory indication of the final performance of an earth dam or a highway embankment.

7.9 A compaction specification for a highway fill project requires a dry unit weight equal to 100% of standard AASHTO for soil having the following properties: Retained on #4 US sieve = 15% (G_s = 2.64), Passing #4 US sieve = 85% (G_s = 2.72), Liquid limit (ω_L) = 35, Plastic limit (ω_P) = 18, Field moisture content = 12% and, Bulk unit weight = 120 pcf. Does the fill meet the specifications? Explain.

7.10 A certain soil has a solids specific gravity, G_s = 2.67. A 1000 cm^3 container is just filled with this soil in its loosest possible state, and later the same size container is filled at the densest possible state. The dry weight of the soil for the two samples were measured to be 1550 g and 1700 g, respectively. The soil in nature is known to have a void ratio of 0.61. Determine the relative density of the soil.

Chapter 8

Cracking–fracture–tensile behavior of soils

8.1 Introduction

8.1.1 General discussion

Soil cracking is a natural phenomenon and is frequently observed in many natural and man-made earth structures. These cracks are a result of an internal energy imbalance in the soil mass caused by nonuniform moisture and temperature distribution or compaction energy during construction. Local environmental changes such as pollution intrusion (Sec. 1.3), wetting–drying, and freezing–thawing cycles (Sec. 6.7) will also affect soil cracking behavior. In many cases, overconsolidated clay (Sec. 9.4) exhibits cracking and fissures due to natural desiccating processes. These small unnoticed cracks create many premature or progressive failures in excavations (Sec. 13.12), earth slopes (Sec. 14.3), dam, highway embankments, hydraulic barriers, landfill's top covers, and bottom seals or clay liners (Sec. 16.9).

Due to rainfall, flood, melting snow, or groundwater fluctuations, where water fills the cracks, there is a softening of the bonding strength between soil particles. Crack growth depends on local environmental conditions and is assisted by the action of capillary tension (Sec. 5.3), thermal or proton migration (Sec. 6.6), electroviscous effects, and/or thermal–electric–magnetic effects (Sec. 6.10).

8.1.2 Cracking–fracture relationship

For a given material, the cracking and fracture behaviors belong to one system. Cracking represents the prefailure phenomena of a material and fracture reflects the behavior of a material at the failure condition. To illustrate this relationship, a typical deformation versus loading curve for a given soil is presented in Figure 8.1. This figure shows the interrelationship between the prefailure and failure conditions of a soil from a geotechnical engineering point of view. At the early stage (points from a to b) of deformation versus loading curve, the cracks develop due to various natural or man-made causes. At this stage, the soil behavior is controlled by multimedia energies such as thermal, electric, and magnetic energies. However, at the failure stage (point c), it is dominated only by mechanical energy, which is the loading itself. Of course, it is important to assess the behavior of soil before failure. As such, to evaluate the useful life of soil under a given loading condition, it is necessary to investigate cracking-fracture mechanisms and their interaction. Fracture load testing

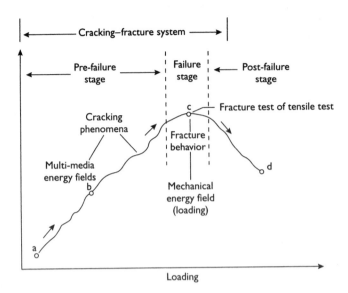

Figure 8.1 Interrelationship between prefailure and failure conditions of a soil in the cracking–fracture system.

(Sec. 8.7) and tensile strength testing (Sec. 8.9) are the quantitative measurements of cracking-fracture behavior of soil as indicated in Figure 8.1.

8.2 Soil cracking mechanisms and types

8.2.1 Soil cracking mechanisms

The natural formation of soil structure on the microscopic scale is always related to water loss and concomitant shrinkage in cohesive soils as discussed in Section 4.4. In this chapter, further discussion of shrinkage behavior of soil as it relates to the cracking pattern is presented. Soil cracking mechanisms can be explained by the following principles: (a) the development of surface cracks, which can be explained by the law of the triple-angle – a special form of the law of least energy. This means, cracks are developed when a minimum amount of surface energy is applied; (b) the linear elastic fracture mechanics (LEFM) concept, which states that cracks will follow the shortest distance to release the strain energy. The cracking pattern produces the greatest stress relief with the least amount of work involved. The concept of strain energy will be discussed in Section 8.6; and (c) the particle-energy-field theory as discussed in Sections 1.6 and 4.3 will provide additional information to understand the mechanisms of cracking and fracture behavior of soils.

8.2.2 Soil cracking types and causes

As indicated in Figure 8.1, to study the fracture or failure behavior of soil, it is necessary to understand the characteristics of soil cracking causes and mechanisms. The following four basic types of cracking frequently exist in a soil mass.

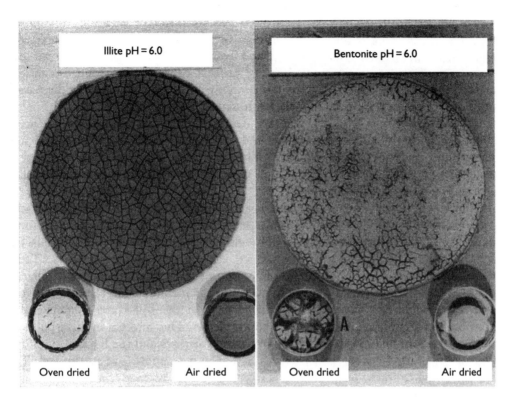

Figure 8.2 Photo shows the cracking patterns of mud flat of illite and bentonite clays.

1 Shrinkage cracking: It is the most common cracking found in earth structures and the phenomenon can be observed in drying mudflats (Fig. 8.2). Two types of clay minerals illite and bentonite (Sec. 3.9) are used. As water is lost from the surface soil mass, tensile forces are established in the drying surface layer. Because of the water loss, soil also loses its ability to relieve these tensile forces by plastic flow. These stresses are finally relieved by the formation of shrinkage cracks that break up the surface layer into pieces of more or less distinct geometric shapes. The geometric shape of the cracks depends on the clay mineral composition, the heating process, pore fluids, and more.

2 Thermal cracking: Thermal cracks are caused by a change in the thermal stresses of a material. The stresses develop when a material is heated and then suddenly cooled. Such stresses may be induced by cycles of freezing–thawing or wetting–drying as well as pollution intrusion in soil. Thermal cracks for soil are somewhat different than with other construction materials because temperature and moisture in the soil mass are closely related and the state of stress of soil is extremely sensitive to both temperature and moisture behavior. Plastic deformation occurs in different zones of the soil mass as a result of differential temperature with accompanying volume changes arising during the cooling cycle. If, deformations are within the elastic strength range of the soil mass, then no stress remains after temperature equilibrium is reached.

Cracking patterns are also influenced by the drying process. The cracking patterns between air, oven, or, microwave oven dried soil are different with the same initial moisture content. This is due to the rate of the drying process. For example, the drying process in an oven is more intense and less uniform as compared to an air–dry process (Fig. 8.2). The microwave heating process causes soil particle excitation; therefore, the cracking pattern appears more intense and significant. Further discussion on this aspect of cracking patterns will be presented in Section 8.3.

3   *Tensile cracking*: Tensile cracks are caused mainly by overburden pressures including structural loading, rainfall, ice and snow loads, trees, vegetation, and seasonal ground surface creep loads. Sometimes, it is also associated with changes in moisture or thermal stress as discussed in Case (2) and related with fracture loads which will be discussed in Case (4).

4   *Fracture cracking*: In a soil mass either man-made or in the natural state, cracks always exist. This is due to daily moisture, temperature changes, caused by seasonal groundwater table fluctuations, rainfall, or melting snow, which will fill water into the cracks or voids, and consequently, produce variable porewater pressures. These porewater pressures vary with changes of environmental condition as does capillarity tension (Sec. 5.3). When a saturated soil mass dries, a meniscus develops in each void of the soil structure, which produces the tension in the soil–water system and a corresponding compression force in the soil skeleton. This internal cyclic-type load, caused by the combination of shrinkage or thermal stresses and the fluctuation of the porewater pressure between soil particles, is called a *fracture load* in the soil. The crack produced from the fracture load is referred to as a *fracture crack*.

8.3   Soil cracking patterns

8.3.1   Characteristics of soil cracking patterns

1   *Homogeneous soil system*: If the soil material is homogeneous, the cracking pattern will be hexagonal according to the law of the triple-angle, which is a special form of the law of least energy. This phenomenon can also be explained by the linear fracture mechanics concepts in Section 8.4, in which cracks will follow the shortest distance to release the strain energy. The cracking pattern produces the greatest stress relief with the least amount of work involved.

2   *Heterogeneous soil system*: If the soil system is heterogeneous, as in the case of organic matter with varying water affinity, the hexagons will tend to become rounded and the organic matter concentrates at the surface of the fissures. In the case of soil particles that are not uniformly mixed, most cracks will occur in the vicinity of the larger soil particles.

3   *Contaminated soil system*: The cracking patterns of contaminated soil are variable, depending on the chemistry of the pollutant. Color changes are also noted in contaminated systems. Most of the colors are dark blue, brown, and red. The color depends on the types of clay minerals and their interactions with the pollutants.

8.3.2   Specimen preparation for laboratory cracking pattern test

Soil cracking measurements, in general, are observed directly from the characteristics of cracking patterns. These patterns are quite sensitive to the way in which a specimen is prepared. In the interest of consistency, the following procedures have been developed (Fang, 1994) and used as a guide for cracking measurements.

1 All soil samples must pass a #40 US standard sieve. Smaller particle sizes or larger surface area per unit volume of soil sample will give more distinct cracking patterns. Approximately 50–100 g of soil are needed.
2 Water content of the specimen must be at the completely saturated condition. For simplicity, the water content of the liquid limit (Sec. 2.5) of soil can be used. Soil–water must be mixed thoroughly.
3 Wet soil paste prepared from step (2) is then uniformly spread on a clean glass plate. A round shaped wet soil paste (mud pad) sample is formed, as shown in Figures 8.3(a) and (b).
4 The maximum height of the wet mud pad must be less than 1/20 in. (1.27 mm) as shown in Figure 8.4(b). The thinner the wet mud pad, the more distinct cracking patterns will develop.
5 The drying process for the soil sample can be determined by three common methods used in the laboratory: (a) air-dry at room temperature (20°C) for 24 h; or (b) oven dry at 110°C for 6 h; or (c) in a microwave oven for 4–5 s.

It must be noted that in using any of these three drying methods for determination of the moisture content of a soil sample, the results will be very close (practically the same), however, the results of the cracking pattern under these three different methods for the same soil will be different. The cracking pattern is closely related to the rate of heat intrusion and intensity of the heat. A soil sample dried by air at room temperature has cracking patterns that are generally uniform. If oven dried, there are usually more cracks around the edges of the specimen. Heat produced from a microwave is much more intensive, due to the induced vibration of water molecules, therefore, the cracking pattern is also more irregular and larger crack openings are observed.

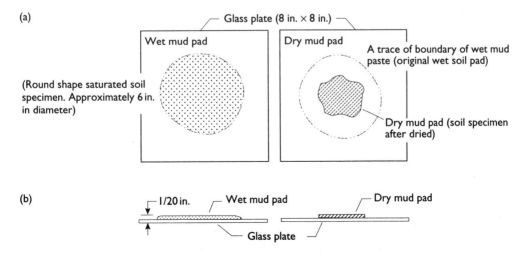

Figure 8.3 Preparation of laboratory soil specimen for cracking pattern test. (a) Top view of wet–dry mud pad, and (b) cross-section of wet–dry mud pad.

8.3.3 Cracking pattern interpretations

Items to be observed from cracking pattern tests include (a) geometry of the cracking pattern for the soil specimen including crack openings, length, depth, and shape; (b) estimation of percent of shrinkage in comparison with the original (wet pad) soil specimen and also determining the shrinkage limit (Sec. 2.5); (c) cracking pattern changes caused by the heating process, such as air-dried soil specimen, oven dried (Fig. 8.2) soil specimen, and microwave dried soil specimen; (d) pattern changes caused by various pore fluids such as inorganic and organic pore fluids; (e) types of clay minerals and ions in the soil sample clay mineral types, and ion types and concentration; and (f) description of the soil color (Sec. 2.10) after it is dried.

Soil color changes are caused by various types of pore fluids in the soil voids. These changes, caused by the physicochemical or biological interaction between soil particle surface and pore fluids, can be explained by the diffuse-double-layer theory (Sec. 6.8). In general, the colors of contaminated soils are more distinct and colorful in comparison with the non-contaminated (water) condition.

8.4 Soil cracking–fracture interaction

8.4.1 General discussion

The mechanism and the interaction between cracking and fracture are complex phenomena. There are numerous factors affecting the cracks and how these cracks relate to the failure condition. In general, all cracks are related to the moisture content in the soil and local environment. Equilibrium conditions of soil–water systems differ when in contact with liquid water and water vapor. Even at the same moisture content, the structure and physicochemical properties of a soil–water system may differ considerably, depending upon the flow path with which a given moisture content has been attained. Since soil cracking mechanisms are related to the moisture content in all cases, the flow directions or paths are also related to the cracking patterns. Under such condition, there are three basic mechanisms of soil cracking and fractures: (a) drying soil, (b) saturated soil, and (c) contaminated soil. Phenomena and reactions of each case are presented as follows:

8.4.2 Cracking mechanism when soil is drying

As a wet or moist cohesive soil system loses water, the soil particles move closer and closer together. If the drying proceeds from the surface downward as in a mud flat, the dehydrated surface layer shrinks while the water resistance between the upper and lower layers and in the layers themselves prevents an adjustment to the volume decrease of the surface layer. As a result, tensile stresses are developed in the surface layer. This cracking pattern produces the greatest stress release with the least amount of work. In other words, cracking and fractures produced by drying are controlled mainly by thermal energy. A similar process known as the shrinkage process was discussed in Section 4.4.

8.4.3 Cracking mechanism when soil is saturated

The entrance of water into a porous dry soil system can also cause cracking/fracture of a soil mass. In this case, these causes and phenomena are more complicated than in drying conditions, because the cracking mechanisms are controlled by the multimedia energy fields, including the thermal–electric–magnetic energies. An explanation of these phenomena is presented as follows:

1 The internal driving force in the soil mass producing the cracking includes (a) the water–vapor force (Sec. 5.11), (b) the heat of wetting (Sec. 6.2) as dry soil becomes wet, (c) the kinetic dispersive force (Sec. 6.2) produced caused by form factors (Winterkorn, 1958) when thermal energy is involved, and (d) the electromotive force (Sec. 6.9.2) produced when electrical energy is involved.
2 The swelling capacity (Sec. 4.4) increases when soil becomes saturated. This capacity of the clay minerals may result in internal swelling that decreases the permeability.
3 The electroviscous effect (Sec. 6.13) can be explained as water flows under a hydraulic gradient between negatively charged soil surfaces, the exchangeable cations (Sec. 4.7) are then swept downstream and a streaming potential develops (Elton, 1948; Low, 1968), which will create internal cracks.
4 The effects of exchangeable ions or pH values on cracking behavior as reflected on volumetric changes can be seen as the characteristics of pore fluids significantly affecting the volume change or crack patterns.
5 If internal forces stated in Case (1) produced by water enter into the soil mass, it may be large enough to result in surface exfoliation of the systems.

8.5 Cracking–fracture characteristics of contaminated soils

8.5.1 Mechanism of cracking–fracture pattern

The mechanism of the cracking–fracture pattern for a contaminated soil is different than a non-contaminated soil as shown by using water (non-contaminated) or either acid or base (contaminated) as an example. Also, soil cracking patterns vary according to the dielectric constant (Sec. 6.10) and surface tension (Sec. 4.3) of the pore fluid. In order to explain why and how these factors affect soil cracks, it is necessary to review the characteristics of soil, water, and their interaction in the environment as discussed in Section 4.3. However, a brief outline of these effects is noted as (a) interaction between a liquid (water) and a solid (soil) occurs only on the solid's surface as discussed in Section 4.1; (b) since soil is composed of electrically negative mineral surfaces and water is composed of electric water dipoles (Sec. 3.8) and predominantly positively charged ions, it then follows that a soil–water system possesses a highly electric character and will respond to the application of an electric potential; and (c) the interaction of electrically charged components of the system is a function of temperature (Sec. 6.3).

The surface tension and dielectric constant are important controlling parameters in evaluation of prefailure condition of soil–water system related to many geotechnical problems including the cracking–fracture characteristics of soil.

8.5.2 Cracking pattern affects by soil structures

The Law governing the formation of shrinkage cracks has been discussed in detail by Hains (1923) and the effects of freeze–thaw and wet–dry cycles on the soil cracking

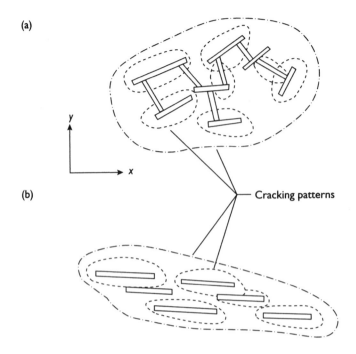

Figure 8.4 Schematic diagram illustrates the effect of soil structures on cracking patterns. (a) Flocculated structure, and (b) dispersed structure.

patterns and fracture behavior have been studied by Czeratzki and Frese (1958). Soil structures such as flocculation or dispersion will affect the cracking pattern as illustrated in Figure 8.4. Andrews *et al.* (1967), Alther *et al.* (1985), and Evans (1991) pointed out that flocculating and dispersive structures and pore fluids create different types of cracking patterns. A flocculated structure produces larger areas of cracking than dispersive structures.

The causes of internal cracking and other cracking phenomena of soil are similar to surface cracking behaviors; however, the measurement of internal cracking is difficult. Recently developed techniques using the computed tomography (CT) technique will be further discussed in Section 8.8.3.

8.6 Application of LEFM

8.6.1 General discussion

As indicated in Figure 8.1, fracture behavior represents the failure condition of a given soil. The fracture behavior of soil is considered in the mechanical energy field. The fracture load test is a quantitative measurement of the cracking behavior of soil, and as such these two characteristics are interrelated. In order to use LEFM to evaluate soil cracks, the basic concept of existing fracture mechanics must be re-examined together with justifications and limitations. In the following section a review of LEFM is examined based on a geotechnical engineering viewpoint. The basic concepts of fracture mechanics theory were developed in 1921 by Griffith and it has acquired a considerable amount of success in predicting failure caused by crack propagation for metals. More recently, it has been applied for predicting fracture behavior in rocks, concrete, soil cement, asphalt pavement, marine clay, adobe bricks, stabilized fine-grained soils, and solidified sludge refuse material.

8.6.2 Justification of LEFM for use in geotechnical problems

The basic concept of the LEFM theory is that crack-like imperfections are inherent in engineering materials. These flaws act as stress raisers and can trigger fracture when subjected to a critical state load or when damage is done cumulatively under cyclic loading. State-of-the-art of the fracture mechanics theory have been summarized by Irwin (1960) and Sih and Liebowitz (1968). Strictly speaking, the theory is limited to linear-elastic materials in spite of the fact that plastic or nonlinear strains unavoidably prevail in the vicinity of flaws or cracks. Nevertheless, it does provide an ideal and simple way of estimating the amount of energy required to create free surface in the material. The degree of simplicity is achieved by focusing attention on the leading edge of the crack where nonlinear strains exist but are regarded as localized within a zone of negligibly small dimensions. Although the stresses and strains within this zone cannot be analyzed in fine detail, the LEFM theory is able to give an adequate description of the gross feature of the stresses and strains near the crack.

8.6.3 Strain energy release rate and stress intensity factor

1 Strain energy release rate, G: The strain energy release rate, G, concept proposed by Griffith (1921), is in a formal sense the force driving the fracture process. At present, plasticity and nonlinear strains can be included in the fracture mechanics treatment only in a rather superficial way without incurring serious losses of clarity and simplicity. Previous work on fracture mechanics has been mostly focused on crack-toughness testing studies of common metals in an effort to characterize the fracture toughness of a material by the critical value of energy release, that is, G_c. In this way, one can compute values of nominal stress necessary for crack propagation for various size-specific cracks. Many of these computations have been carried out in the past for laboratory specimens with simple configurations by assuming material isotropy, homogeneity, etc. However, application of fracture mechanics principle is no longer a simple matter when applied to a complex system such as a natural, multilayered soil.

2 Stress intensity factor: Fracture analysis has been significantly enhanced by the introduction of the concept with a *stress intensity factor, K*, as initiated by Irwin (1960) and later used by Sih and Liebowitz (1968) and others for analyzing various structural problems. This factor, often referred to as K, is related to the amount of energy, G, required to create a new fracture surface used in the Griffith theory. The main idea behind the stress intensity factor, K, is that all the information relating to crack loading and geometry is contained therein. While the factor, K, changes from problem to problem, its association with the stress state near the crack tip is always the same. In particular, one can determine the critical strain energy release rate, G_c, or fracture toughness, K_c, which is the conjugate to the force driving the fracture process. Detailed procedures are discussed in the following sections.

8.7 Laboratory fracture load tests

8.7.1 Fracture load determined by tensile (pull) test

1 *General discussion*: The basic idea for laboratory fracture load testing for soil is borrowed from fracture testing of metals; however, some modifications are necessary for soil specimen preparation. In practice, the value, G_c, is measured in a simple laboratory test (ASTM E399). It uses a cracked specimen, which is pulled apart by a load, P, as shown in Figure 8.5(a). The fracture load, P, can also be calculated from mathematical equations (Sih and Liebowitz) once the crack dimensions and geometry and, G_c, or fracture toughness, K_c, are known. For geotechnical applications, it is found that measuring the fracture load, P, and fracture toughness, K_c, are most useful.

2 *Specimen mold and test specimen*: The specimen mold consists of a steel base with removable sides and interior walls. It can accommodate up to four specimens at the same time. The size of the specimen is 3 in.× 3 in. × 0.25 in. (7.62 cm × 7.62 cm × 6.35 cm) as shown in Figure 8.5(a). The length of the notch is 1 in. (2.54 cm). This notch is used for creating the cracks when the fracture load, P, is applied. The interior of the mold was coated with a lubricating agent to facilitate the removal of the specimens. After the specimens were extracted from the mold, they were air-dried. A glass plate cover was placed over them to prevent warping. The size of mold is modified when larger soil particles are involved. The dimension is proportional to the grain size distribution. For stabilized soil that measures 6 in. × 6 in. × 0.50 in. (15.24 cm × 15.24 cm × 1.27 cm) a modified mold was used. The test procedure followed American Society of Testing and Materials (ASTM E399). However, the mechanical strain gauge – a simple recording device – has been modified for determination of the fracture load, P.

8.7.2 Fracture load test on fine-grained soils and stabilized sludge

Experimental studies are useful in illustrating fracture loading concepts. Two such studies were carried out on soft marine clay and sludge waste. These are discussed as follows.

1 *Fracture load results on soft marine clay*: All samples were passed through a US sieve #40. Molding water content for all samples was constant at 25%, while varying the molded dry density. All samples were air-dried before testing. The fracture load, P, and the corresponding change in gauge length, L, were recorded. The crack growth, a, was also determined. In this study, all samples failed rapidly, therefore, the crack growth, a, value was assumed as the length between the line of application to the total length of the specimen ($= 2.5$ in.). The maximum fracture load before failure was 6, 11.5, and 12.5 for specimens molded at 91.4,

(a)

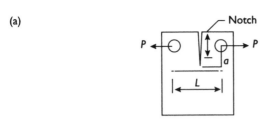

Specimen size 3 in. × 3 in. × 1/4 in.

L = Gage length
P = Fracture load
a = Crack length

(b)

Figure 8.5 Determination of the fracture load of compacted soil: (a) dimension of soil specimen, and (b) soil sample and failure mode.

95.2, and 100 pcf. Similar to compressive strength, tensile strength appears to increase with increasing dry density.

2   *Experimental results on sludge wastes*: Experimental results on solidified sludge wastes (Pamukcu and Topcu, 1991) are summarized. The results of the fracture load tests appear to support the assessment of other test results such as unconfined compressive strength and tensile strength. All these tests were measured over a range of water contents and curing temperatures. The principle aim for these measurements was to confirm that strength development and fracture load, *P*, are correlated under similar environmental conditions.

8.8   Applications of cracking–fracture data

In addition to utilizing the cracking patterns for identification and characterization of ground soil behavior as presented in the previous section, there are other applications in geotechnical engineering. Brief discussions are presented in the following sections:

8.8.1 Use as a guideline in ground improvement programs

A diagram illustrating the cracking-time relationship at various cracking stages is shown in Figure 8.6. This diagram is used as a basic guideline in ground improvement programs for estimating crack growth in brittle construction materials. In examining Figure 8.6, it is seen that an initial crack size, C_i, point a can remain constant for a large number of load cycles before an increase is noted. Between points b to c is called the *ground improvement stage* (degradation stage). In that period, the ground improvement must be made, in order to extend the useful life of the material or structure. Shortly thereafter the useful life of the structure is met (point c) and if no remedial work is done, the fatigue life (point d) will rapidly manifest. Also, it can be seen in Figure 8.6, that shrinkage, thermal, tensile, and fracture caused cracks are interrelated with the useful life of any geo-structure, especially in progressive failure of slopes, erosion, and landslide problems.

8.8.2 Cracking patterns used for identification of soil behavior

Cracking patterns can be used for identification or characterization of soil behavior. Also, the color and patterns together can be used for identifying contaminated ground soil.

8.8.3 Internal soil cracking measurement

Characteristics of internal soil cracking can be used for estimation of progressive erosion or predicting potential landslides. This type of cracking measurement can

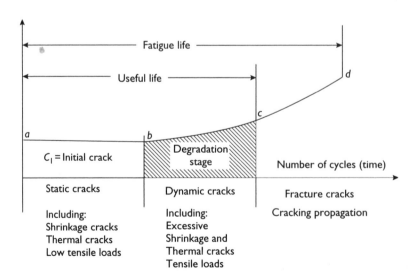

Figure 8.6 Cracking-time relationship.

Source: Fang, 1997 with modification.

be made by the use of CT techniques. CT is a relatively new X-ray method for nondestructive testing and evaluation. CT measures point-by-point density values in thin cross sections of an object, thus allowing three-dimensional imaging of the internal structure when successive transverse sections are compared. In X-ray CT, an X-ray beam passing through the test object is measured by an array of detectors located on the other side of the object. The object is rotated slightly or in some cases, translated and rotated, and a new set of measurements is made. This process is repeated until the object has been fully rotated. The measurement projections are controlled by a computer, which is the major advantage of this technique over conventional radiography.

Soil cracking pattern tests are a qualitative indicator property, which should be correlated with other soil performance measures in order to have a significant meaning. These performance tests include tensile strength, fracture loading, clay mineral structure analysis, and pore fluid chemistry analysis. Other related factors include soil types, unit weight, grain size distribution, void ratio, and geometry of the primary and secondary drying surfaces.

8.9 Tensile strength of soil

Many failures of earthen structures are related to the tensile characteristics of soil itself. Specifically, crack growth at ground surfaces often initiates failure. Low tensile strength may indicate low bearing capacity in shallow foundations (Sec. 12.2). Tensile cracks behind retaining walls, earth dams, etc. will create negative lateral active earth pressure (Sec. 13.5) and initially trigger landslides (Sec. 14.3).

8.9.1 Mechanism of tensile strength of soil

Soil tensile strength mechanisms are complex. The tensile strength varies with the moisture content of soil. Basically, there are three stages of moisture content of soil: the dry stage, partially saturated stage, and saturated stage. Mechanisms of these three stages are summarized in the Table 8.1. The explanation of each stage follows.

1   When soil is dry, individual particles are held together by cementing agents and adhesive films. The cementing power of these films is a function of their own

Table 8.1 Mechanism of tensile strength of soil at various stages of moisture content

Dry soil		Controlled by cohesion, friction, and some tension
Partially saturated soil	Mainly controlled by tensile strength	Caused by (1) Surface tension of soil moisture (2) Oriented dipoles linking a positive charge on one particle with a negative one on the neighboring particles (3) Formation of an electronic field by ions (4) Diffuse double layer thickness expands (5) Permits a more orderly arrangement of particles (6) Electrolyte concentration reduces
Saturated soil		Tensile strength very small

physical, chemical, and physicochemical properties. The adhesive force between soil properties is also dependent on the soil particle size distribution and the surface area.

2 If soil gradually becomes wet, tensile strength develops due to the surface tension (Sec. 4.3) and other forces as listed in Table 8.1. From a physicochemical point of view, the tensile strength for partially saturated soil is caused by dipoles (Sec. 3.7) linking a positive charge on one particle with a negative one of the neighboring particles.

3 As the diffuse-double-layer thickness (Sec. 6.12.3) expands and the electrolyte concentration reduces, the degree of flocculation also decreases, which permits a more orderly arrangement of particles. If there is a further increase in moisture content, which dilutes the concentration of soil particles per contact, the net attractive force (Sec. 3.7) between particles is reduced.

4 As the soil becomes saturated, there is a very low tensile strength with no engineering significance. Therefore, the tensile strength is primarily relevant to partially saturated soil. When soil is dry, it is held together not only by tensile strength but with cohesion (Sec. 10.3) and frictional forces (Sec. 10.8) as well.

8.9.2 Tensile strength measurements

1 *General discussion*: There are numerous test procedures to measure the tensile strength of soil. Satyanarayana and Rao (1972) used seven different tensile tests including beam test, prism split test, cubic diagonal split test, direct tensile test, briquette test, cubic side split test, and cylinder split test to determine the tensile strength of fine-grained soil compacted at modified AASHTO compaction (Sec. 7.2). Figure 8.7 summarizes these laboratory test results. Based on these test results, indications are that tensile strength determined by the beam type tensile test gives the highest results. A comprehensive review of tensile and shear strengths has been reported by Yong and Townsend (1981). Among these test methods, the split tensile test is most commonly used.

2 *Split-tensile test*: The conventional split tensile test equation is presented as Equation (8.1). Equation (8.1) has been derived from the theory of linear elasticity (Timoshonko, 1934) and has a simple form. A schematic diagram of a split test is presented in Figure 8.8.

$$\sigma_t = \frac{2P}{\pi L d} \tag{8.1}$$

where σ_t = simple tensile strength (psi, or kN/m^2), P = applied load (lb or N), π = constant = 3.1416, L = length of specimen (in. or cm), and d = diameter of specimen (in. cm).

8.9.3 Soil tensile strength determined by unconfined-penetration test

1 *General discussion*: The unconfined-penetration test (UP test) for determination of indirect tensile strength of soil was developed at Lehigh University in early 1970 (Fang and Chen, 1971). The UP test has an advantage in that the test can

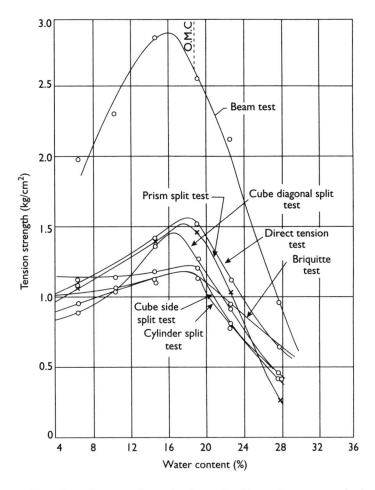

Figure 8.7 Comparison of tensile strength results determined by various test methods.

Source: Satyanarayana and Rao (1972).

conveniently be performed in conjunction with routine compaction tests (Sec. 7.2) and California Bearing Ratio (CBR) (Sec. 12.9) tests. Various sizes and shapes of specimens can be used for the UP tests as well. The conventional split-tensile test method (Fig. 8.8) measures the tensile strength across a predetermined failure plane, whereas the UP test always causes failure on the weakest plane, which results in the measurement of the true tensile strength.

   2 Theoretical considerations: The theoretical consideration for the UP test is based on the plasticity theory. This theory, which was previously used for computing the bearing capacity of concrete blocks or rocks proposed by Chen and Drucker (1969), has been extended to an application of soils and other nonmetallic construction materials. Two major assumptions are made in the theory. The first is that sufficient local deformability of soils in tension and in compression does exist to permit the application of generalized theorems of limit analysis (Sec. 12.2) to soils idealized as a perfectly plastic material. The second is that a modified Mohr-Coulomb failure

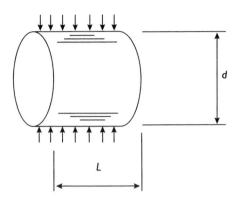

Figure 8.8 Conventional split-tensile test.

surface (Fig. 10.3) is postulated as a yield surface for soils. Figure 8.9(a) shows an ideal failure mechanism for a UP test on a cylindrical specimen. It consists of many simple tension cracks along the radial direction and two cone-shaped rupture surfaces directly beneath the punches. Detailed derivation of tensile strength equations can be obtained from Chen and Drucker (1969).

3   *Tensile test apparatus and set-up*: The apparatus for determining tensile strength is the same as the unconfined compression test (Sec. 10.7) with the exception that punches are utilized as shown in Figure 8.9(a) and (b). The laboratory setup of the UP test is shown in Figure 8.9(b), and Figure 8.9(c) shows the failure modes of soil specimens. The cone-shaped formation with 2- or 3-pierce cracks is generally observed for the soils.

4   *Tensile test procedure and calculations*: (a) By using two steel disks (punches) centered on both the top and bottom surfaces of a cylindrical soil specimen, (b) a vertical load, P, is applied on the disks until the soil specimen reaches failure, (c) the calculation of tensile strength follows as Equation (8.2).

$$\sigma_t = \frac{P}{\pi\,(KbH - a^2)} \tag{8.2}$$

where σ_t = tensile strength (psi or kN/m^2), P = load (lb. or N), π = constant = 3.1416, K = constant, depending on specimen-punch size, and soil type (Table 8.2), b = radius of specimen (cm), H = height of specimen, (cm), and a = radius of disk (punch) (cm).

5   *Typical soil tensile strength test results presentation*: For illustrating purposes, consider a medium rated plasticity soil (liquid limit = 31, plasticity index = 10). Soil samples passed a #10 US. Standard sieve and were air-dried. A 4 in. × 4.6 in. (10.16 cm × 11.68 cm) Proctor mold (Sec. 7.3) was used for preparation of the soil specimen. Specimens were compacted in three layers with a 5.5-lb (2.49 kg) hammer and 12-in. (0.305 m) drop; 15, 25, and 55 blows per layer were applied (Sec. 7.2). Results of molded dry density (unit weight) versus molding moisture content are shown in Figure 8.10(a). Strength was computed from Equation (8.2), where b = 2 in. (5.08 cm), H = 4.6 in. (11.68 cm), and a = 0.5 in. (1.27 cm). The results of tensile strength versus molding moisture content (number of blows) are presented in Figure 8.10(b). For these results, maximum tensile strength exists on the dry-side

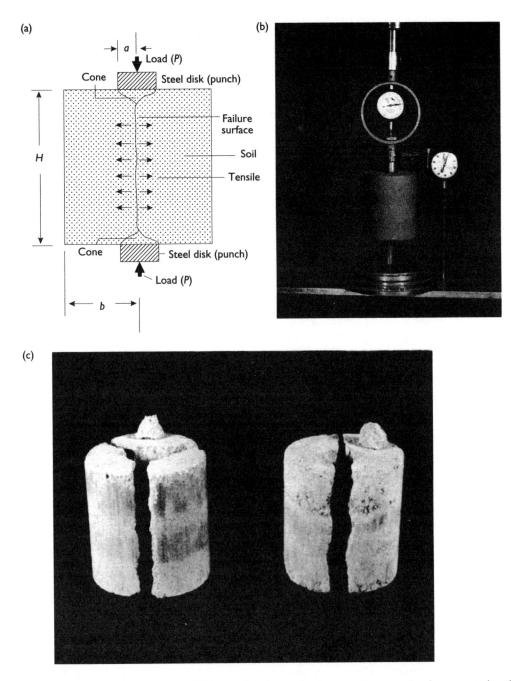

Figure 8.9 Unconfined-penetration (UP) test for determination tensile strength of compacted soil. (a) geometry of test specimen and terminology, (b) test set-up, and (c) failure mode of soil specimen.

of the optimum moisture content, OMC. Figure 8.10(c) was interpreted from Figures 8.10(a) and (b) and indicates that at higher moisture content, as density (unit weight) increases the tensile strength increases slightly, however, at lower moisture content, as density (unit weight) increases, the tensile strength increases sharply.

Table 8.2 Recommended values for parameter, K, specimen-punch size, specimen height–diameter ratio, and rate of loading

Specimen size	K value	
	Soil	Stabilized materials
Proctor mold	1.0	1.2
4 in. × 4.6 in. (10.2 cm × 11.43 cm)		
CBR hold	0.8	1.0
6 in. × 7 in. (15.24 cm × 17.78 cm)		
Larger than 12 in. (0.305 cm)	0.8	
Specimen-punch size:	0.2–0.3	
Height to diameter ratio of specimen:	0.8–1.2	
Rate of loading:	ASTM recommendation for the axial strain at a ratio of 0.5–2% of height per minute	

EXAMPLE 8.1

Given a 4 in. × 4.6 in. compacted soil specimen prepared from standard AASHTO compaction procedure.

Determining the tensile strength by use of the UP test.

SOLUTION

From Equation (8.2) where

Diameter of soil specimen = 4 in.; height of specimen, H = 4.6 in.

b = 4/2 = 2 in. (Fig. 8.10), a = diameter of punch = 0.5 in.

K = constant depend on specimen size = 1.2 (Table 8.2)

P = maximum load obtained from test at failure condition = 200 lb then

$$\sigma_t = P / \pi (KbH - a^2) = 200 / 3.1416 [1.2 \times 2 \times 4.6 - (0.5)^2]$$
$$= 5.9 \text{ psi} = 40.6 \text{ kN/m}^2$$

8.9.4 Comparisons of test result between UP and split tensile test

1   *Comparison of load-deflection curves*: Figure 8.11 shows the typical load-deflection curves for both conventional split-tensile test (Eq. (8.1) and Fig. 8.8), and UP test (Eq. (8.2) and Fig. 8.9). For all the cases, the similar load-deflection curve patterns are found for both split tensile and UP tests.

2   *Comparison of tensile strength test results*: A comparison of tensile strength determined from both conventional split tensile and UP tests is presented in Figure 8.12. Test results include soil, concrete, mortar, bitumen- and cement-treated highway base, subbase materials, and rock. Good agreement is found between the two methods.

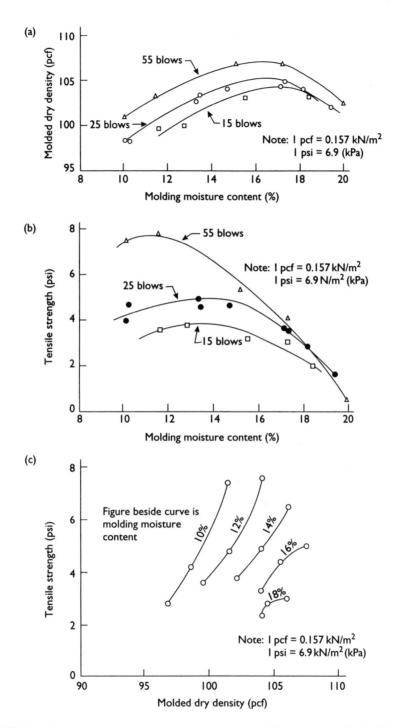

Figure 8.10 Typical laboratory tensile test results of compacted fine-grained soil. (a) Molded dry unit weight versus molding moisture content, (b) tensile strength versus molded moisture content, and (c) tensile strength versus molded dry unit weight.

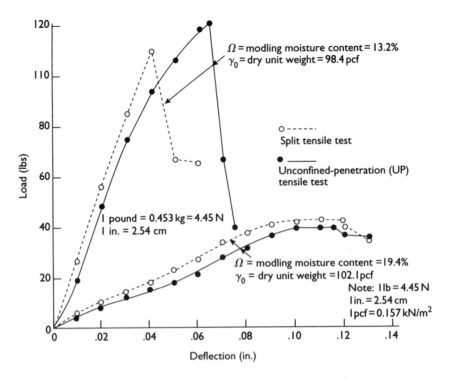

Figure 8.11 Comparison of load–deflection curves from tensile tests.

8.10 Tensile characteristics of compacted soil

8.10.1 General discussion

The tensile strength of soil is more sensitive to the moisture content than other strength parameters of soil such as cohesion, c, friction angle, ϕ, and unconfined compressive strength, q_u. A comprehensive laboratory experiment is summarized in Table 8.3. Six soil types were used and routine soil classifications were made, based on Atterberg limits, gradations, compaction test, and tensile and unconfined compression strength tests.

8.10.2 Relationship of tensile strength with soil constants

1 *Tensile strength related to soil types*: The relationship between tensile strength and plasticity index is presented in Figure 8.13. The tensile strength increases as the plasticity index and liquid limit increase. The split tensile strength data was obtained from Narain and Rawat (1970). All tensile strength tests were conducted for soil prepared at the OMC.

2 *Tensile strength related to activity*: The activity (Sec. 2.5) is defined as the plasticity index divided by the percent of clay. In general, the activity value increases

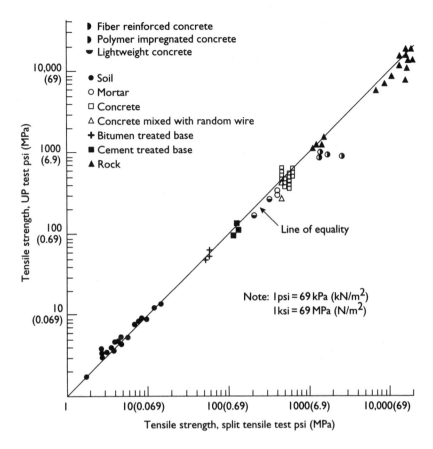

Figure 8.12 Comparison of test results determined by UP and split tensile test.

as the plasticity index and liquid limit increase. The tensile strength tests were performed at both OMC (Sec. 7.2) and air-dried conditions. Six types of soil were tested. The first group of soil specimens was molded at OMC and the tension test was performed at OMC condition. The second group of specimens was also molded at OMC, but was air-dried at room temperature for 40 h, then, the tension tests were performed. Tension test results and activity values for both cases are plotted in Figure 8.14(a). It is clearly indicated that the tensile strength increases significantly for all types of soil when they are air-dried. However, the amount and rate of increase in strength depends on the soil types. For low plasticity soil, the differences between two moisture conditions are more pronounced than for a high plasticity soil.

3   *Tensile strength related to the toughness index*: The toughness index (Sec. 2.5) is frequently used as an indicator for the evaluation of soil stabilizing materials. Figure 8.14(b) shows the tensile strength results versus the toughness index for both moisture conditions. Similar trends are indicated in both cases in Figures 8.14(a) and (b). From these results, it can be projected that there will be greater differences in strength between the two moisture conditions for a low plasticity soil than for a high plasticity soil.

Table 8.3 Summary of experimental results on gradation, Atterberg limit, compaction, tensile, and unconfined compression tests

Soil number	% Pass #200	Compaction text OMC %	Compaction text γ_d pcf	W_L	I_p	A	I_f	I_t	σ_t psi OMC	σ_t psi Air dried	q_u psi OMC	q_u psi Air dried	q_u/σ_t OMC	q_u/σ_t Air dried
A	14.1	12.9	121.7	24.6	4.9	0.34	11.4	0.43	2.1	42.7	58.7	170.2	28.0	4.0
B	21.4	15.0	115.9	27.7	5.0	0.24	9.5	0.53	3.2	45.5	48.7	142.0	15.2	3.1
C	17.0	13.3	120.8	25.5	5.2	0.30	7.5	0.68	3.6	39.5	73.8	212.6	20.5	5.4
D	18.3	14.1	118.0	28.1	7.2	0.39	8.3	0.86	4.9	36.5	73.5	151.2	15.0	4.1
E	25.8	16.5	113.1	73.2	34.2	1.32	23.5	1.46	4.8	27.3	67.0	184.0	14.0	6.7
F	32.2	17.2	107.6	123.5	78.1	2.42	46.3	1.69	6.1	25.1	92.0	114.8	15.1	4.6

Source: Data from Fang and Fernandez, 1981 with additional data.

Notes
γ_d = Maximum dry unit weight; W_L = Liquid limit; I_t = Toughness index; OMC = Optimum moisture; content I_p = Plasticity index; σ_t = Tensile strength; I_f = Flow index; A = Activity; q_u = Confined compressive strength.

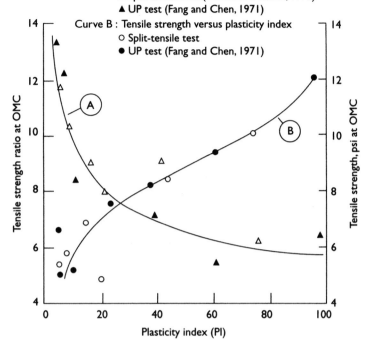

Figure 8.13 Relationship between plasticity index versus tensile strength and compressive–tensile strength ratio.

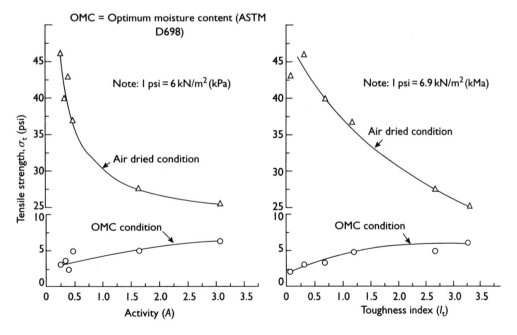

Figure 8.14 Tensile strength versus soil constants of two moisture contents during tensile test. (a) Activity and (b) toughness index.

Source: Data from Table 8.3.

8.10.3 Relationship of tensile strength with other strength parameters

1 Tensile strength related to unconfined compressive strength: The ratio of unconfined compressive strength to tensile strength of materials is of interest to all design engineers because of its practical uses. The result of the ratio of the unconfined compressive strength to tensile is shown in Figure 8.15. It is evident that for air-dried specimens, the compressive–tensile ratios for the six types of soils from Table 8.3 were relatively constant. However, for higher moisture contents, the ratios increased significantly for all types of soil, especially for the low plasticity soils. In Region I of Figure 8.15, when the plasticity index is less than twenty, the effects of moisture content and soil type on the compressive–tensile strength ratio are equally important. However, in Region II, the compressive–tensile strength ratio is insensitive to the variation in percent of clay and soil types as reflected in the plasticity index.

Figure 8.16 presents the relationship between compression–tension ratios versus molding moisture content for various soil types as reflected by the plasticity index, I_P. In examining Figure 8.16, the compression–tension ratios were sharply increasing for the low plasticity soil ($I_P = 4.9$). However, for the higher plasticity index, the ratio change was quite small.

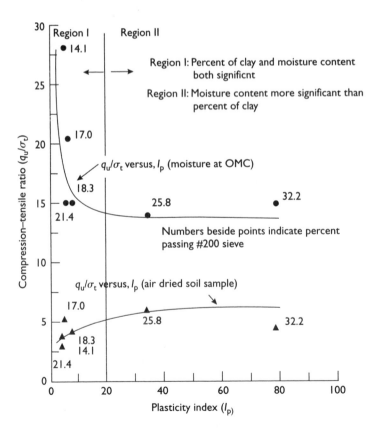

Figure 8.15 Compressive–tensile ratio versus plasticity index of two moisture contents during compression and tension tests.

Source: Data from Table 8.3.

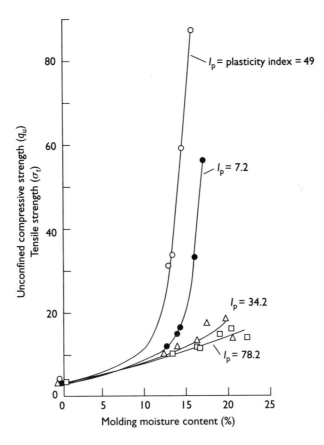

Figure 8.16 Compression–tension ratio versus molding moisture content with various soil type as reflected by plasticity indexes.

Source: Data from Table 8.3.

2 *Tensile strength related to cohesion*: The relationship between cohesion and tensile strength is useful to the practitioner. When soil is dry, the ratio of cohesion and tensile strength is influenced only by soil type. Based on data presented by Fang and Fernandez (1981) and Table 8.3, plots are made, showing the cohesion–tensile ratio versus molding moisture content for three different soil types as reflected by the plasticity index. A curve with a plasticity index equaling 4.9 shows that a slight increase of the molding moisture content, sharply affects the cohesion–tensile ratio, while a curve with a plasticity index equaling 78.2, indicates that molding moisture has a lower effect on the cohesion–tensile ratio. Figure 8.17 presents the relationship between tensile strength and cohesion.

3 *Tensile strength related to the internal friction angle*: Based on data from Table 8.3, a curvilinear relationship between friction angle, ϕ_t, and tensile strength was observed for all three types of soils as indicated in Figure 8.18. It indicates that soils with higher plasticity indexes have lower friction angles. It can also be seen that at lower tensile strengths the rate of increase of the friction angle is much higher than at higher tensile strength.

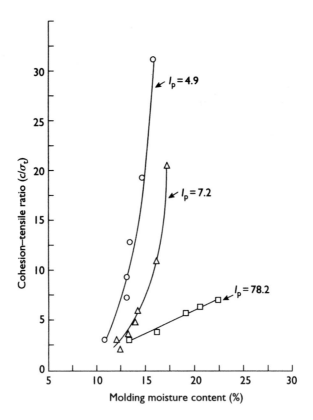

Figure 8.17 Cohesion–tensile ratio versus molding moisture content with various soil type as reflected by plasticity indexes.

8.10.4 Tensile characteristics of rocks and stabilized materials

1 *Rocks*: As with soil, the tensile strength of rock is considerably less than the compressive strength. In general and depending on the mineralogy, the tensile strength of most rock material is about 10–20% of its compressive strength. For example, Dismuke *et al.* (1972) observed compressive strengths ranging from about 5 to 20 kips per square inch (ksi) (1 kip = 1000 lb). The corresponding tensile strengths ranged from 1 to 3 ksi. Strength is also related to unit weight, and the observed increases in tensile or compressive strength correlated with unit weights ranging from 160 to 220 pcf.

2 *Stabilized materials*: The variation of tensile strength with the additives, including lime, fly ash, and cement was studied. The results are shown in Figures 8.19(a) and (b). Figure 8.19(a) shows the variation of tensile strength with different percentages of additive. The percentage by weight of optimum additives for each stabilizing agent can be determined from these curves. Figure 8.19(b) shows the variation of tensile strength with molding water content of optimum additives. It indicates that the maximum tensile strength occurs on the dry-side of OMC conditions for all three cases. Density tests were determined by the standard AASHTO compaction method (Sec. 7.2).

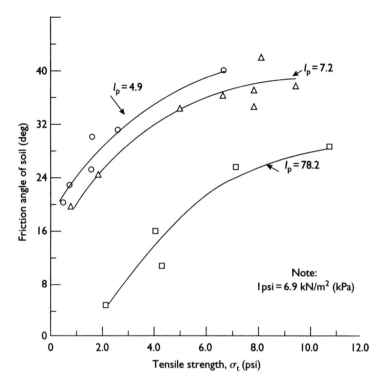

Figure 8.18 Friction angle versus tensile strength for various types of soil as reflected by plasticity indexes.

8.11   Environmental factors affecting tensile strength

8.11.1   Type of exchangeable ions in soil

As discussed in Section 8.10.3, the ratio of unconfined compressive strength to tensile strength of materials is of interest to all design engineers because of its practical uses. In this section, the effect of exchangeable ion on the ratio of unconfined compressive strength to the tensile strength of soil is discussed. Table 8.4 data are presented for tensile and compressive strengths of specimens of nine ionic modifications (Sec. 3.10) of two different clay soils. The specimens were molded in the plastic range and dried. In examining Figure 8.20, it is noted that for Hagerstown soil (New Jersey silty clay) consistently exhibits more variation in all index tests.

8.11.2   Freezing–thawing cycles effects on tensile Strength

Freezing–thawing cycles reduce tensile strength significantly as indicated in Figure 8.21. Most of the damage appears after the first cycle. It is especially true for the expansive bentonite clay. Of course, lower tensile strength translates into a greater cracking potential. A practical example of where such knowledge is important is in the design of waste containment facilities. Often, such facilities use compacted clay to

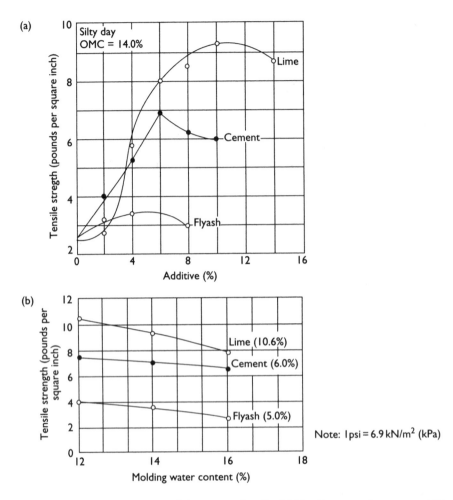

Figure 8.19 Tensile strength compacted fine-grained soils. (a) Tensile strength versus additive as a percentage at optimum moisture content, and (b) tensile strength versus water content at optimum additive (figures in parenthesis are optimum additive percentage).

Source: Fang and Deutsch (1976).

prevent rainwater from infiltrating into the waste, and the compacted clay must be protected from frost penetration.

8.12   Summary

The characteristics of cracking and fracture belong to one system. Cracking represents the prefailure phenomena of a material, and fracture is the behavior of a material at the failure condition. Soil cracking is the major cause of the progressive failure of earth slopes, erosion, and premature failure of all types of earth structures.

Soil cracking is a natural phenomena due to internal energy imbalance in the soil mass and changes in local environments; four types of cracking are examined namely shrinkage, thermal, tensile, and fracture.

Table 8.4 Tensile–compressive strength and tensil/compressive strength ratio of natural and homoionic soils of cecil and Hagerstown soils

Soil type	Exchange ions	Strength, psi		Compressive tensile
		Tensile	Compressive	
Cecil subsoil	H	61	73	1.20
	Na	52	86	1.65
	K	63	110	1.74
	Mg	70	102	1.45
	Ca	51	125	2.45
	Ba	75	121	1.61
	Al	72	147	2.04
	Fe	60	142	2.37
	Natural	65	124	1.91
Hagerstown soil	H	105	445	4.24
	Na	135	342	2.53
	K	88	341	3.88
	Mg	110	575	5.23
	Ca	182	357	1.96
	Ba	153	341	2.23
	Al	93	368	3.96
	Fe	98	302	3.08
	Natural	160	312	3.01

Source: Based on Winterkorn, 1995 with additional data.

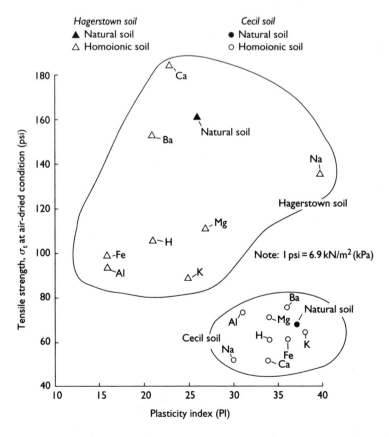

Figure 8.20 Effects of exchangeable ions on tensile strength.

Source: Date from Table 8.4.

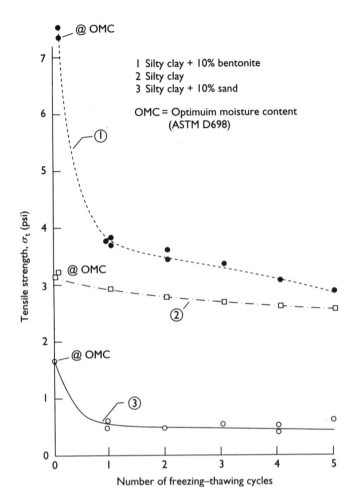

Figure 8.21 Effects of freezing–thawing cycles on tensile strength of sand bentonite liner material.

Source: Fang, H. Y., Daniels, J. L., and Work, D. V. Soil contamination and decontamination mechanisms under wet–dry and freeze–thaw conditions. *Proceedings of the 4th International Symposium on Environmental Geotechnology and Global Sustainable Development*, Danvers, MA, pp. 1158–1171. © 1998 ASCE. Reproduced by permission of the American Society of Civil Engineers.

LEFM is limited to linear elastic materials. The concept has been expanded to include materials such as concrete, bituminous mixture, stabilized soils, as well as timber; LEFM cannot be applied to expansive clays (such as bentonite clay) due to the complicated soil–water interactions. Further research is needed in applications in the various geotechnical areas such as landslides, progressive failure, erosion, etc., and in the area of correlation between cracking pattern and fracture load and characterization of soil behavior based on fracture loads or strain energy rate.

For quantitative measurement of the cracking behavior of soil, determining the tensile strength of soil is one of the simple approaches. Tensile strength determined by UP testing has been described with typical test results for various soil types.

The UP test is simple and easy to perform. No additional equipment is needed for the test, which could be tied in with routine CBR or compaction tests. The tensile test can be used for both laboratory and in situ construction control.

PROBLEMS

8.1 Explain the relationship between cracking, fracture, and tensile strength of a soil.

8.2 Prove that the shrinkage limit is the function of unit weight of water, soil, and specific gravity of solids.

8.3 Explain why cracking patterns form, and how to measure the cracking patterns of soil.

8.4 Discuss the limitations for applications of linear fracture mechanics. Also explain, why the fracture mechanics concept cannot be applied for evaluation of fracture behavior of expansive clays?

8.5 Explain why fracture and tensile tests are the quantitative measurements of the cracking behavior of a soil mass?

8.6 Discuss why the UP test is better than other methods used for measuring tensile strength of soil? Explain the significant difference between the split tensile test and the UP test.

8.7 Derive from Equation (8.4) that tensile strength is the function of load and geometric shape of a soil specimen.

8.8 Explain the differences between ASTM standard split-tensile test and UP test for the determination of tensile strength of concrete block, rock, and compacted clay including failure modes.

8.9 Explain why the moisture content is so sensitive to tensile strength over compressive strength. Also, explain why there is very little tensile strength for saturated soil.

8.10 When soil is dry, the ratio of cohesion–tensile strength influences only the soil types (not the others). Explain why?

8.11 A low plasticity index soil, as shown in Figure 8.18, behaves such that a slight increase of the molding moisture content sharply affects the cohesion–tensile strength ratio, while with a higher plasticity index, the molding moisture content has a lower effect on the cohesion–tensile strength ratio. Explain why?

Chapter 9

Consolidation, stress distribution, and settlement

9.1 Introduction

When a load is applied to a soil mass, deformation may result from (1) immediate elastic and inelastic deformation of the soil structure, (2) porewater drained from the soil mass, (3) continuous time dependent or viscous flow under shear stress resulting in reorientation of the soil particles, and (4) a combination of all the above, which in most cases occurs simultaneously. However, it will depend upon soil types, properties, drainage conditions, stress history, and environmental conditions. Case (1) is generally referred to as compaction or densification (Ch. 7), Case (2) is referred to as consolidation or settlement, Case (3) is referred to as creep (Sec. 10.9); and Case (4) is referred to as subsidence (Sec. 16.6). In this chapter, only Case (2), the consolidation and settlement are discussed.

Consolidation processes tend to control the engineering properties of compressible soils, have a dominating influence on their strength, and govern the rate and magnitude of settlement that occurs when such deposits are subjected to load. There are three general types of consolidation phenomena in clay deposits existing in the natural condition namely (a) normally consolidation, (b) overconsolidation, and (c) underconsolidation. In most case, the soil deposit is normally consolidated. Figure 9.1 shows the difference between normally and overconsolidated soil

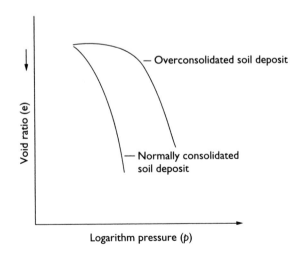

Figure 9.1 Void ratios, e versus logarithm of effective pressure, p curves: (a) Normally consolidated soil deposit; and (b) Overconsolidated soil deposit.

deposits in laboratory standard consolidation test. An overconsolidated soil deposit has a complex failure mechanism but is generally desirable from a construction perspective. In the following sections all three types of consolidation phenomena will be discussed with emphasis on the normally and overconsolidated consolidation soil deposits.

In addition, the determination of vertical stress or pressure distribution in the soil mass caused by various types of surface loading, and the settlement analysis are discussed in detail. The section on settlement is divided into sections which deal with cohesive and cohesionless soils. This subdivision is not entirely satisfactory since natural soils frequently do not fit completely into either category. However, some additional information of settlement in various environmental conditions is added such as settlement analysis caused by dewatering, deep excavation, and decomposition at landfill sites.

9.2 Consolidation phenomena and mechanisms

9.2.1 General discussion

When a load is applied to a saturated compressible soil mass, the load is carried initially by the water in the pores because the water is relatively incompressible when compared with the soil structure. The pressure, which results in the water because of the load increment, is called *hydrostatic excess pressure* because it is in excess of that pressure due to the weight of water. If the water drains from the soil pores, the hydrostatic excess pressure and their gradients gradually decrease, and the load increment is shifted to the soil structure. In other words, the transfer of load is accompanied by a change in volume of the soil mass equal to the volume of water drained. This process is known in soil mechanics as consolidation. A theory relating to loading, time, and volume change was proposed by K. Terzaghi in the early 1920s and has been become known as the Terzaghi Theory of Consolidation. One of the major assumptions in the theory is that volume change and the outflow of porewater occur in one direction only. For this reason it is sometimes referred to as *one-dimensional consolidation theory*.

9.2.2 One-dimensional and three-dimensional consolidation phenomena

1 One-dimensional consolidation phenomenon: In the one-dimensional consolidation concept, it is assumed that strains occur in the vertical direction only and are controlled solely by the magnitude of the vertical normal stress. If the compressible stratum is at a considerable distance below the ground surface, it is generally conceded that very little lateral movement can occur in the clay during its compression; thus, this requirement is essentially satisfied. However, when a structure (footing) rests directly on compressible soil, lateral movement takes place. Consequently, the consolidation is not one-dimensional and the one-dimensional consolidation theory is not strictly applicable. There are three types of one-dimensional consolidation phenomena namely, initial, primary, and secondary consolidations as discussed in Section 9.2.4.

2 Three-dimensional consolidation phenomenon: Natural soil deposits are consolidated three-dimensionally, and laboratory studies of their strength characteristics, accordingly, attempt to duplicate field conditions. In such studies the Terzaghi theory has been presumed to hold, but there are insufficient data to substantiate this presumption. A three-dimensional theory is available (Biot, 1941) but thus far has not been practically applied. However, a brief presentation of Biot's theory is as follows.

9.2.3 Biot's three-dimensional consolidation theory

Assumptions for Biot's theory include (a) isotropy of the material: the elastic properties of soil are the same in any direction at a point. Due to the fact that many compressible clays are of sedimentary origin, this assumption is frequently violated in practice; (b) reversibility of stress–strain relations under final equilibrium condition; (c) linearity of stress–strain relations: the relationship between void ratio and effective pressure is a straight line during three-dimensional consolidation; (d) small strain: the soils only endure a limited deformation during the three-dimensional consolidation; (e) water may contain air bubbles; (f) solids and liquid are incompressible: compressibility of soil grains and water is assumed to be negligible. Thus the total volume change is equal to the change in the void space; and (g) validity of Darcy's Law: this means that the flow is laminar.

The above assumptions constitute the basics of Biot's general three-dimensional theory. The assumption of isotropy is not essential, and anisotropy can easily be introduced as a refinement. Assumption (c) is subject to criticism but does not introduce serious errors when the stresses are small. Assumption (d) requires that the grain pattern to be undisturbed, which is reasonable for primary consolidation only. It is known from one-dimensional tests that the stress–strain relations in clay soils are not reversible; thus, assumption (b) is probably not even approximately true in practice. For this reason, Biot's theory has not been widely accepted.

9.2.4 Initial and primary consolidations

1 *Initial consolidation*: Initial consolidation is defined as a comparatively sudden reduction in volume of a soil mass under an applied load due principally to expulsion and compression of gas in the soil's voids preceding primary consolidation. This term also is referred to as *initial compression*.

2 *Primary consolidation*: Primary consolidation is the reduction in volume of a soil mass caused by the application of a sustained load to the mass and due principally to a squeezing out of water from the void spaces of the mass and accompanied by a transfer of the load from the soil-water to the soil-solids. This term also is referred to as *primary compression* or called *primary time effect*. Terzaghi consolidation theory is solely based on the primary consolidation phenomenon, which will be discussed in detail in Section 9.3.

9.2.5 Secondary consolidation

1 *General discussion*: Secondary consolidation is defined by ASCE (1958) as the reduction in volume of a soil mass caused by the application of a sustained load to the mass and due principally to the adjustment of the internal structure of the soil mass after most of the load has been transferred from soil-water to soil-solids.

This term is also referred to as *secondary compression* or called *secondary time effect*. Leonards and Ramiah (1959) reported further observations of the consolidation process, both in the field and in the laboratory which demonstrated that volume changes continue to occur after excess hydrostatic pressures had essentially dissipated. This process has been termed *secondary* or *secular consolidation*. A review of the development of secondary consolidation process is presented as follows:

2   *Casagrande method*: Casagrande (1936) appears to be among the first to clearly identify the phenomenon of secondary compression in connection with his efforts to develop improved "curve fitting" procedures to obtain better agreement between values of the coefficient of permeability measured directly and those computed from consolidation test data.

3   *Buisman method*: Buisman (1936) proposed a semi-empirical formula for estimating the amount of secondary compression on the basis of laboratory test data as

$$\Delta h = h \, q \, (\alpha_p + \alpha_s \log t) \qquad (9.1)$$

where Δh = settlement, h = thickness of the stratum, q = loading intensity, α_p = coefficient of primary compression, α_s = coefficient of secondary compression, and t = time. As α_p and α_s were considered constant, Buisman (1936) assumed that the secondary compression per unit thickness, per unit load intensity was proportional to the logarithm of time, t.

4   *Taylor and Merchant work*: The first rational theory to account for secondary compression was proposed by Taylor and Merchant (1940), who assumed that secondary compression began after the primary ceased and that their rate was proportional to the undeveloped secondary compression. Later, Taylor (1942) supplemented Terzaghi theory (1925) to include secondary effects during primary consolidation.

5   *Tan's work*: Tan (1957) objected to the Taylor theory (1942) on the grounds that the ultimate settlement could not exceed that predicted from Terzaghi's theory. Tan proposed his own theory assuming that the secondary compressions were due to a combined viscous and elastic action.

9.2.6   Consolidation mechanisms and models

1   *General Discussion*: The basic consolidation theory was developed by Terzaghi; however, the first rational theory to account for secondary consolidation (compression) was proposed by Taylor and Merchant in 1940. Biot (1941) proposed a general theory of three-dimensional consolidation to include the volume changes during the consolidation load under three directions as discussed in Section 9.2.2. Gray (1936) and Taylor (1942) carried out comprehensive laboratory studies on various aspects including drainage systems, secondary compression, and standardization of test procedures. From this effort, several models of behavior have been developed. These models are useful for providing insight into basic processes as well as serving as a basis for subsequent numerical and computational modeling of soil behavior.

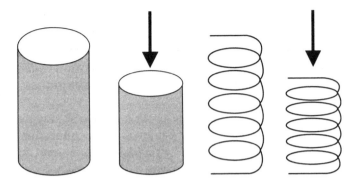

Figure 9.2 Spring analogy for consolidation.

2   *Terzaghi model*: The Terzaghi model considers the consolidation process to be analogous to a spring loading, as shown in Figure 9.2. As the spring, which is defined by an associated spring constant, *k*, is compressed, porewater escapes. The mechanism is presented in Section 9.2.1.

3   *Taylor model*: The Taylor model (1948) considers the consolidation process to be similar to the simultaneous loading of a spring and a viscous fluid. The addition of a viscous fluid helps to better account for secondary consolidation.

4   *Tan model*: Tan (1957) proposed a new theory, which assumed that the secondary compression was due to a combined viscous and elastic action. In his theory, which accounts for both primary and secondary compression, it is assumed that secondary compression continues until spring, *S*, carries all of the external load.

In these models, primary consolidation ceases when the shortening of the spring is complete. Thus, in the modified Taylor theory, primary and secondary consolidation develop and cease simultaneously, while Leonards and Ramiah (1959) commented on the consolidation phenomena based on the results of laboratory studies from Taylor in 1942 and Fang in 1956 (cited by Leonards and Ramiah, 1959), which show substantially reduced secondary compression during triaxial consolidation (volume change) as compared with standard one-dimensional tests (ASTM D2435).

9.3   Terzaghi's one-dimensional consolidation theory

9.3.1   Assumptions

The assumptions of consolidation theory include (a) soil is homogeneous (although many soils are heterogeneous); (b) soil is completely saturated (although most naturally deposited clays below the phreatic line approach saturation, earth fills are usually less than 90% saturated); (c) consolidation is one-dimensional (this is one of the major assumptions in the theory, that the strains occur in the vertical direction only); (d) flow is one-dimensional and consolidation is due to outflow of porewater in one direction only; (e) effect of primary consolidation only; (f) assumed that solids and fluid are incompressible. The compressibility of soil grains and water is assumed to be negligible. Thus, the change in volume is due exclusively to a change of the void ratio; (g) voids in soil element remain saturated, which means the flow is continuous; (h) Darcy's Law is valid, which means that the flow is laminar; (i) relationship between void ratio and effective pressure is a straight line during consolidation under a given pressure increment. This assumption is probably not satisfied in practice, but since the actual relationship has not been established, it is customary to assume linear variation; (j) a constant value for the ratio of certain soil properties. The individual soil properties vary considerably with pressure, but their ratio usually remains sensibly constant.

9.3.2   Laboratory consolidation tests

The main purpose of a consolidation test is to establish the deformation versus time relationship. A brief review of the one-dimensional consolidation apparatus is presented as follows. The first apparatus for determining the compressibility characteristics of soil was built by Terzaghi around 1925–1926 and was called an *Oedometer*. The first consolidation apparatus for large diameter samples was designed and prepared by A. Casagrande in 1932 or 1933 (Gilboy, 1936). At present, the two methods most widely used for applying sustained loads are the jack with load measurement by platform scale, and the wheel or lever system on which weights of known magnitudes are placed. The former apparatus is more portable, but the latter one is more compact and tends to hold the applied loads more nearly constant.

Two types of soil containers are in common use: the *fixed-ring* container and the *floating-ring* container. In the fixed-ring container, specimen movement relative to the container is downward while in the floating-ring container, compression occurs toward the middle from both top and bottom directions. When using the floating-ring container, the effect of friction between the container wall and the soil specimen is minimized; on the other hand, only the fixed-ring container can be adapted for permeability tests (ASTM D2435).

9.3.3   Fundamental differential equations and its solutions

1   General discussion: Consider an elemental volume of clay of height, ds, and area, A, with its top surface at distance, z, below some reference elevation. The porewater is squeezed in a vertical direction from the voids during one-dimensional consolidation. Since the solids and the water are incompressible, the change in volume is equal to the change in volume of the voids. Based on this point of view, and using the assumptions previously listed, the following fundamental equation was derived by Terzaghi (1925):

2   Fundamental Differential Equation and Its Solution:

$$\frac{\partial u}{\partial t} = c_v \frac{\partial^2 u}{\partial z^2} \tag{9.2}$$

where u = hydrostatic excess pressure, t = time, c_v = coefficient of consolidation, and z = depth from ground surface. For the particular case of a linear initial pressure distribution throughout the depth of the compressible stratum and with boundary conditions $u = 0$ for $z = 0$ or $z = 2H$ for all values of the t (time) except $t = 0$, the solution is

$$U\% = 1 - \sum_{N=0}^{N=\infty} \frac{2e^{-M^2 T}}{M^2} \tag{9.3}$$

where $M = \pi/2 \, (2N + 1)$, N = any integer, include zero, e = base of natural logarithm. Equation (9.3) expresses the relationship between time and the average state of consolidation over the height of the stratum. In the application of the consolidation theory to the prediction of settlements, only the average consolidation need be considered. It has been found that this equation may be represented with high precision by the following expressions:

When $U < 60\%$, $T = \pi /4 \, U^2$ \hfill (9.4)

and when $U > 60\%$, $T = -0.9332 - \log_{10} (1 - U) - 0.0851$ \hfill (9.5)

3 *Relationship between void ratio and pressure*: For the determination of the void ratio of soil under one-dimensional consolidation, since it is assumed that the volume change occur is one dimension only, the area of specimen under the vertical load remains constant. The relationships of void ratio, e, at any pressure, p, is given by Equation (9.6):

$$e_1 = e_o - \frac{\Delta h}{h_s} = e_o - \frac{R}{h_s} \tag{9.6}$$

where e_f = final void ratio, e_o = initial void ratio, $R = \Delta h$ = total change in height of the specimen, h = height of the original specimen, h_s = reduced height of solids:

$$h_s = W_s /G_s \, \gamma_w \, A = h / 1 + e_o \tag{9.7}$$

where W_s = weight of the solids, G_s = specific gravity of solids, γ_w = unit weight of water, A = cross-sectional area of the specimen, and h = initial height of soil specimen.

4 *Computation of consolidation settlement*: The major conclusion of Terzaghi's consolidation theory is that it produced two main equations for the prediction of magnitude and time rate of consolidation. Then use the degree of consolidation to link settlement and the time rate relationship as shown in following equations:

a   *Total (Maximum) consolidation settlement*

$$\Delta H = \frac{H}{1 + e_o} c_c \log_{10} \frac{p_1 + \Delta p}{p_1} \tag{9.8}$$

b   *Consolidation time rate*

$$t = \frac{TH^2}{c_v} \tag{9.9}$$

where ΔH = total settlement, H = thickness of clay layer, e_o = initial void ratio, c_c = compression index p_1 = overburden pressure, Δp = added stress, t = time, T = time factor, dimensionless, and c_v = coefficient of consolidation.

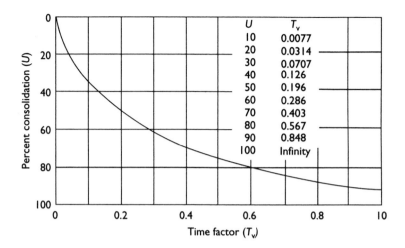

U	T_v
10	0.0077
20	0.0314
30	0.0707
40	0.126
50	0.196
60	0.286
70	0.403
80	0.567
90	0.848
100	Infinity

Figure 9.3 Degree of consolidation versus time factor.

c   *Degree of consolidation and time factor*: The *degree of consolidation* or *percent of consolidation* is defined as the ratio, expressed as a percentage of the amount of consolidation at a given time within a soil mass to the total amount of consolidation obtainable under a given stress condition. Based on Terzaghi consolidation theory, it can be computed from Equation (9.3). Figure 9.3 presents the degree of consolidation or percentage of consolidation, U, versus time factor, T. The time factor, T, is a dimensionless number derived from Terzaghi consolidation theory. This relationship is controlled by drainage pattern during the consolidation. Several approaches have been proposed (Taylor, 1948), however, Figure 9.3 is a simple approach and commonly used when the drainage pattern is assumed linearly versus the depth of soil layer. Figure 9.3 also referred to as *consolidation–time curve* or *theoretical time curve* or simply the *time curve*.

9.3.4   Terms used in Terzaghi's consolidation equations

Various terms used in Terzaghi's one-dimensional consolidation equation are explained as follows. In addition to the standard method some time saving short-cut methods are also presented.

1   Compression index, c_c: Compression index, c_c, is the slope of the linear portion of the pressure–void ratio curve on a semi-logarithm plot. This index is used to indicate the degree of compressibility of clays. For normally consolidated clay, this index can be estimated roughly from liquid limit, or initial moisture content, and initial void ratio. Goldberg *et al.* (1979) proposed an experimental equation that can be estimated from initial void ratio, Atterberg limits, and preconsolidation pressure as follows:

$$c_c = 0.568(e_o + 0.0033\, \omega_L - 0.0082\, \omega_P + 0.0329\, p_c - 0.4322) \qquad (9.10)$$

where c_c = compression index, ω_L = liquid limit, ω_P = plastic limit, and p_c = preconsolidation pressure. Equation (9.10) is developed based on 96 soil samples obtained from alluvial deposits, Wabash lowland and Crawford upland soils from the State of Indiana. The equation can be used for both normally and preconsolidated clay deposits. For normally consolidated deposits, the p_c value is equal to 1 in Equation (9.10), and for preconsolidated clay, an averaged value of the entire deposit is recommended. The c_c also can be estimated from dry unit weight, γ_o, of soil and density of water, γ_ω, as proposed by Rendon-Herrero (1980) as

$$c_c = 0.5 \, (\gamma_o/\gamma_\omega)^{12/5} \tag{9.11}$$

where γ_o = dry unit weight of soil, and γ_ω = unit weight of water. The application of Equation (9.11) requires that an estimate first be made of a soil materials stress history for determining an appropriate value of γ_o. Comparison of the in situ moisture content with liquid and plastic limits, ω_P, will generally suffice in estimating a soil material's stress history. If a soil deposit is normally consolidated, it has a suggested value of 95% of the initial void ratio, e_o, which should be used for computing the dry unit weight, γ_o.

If the soil deposit is overconsolidated however, the in situ soil index properties will aid only in computing a dry unit weight approximately corresponding to the recompression portion of the laboratory compression curve. Equation (9.11) was developed based on 94 consolidation tests where approximately 85% were for fine-grained soil with compression indices between 0.04 and 0.50. For practical uses, Koppula (1981) pointed out that there is a high correlation existing between initial moisture content and compression index, and it is postulated that the initial moisture content of the soil may yield a good indication of the magnitude of compression index. A statistical approach based on 134 consolidation tests results from the Province of Alberta, Canada were used. A simple correlation equation between c_c and ω_o is given as

$$c_c = 0.01 \, \omega_o \tag{9.12}$$

where c_c = compression index, and ω_o = initial moisture content. The range of values for c_c in Equation (9.12) varied between 0.0075 and 0.0111 for cohesive soils. Other such regression analysis equations for various soil types are given by Azzouz et al. (1976). The value of c_c is required for computing the maximum settlement of a soil deposit under the structural loads and will be further discussed in Section 9.4. The typical values of c_c in natural condition are Alluvial deposit = 0.42, Boston blue clay = 0.35, Chicago clay = 0.42, loess = 0.09–0.23. For remolded condition of soil, the c_c value is smaller.

2 *Coefficient of compressibility, a_v:* The coefficient of compressibility, a_v, is the ratio of the change in void ratio, e, to the corresponding change in pressure, p. In other words,

$$a_v = de/dp \approx (e_1 - e_2) \, / \, (p_2 - p_1) \tag{9.13}$$

Also, a_v is the slope of the void ratio, e, versus pressure, p curve, and can be computed from c_c as

$$a_v = \frac{0.435 \, c_c}{p_{(av)}} \tag{9.14}$$

where $p_{(av)}$ = the average pressure for the increment during the consolidation test. The a_v value is used to indicate the degree of compressibility of clay deposits.

3 *Coefficient of consolidation, c_v*: The coefficient of consolidation, c_v, is used to indicate the combined effects of permeability and compressibility for the given void ratio range. It is an important parameter for predicting the time rate in the settlement analysis. This value can be determined from consolidation test data, method as proposed by Casagrande (1936), a square-root method by Taylor (1948), and inflection point method by Cour (1971). The c_v value based on the inflection point method as shown in Figure 9.4 can be computed from following equation (Eq. 9.15):

$$c_v = 0.405 \, (H^2 \, / \, t_i) \tag{9.15}$$

where H = the average thickness of soil specimen or the longest drainage path during the given load increment, and t_i = the inflection point occurs in a plot of deformation (dial reading) versus the logarithm of time. If the time curve is plotted, the t_i value can easily be recognized with a reasonable degree of accuracy. If the time curve is not plotted, or if a more precise location of the inflection point is desired, it can be defined as the point at which the absolute value of the tangent to the time curve on a semi-logarithmic plot reaches a maximum.

A curvilinear relationship between c_v and liquid limit ω_L was observed (Hough, 1957) as c_v decreases the ω_L value increases. Typical values for c_v vary from 10^{-2} to 10^{-4} cm^2/s. However, for clay minerals such as illite and montmorillonite, the c_v value can reach 3×10^{-5} cm^2/s.

4 *Coefficient of permeability, k*: The coefficient of permeability, k (Sec. 5.4), can also be computed from consolidation test result as

$$k = (c_v) \, (m_v) \, (\gamma_\omega) \tag{9.16}$$

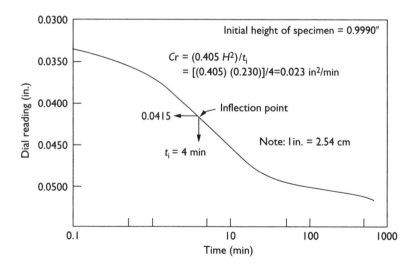

Figure 9.4 Determination of coefficient of consolidation, c_v, by the inflection point method.

Source: Cour, F. R. Inflection point method for computing c_v, *Journal of the Soil Mechanics and Foundations Division Proceeding of the ASCE*, v. 97, no. SM5, pp. 827-831. (c) 1971 ASCE. Reproduced Permission of the American society of Civil Engineers.

where c_v = coefficient of consolidation, and can be obtained from a standard consolidation test, and m_v is known as the coefficient of volume compressibility, and γ_ω = unit weight of water.

5 *Coefficient of volume compressibility, m_v:* The coefficient of volume compressibility, m_v, is defined as illustrated in Equation (9.16). The m_v value is used to indicate the compression of the clay per unit of original thickness due to a unit increase of the pressure. The units of m_v are the same as a_v (cm²/kg), using the average void ratio for the increment during the consolidation test. Zeevaert (1983) reported that the k value presented in Equation (9.16) is a linear relationship with the coefficient of consolidation, c_v, when m_v varies from 0.1 to 0.001 cm²/kg.

$$m_v = \frac{a_v}{1 + e_{ave}} \tag{9.17}$$

where m_v = coefficient of volume compressibility (cm²/kg), a_v = coefficient of compressibility, and e_{ave} = average void ratio for the increment during the consolidation test.

6 *Swell Index, c_s:* To express the expansion which may occur upon the unloading of a soil sample, one commonly uses the swell index, c_s, as shown in Figure 9.5. The c_s values are always much smaller than the c_c values for a virgin compression. The swell index can be expressed empirically as follows as proposed by Nagaraj and Murty (1985):

$$c_s = 0.0463 \left(\frac{\omega_L}{100} \right) G_s \tag{9.18}$$

where c_s = swell index, ω_L = liquid limit (%), and G_s = specific gravity of solids. The typical value for Boston blue clay (Sec. 2.11) ranges from 0.07 to 0.09 and for soft silt from Shanghai (Sec. 2.11) varying from 0.02 to 0.04.

9.4 Overconsolidated clays

9.4.1 *Definition and classification of overconsolidated soil deposit*

1 *Definition of overconsolidated soil:* If the maximum past intergranular pressure, p, is larger than the present overburden pressure, p_o, ($p > p_o$), then this intergranular pressure is called *preconsolidation pressure*, p_c, ($p = p_c$) or *past pressure*. The ratio between preconsolidation pressure, p_c, and overburden pressure, p_o, is called *overconsolidation ratio* (OCR):

$$\text{OCR} = p_c/p_o \tag{9.19}$$

where OCR = overconsolidation ratio; p_c = preconsolidated pressure; and p_0 = overburden pressure. (a) If OCR = 1, then the soil deposit is called *normally consolidated*; (b) If OCR > 1, then it is called *overconsolidated*, and (c) If OCR < 1, it is called *underconsolidated*.

2 *Identification of overconsolidation clay deposits:* By visual observation, there is no difference between normally consolidated and overconsolidated clay deposits.

However, it can be identified by standard laboratory consolidation test results on void ratio versus logarithm pressure curve as indicated in Figure 9.5. It also can be identified by simple relationships of the following equation. A liquidity index, I_L, value less than 0.4 may also imply that the clay deposit is overconsolidated (Fang, 1997):

$$I_L < 0.4 \tag{9.20}$$

where I_L = liquid index (Sec. 2.5.3).

3   *Classification of degree of overconsolidation*: The degree of overconsolidation can be classified based OCR such as:

  a   Lightly overconsolidated clay: $1.0 < OCR < 2.5$;
  b   Heavily overconsolidated clay: $8.0 < OCR$.

An OCR value reaching as high as 32.0 has been reported. OCR values can be used to indicate the stress history of a given soil deposit. For example, a plot of the OCR for Shanghai soft clay (Sec. 2.11) versus depth clearly indicates that there are three distinct subsurface layers with various OCR values which reflect the stress history of the clay deposit.

9.4.2   Causes of preconsolidation pressure

There are numerous reasons and factors causing preconsolidation, including geological features, changes in pore pressures, changes in soil structure, and changes in environmental conditions. A list of such factors causing the preconsolidation pressure may be dichotomized according to mechanical or multimedia energy causes: (a) mechanical: loading or surcharge, and change in porewater pressure; and (b) multimedia: acid rain or acid drainage, freezing–thawing process, ion exchange reaction, and wetting–drying cycles, as shown in Table 9.1.

9.4.3   Procedures for estimation of preconsolidation pressure

1   *Casagrande graphical procedures*: The Casagrande procedure (1936) involves selecting the point [A] corresponding to the minimum radius of curvature on the e versus $\log p$ curve (Fig. 9.5);

  a   At this point [A] horizontal and tangent lines are drawn;
  b   The angle between them (horizontal and tangent lines) is bisected;

Table 9.1 Causes of preconsolidation pressure

1   *Caused by mechanical energy*
(a) Structural loading
(b) Surcharge loading
2   *Caused by multimedia energy*
(a) Porewater pressure
(b) Ion exchange reaction
(c) Freezing–thawing process
(d) Wet–dry, hot–cold cycles
(e) Pollution intrusion

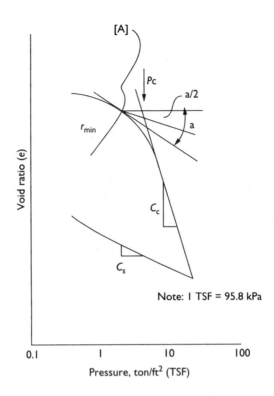

Figure 9.5 Casagrande's graphical procedure for determination of preconsolidation pressure.
Source: After Casagrande (1936) and personal communication.

c   Then the straight-line portion of the curve is projected back to intersect the bisector of the angle;

d   The pressure corresponding to this point [**A**] of intersection is equal to the maximum preconsolidation pressure.

2   *Schmertmann and Leonards approaches*: The techniques for estimating p_c that have been proposed by Casagrande (1936), Burmister (1951), Schmertmann (1955), and Leonards (1962) can be explained by the combined system as follows:

3   *Combined system*: A range of the preconsolidation pressure, p_c, can be estimated as follows:

a   Establish void ratio, e, versus logarithm pressure, p, curve (heavy line in Fig. 9.6) from standard laboratory consolidation test (ASTM D2435);

b   Point c is the beginning of straight-line portion of the e versus log p curve;

c   Point d is equal to 0.40 to 0.42 e_o suggested by Schmertmann (1955) on e versus log p curve;

d   Point a is a straight-line extension from point c;

e   Point b is vertical line from point c;

f   Distance ab is the range of preconsolidation pressure p_c suggested by Leonards (1962);

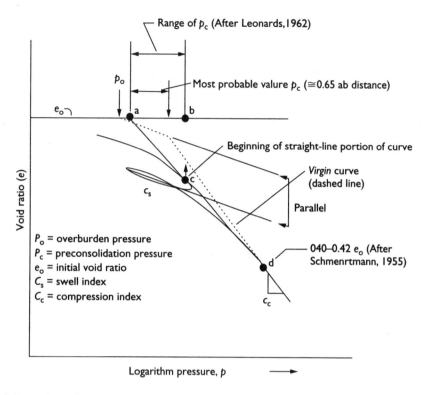

Figure 9.6 Procedure for interpretation of maximum preconsolidated pressure – a combined approach.

(g) The most probable value of p_c is approximately equal to 0.65 to 0.70 in the distance of **ab** as shown in Figure 9.6. If the first loading curve (Fig. 9.5) indicates a well-defined break, the Casagrande construction procedure is applied.

9.4.4 *Engineering problems of overconsolidated deposit*

Overconsolidated soils tend to have reduced settlement and greater strength than soils which are normally consolidated, which is desirable. However, overconsolidated soils are more complicated than normally consolidated soils due to (a) the variable causes of preconsolidated pressure; (b) mechanisms of failure are not clearly understood; and (c) the difficulty in predicting the failure, especially with respect to slope stability. Further discussions of engineering properties and practice in overconsolidated deposits are given by Transportation Research Board (TRB, 1995) including sampling, in situ and laboratory strength tests, and design practice and case studies. Slope stability and landslide problems in overconsolidated soil will be discussed in Section 14.10.

9.5 Consolidation characteristics of contaminated soil deposits

Many natural soil deposits are contaminated. The term contamination is used loosely, which includes numerous factors such as temperature, pore fluid, and types and concentrations of exchangeable ions. Most studies used to determine the consolidation characteristics of these deposits were carried out in the laboratory environment, discussed as follows.

9.5.1 Environmental factors affecting consolidation test results

1 *Temperature effects:* Gray (1936) reported an increase in compressibility with an increase in temperature, with the greatest effects being observed in the range of secondary consolidation (Sec. 9.2). Lo (1961) demonstrated that even relatively small changes in temperature can cause a marked change in compressibility in the secondary consolidation range. Paaswell (1967) found that increase in temperature caused immediate volume changes, the magnitudes dependent upon the magnitude of the temperature change. The magnitude of the initial stress (excess pore pressure) was seen to have a secondary effect on the magnitude of the volume change. These volume changes are attributed to a transfer of stress between the pore fluid and the matrix, as increase in temperature increases the pore pressure but decreases the matrix strength. Plum and Esrig (1969) found that the amount of temperature induced consolidation is related to soil compressibility; the higher the compressibility the greater the consolidations for a given temperature increase. A given increment in temperature at constant effective stress had an effective equivalent to some increment of pressure at constant temperature. It also showed the volumetric strain of an illite–water system as a function of overconsolidation ratio, for a temperature increase from 24°C to 50°C.

2 *Effect of pore fluid:* Standard laboratory consolidation tests have been performed to determine the effect of pore fluid on the compression index, c_c, and expansion index, c_s. Two types of soils, sodium bentonite and an acid kaolinite, were tested. Seven types of pore fluids including water, aniline, nitrobenzene, methyl alcohol, ethyl alcohol, butyl alcohol, and propyl alcohol were used. Results of these consolidation tests are summarized in Figure 9.7.

3 *Effect by exchangeable ions:* Figure 9.8 shows the relationship between coefficients of consolidation, c_v, and compressibility, a_v, versus void ratio, e, for Putnam clay. Five homoionic modifications soils including K, H, Ca, Mg, and Na were used. Since the compressibility indicates the intensity of the forces holding the water films between soil and water, this figure gives a picture of the water-fixing ability of the Putnam clay as a function of the exchangeable ions.

9.5.2 Underconsolidated soils

In addition to normally and overconsolidated soils, there are underconsolidated soils as well. Underconsolidated soils are those in which a stratum of clay deposit is found to exhibit a preconsolidation pressure less than the calculated existing overburden pressure. This is the case where a given deposit is undergoing consolidation from a previously applied load. The deposit has not yet reached an equilibrium condition under the applied overburden stresses. This situation occurs in areas of newly established landfills (Sec. 16.9). When analyzing the settlement in an underconsolidated deposit, both the previous and current load applications must be considered.

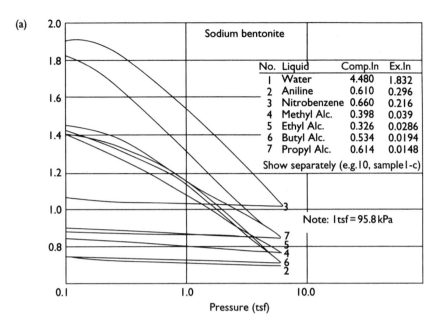

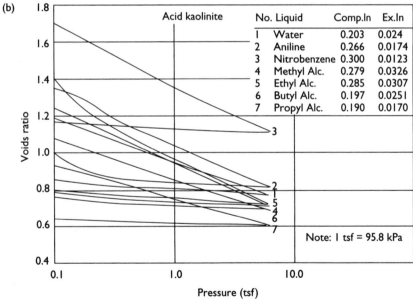

Figure 9.7 Effect of pore fluid on consolidation test results (a) Sodium bentonite, and (b) Acid kaolinite.

Source: Waidelich, W. C., Influence of liquid and clay mineral type on consolidation of clay-liquid systems. In *Highway Research Board Special Report No. 40*, Highway Research Board. National Research Council, Washington DC, 1958, pp. 24–42. Reproduced with permission of the Transportation Research Board.

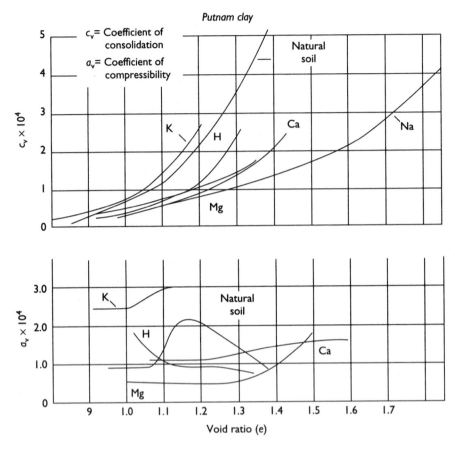

Figure 9.8 Effect of exchangeable ions on coefficient of consolidation and compressibility (a) Coefficient of consolidation, c_v, and (b) Coefficient of compressibility, a_v.

Source: Winterkorn, H. F. and Moorman, R. B. B., A study of changes in physical properties of Putnam soil induced by ionic substitution. In *Proceedings of the 21st Annual Meeting, Highway Research Board, Highway Research Board.* National Research Council, Washington DC, 1941, pp. 415–434. Reproduced with permission of the Transportation Research Board.

9.6   Vertical stress and pressure distribution

9.6.1   *General discussion*

In order to discuss stress or pressure distribution and the resulting settlement, it is necessary to recognize the sources of the pressure. Besides the self-weight of the soil, there are primarily dead and live loads of the structure plus wind loads, which can produce moments, as well as lateral and vertical loads. One of the important aspects of wind loading is the time variation of the load. This chapter will restrict itself to a discussion of those vertical loads which may be considered time invariant. These are the conventional loads used in most settlement calculations.

If a surface load is applied to a unit area, the vertical stress will decrease as the depth of the soil below the ground surface increases. There are two general methods for calculating vertical stress below a concentrated (point) load, namely Boussinesq (1883) and Westergaard (1938) equations. Both of these equations result from the theory of elasticity, which assumes that the stress is proportional to strain. In general, vertical stress is grouped into five types of surface loading conditions as: (a) *point load*: such as monument, silo, TV towers, sign post; (b) *line load*: such as pipe line, sewerage pipe; (c) *strip load*: such as wall footings (d) *uniform load*: such as footings, mat foundation, swimming pool; and (e) *embankment load*: such as highway embankment, earth dams, retaining walls. In this chapter, only the point, uniform, and embankment loads are emphasized.

9.6.2 *Vertical stress under point (concentrated) load*

1 General discussion: The most common approach to determining the distribution of vertical stress in a homogeneous soil beneath a foundation is through the use of some form of linear elastic theory. The justification for this lies in both laboratory and in situ tests, which have utilized in situ pressure cells to measure stresses. The measurements suggest that when the boundary conditions of the analytical model approximate the in situ boundary conditions, the computed stress distribution will correspond reasonably with the in situ conditions.

In the case of layered soils, there is not full agreement on what approach should be used. This depends to some extent on the manner in which the soils are layered. Sowers and Vesic (1962) suggest that the distribution of vertical stress in a material overlain by a much stiffer layer may be better approximated by assuming homogeneity than by using a layered elastic theory. In general, Boussinesq and Westergaad theories are used for computing the vertical stress distribution in soil layers. Figure 9.9 is stress in elastic half-space due to point at the surface.

2 Boussinesq equation: Boussinesq provided a solution to the distribution of stresses within a linear elastic half-space under the influence of a surface point load (Fig. 9.9).

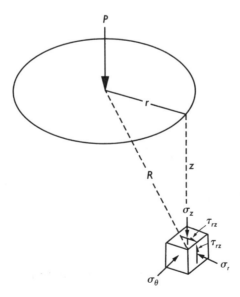

Figure 9.9 Stress in elastic half-space due to point load at the surface.

It can be used to solve the problem of stresses produced at any point at the surface in a homogeneous, elastic, and isotropic medium. Equation (9.21) gives stress, σ_z, as a function of both, the vertical distance, z, and the horizontal distance, r:

$$\sigma_z = I_z \, P / \, z^2 \tag{9.21}$$

$$I_z = (3/2\pi) \, / \, [1 + (r/z)^2]^{\,5/2} \tag{9.22}$$

where σ_z = vertical stress at depth z, P = point (concentrated) load, z = depth from ground surface, r = horizontal distance from point of application of P to point at which σ_z is desired, I_z = influence value which depends only upon the geometry; that is, the location of the point at which the stress is desired relative to the point load P. This influence value is plotted in Figure 9.10 (Perloff, 1975).

3   *Westergaard equation*: Westergaard (1938) proposed an equation for determination of the vertical stress caused by a point load in an elastic solid medium in which layers alternate with thin rigid reinforcements as shown in Equation (9.23):

$$I_z = (1/\pi) / \, [1 + 2(r/z)^2]^{3/2} \tag{9.23}$$

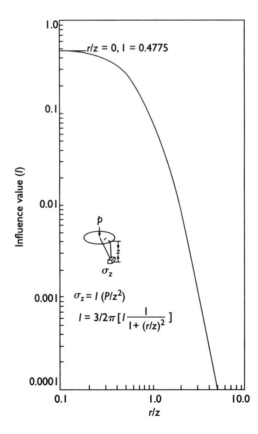

Figure 9.10 Influence diagram for vertical normal stress due to point load on surface of elastic half-space.

Source: Perloff, W. H., Baladi, G. Y. and Harr, M. E., Stress distribution within and under long elastic embankment. In *Highway Research Record* No. 181, Highway Research Board. National Research Council, Washington DC, 1967, pp. 12–40. Reproduced with permission of the Transportation Research Board.

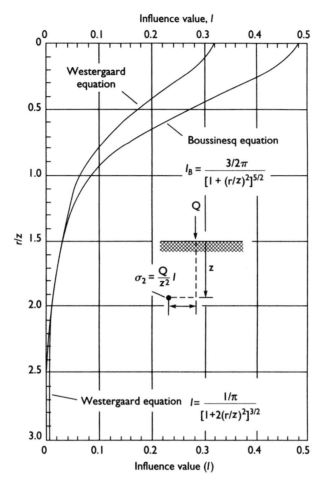

Figure 9.11 Comparison of influencing values determined by Boussinesq and Westergaard equations: (a) Boussinesq equation, and (b) Westergaard equation.

Source: Fenske (1954). Reprinted with permission.

The parameters are as defined in Equation (9.22). A comparison of influence values for point loads determined by the Boussinesq and Westergaard equations is presented in Figure 9.11.

9.6.3 Vertical stress under uniform footing loads

Equations (9.22) and (9.23) apply when the surcharge is applied as a point load at the surface. If the load acts on a footing, several methods are proposed. There are various forms of uniform loaded area such as rectangular, circular loaded areas, and numerous chart or tables are available for specific applications. However, in the following section only the most fundamental forms are presented.

1 Newmark chart (Newmark–Boussinesq method): As indicated in Equation (9.21), vertical stress is independent of the material properties; thus, I_z is a function of geometry only. By the principle of superposition, the point load ban is integrated over a finite area to produce the stress distribution resulting from a uniform stress applied to the surface. Based on the principle of superposition, Newmark (1942) developed a chart for estimation of vertical stress under a uniform footing load. The procedures for determination of vertical stress by the Newmark chart are presented as follows: (a) it is necessary to determine the scale for converting the actual distance or length into model or scaling distance; (b) draw a plan of the building or footing on the tracing paper, to such a scale that the depth, Z, is equal to the distance, AB, (Fig. 9.11), and (c) then place the tracing paper over the influence chart, so that the point on the tracing paper (which is selected as the point for which the vertical stress is desired) is directly above the point in the center of the chart, then count the number of elements enclosed by the outline of the loaded area.

$$p = qIA \tag{9.24}$$

where p = vertical stress, q = surface loading, I = influence value (provided by each chart, see Fig. 9.12), and A = influence area determined from Newmark chart. With the advantage of computers, there have been a very large number of influence charts developed which cover a wide range of surface loading geometries (e.g. Holtz, 1991).

2 *Fenske chart (Fenske-Westergaard method)*: A similar influence chart for determination of vertical stress under uniform footing load based on Westergaard equation (Eq. 9.23) was developed by Fenske (1954) as shown in Figure 9.13. The procedures for the Fenske or Newmark's charts are similar.

3 *Discussion of Boussinesq and Westergaard methods*: The Westergaard theory (1938) is an improvement of Boussinesq (1883) theory. Many geotechnical engineers believe that the use of the Westergaard equations will give a better estimate of the vertical pressure distribution in clay strata. Taylor (1948) has presented a method for determining the vertical stress distribution under a corner of a uniformly loaded rectangular area, using Westergaard's equation. Fenske (1954) pointed out that the Westergaard equations will also have to be integrated to determine the distribution of the vertical stress under uniformly loaded areas, and it would be convenient to have simplified methods such as those Newmark devised for the Boussinesq equations.

4 *Pressure-bulb procedure*: The term pressure-bulb refers to a zone within which appreciable stresses are caused by an applied load. The stress in a soil due to an applied load is dependent upon factors such as the type of soil deposit, loading intensity, shape, and size of the loaded area. If vertical stresses are considered to be of negligible magnitude when they are smaller than $0.2q$, the bulb line shown in Figure 9.14 represents the loaded areas. In this case, the depth, D, of a pressure-bulb is approximately the same as the width, B, of the square loaded area. For a rectangular loaded area, the depth of the pressure-bulb, expressed in terms of its width, is slightly greater than that for a square loaded area. In general, the pressure-bulb is often considered to have a depth, D, in the range of 1–1.5 times the width, B, of the loaded area.

5 *2:1 Method*: The 2:1 method is an approximate method to determine the increase of stress with depth, z, caused by the construction of a foundation structure

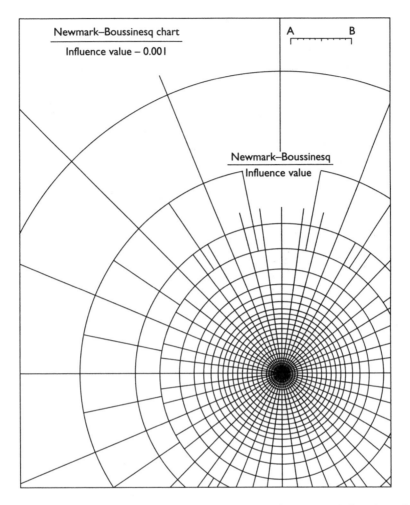

Figure 9.12 Influence values for vertical stress under uniform footing loads based on Boussinesq equation.

Source: Fenske (1954). Reprinted with permission.

such as a footing as illustrated in Figure 9.15. The increase of stress at a depth, z, can be computed by following Equation (9.25):

$$\Delta p = \frac{qBL}{(B + z)(L + z)} \tag{9.25}$$

where Δp = stress increase, q = surface load, **B** = width of the footing, **L** = length of the footing, and z = depth from ground surface.

9.6.4 Embankment loading

1 *Osterberg method*: A method for computing the vertical stress under an embankment load of finite length was developed by Osterberg (1957). The chart was based on

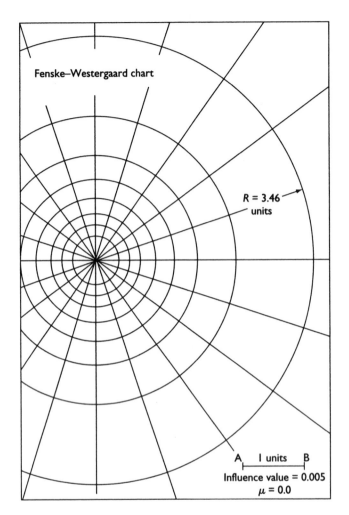

Figure 9.13 Influence values for vertical stress under infinite footing loads based on Westergaard equation.

Source: Fenske (1954). Reprinted with permission.

the Boussinesq solution for a triangular load of infinite extent (Newmark, 1942). By superposition procedures, the vertical stress at the location shown is a function of a/z, b/z, and the unit load, q. From this relationship an influence chart shown in Figure 9.16 has been constructed, which simplifies the computation of vertical stresses beneath an embankment loading. The stress given by the chart is the vertical stress directly under the vertical face of a portion of an embankment of infinite extent. Vertical stresses for any point in the ground foundation can be found by superposition. For stresses under a corner such as under the vertical face of an embankment ending abruptly against a wall, the stresses are one half of those given in the chart.

2   *Perloff–Baladi–Harr method*: The embankment load method developed by Perloff *et al.* (1967) is the determination of the distribution of stresses within and under an embankment resulting from the self-weight of the embankment. Stresses due

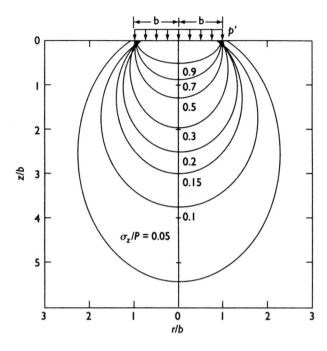

Figure 9.14 Pressure-bulb for determination of vertical stress of soil.

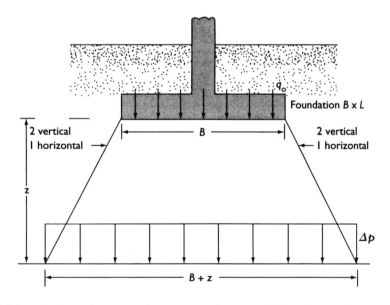

Figure 9.15 2:1 methods to determine the increase of stress with depth caused by the construction of a foundation.

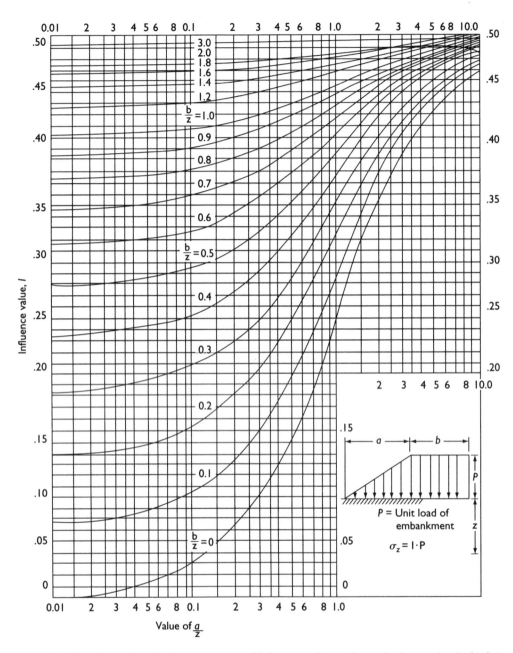

Figure 9.16 Osterberg chart for determination of Influence value under embankment load of infinite length. (a) Problem considered; and (b) Influence values.

Source: Fenske (1954). Reprinted with permission.

to the weight of material underlying the embankment must be superimposed to obtain the total stress. The stresses are expressed in dimensionless form. The coordinates are also in dimensionless form. For convenience in the semi-logarithmic plot, the depth is measured from the top of the embankment. A series of design curves are developed and are shown by Perloff *et al.* (1967) and Perloff (1975).

9.7 Settlement analysis

9.7.1 General discussion

Excessive foundation settlement may result in failure of or damage to structures regardless of the margin of safety against shear failure of the foundation soil. Settlement analysis is, therefore, an essential step in the design of shallow foundations. The purpose of settlement analysis of soil is to predict the magnitudes of settlements and times required for their occurrence. It is generally recognized that settlement due to a change in vertical stress is made of three components as illustrated in Equation (9.26) and Figure 9.17.

$$S = S_d + S_c + S_s \tag{9.26}$$

where S = total settlement, S_d = immediate or distortion settlement, S_c = consolidation settlement, and S_s = secondary consolidation settlement. These are shown schematically in Figure 9.17. Although it is convenient to separate each of these components for analysis, it is important to recognize that in nature all three components can occur simultaneously, and the rate at which S_c and S_s occur is determined by the type of material and drainage boundary conditions.

9.7.2 Causes and types of settlement

There are different methods for computation of the various settlement types. These settlement types and methods can be summarized as (a) the immediate settlement, which is caused by the elastic deformation, and calculations are generally based on equations derived from elastic theory; (b) the consolidation settlement, which is result of a volume change in saturated cohesive soils. Calculations are based on the Terzaghi consolidation theory; and (c) the secondary consolidation settlement, which is observed in saturated cohesive soils and is the result of the plastic adjustment of soil particles.

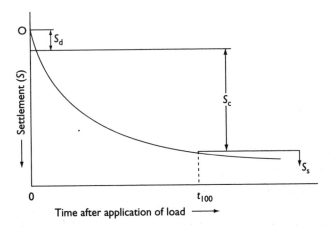

Figure 9.17 Schematic diagram illustrating various types of settlement.

9.8 Immediate settlement

The *immediate settlement* is also called *initial settlement, contact settlement, shear strain settlement,* or *elastic settlement*. It is primarily due to the change in shape (distortion) of the soil elements underneath the foundation. In designing shallow foundations, the immediate settlement can be checked as follows:

9.8.1 Immediate settlement determined by elastic theory

The immediate settlement can be computed on the basis of the theory of elasticity if the soil below the foundation is homogeneous, elastic, and isotropic. The general equation for the estimation of immediate settlement of soil is based on the elastic theory proposed by Janbu *et al.* (1956).

$$S = \frac{qB(1 - \mu^2)}{E} I_s \qquad (9.27)$$

where S = settlement, q = contact pressure between foundation and soil, B = width of the foundation, E = modulus of elasticity of the soil, μ = Poisson's ratio of the soil, and I_s = influence coefficient. This value varies according to the shape of the foundation: for circular foundation, $I_s = 0.85$; for square foundation, $I_s = 0.95$ (Bowles, 1988). Equation (9.27) is based on the assumption of uniform contact pressure between the foundation and the soil underneath. In actuality, however, the contact pressure under rigid foundations will vary. It is noted that in engineering practice the structural design of footings is often based on the same assumptions of uniform contact pressure. This is a conservative assumption for footings on cohesionless soils. In the case of highly cohesive soils, this may be unsafe due to the increase in contact pressure toward the outside edges of the footing. Ordinarily, the safety factor used in structural design is likely to be adequate for covering this condition. For very wide foundations, such as mats, the variation in contact pressure requires greater attention.

While the nonuniformity of contact pressure discussed above results in more complicated settlement computations than that indicated by Equation (9.26), there are other factors which contribute to the complexity of settlement analysis. The assumptions of the soil being homogeneous, elastic, and isotropic referred are often far from the actual conditions. In addition, it is very difficult to evaluate the in situ stress–strain characteristics of soils below proposed foundations. For cohesionless soils, the modulus of elasticity normally increases with depth. This is another significant factor causing the difficulties in estimating the immediate settlement of foundations on cohesionless soils. For this type of soil, the empirical procedures described in the following section are often employed for settlement analysis.

9.8.2 Immediate settlement for cohesionless soil (Sand)

A commonly used method for estimating the immediate settlement of foundations on cohesionless soils is based on a correlation of the standard or cone penetration resistance (Ch. 10) of foundation soils with the settlement. In particular, the modulus of elasticity, E, may be estimated and used in Equation (9.27). Table 9.2 lists several equations which relate either standard penetration (SPT) N values or cone penetration (CPT) q_c values to the value of E. More details may be found in Bowles (1988).

Table 9.2 Equations used to estimate the modulus E from SPT or CPT data for use in computing immediate settlements

Soil type	SPT (N-value, E in kPa)	CPT (E in same units as q_c)	Undrained Shear strength (E in same units as s_u)
Sand (normally consolidated)	$E = 500 (N + 15)$	$E = 2–4 (q_c)$	—
Sand (saturated)	$E = 250 (N + 15)$	—	—
Sand (overconsolidated)	$E = 18{,}000 + 750 (N)$	$E = 6–30 (q_c)$	—
Gravelly sand and gravel	$E = 1200 (N + 6)$	—	—
Clayey sand	$E = 320 (N + 15)$	$E = 3–6 (q_c)$	—
Silty sand	$E = 300 (N + 6)$	$E = 1–2 (q_c)$	—
Soft clay	—	$E = 3–8 (q_c)$	—
Clay, $I_p > 30$ or organic	—	—	$100–500 (s_u)$
Clay, $I_p < 30$ or stiff	—	—	$500–1500 (s_u)$

Source: After Bowles, 1988.

9.8.3 *Immediate settlement on highly cohesive soils (Clay)*

For highly cohesive soils, the variation in modulus of elasticity with the depth below ground surface is often much less than that in the case of cohesionless soils. If the assumption of constant modulus of elasticity is made, the immediate settlement of foundations on highly cohesive soils may be estimated as described by Bowles (1988). Alternatively, Table 9.2 lists several equations which can be used to relate SPT, CPT, or undrained shear strength data to the modulus of elasticity, *E*. Again, Equation (9.27) can be used to estimate the settlement. Immediate settlements may also be computed using finite element techniques. It can be particularly helpful in problems with significant heterogeneity or complicated boundary conditions.

9.9 Consolidation settlement

Consolidation settlement, also called *compression settlement* or *long-term settlement*, of a foundation is caused mainly by a reduction in the void ratio of compressible soils below the foundation. It is based on Terzaghi's consolidation theory as discussed in Section 9.3. Since that time, progress has been made in improving testing procedures and refining computational methods. The importance of testing procedure and sample disturbance is well-recognized. It is known that sample disturbance tends to underestimate the compressibility of normally consolidated soils and overestimate the compressibility of overconsolidated clays.

9.9.1 *Consolidation settlement-based on Terzaghi consolidation theory*

1   *Normally consolidated soil deposit*: The computation of consolidation settlement in normally consolidated soil deposits based on Terzaghi consolidation theory, previously given by Equation (9.8), again may be written as

$$S_c = (\Delta e\, /\, 1 + e_o)\, H \qquad (9.28)$$

$$\Delta e = -c_c \log_{10}\left(\frac{p_z}{p_o}\right) \tag{9.29}$$

where S_c = consolidation settlement, Δe = change in void ratio at middle of soil layer, e_o = initial void ratio at middle of soil layer, H = thickness of soil layer, c_c = the compression index, p_z = the final effective pressure (stress) at middle of layer and p_o = the initial effective pressure at middle of layer.

9.9.2 Secondary consolidation settlement

Secondary consolidation is conventionally assumed to occur under constant effective stress, that is, following completion of primary consolidation. However, secondary consolidation likely occurs during primary consolidation as well. Although much research has been conducted on secondary consolidation, no reliable methods are available for calculating the magnitude and rate of consolidation. Approximate estimates of secondary consolidation have been made using Equation (9.30):

$$\Delta e = -C_\alpha \log \frac{t_2}{t_1} \tag{9.30}$$

9.9.3 Discussions and comments on settlement analysis

The routine method for estimating the consolidation settlement of a foundation is (a) conduct subsurface exploration for determining the thickness of the compressible stratum; (b) obtain undisturbed samples representing the compressible soil; (c) perform standard laboratory consolidation tests (ASTM D2435); and (d) then compute the ultimate settlement by assuming one-dimensional compression of the soil (Eq. (9.28)). The data from the consolidation tests and the soil profile information are also used for evaluating the time rate of settlement (Eq. (9.9)). Although the basic principles of settlement calculations have been known for some time, continual refinements in test procedures and analytical methods are being introduced. However, there are still many uncertainties. Settlement in clay predictions with accuracy of ±20% may be achievable in some situations where knowledge and past experience with similar structures are available. However, the literature attests to the fact that predictive accuracy as low as ±50–200% are still quite common. Difficulties remain in estimating correct foundation stresses, realistic stress–strain relationships, and horizontal and vertical variability of soil profiles.

9.10 Settlement estimation under environmental conditions

In general, settlement may be initiated by factors other than loading alone. These environmental conditions may be broadly attributed to either natural or anthropogenic causes. Natural causes include climatic variations that lead to wet/dry and freeze/thaw cycling, as well as flooding, sinkholes, and subsidence. Anthropogenic sources include dewatering, excavation, and landfill sites. Landfill settlement is particularly difficult to control, given a nature which is even more variable than soil

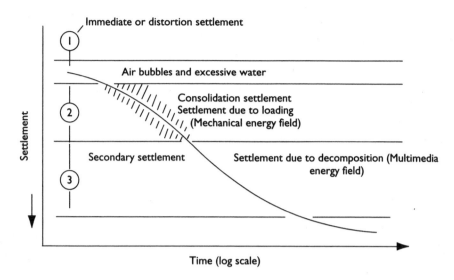

Figure 9.18 Schematic diagram illustrating settlement versus time for a degradable material.

itself. A schematic diagram illustrating the settlement potential versus time in a landfill site is presented in Figure 9.18. Terzaghi's one-dimensional consolidation theory covers only a part of this settlement behavior, because the theory is based on loading, that is, mechanical energy. In landfill areas, the decomposition process involves multimedia energy fields including biological and physicochemical processes.

9.10.1 Settlement analysis and prediction at landfill Site

There are two general methods for prediction of settlement in landfill sites: (a) Sower method, and (b) Yen and Scanlon method. These two methods are generally based on Terzaghi consolidation theory as discussed in Section 9.2.

1 *Sower method*: Sowers (1973) proposed that continuing settlement of sanitary landfill is analogous to secondary compression of soil. The settlement with respect to time and depth of fill was described as follows:

$$\Delta H = -\alpha \frac{H}{1+e} \log \frac{t_2}{t_1} \tag{9.31}$$

where ΔH = total settlement, α = coefficient which depends on field conditions: $\alpha = 0.9e$, for conditions favorable to decomposition, $\alpha = 0.3e$, for unfavorable conditions; e = initial void ratio; H = fill height; and t = time.

2 *Yen and Scanlon method*: Yen and Scanlon (1975) proposed that settlement rate can be computed by the following equation:

$$m = \frac{\alpha}{1+e} \frac{H}{t-1} \log(t) \tag{9.32}$$

where m = settlement rate (ft/month), t = time elapsed (month). Other notations in Equation (9.32) are the same as defined in Equation (9.31). Due to the nature of the problem, settlement analysis of landfill cannot be solved by mathematical equations. These semi-empirical methods may be the best approach at present time. In general, precompaction (Sec. 16.9) of landfill materials can eliminate some of the initial settlement. Controlled addition of water, leachate and/or to landfills also serves to accelerate decomposition and settlement. The rate of settlement in landfill site is an important factor on which numerous investigators have reported include modeling, laboratory testing, field observations and case studies. However, field in situ settlement measurements tend to be the most reliable.

9.11 Summary

Settlement calculations are of central importance in geotechnical engineering, as the Leaning Tower of Pisa might suggest. Settlement is broadly categorized into initial/immediate settlement, consolidation settlement, and secondary settlement. Total settlement at the ground surface will reflect all of these components, as a function of time. Terzaghi's one-dimensional consolidation theory is used to quantify the rate at which the consolidation process takes place. This theory was a major breakthrough in soil mechanics, in part because it is based on rational consideration of the physical problem, not simply an empirical equation fitted to field data. Consolidation processes are often initiated by a surcharge placed at the ground surface. The transmission of stresses from the surface to a given depth may be assessed using methodologies developed by Boussinesq, Westergaard, Newmark, and others. The approach varies depending on the geometry of the load, for example, whether a point load, strip load, rectangular, or circular in plan view. This chapter also reviewed the influence of environmental conditions on the consolidation process. In particular, changes in pore fluid, mineralogy, temperature, and moisture content can influence results.

PROBLEMS

9.1 Comment on Terzaghi's one-dimensional consolidation theory. What are the limitations? Name and define two different terms, which are used to indicate the compressibility of soil.

9.2 The coordinates of two points on a straight-line section of a semi-logarithmic compression diagram are $e_1 = 2.0$; $p_1 = 1.2$ ton/ft^2 and $e_2 = 1.4$; $p_2 = 3.8$ ton/ft^2. Calculate the compression index, c_c.

9.3 Given weight of solid = 108.73 g, area of sample = 31.67 cm^2 and specific gravity of solid, $G_s = 2.82$, the initial sample thickness = 1.000 in., initial void ratio, $e_0 = 1.0848$. Dial reading at $e_o = 0.0000$ inch, Compute: (a) Coefficient of consolidation, c_v; (b) Compression index, c_c; (c) Coefficient of permeability, k, (d) Coefficient of compressibility, a_v; and (e) Overconsolidated pressure, p_c.

9.4 A soil profile consists of 20 ft (6.1 m) of fine sand underlain by 50 ft (15.25 m) soft clay on rock. The clay was normally consolidated. The free water level was at the surface of the clay. The unit weight of sand is 120 pcf (18.8 kN/m^3) and the unit weight of clay is 119 pcf (18.7 kN/m^3). The clay had a liquid limit = 60, plastic limit = 38, and a natural water content = 50%. What is the estimated maximum settlement if 10 ft of embankment (unit weight = 110 pcf) is placed on the ground surface? If the free water level were to drop 15 ft (4.575 m), how much would your estimate change?

9.5   It is predicted from laboratory tests that a building above a 7.6 m thick layer of compressible soil will change the average void ratio from the initial value of 1.110 to a final value of 1.032. What will the ultimate settlement of the building be?

9.6   A uniform load of 1 kg/cm² is distributed over a very large area at the ground surface. The soil profile consists of dense sand with two inter-bedded strata of clay. The clay strata extend from 4.6 to 7.6 m and from 20 to 23 m below the surface, and the water table is 1.5 m below the surface. The soils have the following properties:

Sand: G_s = 2.67, e = 0.67, ω = 8% above water table.
Clay: G_s = 2.75, Unit weight = 2.10 g/cm³, ω = 35% and ω_L = 45.

What is the total settlement under this load?

9.7   Under a given loading, a fully saturated clay layer, 10 ft (3.05 m) thick, resting directly on sound unweathered rock, was expected to change its thickness by 6 in. (15.24 cm). Subsequent investigation indicated the clay layer was 20% thicker than originally estimated. What ultimate settlement may now be expected?

9.8   A settlement of 1 in. (2.54 cm) occurred during the first two months after application of the load. How many months will be required for settlement to reach 50% of its ultimate value?

9.9   A flagpole that weighs 3500 lb is to be erected 10 ft away from one side of a structure. Compute the vertical stress increment at depths of 1 ft, 5 ft, 10 ft, 20 ft, 30 ft and 50 ft below the (a) the flagpole and (b) the structure (10 ft away from flagpole).

Chapter 10

Stress–strain–strength of soil

10.1 Introduction

10.1.1 General discussion

When a load is applied to a soil mass, soil deformation can be either elastic or nonelastic until the deformations become unacceptably large. Then, it is said that the soil mass has failed. Therefore, the deformation or resistance (strength) of the soil is an important criterion in estimating the extent to which soil can be loaded. While compressive forces are typically applied to soil, the mode of failure is almost always in shear. There are two general types of shear failures: the general and local shear failures. (a) The *general shear failure* in which the ultimate strength of the soil is mobilized along the entire potential surface of sliding before the structure supported by the soil is impaired by excessive movement; and (b) The *local shear failure* in which the ultimate shearing strength of the soil is mobilized only locally along the potential surface of sliding at the time the structure supported by the soil is impaired by excessive movement.

10.1.2 Mechanics of shear strength

Shear strength is the maximum resistance a soil has to shearing stresses. It can be determined from both theoretical and experimental approaches. The theoretical approach assumes that soil has elasticity, plasticity, or viscoelasticity properties, and the experimental approach relies on direct measurement of strength for a given set of laboratory and/or in situ soil conditions. A brief discussion of theoretical, laboratory, and in situ conditions for measuring shear stress are presented as follows.

10.2 Constitutive modeling of soils

10.2.1 General discussion

The stress–strain relationship for any material is used for analyzing its stability as part of an engineered system. Numerous mathematical models have been developed to predict soil behavior including elastic, plastic, viscoelastic, elastic-plastic, and work hardening and softening stress–strain behavior as illustrated in Figure 10.1.

These various proposed theories, mathematical models, and their limitations are discussed by many investigators. In general, the elasticity- and plasticity-based models

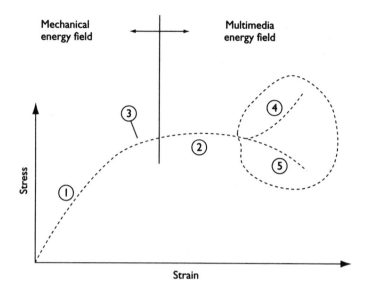

Figure 10.1 Typical stress–strain relationship of soil.

Notes
(1) Elastic; (2) Plastic; (3) Elastic-plastic; (4) Hardening; (5) Softening.

are commonly used for analysis of soils and rocks. A summary of the advantages and limitations of elasticity- and plasticity-based models of the constitutive modeling of soils may be found in saleeb and Chen (1982).

10.2.2 Elasticity-based and plasticity-based models

1 Elasticity-based model: The elasticity-based model is divided into secant models and tangential models. Most nonlinear models used in soil mechanics are based on the linear model. To formulate such nonlinear models you must replace the elastic constants in the linear stress–strain relations with the appropriate secant modulus dependent on the stress and/or strain invariant. The hyperelasticity theory provides a more rational approach in formulating secant stress–strain models for soils. Hence, the constitutive relations are based on the assumption of the existence of a strain energy function or a complementary energy function. Hypoelastic models are formulated directly as a simple extension of the isotropic linear elastic model with the elastic constants replaced by variable tangential moduli, which are taken to be functions of the stress and/or strain invariants. This approach has been used in many geotechnical applications.

2 Plasticity-based model: The plasticity-based models are also frequently used for many geotechnical applications, especially for computing the bearing capacity (Ch. 12) of ground soil. They have been extended to include earth pressure (Ch. 13) and slope stability (Ch. 14). The basic difference between the elasticity and plasticity models lies in the treatment of loading and unloading in plasticity theories achieved by introducing the concept of a "loading function." The Drucker–Prager type of elastic-perfectly plastic models (Drucker and Prager, 1952) is the major thrust for the plastic models used in geotechnical engineering as discussed and evaluated by Chen (1994). With the proper selection of the material constants, the Drucker–Prager criterion can be matched with the Mohr–Coulomb condition. The models reflect some of the important characteristics of soil behavior such as elastic response at lower loads, small material stiffness near failure, failure condition, and elastic unloading after yielding. However, the main disadvantages of the models are the excessive amount of plastic dilatation at yielding as a result of the normality rule used and the inability to describe hysteretic behavior within the failure surface.

The critical state or capped plasticity, also called *CAP type* of plasticity models, is based on the elastic-plastic strain-hardening material (Drucker *et al.* 1957). They introduced a spherical end CAP to the Drucker–Prager model in order to control the plastic volumetric change of soil or dilatancy. This model introduces soil density as the strain-hardening parameter, which determines the change in successive loading surfaces and CAPs. This work has been refined and expanded at Cambridge University, based on experimental data from triaxial compression testing (Roscoe *et al.*, 1958; Schofield and Wroth, 1968). Sometimes it is referred to as the *Cambridge model*.

10.3 Failure criteria

10.3.1 General discussion

Numerous failure criteria have been proposed for the stability analysis of soil mass, but most of them are borrowed from basic engineering mechanics. A summary of advantages and limitations of these criteria is provided in Chen and Saleeb (1994). Each of these criteria reflects some important features of soil strength. In general, failure criteria are classified as (a) *one-parameter model*, and (b) *two-parameter model*. One-parameter models include Tresca, von Mises, and Lade–Duncan (1975) criteria; and two-parameter models include the classic Mohr–Coulomb criterion, extended Tresca, and Drucker–Prager–Lade models. In all these models, two basic postulates are assumed: isotropy and convexity in the principal stress space. Detailed discussion of each failure criteria with various types of soil behavior has been given by Chen (1994), Chen and Saleeb (1994). The differences among these models may at first seem academic; however, the increasing use of finite element modeling techniques in professional practice lends greater importance. In particular, every finite element model uses some constitutive model to relate stress to strain, settlement, and so on. Certain models are more applicable to a certain soil type and set of boundary/initial conditions, and so one must choose a model with care.

10.3.2 Mohr–Coulomb failure criteria

1 *Classic Mohr–Coulomb failure criteria:* The classic Mohr–Coulomb criteria representing stress conditions at failure for a given material is determined by Mohr's rupture hypothesis. A rupture envelope is the locus of points – the coordinates of which represent the combinations of normal and shearing stresses – that will cause a given material to fail. Since soil is a complicated material, some stress–strain–time behavior is highly

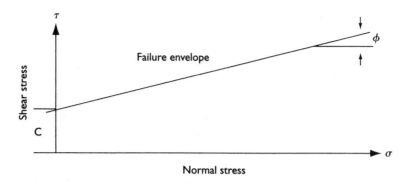

Figure 10.2 Classical Mohr–Coulomb failure criteria.

nonlinear. However, for practical uses, the linear elastic model and Mohr–Coulomb criteria (Fig. 10.2) and their shear equations are commonly used as

$$s = c + \sigma \tan \phi \tag{10.1}$$

where s = shear strength, c = cohesion, σ = normal stress on shear plan, and ϕ = internal friction angle (degree).

   2 Terzaghi modification: In 1925, Terzaghi introduced the effective stress concept to include the porewater pressure effects during the shear tests. The effective stress, also called *effective pressure* or intergranular *pressure*, is the average normal force per unit area transmitted from grain to grain of a soil mass. Also, it is the stress that is effective in mobilizing internal friction together with neutral stress as discussed in Chapter 5. Equation (10.1) has been modified into the following form:

$$s = c' + \sigma' \tan \phi' \tag{10.2}$$

where s' = effective shear strength, c' = effective cohesion, σ' = effective normal stress on shear plane, and ϕ' = effective internal friction angle (degree). The friction force or internal friction angle stated in both Equations (10.1) and (10.2) will be discussed further in this chapter.

10.3.3 Chen–Drucker modification

1 Assumptions: The Chen–Drucker modified Mohr–Coulomb criteria (Chen and Drucker, 1969) is shown in Figure 10.3. The modification is based on two assumptions: (a) Sufficient local deformability of soils in tension and in compression does permit the application of the generalized theorems of limit analysis to soils idealized as a perfectly plastic material; and (b) A modified Mohr–Coulomb failure surface in compression and a small but nonzero tension cutoff is postulated as a yield surface for soils.

   2 Construction of Chen–Drucker modified Mohr–Coulomb circles: The construction procedure for developing a modified a Mohr–Coulomb circle is given in Figure 10.3. The failure envelope is denoted by **AG′H** where **AG′** is part of the circle and **G′H** is a straight line. The distance **AB** is equal to the magnitude of the tensile

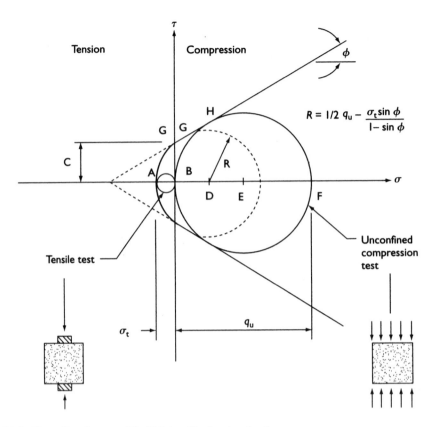

Figure 10.3 Chen–Drucker modified Mohr–Coulomb criterion.

Source: Chen, W. F. and Drucker, D. C., Bearing capacity of concrete blocks and rock, *Journal of the Engineering Mechanics Division, Proceedings of the ASCE*, v. 95, no. EM4, pp. 955–978. © 1969 ASCE. Reproduced by permission of the American Society of Civil Engineers.

strength. **BE** is equal to the radius of the unconfined compressive strength (Sec. 10.6), and distance **BG** is equal to the cohesion, *c*. Angle, ϕ, is the slope of the line **GH**. Finally, the unconfined compressive strength, ϕ, and the radius of circle can be computed from the following equations.

$$q_u = 2\,c\,\tan\,(45° + \phi/2) \tag{10.3}$$

Rearranging Equation (10.3) gives

$$\phi = 2\,\tan^{-1}\frac{q_u}{c} - \frac{\pi}{4} \tag{10.4}$$

To establish the failure envelope, we should know the curve distance **AG′**, in as much as **AG′** is part of the circle whose center is **D**, and whose radius is **R**. The radius may be determined from the following equation:

$$R = \frac{1}{2}q_u - \frac{\sigma_t\,\sin\,\phi}{1 - \sin\,\phi} \tag{10.5}$$

where R = radius of Mohr–Coulomb circle, q_u = unconfined compressive strength, and ϕ = internal friction angle. The circle shown in Figure 10.3 must pass through point **A**, and be tangent to line **GH**, at point **G'**. **AG'H**, therefore, represents the failure envelope of the material. Two applications by use of Chen–Drucker modified Mohr–Coulomb circle are presented in the text; one is the tensile strength determination as described in Section 8.9, and the other one is the short-cut method for determination of undrained strength parameters of soil and stabilized construction materials, presented in Section 10.6.

10.4   Prefailure characteristics of soils

In the previous section, we discussed the failure load and failure criteria. However, this failure load is related to the behavior of prefailure conditions of the same soil. At present, most engineers are interested only in the failure load of the soil, not the prefailure soil conditions. Unfortunately, many premature and progressive failures frequently occur, without any explanations or without effective methods to control these failures because the prefailure and failure conditions are interrelated. In other words, the stability of a soil mass is not only affected by the applied load, but also by prefailure conditions and/or the genetic and/or past stress history of soil itself.

10.4.1   Behavior of soil at prefailure stage

Figure 10.4 illustrates in a schematic diagram the three basic failure conditions when soil is subjected to an applied load, namely prefailure stage, failure stage, and post-failure stage. Characteristics of each individual failure stage controlled by various types of the energy fields are also shown. Failure (point's **cde**) and post-failure (points **ef**) conditions are controlled by mechanical energy. The R value indicated in the Figure 10.4 is called *residual stress*, and $(\Delta\sigma)$ is called *deviator stress*. These two parameters are commonly used in the stress–strain relationship. The values of R and $\Delta\sigma$ of soil at the failure are closely related to the prefailure condition of the same soil. The characteristics of material as indicated by peak point **d** in Figure 10.4 is influenced by the behavior of material (soil) as indicated by point **ab** and point **bc**. The prefailure stage (points **a** to **b**) can be further divided into two substages:

1   Points **a** to **b** is controlled by multimedia energies, which are caused by environmental factors.
2   Point **b** to **c** is controlled by both mechanical and multimedia energy fields. Basically, it includes two stages: (a) soil–water interaction at various water contents without external loads, and (b) soil–water interaction with external loads.
3   Failure and post-failure conditions are also indicated in Figure 10.4 as point **c** to point **d**. Both failure and post-failure conditions are under the mechanical energy field.

10.5   Laboratory shear tests

To determine the strength parameters c, c', ϕ, and ϕ' in Equations (10.1) and (10.2), shear tests must be performed either in the laboratory or in the field. Currently,

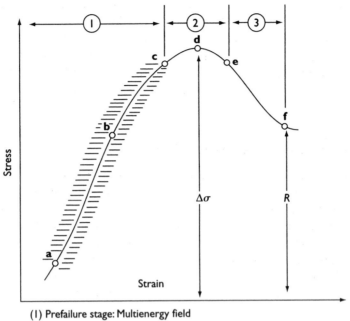

(1) Prefailure stage: Multienergy field
(2) Failure stage: Mechanical energy field
(3) Post-failure stage: Mechanical energy field

• Points **a** to **b**: Controlled by environmental factors (Multimedia energies)
• Points **b** to **c**: Controlled by environmental load factors.

$\Delta\sigma$ = Deviator stress R = Residual stress

Figure 10.4 Schematic diagram illustrating the basic failure stages when soil is subjected to an applied load. (a) Prefailure stage; (b) Failure stage; and (c) Post-failure stage.

various test methods used in the laboratory are available, including (a) Direct shear test, (b) Hollow cylindrical shear test, (c) Ring shear test, (d) Triaxial shear test, (e) Unconfined compression test, and (f) Vane shear and cone penetration tests (also can be used in the field). Field (in situ) measurements will be discussed later in this chapter.

10.5.1   Direct shear test

The *direct shear test*, also called the *direct shear box test*, was one of the earliest tests developed. It provides a measure of the shearing resistance of cohesive or cohesionless soils across a predetermined failure plane. It has been criticized because the failure plane is determined by the test method and not by the soil properties and because of difficulties associated with controlling the sample volume. However, this simple test provides repeatable measurements that have been used for many years.

The soil specimen is enclosed in a box consisting of an upper and lower half. The lower half can slide underneath the upper half of the box, which is free to move vertically. The normal load is applied on the top of the upper half of the box. A horizontal force is applied to the lower half of the box. In the case of wet cohesive soils, porous stones are used to permit drainage of water from the specimen. A stress–strain curve is obtained by plotting the shear stress versus shear displacement. To obtain the failure envelope, at least three tests using various normal stresses are performed on specimens of the same soil. The cohesion, c, and the friction angle, ϕ, also can be obtained graphically from this plot. Details of the test, apparatus, and procedure are provided by ASTM (D3080).

10.5.2 Hollow cylindrical, ring and cubical shear tests

1 Hollow cylindrical shear test: Cooling and Smith (1935) used a hollow cylinder, laterally unconfined and subjected to torque, to obtain the resistance of soils in pure shear. Geuze and Tan (1953) studied the rheology of clays on thin, long hollow cylinders subjected to torque. Later, hollow cylinders were placed in a triaxial cell and pressurized in an effort to generate a wide variety of stress paths. Two approaches were used: (a) The inner and outer pressures are different and thus the specimen is subjected to axial loading; and (b) The inner and outer pressures are identical and thus the specimen is subjected to axial and torsional loading. If the internal and external pressures are different, the stress distribution across the thickness is, by definition, nonuniform. If the internal and external pressures are equal, the stress distribution across the thickness is uniform, provided there are no "end effects." In the hollow cylinder, geometry affects the uniformity of the stress distribution. In the triaxial test on solid circular cylinders, it is customary to consider that a length-to-diameter ratio of 2.5 to 1 is adequate for routine testing.

2 Ring-shear test: The major purpose of the ring-shear test is to determine the residual strength of soil, as, for example, is necessary when analyzing landslide potential in an overconsolidated clay deposit (Sec. 14.10). The concept of residual strength will be discussed in Section 10.13. There are several types of ring-shear devices commercially available. Most soil shear strength testing devices, such as the direct shear and triaxial test, are designed for measuring the peak strength as shown on point **d** in Figure 10.4. Without modification, these instruments cannot deform the sample enough to assess the residual strength. By shearing a ring of soil about a given axis, however, we can keep turning (and shearing) until the residual condition is reached. Further discussion on the test results and applications are presented in Section 10.13.

3 Cubical shear test: A cubical box type apparatus for testing the shearing behavior of soil was developed by Ko and Scott (1967). This type of apparatus is capable of applying any combination of the three principal stresses. The principal stresses can be varied independently or by means of a stress control device. This device is a mechanical-hydraulic analog of an octahedral plane in the principal stress space. The test box has internal dimensions of 4 in. × 4 in. × 4 in. (10.2 cm × 10.2 cm × 10.2 cm) and has six sides built of 5/8 in. (1.5875 cm) thick aluminum plate. The vertical walls are identical so that the box has a mutually perpendicular vertical plane of symmetry. The three principal strains are measured together with an independent measurement of the volume change of the soil sample.

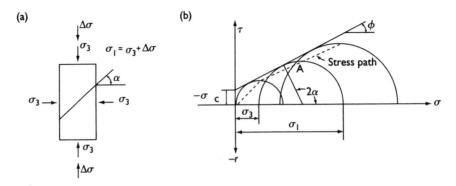

Figure 10.5 Laboratory triaxial shear test on soil. (a) Soil sample under load, and (b) Mohr circles and stress path.

10.6 Triaxial shear test

10.6.1 Test equipment and procedure

A cylindrical soil specimen is encased in a thin rubber membrane with rigid caps, pistons on both ends, and placed inside a triaxial cell. The cell is then filled with a fluid, such as glycerine or water (Sec. 4.3.2), and by an application of pressure to the fluid, the specimen may be subjected to hydrostatic compressive stress. Applying the additional vertical stress creates shear stress in the specimen. This additional vertical stress is called *deviator stress*, $\Delta\sigma$. The deviator stress is steadily increased until failure of the specimen occurs. Drainage of water from the specimen is measured by a burette. To obtain Mohr's envelope, several triaxial tests should be performed on specimens of the same soil using various confining (cell) pressures, σ_3 (Fig. 10.5). From Figure 10.5 the stresses and ϕ can be obtained either graphically or by means of the formulas. Detailed test apparatuses and procedures are provided by ASTM (D2850) and in a standard laboratory manual.

10.6.2 Test results presentation

1   *Construction procedure of Mohr circles*: From Figure 10.5 the stresses, s, and ϕ, can be obtained either graphically or by means of the formulas:

$$\tau = \frac{\sigma_1 - \sigma_3}{2} \sin 2\alpha \tag{10.6}$$

$$\sigma = \frac{\sigma_1 + \sigma_3}{2} + \frac{\sigma_1 - \sigma_3}{2} \cos 2\alpha \tag{10.7}$$

From a simultaneous solution of Equations (10.6) and (10.7), we have

$$\sigma_1 = \sigma_3 \tan^2\left(45° + \frac{\phi}{2}\right) + 2c \tan\left(45° + \frac{\phi}{2}\right) \tag{10.8}$$

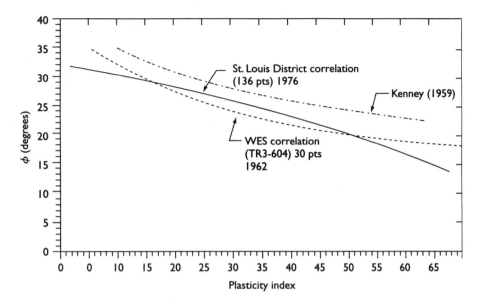

Figure 10.6 Drained friction angle, ϕ' versus soil type as reflected by plasticity index.

Source: ASCE, *Retaining and Flood Walls*, Technical Engineering and Design Guides. Adapted from the US Army Corps of Engineers, no. 4, 313p. © 1994 ASCE. Reproduced by permission of the American Society of Civil Engineers.

$$\sigma_3 = \sigma_1 \tan^2 (45° - \frac{\phi}{2}) - 2c \tan (45° - \frac{\phi}{2}) \tag{10.9}$$

where τ = shear stress, σ_1 = major principal stress ($= \sigma_3 + \Delta\sigma$), σ_3 = minor principal stress (confined pressure, σ_3), and α = angle between normal stress and major principal stress.

2   *Construction of stress path*: A *stress path* is a graphical representation on Mohr circles as it represents a state of stress – point A, in Figure 10.5(b). It has the coordinates

$$x = \frac{\sigma_1 + \sigma_3}{2}; \qquad y = \frac{\sigma_1 - \sigma_3}{2} \tag{10.10}$$

The line or curve connecting these points as illustrated in Figure 10.5(b) is called a *stress path*. In other words, a stress path is a simplified representation of Mohr circles.

10.6.3   Effects of loading and drainage conditions on shear strength

There are three types of loading and drainage conditions under which shear tests may be performed, namely unconsolidated undrained, and consolidated undrained and consolidated drained. These conditions are used to evaluate the effects of loading and porewater pressure. Basically, the intent of the different test methods is to match the anticipated loading conditions in the field. Brief discussions are presented as follows:

1   *Unconsolidated undrained test* (UU or Quick test): No drainage is allowed during the application of confined pressure, σ_3, or normal load in direct shear;

2   *Consolidated undrained test* (CU Test): Drainage is allowed during the application of the confining or normal load. The sample is consolidated with respect to the applied pressure as observed via drainage (or vertical deformation in the direct shear test). No drainage is allowed during the shear test; and

3   *Consolidated drained test* (CD or slow test): The difference between the CD test and the CU test is that drainage takes place during the test and the test is slow enough that porewater pressure does not build up.

The same soil under different loading and drainage conditions presents a different stress–strain relationship. Furthermore, various shear test methods are used which yield various shear strength results. Therefore, careful examination for use of the proper test method and proper selection of strength parameters is important. Table 10.1 presents some guidelines to assist selecting the proper shear test method for preliminary analysis and design of geotechnical engineering projects.

When fine-grained material such as silts and clays are subjected to stress changes, excess pore pressures are induced, because their low permeability precludes an instantaneous water content change. Shear strength envelopes for undrained tests plotted in

Table 10.1 Guideline to assist in selecting the proper shear test

Strength parameters and type of shear tests	Applications
Unconsolidated undrained test (UU or Quick test)	• Embankment constructed rapidly over a soft clay deposit • A strip loading placed rapid on a clay deposit • Short-term slope stability analysis
Consolidated undrained test (CU test)	• Rapid drawdown behind an earth dam • Rapid construction of an embankment on a natural slope
Consolidated drained test (CD or Slow test)	• Embankment constructed very slowly in layers over a soft clay deposit • An earth dam with steady-state seepage • A strip footing on a clay deposit a long time after construction
Residual strength	• Slope stability analysis on over-consolidated clay
Tension test (see text)	• Short-term stability analysis • Analysis of progressive failure • Failed control of stabilized materials
Unconfined compression test	• Short-term stability analysis • Field control of stabilized materials

terms of total stresses exhibit a nonzero cohesion, c, parameter. However, if plotted in terms of effective stresses, the cohesion, c, parameter is small and the friction angle will be essentially equal to that from a drained test. An estimate of the drained friction angle, ϕ', can be made from Figure 10.6.

10.6.4 Pore pressure and parameters A and B

In Sections 5.5 and 10.2, the effective stress is equal to the total stress minus the pore water pressure. For measuring the pore water pressure in the triaxial shear test in laboratory, Skempton (1954) proposed a method for evaluating the pore pressure for both saturated and partially saturated soil as

$$\Delta u = \mathbf{B}\,[\Delta\sigma_3 + \mathbf{A}\,(\Delta\sigma_1 - \Delta\sigma_3)] \tag{10.11}$$

where Δu = change in pore pressure due to increased stresses, $\mathbf{A}$, $\mathbf{B}$ = pore pressure parameters, $\Delta\sigma_1$, $\Delta\sigma_3$ = change in major and minor principal stresses, and $\Delta\sigma_1 - \Delta\sigma_3$ = deviator stress. The parameter $\mathbf{B}$ for dry soil is zero and for saturated soil is 1.0. The parameter $\mathbf{A}$ is dependent on the type of soil and its past stress history. These values can be measured experimentally in both laboratory and field. Typical values and ranges of parameter $\mathbf{A}$ at failure are given: normally consolidated clay = 0.7–1.3; overconsolidated clays = 0.3–0.7; sensitive clays = 1.2–2.5; dense fine sand = –0.3; medium fine sand = 0.0; loose fine sand = 2.0–3.0; and loess = –0.2.

10.7 Unconfined compression test and undrained shear strength

10.7.1 Unconfined compressive strength

The unconfined compression test is considered as a special case of the UU test with a confined pressure equal to zero. The deviator stress at failure is called *unconfined compressive strength* and denoted as q_u. The standard test procedure is provided by ASTM (D2166). For soft clay when the internal friction angle is small or equal to zero, the shear strength can be computed from the unconfined compressive strength as

$$s = c = \tfrac{1}{2}\,q_u \tag{10.12}$$

where s = shear strength, c = cohesion, and q_u = unconfined compressive strength. Unconfined compressive strength is easily obtained in comparison with other shear strength tests and commonly used to evaluate undrained shear for various strength and bearing capacity related problems. Typical unconfined compressive strength ranges for various soil types are very soft clay <0.25 tons/ft^2 (<23.9 kPa), soft clay = 0.25–0.50 tons/ft^2 (23.9–47.9 kPa), medium clay = 0.50–1.00 tons/ft^2 (47.9–95.8 kPa), stiff clay = 1.00–2.00 tons/ft^2 (95.8–191.5 kPa), very stiff clay = 2.00–4.00 tons/ft^2 (191.5–383.0 kPa) and hard clay >4.00 tons/ft^2 (>383.0 kPa).

10.7.2   Undrained shear strength computed from tension and compression test results

Undrained shear strength parameters, cohesion, c, and internal friction angles, ϕ, are commonly determined by direct and triaxial shear tests. These test methods are generally time consuming, expensive and particularly poorly suited to test stabilized soils, highway base, or subbase materials. A simple method for determining the cohesion and friction angle of soil and stabilized materials, which requires knowledge of only the unconfined compressive strength and tensile strength, are proposed by Fang and Hirst (1973).

1 Theoretical considerations: The method described herein is based on the Chen–Drucker modified Coulomb failure envelope as discussed in Section 10.3 and Figure 10.3. The failure envelope is denoted by **AG′H**, where **AG′** is part of the circle and **G′H** is a straight line. The distance **AB** is equal to the magnitude of the tensile strength. **BE** is equal to the radius of the unconfined compressive strength, q_u, and the distance **BG** is equal to the cohesion, c. The friction angle, ϕ, is the slope of the line **GH**. To establish the failure envelope, at least three points on the envelope in Figure 10.3 are required.

2 Procedures for determination of undrained shear strength: **AB** can be determined from tensile tests as described in Section 8.9. Distance **BF** is equal to the compressive strength, q_u, and can be determined by a standard unconfined compression test. This information provides 2 of the 3 points necessary to define the failure envelope. Experimental data indicates that the cohesion, c, is related to the tensile strength, σ_t, of the material as discussed in Section 8.9. Then let

$$\xi = c/\sigma_t \quad \text{or} \quad c = \xi\, \sigma_t \tag{10.13}$$

ξ in Equation (10.13) is the ratio between cohesion and tensile strength of a material. If the soil is in a dry condition, the ξ is relatively constant, and it can be estimated from the following equation:

(a) Cohesion/tension ratio for dry soil

$$\xi = 1/\,[0.34 + 0.01\ I_P] \tag{10.14}$$

(b) Cohesion/tensile ratio for rock

$$\xi = 2.0 \tag{10.15}$$

(c) Cohesion/tensile ratio for general type of soil

$$\xi = 0.36\ q_u/\sigma_t \tag{10.16}$$

where ξ = coefficient between cohesion and tensile strength, I_P = plasticity index, and q_u = unconfined compressive strength. Based on the geometric relations from Figure 10.3, the following interrelationships of the shear strength parameters c and ϕ

can be established as

$$q_u = 2 \, c \tan (45° + \phi /2) \tag{10.17}$$

$$\phi = 2 \tan^{-1} q_u/c - \pi /4 \tag{10.18}$$

To establish the failure envelop, we should know the curve distance **AG'**, inasmuch as **AG'** is part of **AGG'H** in Figure 10.3.

10.8 Friction force and internal friction angle

When a soil mass is subjected to a loading, whether static or dynamic, the increase in density or volume change will change the total internal energy of the soil. It is a process involving potential energy (energy of position), kinetic energy (energy of motion), heat energy (such as change in moisture content of soil) as discussed in Section 1.6. These energies can cause the reorientation of soil particles, which possess forces of repulsion or attraction due to their adsorbed ions and absorbed water molecules (Sec. 4.5). The change of physicochemical properties during the geomorphic process (Sec. 4.11) may be the compression of gases in the soil's voids, thereby, increasing the amount of dissolved gases in porewater, the elastic strain of solid particles, and the characteristics of electric-double-layer thickness (Sec. 6.12).

10.8.1 Mechanism of friction force

During the load-deformation process, there must be a relative motion of soil particles. Force is required to overcome the friction resistance developed between particles during the motion. Energy spent to overcome the frictional resistance is heat energy. There are three basic types of friction resistance which characterize the relative motion between soil particles during loading process: (a) dry condition: when the particle surface is dry; (b) partially saturated condition: partially saturated is also known as hydrodynamic or thick-film lubricated surface friction; and (c) saturated condition: saturated surface-boundary or thin film lubricated surface friction. The force required to produce the relative motion may be expressed by the following basic equation:

$$F = \eta Av / h \tag{10.19}$$

where F = force required to cause the motion, η = coefficient of viscosity of lubricant, A = area of the surface of motion, v = relative velocity of the surface of motion, and h = distance between the surfaces of motion. In Equation (10.19), the surface condition exists only in microparticle systems such as fine silt or clay-like soils.

10.8.2 Angle of repose and friction between soil and wall

1 *Angle of repose*: The angle of repose is the angle between a horizontal and the maximum slope that a soil assumes through the natural processes. For dry granular soils such as sand the effect of the height of slope is negligible; for cohesive soils such as clay the effect of height of slope is so great that the angle of repose is meaningless.

2 *Contact angle*: The contact angle is defined as friction between soil and the wall. It is also referred to as the *angle of external friction* or called *angle of wall friction*. For practical application this friction force or resistance is referred to as *skin resistance*. Such skin resistance includes the force between the soil and retaining walls, between various types of pile foundations, caisson, anchors, etc.

3 *Cohesion and adhesion*: *Cohesion* is the portion of the shear strength of a soil indicated by the term c in Coulomb's equation (Eq. (10.1)). Cohesion is the attractive force between soil particles. *Adhesion* is part of the shearing resistance between soil and another material under zero externally applied pressure. Consideration of adhesion becomes important in geosynthetic applications whereby different materials are placed against soil and each other.

10.9 Sensitivity, creep, thixotropy, and other shear phenomena of soil

10.9.1 Sensitivity of soil

1 *General discussion*: Most clays lose a portion of their strength when remolded. To measure this phenomenon, the term *sensitivity* was introduced by Terzaghi (1943). The sensitivity of clay to remolding and its possible causes are explained, based on physicochemical characteristics of soil, by Winterkorn and Tschebotarioff (1947). Sensitivity is the ratio of the strength of an undisturbed soil sample to the strength of the same soil after remolding. From the point of view of the sensitivity to remolding, clay may be classified as

$$S = S_1/S_2 \qquad\qquad (10.20)$$

where S = sensitivity, S_1 = strength of soil at undisturbed condition, and S_2 = strength of soil at remolded (disturbed) condition. The sensitivity of most clays ranges from 2 to 4, while highly sensitive clays are referred to as "quick." For peat soil (Sec. 2.11), the value ranges from 1.5 to 10 and for marine deposits (Sec. 16.3) it ranges from 1.6 to 26. Classification of clay based on sensitivity as proposed by Skempton and Northey (1952) and Bjerrum (1954) is given as $S < 2$ is classified as insensitive clay; $2 < S < 4$, low sensitivity; $4 < S < 8$, medium sensitivity; $8 < S < 16$, high sensitivity; >16, extra quick.

2 *Sensitivity based on unconfined compressive strength*: Commonly, the unconfined compression tests are used for determination of strength as shown in Equation (10.21). Sometime this term is referred to as the *sensitivity ratio*.

$$S_c = \frac{q_{u(1)}}{q_{u(2)}} \qquad\qquad (10.21)$$

where S_c = sensitivity based on compressive strength, $q_{u(1)}$ = unconfined compressive strength of soil at undisturbed condition, and $q_{u(2)}$ = unconfined compressive strength of soil at remolded condition.

3 *Sensitivity based on tensile strength*: A study indicates that unconfined compression test results are not sensitive to contaminated pore fluids as discussed in Section 4.9, and it is suggested that using tensile tests as discussed in Section 8.9 are more

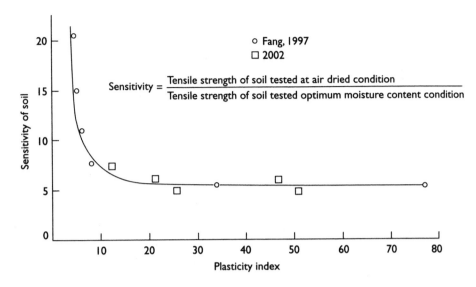

Figure 10.7 Sensitivity versus soil types as reflected by the plasticity index.

effective for evaluation of sensitivity of a contaminated soil–water system. Sensitivity values computed based on tensile strength is suggested by Fang (1997) as illustrated in the following equation.

$$S_t = \frac{\sigma_{t(1)}}{\sigma_{t(2)}} \tag{10.22}$$

where S_t = sensitivity, $\sigma_{t(1)}$ = tensile strength at air dried soil sample, and $\sigma_{t(2)}$ = tensile strength at optimum moisture content soil sample (Sec. 7.2). Figure 10.7 shows the relationship of the sensitivity, S_t, versus soil types as reflected by the plasticity index, I_p. A curvilinear relationship is found. The data used for Figure 10.7 is obtained from Table 8.3 in Chapter 8.

10.9.2   Creep, relaxation phenomena and fatigue behavior

The physical–mechanical behavior of soil changes with time. Moreover, this rate of change is also a function of time. The stress–strain behavior of idealized materials has been investigated at both constant stress ratios and the constant strain rate. Characteristic behavior is often described in terms of creep, relaxation, and fatigue.

1   Creep phenomenon: Creep may be defined as the continued deformation of a material when subjected to a constant stress. The term creep is also used in a more general sense to indicate any inelastic deformation that occurs with time. The study of the creep phenomenon requires an investigation of the change in deformation of a member with respect to time when the material is subjected to a constant stress. Many materials creep at room temperature whereas others require an elevated temperature before significant deformation will occur. Figure 10.8 illustrates typical creep types for a material. Figure 10.9 shows creep curves for various soil types.

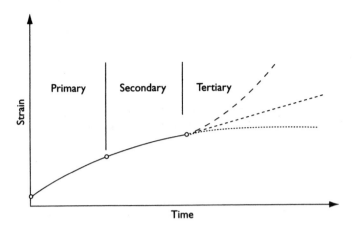

Figure 10.8 Types and characteristics of creep phenomena.

Source: Based on Mitchell (1969) and Fang (1997).

Notes
Primary creep: An initial transition stage where the strain rate decreases continuously.
Secondary creep: A zone where the material creeps at a constant rate.
Tertiary creep: A final transition stage where the strain either decreases to zero or increases to cause creep rupture.

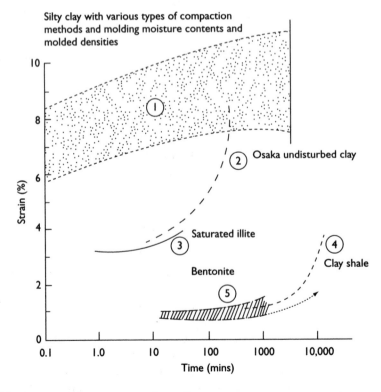

Figure 10.9 Various types of creep curves for soils.

Source: Data from Mitchell (1969) and others, Copyright Thomas Telford Ltd. Reprinted with permission.

2 Relaxation phenomena: Relaxation is the continuous decrease of stress occurring when a material is subjected to constant strain. The same general properties of a material that influence the creep will influence its relaxation.

3 Fatigue behavior: Fatigue properties deal with the behavior of materials when subjected to repeated loads. The cyclic stressing of materials usually leads to a brittle type fracture (Ch. 8) if the magnitude of cyclic stress is sufficient. (a) The fatigue strength of a material is the maximum alternating value of stress that a material can withstand for a specific number of cycles without failure. Fatigue strength will usually be applied to materials that do not exhibit an endurance limit; (b) The fatigue limit, sometime referred to as *endurance limit,* of a material is the maximum value of alternating stress that a material can withstand an infinite number of times without failure, and (c) Fatigue life is the number of cycles of alternating stress of a specified magnitude that are required to fracture a material.

10.9.3   Thixotropy phenomena

1 Thixotropic causes and phenomena: The term *thixotropy* was proposed by Freundlich in 1935 to describe isothermal reversible sol-gel transformation in colloidal suspensions. More recently, the meaning of thixotropy has expanded as defined by ASCE (1958) as the property of a material that enables it to stiffen in a relatively short time on standing, but upon agitation or manipulation it will change to a very soft consistency or to a fluid of high viscosity, the process being completely reversible. Mitchell (1993) defined it as a softening or thinning of a flocculated suspension caused by stirring, followed by a time-dependent return to the original stiffer state. It is further explained by Nalezny and Li (1967) if a flocculated clay (Sec. 3.7) suspension is disturbed by mixing, the edge-to-face contacts (Fig. 3.7) between the particles will be disrupted and the relatively still "gel" will be transformed into a dispersed suspension, which is very fluid. As soon as the mixing is stopped, the suspension will begin to flocculate at a rate depending upon the physicochemical properties of the suspension. Gray and Kashmeeri (1971) studied the thixotropic behavior of compacted soil.

2 Thixotropic hardening and thixotropic strength ratio: The properties of an ideal thixotropic material are shown in Figure 10.10. The *thixotropic strength ratio* (TSR) is the same principal of sensitivity as shown in Equation (10.23). The TSR is the ratio of the aged strength to that of the remolded strength as shown in Equation (10.23):

$$\text{TSR} = S_a/S_r \tag{10.23}$$

where TSR = thixotropic strength ratio, S_a = aged strength of soil, and S_r = remolded strength of soil.

3 Stress hardening and softening: Stress hardening as indicated in Figure 10.1 is the increase in strength of clays with time. It has been found to increase with moisture content up to a certain moisture content, after which it decreases with further increases in moisture. The reason for this phenomenon is that at low moisture content, a particle's mobility is low and the rate of cross link formation is also low. As moisture content increases, particle mobility and the rate of cross link formation increases. However, increased moisture content also increases the average particle separation and net energy barrier to be crossed.

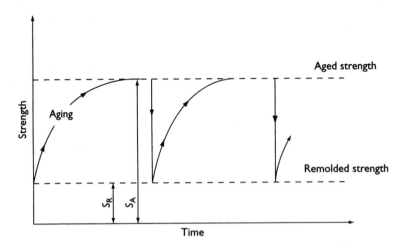

Figure 10.10 Properties of a purely thixotropic material.

10.10 Field shear strength tests

All in situ shear tests measure the soil's undrained shear strength. Commonly used techniques such as the california bearing ratio (CBR) test and standard penetration test (SPT) of bearing capacity of soil are discussed further in Chapter 12. Other tests will be presented in the following sections. Most of these in situ shear tests discussed in this section are available commercially. These in situ shear tests are (a) shear caused by moment or torsion forces including vane shear and dilatometer; (b) shear caused by friction force including cone penetrometer, Burggraf shear apparatus, SPT Test (Sec. 12.9.3), and CBR Test (Sections 7.4 and 12.9.1); and (c) shear caused by both compression and friction forces including pressure meter and self-boring pressure meter.

10.10.1 Shear caused by moment or torsion forces

1 *Vane shear test*: The vane shear test is commonly used for determination of the undrained shear strength of clay. The test basically consists of placing a four-bladed vane in the undisturbed soil and rotating it from the surface to determine the torsion force required to cause a cylindrical surface to be sheared by the vane; the force is then converted to a unit shearing resistance of the cylindrical surface. For computing the shear strength from vane test, the following equation is used:

$$S_u = \frac{6M}{7\pi D^3} \tag{10.24}$$

where S_u = undrained shear strength, M = total resisting moment at failure (resisting moment from the circumference of the cylinder plus resisting moment from the end of the cylinder), D = diameter of the vane, and π = constant = 3.1416. Vane shear can be used for both laboratory and field (in situ) conditions. The field procedure of the vane shear for the cohesive soil has been standardized by the ASTM (D2573). This method covers the field vane test in soft, saturated, and cohesive soils. Knowledge of the nature of the soil in which each vane test is to be made is necessary for assessment of the applicability and interpretation of the test.

2 *Dilatometer test*: The flat plate dilatometer was developed by Marchetti in 1980. Results from the dilatometer are expressed in terms of the three index parameters (a) Material index, I_D, related to the soil type; (b) Horizontal stress index, K_p, related to the in situ coefficient of earth pressure at rest, K_o (Sec. 13.10), and (c) Dilatometer modulus, D_D, a parameter related to soil stiffness. Quantitative estimates of K_o, S_u (undrained shear strength), ϕ (friction angle for sand), OCR (overconsolidation ratio), and M (the constrained modulus) can be obtained from the empirical correlation with the dilatometer's I_D, K_o, and E_D. If the soils are too stiff, a drill rig may not penetrate a dilatometer-blade successfully without pre-boring a "pilot" hole.

10.10.2 Shear resistance caused by friction force

1 *Cone penetration Test (Cone penetrometer)*: The cone penetration test, sometimes called the *Dutch cone test*, is used to estimate the in situ undrained strength of soils. This test, standardized by ASTM (D3441), consists of advancing a cone-tipped probe or penetrometer into the ground and measuring the forces developed. The cone tip has an apex angle of 60° and a projected cross-sectional area of 10 cm². Behind the tip is a cylindrical sleeve. By mechanical or electronic means, the capability is provided to separately measure the point resistance, q_c, developed on the tip and the side resistance, f_s, developed along the sleeve as the cone penetrometer is advanced. Both of these measures have units of force per length squared, with q_c being a bearing pressure and f_s being a shear stress. The ratio of the side resistance to point resistance is termed the *friction ratio* as shown in Equation (10.25).

$$F_r = q_c/f_s \qquad (10.25)$$

where F_r = friction ratio of cone, q_c = point resistance of cone, and f_s = side resistance of cone. For coarse-grained soils, the cone resistance, q_c, has been empirically correlated with the standard penetration resistance, N, value (Sec. 2.7.5). The ratio (q_c/N) is typically in the range of 2–6. This ratio is also related to the grain size of soil as reflected by D_{50} (Sec. 3.3) as shown in Figure 10.11. The undrained shear strength of fine-grained soils is determined from the cone penetration test as suggested by Robertson and Campanella (1983) as

$$S_u = \frac{q_c - p_o}{N_k} \qquad (10.26)$$

where S_u = undrained shear strength, q_c = cone resistance, p_o = the in situ total overburden pressure, and N_k = empirical cone factor typically in the range of 10–20. Figure 10.12 presents soil classification from a cone penetrometer proposed by Robertson and Campanella (1983).

The *Dutch cone penetrometer* is also called the *static penetration test*. It is a popular device used for in situ subsurface soil investigations. Disposable cones are

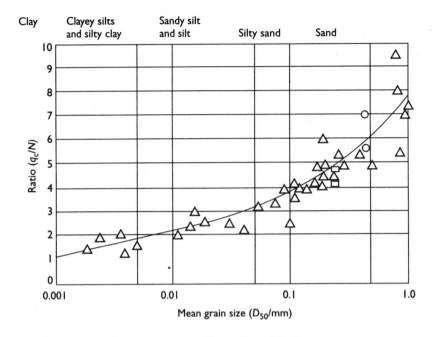

Figure 10.11 q_c/N versus median grain size of soil as reflected by D_{50}.

Source: Robertson, P. K. and Campanella, R. G. (1983). Interpretation of cone penetration tests; Part I and II, *Canadian Geotechnical Journal*, v. 20, no. 4, pp. 718–745. Reprinted with permission by National Research Council of Canada.

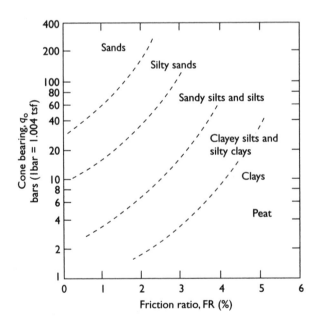

Figure 10.12 Soil classification from cone penetrometer value.

Source: Robertson, P. K. and Campanella, R. G. (1983). Interpretation of cone penetration tests; Part I and II, *Canadian Geotechnical Journal*, v. 20, no. 4, pp. 718–745. Reprinted with permission by National Research Council of Canada.

available commercially. The Dutch cone penetrometer is a technique by which a 60° cone with a base area of 10 cm² is pushed into the ground surface at a steady rate of about 20 mm/s. They are simple, hand-operated instrument, which save time and are low-cost devices.

   2   Burggraf shear resistance apparatus: The Burggraf shear resistance apparatus developed by F. Burggraf in 1939 is used to determine the horizontal shear resistance of highway material in situ condition. With the Burggraf shear apparatus, a hole about 10 in. × 10 in. (25.4 cm × 25.4 cm) is dug in the layer to be tested to a sufficient depth. A vertical face against which the test is to be made is carefully cut to receive a standard compression plate connected to the thrust cylinder. The horizontal thrust is then applied by turning a hand wheel operating the screw-propelled plunger-type pump at a uniform rate to force the compression plate against the soil–aggregate layer until the material ahead of the plate fails. The area of the surface on which the failure occurs is measured, and the strength value is determined by dividing the maximum horizontal thrust by the sheared area. The angle of failure, ϕ, is determined by measuring its tangent, which consists of one measurement from the top of the compression plate to the bottom of the cavity divided by the distance from the face of the compression plate to the most remote edge of the sheared surface.

   3   Standard penetration test (SPT): The SPT has a wide variety of uses. The basic principle of SPT has been discussed in Section 2.6.5. Applications for determination of relative density for controlling the field density during field compaction have been discussed in Chapter 7. The bearing capacity of ground soil will be further discussed in Section 12.9.1.

   4   In situ california bearing ratio (CBR) test: The CBR test can be used for both laboratory and field condition. The basic concept and laboratory test procedures together with typical results have been discussed in details in Chapters 7 and 12. In this section only the field test procedure is discussed.

10.10.3   Shear resistance caused by compression and friction forces

1   Pressure meter test: The pressure meter probe is a cylindrical metal assembly with rubber membranes attached to three independent cells. The central cell contains water under gas pressure so that the increase in volume of this cell is measured by the lowering of the water level in the volumeter at the surface. The lower and upper cells are called *guard cells* and expand under equal gas pressure from the surface to minimize the effects of end restraint on the measuring cell. The volumeter is equipped so that a monitored gas pressure can be used to force water into the measuring cell. There are two types of pressure meters. The Menard type of pressure meter was based on Menard's work for a master's thesis developed at the University of Illinois in 1956 (Menard, 1956). Other types of pressure meters incorporating similar principles were developed independently at the AASHO Road Test in 1956, where the probe was used to measure the internal-deformation characteristics of embankment soil and of base and subbase materials. Both types measure the pressure–volume changes relationship of soil at the in situ condition. It must be noted that the deformation (volume change) measured by the pressure meter or the internal-deformation device reflects both consolidation and shear phenomena in a soil mass. The interrelationships of pressure-deformation phenomena are not clearly understood; however, useful correlation among other shear parameters exist as reported by Higgins (1969).

2 *Strength versus pressure meter results:* Pressure meter results use the bearing capacity of ground soil, compressibility, as well as shear strength. Because it is a measured combination of compression and shear characteristics, the following equation for estimating the undrained shear strength of fine-grained soil may be used as described by Fang (1997).

$$S_{\mathrm{u}} = [P_1 - P'_{\mathrm{ho}}]/2\,K_{\mathrm{b}} \tag{10.27}$$

where S_{u} = undrained shear strength, P_1 = limit pressure, P'_{ho} = effective at rest horizontal pressure, and K_{b} = a coefficient typically in the range of 2.5–3.5 for most clays.

10.11 Shear characteristics of granular soils

10.11.1 General discussion

The shearing strength of a granular system is also expressed in Equation (10.1). The resistance to shear is due to a combination of effects, including sliding friction, rolling friction, and interlocking of particles. In general, the cohesion, c, in Equation (10.1) is small, and in some case, it is assumed as zero. The friction angle controls the shear strength. The typical friction angle, ϕ, under dry condition are as follows: Sand and gravel: loose condition = 32°–36°, dense condition = 46°–50°; Sand, well-graded: loose condition = 30°–32°, dense condition = 40°–46°; Uniform sand: loose condition = 26°–30°, dense condition = 32°–38°; Sand and silt: loose condition = 25°–30°, dense condition = 30°–35°.

10.11.2 Laboratory test methods and characteristics of granular soil

Laboratory test methods for granular soil include the direct shear box and the triaxial shear test, as discussed in Section 10.5. Major characteristics of granular soils are presented as follows:

1 *Critical void ratio (cvr) and dilatancy phenomena:* The dilatancy phenomenon is the volume change (expansion) of cohesionless soils when subject to shearing deformation. This occurs when the initial soil is at a relatively dense state. Shear stresses cause the soil particles to rearrange, resulting in an increase in void ratio.

2 *Tan ϕ and void ratio:* The constant normal stress and tan ϕ increase as the initial void ratio decreases. When a dense granular system is sheared, it expands, because the closely interlocked grains need sufficient space to be able to roll or slide over one another. The increase in volume means that work is done against the normal pressure, which reflects itself in a higher tan ϕ. In a loose system containing many

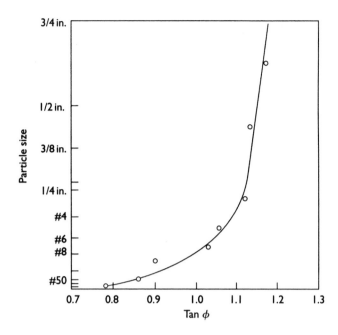

Figure 10.13 Effect of particle size on the tan φ of round gravel.

Source: Hennes (1953). Copyright ASTM INTERNATIONAL. Reprinted with permission.

arches, the volume will decrease during shear as the arches are broken down and the grains rearrange themselves to a denser state. Eventually the CVR is attained, at which shear continues at constant volume (Casagrande, 1936; Taylor, 1948). Winterkorn suggested that in shear, granular systems behave as macromeritic liquids (Sec. 3.5.3).

3   *Tan φ and coefficient of friction, f:* A number of expressions have been derived relating tan φ to the coefficient of friction, f, between the particles. Caquot and Kerisel (1948) considered an irregular system composed of particles of various sizes and shapes in random distribution and found that

$$\tan \phi = (\pi /2)\, f \tag{10.28}$$

4   *Effect of grain sizes on tan φ:* To determine conclusively whether tan φ depends on the grain size, controlled tests on systems with regular arrangements of spherical grains of uniform size and surface characteristics should be performed.

Hennes (1953) measured the shear strength of samples of rounded gravel of approximately uniform size. Results of Hennes's work are shown in Figure 10.13. In examining Figure 10.13, tan φ was found to increase approximately with grain size up to 1/4 in. (0.635 cm), beyond which there was little variation. The shapes of the grains varied considerably within each sample and from sample to sample. The results, therefore, do not show the influence of grain size alone. According to Winterkorn's Macromeritic Theory (Ch. 3), a granular system of identical spheres is in a potentially liquid state when it has a void ratio of 0.62 (the critical value) or higher. The CVR thus corresponds to the melting point of simple chemical substance. Identification of various types of CVR are made by Taylor (1948).

10.11.3 In situ strength test on granular material

Coarse-grained materials such as sands and gravel are sufficiently pervious that excess pore pressures (Sec. 5.5) do not develop when stress conditions are changed. Failure envelopes plotted in terms of total or effective stresses are the same and typically exhibit a zero cohesion, c, value and a friction angle, ϕ, value in the range of 25°–50°. Because of the difficulty of obtaining undisturbed samples of coarse-grained materials, the ϕ value is usually inferred from in situ tests. The SPT (Chs 2 and 10) is one of such in situ tests commonly used. As noted previously, the undrained shear strength of normally consolidated sand is related with relative density, coefficient of uniformity, and effective overburden pressure. The drained friction angle, ϕ', can be estimated from SPT (N) results (ASCE, 1994).

10.12 Shear characteristics of normally and overconsolidated clays

10.12.1 General discussion

As discussed in Chapter 2, due to the past stress history of a natural soil deposit, the deposit can be divided into two distinct characteristics: normally consolidated clay and overconsolidated clay. Most natural soil deposits are normally consolidated; however, the overconsolidated soil deposits give more problems than normally consolidated soil, especially for slope stability problems. In this section, both normally and overconsolidated soil deposits are presented; however, emphasis is placed on overconsolidated soil deposits. Most soil deposits are at the normally consolidated condition. By visual observation, there are no significant differences between normally and overconsolidated soil deposits, except when performing a complete laboratory consolidation test as discussed in Chapter 9.

10.12.2 Shear characteristics of overconsolidated clays

There are no obvious features which can differentiate between normally and overconsolidated soil deposits. However, standard laboratory consolidation tests can identify them as discussed in Chapter 9. Typical overconsolidated clay can also be identified by the liquidity index as discussed in Sections 2.5 and 9.4. The typical stress–strain curves, volume change, and porewater pressure of normally consolidated and overconsolidated clays are shown in Figures 10.14(a), (b), and (c). Figure 10.14(a) shows the stress–strain curves, Figure 10.14(b) plots the pore pressure versus strain (which illustrate what happens to the porewater pressure during shear), and Figure 10.14(c) shows the volume change versus strain for CD triaxial tests. All

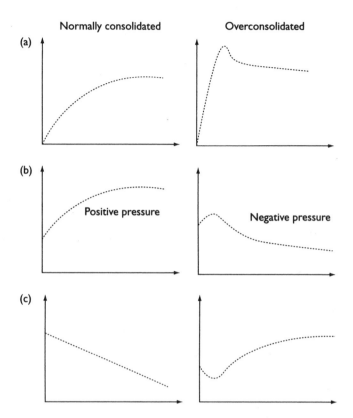

Figure 10.14 Comparisons of shear characteristics of normally and overconsolidated clays. (a) Change in stress versus deformation; (b) Change in pore water pressure versus deformation; and (c) Change in volume versus deformation.

shear tests are performed at the same effective confined stress. In examining these curves in Figures 10.14(a), (b), and (c), several observations may be made, given as follows:

1   In Figure 10.14(a), the overconsolidated specimen has a greater strength than the normally consolidated clay. The maximum occurs at a much lower strain than for the normally consolidated specimen.
2   In Figure 10.14(b) for porewater, the normally consolidated specimen develops positive (+) porewater pressure. However, for the overconsolidated specimen, after a slight initial increase the porewater pressure goes negative (−). In this particular case, the negative porewater pressure goes to the back-pressure, u_0.
3   In Figure 10.14(c), the volume change for overconsolidated clay is expansion during shear, while normally consolidated clay compresses or consolidates during shear.
4   Therefore, the Mohr failure envelopes for total and effective stresses are different for the normally consolidated and overconsolidated clays.

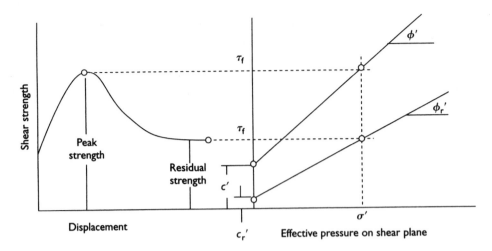

Figure 10.15 Shear characteristics of overconsolidated clay – the residual strength concept.

Source: Skempton, A.W. (1964), Long-term stability of clay slopes, *Geotechnique*, v. 14, no. 2, pp. 77–102. Copyright Thomas Telford Ltd. Reprinted with permission.

10.13   Residual shear strength of clay

10.13.1   Basic concept of residual shear strength

For analysis of shear characteristics of overconsolidated clays in relating to the stability problems, ordinary shear tests (Eq. (10.1)) are not suitable because it overestimates the true shear strength value. Skempton (1964) showed that the strength remaining in laboratory samples after large shearing displacement corresponded closely with the computed strength from actual landslides (Ch. 14); therefore, he proposed a residual strength concept as shown in Figure 10.15 and Equation (10.29) for the analysis of slope stability of overconsolidated clays as

$$S_r = \sigma' \tan \phi_r' \tag{10.29}$$

where S_r = residual shear strength, σ' = effective pressure on shear plane, and ϕ_r' = residual friction angle. In examining Figure 10.15, the peak strength, τ_f, the corresponding effective friction angle, ϕ', and effective cohesion, c', are used for conventional slope stability analysis. However, for overconsolidated clays, the ϕ_r' and c_r' are suggested for the slope stability analysis. The c_r' value is very small or zero.

10.13.2   Residual shear measurements and data interpretations

The residual shear strength in Equation (10.29) can be obtained from slow drained direct shear tests as suggested by Skempton (1964). Since then, several test methods and procedures have been developed including the modified direct shear box, triaxial compression test, and ring-shear tests. Because each test apparatus and test procedure

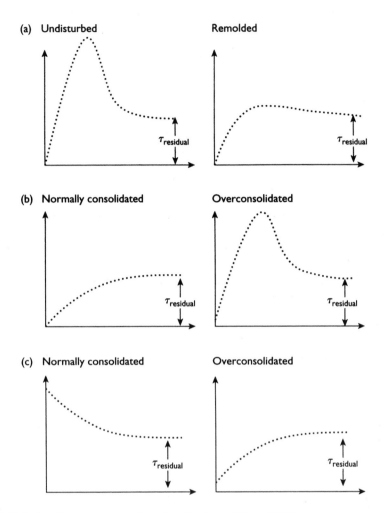

Figure 10.16 Relationship between peak and residual conditions. (a) Shear stress versus displacement for undisturbed and remolded soil specimens; (b) Shear stress versus displacement for over-consolidated and normally consolidated clays; and (c) Water content versus displacement for over consolidated and normally consolidated clays.

is different, some discrepancies or variations are reported. A report presented by Townsend and Gilber (1973) summarizes the residual shear strength of clay shale from five dam sites including Texas, Montana, and South Dakota. Three types of test methods are used including annular shear, repeated direct shear, and rotational shear. The relationship between peak and residual conditions of the soil is presented in Figure 10.16(a), (b), and (c). In examining these three curves, in all cases, the residual conditions remained constant as indicated.

10.13.3 Factors affecting residual shear results

1   *Test methods*: Results of Townsend and Gilber's studies indicate that the effective pressure, σ', along the shear plane varied from 1.5 kg/cm^2 (21.3 psi) to

12 kg/cm^2 (170.7 psi). The results of the effective residual friction angle, $\phi_r{}'$, indicate that there is no significant difference among these methods. Comparison of residual strength data determined by the three laboratory tests (CRRI, 1979) included repeated direct shear, triaxial compression, and ring-shear. Four soil types were used. For all cases, the results from the ring-shear give the highest residual friction angles.

2   *Rate of shear*: Studies including the rate of shear for determination of residual strength (CRRI, 1979) show that the smaller the rate of shear, the higher the residual friction angle, ϕ_r. The rate of shear varies from 0.08 to 0.12 mm/min and ϕ_r varies from 3°36' to 7°31'.

3   *Rubber membrane*: The types of rubber membrane used for the protecting of the soil specimen for determination of residual strength show some significance. As the lateral pressure increases, the correction values for the rubber membrane increase.

10.13.4   *Correlation of residual shear strength to other soil parameters*

The peak or residual shear parameters as shown in Figure 10.15 are relatively difficult to obtain for practical uses. Therefore, some simple experimental equations and correlations for estimating these strength parameters were proposed by various investigators as follows:

1   *Residual strength with clay content*: Skempton (1964) reported that the ultimate ϕ' is related to the clay content ($>2\mu$) for various soil types from England.

2   *Residual strength coefficient (μ_r)*: Voight (1973) summarized various published data from 1967 to 1970 and found a relationship between the plasticity index, I_P, and the residual strength coefficient, μ_r. Results of I_P versus τ_r are shown in Figure 10.17. The μ_r, is defined as

$$\mu_r = \frac{\tau_f - \tau'}{\tau_f - \tau_r} \tag{10.30}$$

where μ_r = residual strength coefficient, τ_f = peak strength (Fig. 10.15), τ' = shear strength at natural sliding surface, and τ_r = residual strength (Fig. 10.15).

3   *Residual strength and normal stress*: A relationship between residual strength and normal stress proposed by a Russian engineer (CRRI, 1976) is as follows:

$$\tau_r = 0.09 + 0.14\,\sigma \tag{10.31}$$

where τ_r = residual strength, and σ = normal stress. Equation (10.31) is based on 200 laboratory tests. The coefficient of correlation, r^2, is equal to 0.78. Equation (10.32) is the correlation of shear strength along the natural sliding surface, τ', with laboratory test results of normal stress, τ_r, as follows:

$$\tau' = \tau_r = 0.06 + 0.15\,\sigma \tag{10.32}$$

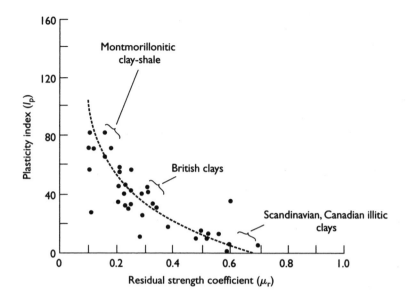

Figure 10.17 Relationship between plasticity index and residual strength coefficient.

Source: Voight, B. (1973), Correlation between Atterberg plasticity limits and residual shear strength of natural soils, *Geotechnique*, v. 23, no. 2, pp. 265–267. Copyright Thomas Telford Ltd. Reprinted with permission.

Equation (10.32) is based on the 50 field observations. The coefficient of correlation, r^2, is equal to 0.82. It was found that laboratory residual strength, τ_r, was very close to the field shear strength along the natural sliding surface.

4 *Friction angle with plasticity index and liquidity index:* Equation (10.33) proposed by the Chinese Railroad Research Institute (CRRI, 1979) is based on 236 laboratory undrained multi-repeated shear tests and 59 field observations on 91 types of sliding soils.

$$\log \phi_r = 2.4278 - 1.2279 \log I_P - 0.1173 \log I_L \tag{10.33}$$

where ϕ_r = residual friction angle, I_P = plasticity index, and I_L = liquidity index. Figure 10.19 presents the graphical form of Equation (10.33). It shows the residual friction angle, ϕ_r, relating to the plasticity index and liquidity index. However, it suggests that the plasticity index has a greater influence than the liquidity index.

5 *Effective residual friction angle with liquid limit and plasticity index:* Equation (10.34) proposed by Jamiolkski and Pasqualini, as reported by CRRI (1979), is based on the weathering of blue and grey clays from Italy. A similar relationship, given as Equation (10.35) was also reported by CRRI (1979).

$$\phi_r' = 453.1 (\omega_L^{-0.85}) \tag{10.34}$$

$$\phi_r' = 46.6/I_P^{0.446} \tag{10.35}$$

where ϕ_r' = effective residual friction angle, ω_L = liquid limit, and I_P = plasticity index. Figure 10.18 presents the graphical forms of Equations (10.34) and (10.35)

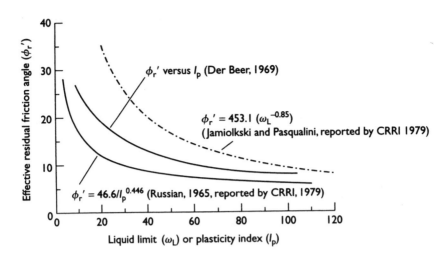

Figure 10.18 Effective residual friction angle versus liquid limit or plasticity index.

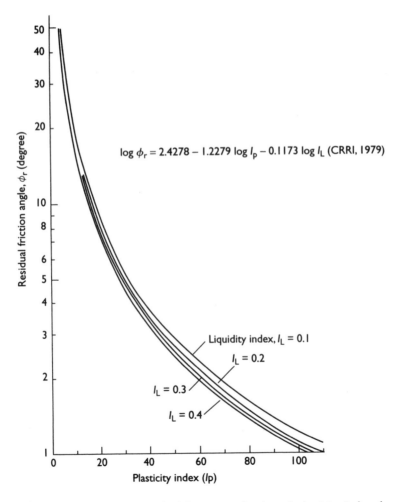

Figure 10.19 Correlation of effective residual friction angle, ϕ_r and plasticity index, I_p with various liquidity indexes.

together with previously published data reported by De Beer (1969) for comparison purposes.

10.14   Genetic diagnosis approach for evaluation of shear strength of soil

10.14.1   General discussion

As discussed in previous sections, the stress–strain–strength relationship of soil can be changed from elastic to plastic behavior, or it can be changed from softening to hardening behavior if certain local environmental conditions change. Therefore, the stress–strain–time relationship cannot be generalized into a mathematical model and it cannot be arbitrarily assumed as elastic or plastic behavior as indicated in Figure 10.1. Therefore, a genetic diagnosis approach is proposed, which may assist in understanding or solving the basic stability of earth structures in the environment. Let us assume the shear strength can be grouped into two components as

$$S = S_P + S_e \tag{10.36}$$

where S = shear strength in the environment; S_P = shear strength inherited from parent material (stress past history); and S_e = shear strength caused by present local environmental conditions.

10.14.2   Strength of soil inherited from parent material

1   *General discussion:* Indications are that an inheritance factor in origin of clay minerals in soil plays an important role. Using halloysite and kaolinite clay minerals (Sec. 3.9) as examples, shear strength is controlled by bonding energy between particles such as (a) mineral structure; (b) particle sizes and specific surface; (c) ion exchange capacity; and (d) sensitivity of soil particle to environment. Because these two types of clay mineral have distinct mineral structure as indicated in Section 3.9, consequently, the engineering behavior of these two types of material will also be different. Kaolinite can produce high density while halloysite only produces a low density. The sensitivity of soil to the environment has been discussed in Section 4.9; however, the effect of types of exchangeable ions on shear strength will be discussed in the following section.

2   *Types of exchangeable ions:* As reported by Vees and Winterkorn (1967), the effects of types of exchangeable ions on shear strength of kaolinite clay are presented in Figure 10.20. Figure 10.20 shows the shear resistance of homoionic modifications of kaolinite clay as a function of normal pressure, which is also the consolidation pressure. Figure 10.20 shows the shear strength for kaolinite as a function of void ratio. At low void ratios, the structural influence is most effective as indicated by the order of the cation effect: Th > Na > Al > Ca. With increasing void ratio, the effect of the Th ion decreases rapidly and falls below those of Na and Al at a void ratio of 1.4 and 1.45, respectively.

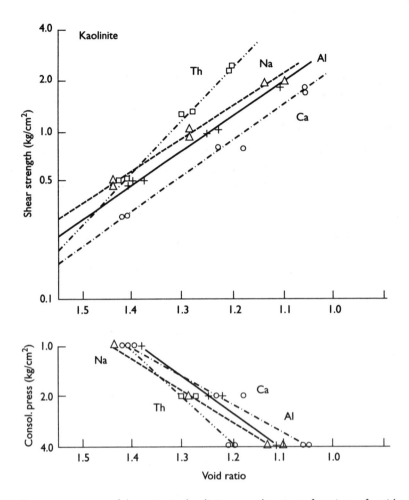

Figure 10.20 Shear resistance of homoionic kaolinite samples as a function of void ratio with relationship of void ratio to consolidation pressure.

Source: Vees, E. and Winterkorn, H. F., Engineering properties of several pure clays as functions of mineral type, exchange ions, and phase composition. In *Highway Research Record* No. 209, Highway Research Board. National Research Council, Washington DC, 1967, pp. 55–65. Reproduced with permission of the Transportation Research Board.

10.14.3 Strength of soil influenced by environment factors

1   *General discussion*: As discussed in Chapter 4, shear strength is influenced by local environmental factors including (a) Wetting–drying, (b) Freezing–thawing, (c) Leaching, (d) Adsorption, (e) Exchangeable ions, and (f) Geomicrobiological factors. Jones (1955) and Daniels and Cherukuri (2005) pointed out the importance of microbiological factors in the characteristics in soil stabilization as well as the strength. Also, discussed in Chapter 4 are the mechanisms of how microbiological factors relate to soil behavior including volume change, compressibility, and shear strength.

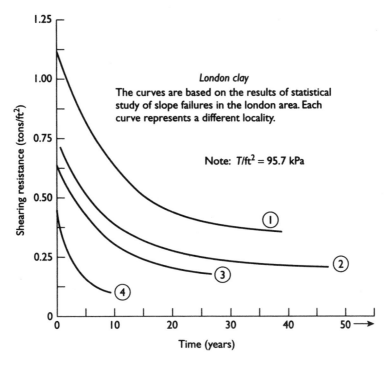

Figure 10.21 Diagram showing gradual decrease of shear resistance of stiff, fissured London clay.

Source: Skempton, A.W. (1964), Long-term stability of clay slopes, *Geotechnique*, v. 14, no. 2, pp. 77–102. Copyright Thomas Telford Ltd. Reprinted with permission.

2   *Effects on time*: Shear strength of soils decreases with the time, mainly caused by local environmental changes such as weathering, wet–dry, freeze–thaw, hot–cold cycles, surface and internal erosions, cracking, creep, and many others. These environmental factors can cause the loss of bonding stress between soil particles, resulting in a gradual loss of shear resistance as indicated in Figure 10.21.

3   *Effect on moisture content*: Moisture content in the soil mass is a major factor for controlling the state of stress of the soil. In general, an increase in the moisture content or degree of saturation in soil will decrease the shear resistance as reflected by cohesion, c, or friction angle, ϕ. This relationship is shown in terms of cohesion by Figures 10.22 and 10.23.

4   *Effect of Temperature*: Murayama (1969) used a rheological model for analysis of the elastic modulus of clay–water systems showing that the modulus decreased as the temperature increased. Mitchell (1969) studied the relationship between initial stress and strain in stress relaxation tests at various temperatures. Considering the straight-line portions of the curves through the plotted data as representative of the elastic modulus of the soil, it is concluded that the modulus decreases with increase in temperature. Many other investigators made similar conclusions regarding the temperature effect on shear strength.

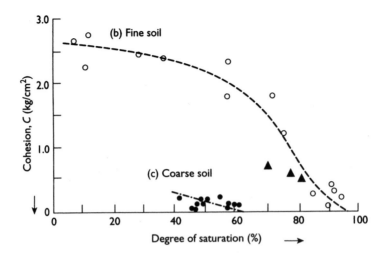

Figure 10.22 Increase in degree of saturation decreases cohesion for weathered residual soils.

Source: Lumb, P. (1962), The properties of decomposed granite, *Geotechnique*, v. 12, no. 3, pp. 226–243. Copyright Thomas Telford Ltd. Reprinted with permission.

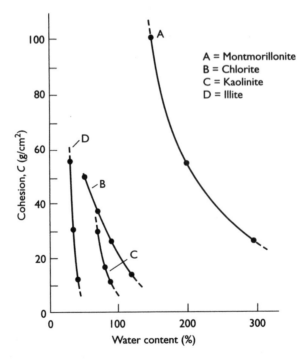

Figure 10.23 Effect of moisture content on cohesion for four basic clay minerals.

Source: Based on Shibuya (1973) and Fang (1997).

Table 10.2 Effect of pore fluid on strength parameter

Shear parameters	Soaking 5-day under	
	Water (H₂O)	Acid (H₂SO₄)
Laboratory Test		
Friction angle (θ), deg.	22.2	15.2
Coefficient of friction, (f)	0.408	0.271
Cohesion (c), kPa	87.5	75.5
In situ measurement		
Static cone penetration (P_S), kPa	3060	1370

Source: Based on Sun (1989) and Fang (1997).

Mitchell (1969) emphasized the role of porewater pressure changes accompanying temperature changes. Determinant factors under drained conditions appear to be the thermal expansion of the porewater, the compressibility of the soil structure, and the initial effective stress. Their experimental results show several clay–water systems in which each change in temperature by 1°F changed the porewater pressure by about 0.75–1.0% of the initial effective stress. For less compressible materials the porewater pressure change was considerably greater. The shear strength of cohesive soils, while a function of phase composition (including water content) and temperature, is not uniquely controlled by these but depends also on the direction from which the water content has been reached. Higher shear strength at a particular water content and higher thermal conductivities (Ch. 6) were obtained when the path was from the wet-side to the dry-side of optimum moisture content (Ch. 7).

5   *Effect of pore fluid*: The effect of pore fluid changes, as reflected by sulfuric acid (H_2SO_4), on laterite soil was reported by Sun (1989). A limited amount of laboratory experimental data is shown in Table 10.2. A significant difference of friction angle, coefficient of friction, and cohesion values are found. In addition, a larger difference of in situ static cone penetration results is found. In all cases, when the pore fluid is composed with H_2SO_4, smaller values are obtained.

The combined effects of both pore fluid and temperature on sand–bentonite mixture are reported by Naik (1986) as indicated in Figure 10.24. The influence of pore fluid is reflected by the pH value. An increase in the pH value will decrease the shear strength in all cases. Also, increasing the temperature will decrease the shear strength.

10.14.4   Genetic diagnosis approach

Based on Equation (10.36) and limited experimental results, it may be concluded that the shear strength of soil must be examined carefully to include types of parent material and stress history together with a degree of sensitivity of soil respect to the local environmental conditions such as load, temperature, and pore fluid. It is especially important for designing with contaminated soil. Arbitrarily assumed shear properties require second thoughts.

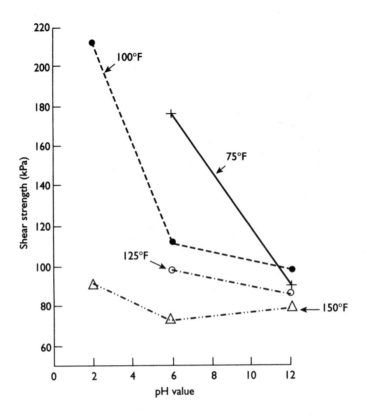

Figure 10.24 Effect of pore fluid and temperature on shear strength of sand–bentonite mixture.
Source: Based on Naik (1986) and Fang (1997).

10.15   Summary

The stress–strain–strength of soil and the failure criteria have been summarized. Failure criteria may be specified in terms of a maximum load or settlement. When considering shear strength, it is also important to distinguish between peak and residual conditions. While many designs are based on the peak value of strength, conditions that involve considerable displacement (e.g. a landslide) warrant use of the residual strength. Correlation of residual strength parameters to other soil constants such as the liquid limit, the plasticity index, and the liquidity index were presented. In terms of stress–strain relationships, soil may be broadly categorized according to whether it is loose or dense (granular soils) or whether it is normally consolidated or overconsolidated (fine-grained soils). Comparisons of shear characteristics of normally consolidated and overconsolidated clays were examined. A simple method for determination of undrained shear parameters, c, and ϕ, from unconfined compressive strength, q_u, and tensile strength, σ_t, is introduced. Genetic diagnosis approach for analysis of shear strength of soil is proposed. In this approach, the shear strength of soil is influenced by both parent material and the prevailing local environment factors.

PROBLEMS

10.1 (a) Define the strength of soil. (b) What is a rational strength test? (c) Explain Mohr's theory of rupture. (d) List two common structural materials for which Mohr's theory is not applicable and indicate where the theory breaks down. (e) If the Mohr rupture envelope is known, show how the theoretical failure planes can be determined when the material is subjected to a torsion test.

10.2 Sketch a stress–strain curve and a sample at failure for a brittle, a plastic, and a semi-plastic failure of a cohesive soil mass tested by unconfined compression tests. Explain why a height–diameter ratio of test specimen not less than 2:1 is desirable for strength testing?

10.3 Certain deposits of clay lose a considerable portion of their strength and stiffness after being remolded, while others do not, why? Define the term: sensitivity of soil. State how it is usually found, and discuss briefly its significance in engineering problems.

10.4 Assuming no volume change during a compression test of soil, derive a formula for the corrected area in terms of percent strain and original area for use in compression test computations. Explain why this formula is needed.

10.5 For a noncohesive granular material having a straight-line strength envelope passing through the origin, derive – by means of Mohr's circle – a relationship between the angle of internal friction of the material and the inclination of the shear planes in a compression test.

10.6 What type of laboratory shear test should be performed in order to obtain the reasonable results for the analysis of the following cases: deep excavation, highway embankment, earth dam, bridge pier, pile foundation, stability of retaining wall, and Shallow foundation. Explain.

10.7 A uniform deposit of fine sand has a void ratio of 0.65 and a specific gravity of solids of 2.65. The top 5 ft (1.53 m) of sand is moist with a water content of 15%; the ground water table is 10 ft (3.1 m) below the ground surface; and the maximum height of capillary rise is 5 ft (1.53 m). If the K_0 equals 0.5 and the ϕ is 30°, how much shear stress could be resisted on a plane inclined at 60° to the horizontal and at a depth of 20 ft (6.1 m) below the ground surface, if the sand remains in an at rest condition?

10.8 A consolidated undrained triaxial test is performed on a specimen of saturated clay. The value of σ_3 is 196.2 kPa (28 psi). At failure, we have $\sigma_1 = 274.7$ kPa (39.8 psi), $u = 186.6$ kPa (27.1 psi). If the failure in this test makes an angle of 57° with the horizontal, calculate the normal and shear stresses on the failure surface and the maximum shear stress in the specimen.

10.9 A given subsoil exploration showed the free groundwater level at a depth of 3.3 m (10.8 ft) below the ground surface. The soil was found to have a void ratio $e = 0.58$ and a specific gravity of solids $G_s = 2.67$ from the ground surface to a depth of 8.2 m (26.9 ft). The water content of the soil above the free water surface was 18%. (a) Determine the effective pressure in units of N/m^2 at 3.3 m (10.8 ft) and 8.2 m (26.9 ft) below the ground surface, and (b) If the groundwater surface is lowered 1.6 m (5.3 ft) and the soil between the original and final position of the free groundwater surface remains saturated by capillary action, by what amount will the effective pressure be changed at 3.3 m (10.8 ft) level and at the 8.2 m (26.9 ft) level?

10.10 For a given compacted stabilized soil, the unconfined compressive strength, q_u, equals to 200 kPa (29 psi) and its tensile strength equals to 90 kPa (13.1 psi). Compute the other shear parameters, cohesion, c, and the friction angle, ϕ.

10.11 The shear strength of clay is known to be directly related to the effective stress acting on the failure plane at the time of failure; however, it is difficult to estimate what this effective stress is likely to be in many situations. Why is this difficult and how is it overcome?

Chapter 11

Dynamic properties of soil

11.1 Introduction

Shear characteristics of the soil element under a static load have been presented in Chapter 10. Since many geotechnical engineering projects are not always under static loading alone, however, knowledge of dynamic properties and characteristics of soil is also important and needed for analysis and design. Example projects are those such as nearshore/offshore structures and facilities in earthquake zone, blasting area, etc. These structures face wind loads, wave action, current, seismic loads, machine vibrations, and the like. Dynamic loading can be grouped according to natural and anthropogenic origins: (a) natural sources of dynamic loads including earthquake, tsunamis, volcanic explosion, wind, rainstorm, waves, ice movement, and current; and (b) anthropogenic sources of dynamic loads including machine vibration, bomb blasts, construction and quarry blasting, construction operations, traffic, ship impact, and landing aircraft. Each of these loadings present unique challenges to the geotechnical engineer.

11.1.1 Kinematics, kinetics, and equivalent static effects

Kinematics deals with motion, that is, with time–displacement relationships and the geometry of movements. Kinetics considers the forces that produce or resist motion. From Newtonian physics the simple definition of mechanical force is equal to mass × acceleration. Mass is the measure of the property of inertia, which is what causes an object to resist change in its state of motion. Mass is weight, which is the force defined as $W = mg$, where g = constant acceleration of gravity 9.81 m/s^2 (32.2 ft/sc^2). The use of equivalent static effects permits simpler analysis and design by obviating the need for a more complicated dynamic analysis. To make this possible the load effects and the structure's responses must be translated into static terms. For example, (a) for earthquake effects the primary translation consists of establishing a hypothetical horizontal static force that is applied to a structure to simulate the effects of sideward motions during ground movements; (b) for wind load the primary translation consists of converting the kinetic energy of the wind into an equivalent static pressure, which is then treated in a manner similar to that for a distributed gravity load; and (c) for moving traffic load it can translate all types of moving vehicle loads into an equivalent 18-kip static load.

11.1.2   Types of waves and stress conditions of soil element under dynamic loading in general

1   *Types of waves*: There are four types of waves namely (a) harmonic; (b) periodic; (c) random; and (d) transient as illustrated in Figure 11.1. Harmonic motion is a special kinematic problem of major concern in structural analysis.

2   *Stress conditions of soil element*: The problems of dynamic loading of soils and soil structure interaction, which will be discussed in subsequent sections include earthquake, blasting, machine vibration, etc. Typical stresses on a small element in an infinitely homogeneous, isotropic, and elastic medium under wave propagation is illustrated in Figure 11.2. Consider the variation in stresses on the

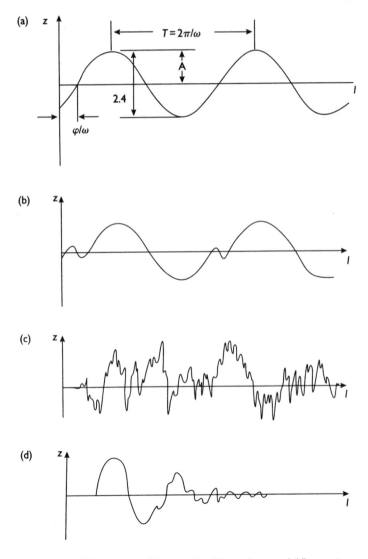

Figure 11.1 Types of waves: (a) harmonic; (b) periodic; (c) random; and (d) transient.

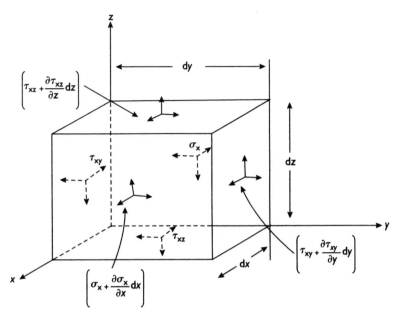

Figure 11.2 General stress conditions under dynamic load.

opposite faces of this element. The stresses on each face of this element are represented by a set of orthogonal vectors. Translational equilibrium of this element can be expressed by writing the sum of forces acting parallel to each axis.

11.2 Earthquake, earthquake loading, and measurements

11.2.1 Nature of earthquake and earthquake terminology

An *earthquake* is a natural phenomenon that can occur virtually anywhere. Man's understanding of the earthquake phenomena has undergone a gradual transition over the centuries. In Greek mythology the god Poseidon was the ruler of the sea and earthquakes. Throughout the centuries, man has been plagued by earthquake superstitions. The early Chinese attributed earthquakes to the rolling over of a huge monster within the earth, and others such as India and Japan also have similar ideas. To a certain extent there still remains with us in this modern time a certain amount of superstition and misconception. There are, however, two main areas, which between them, account for more than 80% of the earthquakes in modern times. The Circum-Pacific Belt, which runs from Chile, north to Alaska, and then down through Japan to New Guinea. The Mediterranean Belt (Alpide Belt) extends from Spain and North Africa, through Italy and the Middle East countries, and joins the Circum-Pacific belt in the East Indies. According to record, one of the strongest earthquakes was in Lisbon, Portugal on November 1, 1755. In terms of damage, the September 1, 1923 earthquake in the Kwanto plain in Japan was one of the worst. The sea-bottom in one area of the Sagami Bay sank more than 1000 feet (305 m). In the cities of Tokyo and Yokohama more than 50,000 buildings were completely destroyed. Of course there have been many notable earthquakes, and further discussion and a historical perspective is given by Scawthorn (2003). In order to better understand the earthquake characteristics and phenomena it is imperative to learn some common earthquake-related terms as used by geologists and engineers briefly explained as follows:

1 *Fault and displacement*: A fault is a crack in the Earth's crust along which there has been displacement of the two sides, relative to one another and parallel to the fracture. The *displacement* is relative movement of two sides of a fault, measured in any specified direction.
2 *Focus*: Focus is the origin of an earthquake in the Earth's crust or upper mantle.
3 *Mohorovicic discontinuity*: The transitional area, below the Earth's crust and above the mantle of the Earth, where earthquake waves travel at different speeds than they do in either adjacent zone. Also known as *Moho*.
4 *Macro- and micro-seismic effects*: Macro-seismic effects are the effects of earthquakes that can be observed on the large scale in the field, without instrumental aid. Micro-seismic effects are small scale, observable with instruments.
5 *Seismic load*: Forces due to a structure's inertia while it is being subjected to earthquake vibrations.
6 *Tectonic earthquake and creep*: Earthquake associated with faulting or other structural processes. Tectonic creep is a continuous displacement along a fault at a slow but varying rate, usually not accompanied by observable earthquakes.
7 *Zonation and microzonation*: Earthquake zonation is based on historical earthquakes and geological faults to distinguish regions or zones of different levels of risk or probability of occurrence. Microzonation involves the determination of pertinent site characteristics in an effort to reduce earthquake damage to an acceptable level. Microzonation is the determination of an earthquake damage potential of an area smaller than that considered in zonation. Other terms such as intensity, magnitude, liquefaction phenomena, P- and S- waves, and Richter scale are discussed in detail in the following sections.

11.2.2 Earthquake theory and related activities

11.2.2.1 Earthquake theory

1 Continental drift theory: According to the continental drift theory, the Earth's surface is composed of perhaps as many as 10 or more great plates that are moving. The movement of these great plates causes deep oceanic trenches and mid-ocean rifts; they crash into one another or slide past, or more appropriately, grind past one another. As a result of these tremendous forces and pressures the crust is subjected to tectonic stresses and strain, which periodically must be relieved. This relief shows up as a geologic fault.

2 Elastic rebound theory: Elastic rebound theory was developed by Harry Fielding Reid about 1906. According to his theory, there are three steps: (a) where a fault line exists, underground rocks accumulate strain as a result of gradual movement of the Earth's crust; (b) the fault is incapable of movement until strain has built up on either side of the fault. The terrain becomes distorted while retaining its original position until the resistance of opposing rocks is overcome; and (c) then the Earth snaps back into an unrestrained position creating an offset, which can be vertical, horizontal, or both. These violent subterranean movements create shock waves called *earthquake*.

11.2.2.2 Earthquake related activities

1 Tsunami: The name of *tsunami* comes from the Japanese term for "harbor wave." It is a special type of earthquake phenomenon, and caused, in general, by seafloor earthquake or volcanic eruption at the ocean-bottom. After a seafloor earthquake, a chain of waves races across the ocean at speeds greater than 500 mph (805 km/h), over the deep ocean. A tsunami wave attack occurred in July 1998 at Papua, New Guinea. A 23 ft (7.015 m) wave crashed into the northern coast of Papua, New Guinea, washing away five villages built on beaches, killing nearly 3000 people, and leaving hundreds of others missing. More devastating still was the tsunami that struck southeast Asia on December 26, 2004, where loss of life, property was severe in Thailand, Indonesia, India, Sri Lanka and elsewhere.

2 Volcanic Explosion: There are 455 active volcanoes in the world. There are more than 80 under the sea. Indonesia has the greatest number of active volcanoes. The greatest volcanic explosion in recent times occurred on August 17, 1883 on Krakatoa an island in the Sunda Strait in Indonesia. In all, 163 villages were wiped out by the tsunami action. The explosion is one of the major dynamic loadings based on record: the explosion force threw rocks 34 miles (54.74 km) up into the air and dust fell as far away as 3000 miles (4830 km) 10 days later.

11.2.3 Intensity, energy, and magnitude of earthquake

1 Earthquake intensity: There are numerous methods for measuring the intensity and magnitude of an earthquake. The seismograph is a device for detecting and recording Earth movements, primarily those originating with earthquakes. Analysis of the data from several seismographs can determine the energy released by the earthquake and its approximate location (latitude, longitude, and depth below the surface). The first crude seismograph was invented in China around 132 BC. As reported by Davidson in 1921 and 1933, 39 different intensity scales were listed. The modified Mercalli scale of 1931 was abridged and rewritten by Wood and Neumann (1931).

The intensity of an earthquake is a measure of earthquake effects based primarily on human reactions. Mercalli Scale range is I–XII in increasing intensity. The magnitude is a rating given to an earthquake independent of the place of observation, calculated from seismographic measurements.

2 Magnitude versus equivalent energy: The magnitude of an earthquake expresses the total amount of energy released as determined from the measurement of the amplitude of the seismic waves produced on seismographs. It is a measure of the absolute size and does not consider the effect on any specified location. During an earthquake, seismographs record the amplitude of the wave as received at the seismograph location. Comparison of Richter scale magnitude versus equivalent energy of TNT is presented in Figure 11.3.

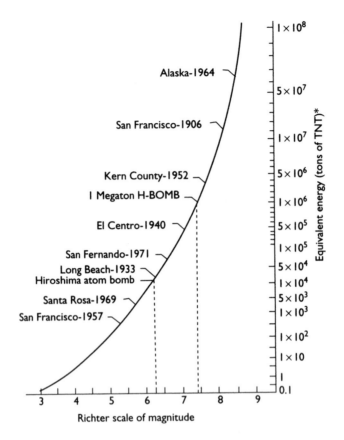

Figure 11.3 Richter scale magnitude versus equivalent energy.

Source: Federal Highway Administration with additional data.

Note
* TNT – abbreviation for 2, 4, 6–trinitrotoluene ($C_7H_5N_3O_6$), a highly explosive compound.

3   *Richter scale*: The Richter scale was devised by C. W. Richter and B. Gutenberg at the California Institute of Technology in 1935 for measuring the magnitude of earthquakes as shown in Table 11.4. It is a numerical scale in which each unit increment involves a logarithmic increase in the size of the ground waves generated at the earthquake's source and is based on the logarithmic scale of base 10. The Richter scale is open-ended; that is, it has no fixed maximum or minimum value. The relationship between Richter scale, magnitude, and energy are explained by Lew *et al.* (1971). Using the 1971 San Fernando Earthquake as an example, one unit change in the scale represents an increase of 10 times in the measured amplitude of the wave and an increase of approximately 32 times in energy release. Thus, a Richter magnitude 8 represents an increase in the seismograph amplitude of 10 over magnitude 7100 times over a magnitude 6, etc. A magnitude of 8 earthquake releases 32 times more energy than a magnitude 7, or 32 to the fourth power times as much energy as a magnitude 4. The approximate correlation between modified Mercalli intensity scale (Table 11.1) and the Richter scale (Table 11.2) is presented in Table 11.3.

Table 11.1 Modified Mercalli intensity scale

Category	Description
I	Not felt except by a very few under especially favorable conditions
II	Felt only by a few persons at rest, especially on upper floors of buildings
III	Felt quite noticeably by persons indoors, especially on upper floors of buildings. Many people do not recognize it as an earthquake. Standing motor cars may rock slightly. Vibrations similar to the passing of a truck. Duration estimated
IV	Felt indoors by many, outdoors by few during the day. At night, some awakened. Dishes, windows, doors disturbed; walls make cracking sound. Sensation like heavy truck striking building. Standing motor cars rocked noticeably
V	Felt by nearly everyone; many awakened. Some dishes, windows broken. Unstable objects overturned. Pendulum clocks may stop
VI	Felt by all, many frightened. Some heavy furniture moved; a few instances of fallen plaster. Damage slight
VII	Damage negligible in buildings of good design and construction; slight to moderate in well-built ordinary structures; considerable damage in poorly built or badly designed structures; some chimneys broken
VIII	Damage slight in specially designed structures; considerable damage in ordinary substantial buildings with partial collapse. Damage great in poorly built structures. Fall of chimneys, factory stacks, columns, monuments, walls. Heavy furniture overturned
IX	Damage considerable in specially designed structures; well-designed frame structures thrown out of plumb. Damage great in substantial buildings, with partial collapse. Buildings shifted off foundations
X	Some well-built wooden structures destroyed; most masonry and frame structures destroyed with foundations. Rails bent
XI	Few, if any (masonry) structures remain standing. Bridges destroyed. Rails bent greatly
XII	Damage total. Lines of sight and level are distorted. Objects thrown into the air

Source: US Geological Survey.

Table 11.2 The Richter magnitude scale

Magnitude	Description
M = 1–3	Recorded on seismograph, but not generally felt
M = 3–4	Often felt, but little to no damage
M = 5	Felt widely, little damage near epicenter
M = 6	Damage to poorly constructed structures within ≈10 km
M = 7	Major earthquake, serious damage within ≈100 km
M = 8	Great earthquake, significant destruction and loss of life >100 km
M = 9	Rare great earthquake, major damage over a large region >1000 km (e.g. Alaska 1964, Chili 1960)

Source: After Richter and Gutenberg, 1954.

11.2.4 Earthquake waves and dynamic shear force measurements

1 *Hypocenter and epicenter*: The origin of an earthquake in the Earth's crust or upper mantle is the "focus," or sometimes referred to as the *hypocenter*. The epicenter is defined as the point on the Earth's surface directly above the focus or hypocenter

Table 11.3 Relationship between modified Mercalli intensity scale and the Richter magnitude scale

Modified Mercalli intensity scale	Richter magnitude scale
I–II	2
III–IV	3
V	4
VI–VII	5
VII–VIII	6
IX–X	7
XI	8

Source: National Bureau of Standards, 1971.

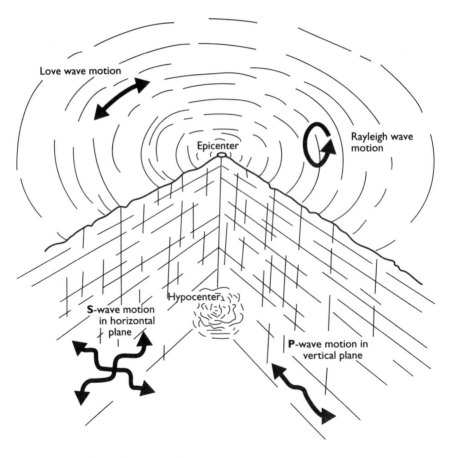

Figure 11.4 Elastic earthwuake generated waves.

Source: National Bureau of Standards, 1971.

as illustrated in Figure 11.4. The epicenter is that point on the Earth's surface that is intersected by a radial line drawn from the center of the Earth and passing through the hypocenter. When an energy release occurs at the hypocenter this energy is dissipated in the form of waves that travel through the crustal media to the surface and create a dynamic oscillatory motion at the surface. Two groups of seismic waves are generally named; (a) body waves and (b) surface waves.

2   *P- and S-waves*: There are two types of body waves, the primary wave and an orthogonal shear or secondary wave, sometimes referred to as the P- and S-waves respectively as illustrated in Figure 11.4. Both waves produce deformation in the direction of propagation. The P-motion is longitudinal and the S-motion is transverse to the direction of propagation (Bolt, 2004). Surface waves include Rayleigh wave and Love wave.

3   *Rayleigh and Love waves*: Lord Rayleigh was the first to prove that it was possible for a special type of elastic wave to propagate along the surface of a bounded elastic solid. The theory indicates that the motion of the particles at the surface is elliptical with the wave motion being retrograde; that is, the particles at the top of the ellipse are moving opposite to the direction of propagation. A. E. H. Love investigated transverse waves, and his theory indicates that in transverse waves the displacement is altogether transverse to the horizontal motion. Both Rayleigh and Love waves are illustrated in Figure 11.4. Detailed discussions are given by Richter (1958).

Both P-wave velocity and S-wave velocity can also be determined by field measurements by several methods, namely: (a) seismic refraction; (b) cross-hole seismic; (c) in situ impulse test; (d) down-hole (up-hole) seismic test; and (e) surface vibration. Detailed discussion of these methods are given by U.S. Army (1998).

11.3   Liquefaction phenomena and characteristics of granular soil

Liquefaction represents the largest seismic induced threat to the stability of most types of geotechnical structures. According to ASCE (1958), liquefaction, also called *spontaneous liquefaction*, is a sudden large decrease of the shearing resistance of a cohesionless soil. It is caused by a collapse of the soil structure by shock or other type of strain and is associated with a sudden but temporary increase of the pore fluid pressure. It involves a temporary transformation of the material into a fluid mass. The increase in pore pressure leads to a reduction in effective stress.

11.3.1   Laboratory studies of liquefaction

The simplest method of studying liquefaction involves placing a saturated soil specimen in a container, on a shaking table. The table is shaken until liquefaction of the soil sample is observed at a given acceleration. However, the results from this type of test tend to reflect the test conditions, that is, the duration and frequency of the shaking, with little information provided on the state of stress. This makes it difficult to apply the results to the field. An improved method of analysis involves cyclic, undrained loading with a triaxial device which provides information on the stress conditions that produce liquefaction. Seed (1976) notes that these types of test generally show that liquefaction can be readily induced in loose to medium dense sands. There are four types of laboratory tests commonly used for measuring the liquefaction potential of a sand mass; that is (a) cyclic triaxial shear test, (b) cyclic simple shear test, (c) cyclic torsion shear test, and (d) shake table test. The cyclic triaxial shear test is singled out for discussion as follows:

Cyclic triaxial shear test The cyclic triaxial shear test was developed by Seed and Lee (1966) to study the factors controlling the liquefaction of a saturated sand under cyclic loading conditions. A saturated cylindrical sample of sand is consolidated under an effective ambient pressure, σ'_o. All drainage is prevented and the sample is subjected to cycles of axial stress change, $\pm \Delta \sigma_d$. In saturated samples this loading procedure creates stress conditions on a plane at 45° through the sample, which corresponds to horizontal planes under the level ground conditions (Seed and Peacock, 1971). As indicated by Lee and Seed (1967), there are four major factors which influence the liquefaction potential of saturated sand. The factors are (a) void ratio or relative density of the sand, (b) intensity of cyclic shear stress, (c) duration of cyclic load or number of cycles of load, and (d) the initial effective confining pressure, σ'_o. The cyclic triaxial test is generally recommended for use in engineering practice, because it is simple to run and the equipment is widely available.

11.3.2 In situ measurements and correlation studies of liquefaction

In situ measurements and correlation studies of liquefaction include both field and laboratory studies. Liquefaction potential may be predicted on the basis of (a) soil types, (b) soil constants, (c) soil standard penetration test (SPT) blow counts, (d) laboratory cyclic stress ratio, and (e) dielectric constant. In this section only soil types, soil size, density, and SPT results are discussed. The use of soil types and particle size is a common procedure for identification of liquefaction potential. In particular, uniform cohesionless materials such as sands are more likely to experience liquefaction during earthquake action. In terms of soil density and SPT, Yegian and Whitman (1978) have shown a criterion which relates the possibility of liquefaction and soil density expressed in terms of SPT blow counts. One of the earliest procedures from Ohsaki (1970) suggested a criterion that if the SPT blow count, N, exceeds twice the depth in meters, then liquefaction will not occur. Many more methods have since emerged, including those which incorporate specific soil properties, depth to groundwater, and the earthquake magnitude. More details may be found in Seed *et al.* (1991), while Youd (1988) provides a rough liquefaction map for the contiguous United States, presented as Figure 11.5.

Liquefaction opportunity and liquefaction susceptibility can also be explained in terms of the general cyclic stress ratio equation as proposed by Seed *et al.* (1983) as follows:

$$\frac{t}{S_c} = \frac{0.65 \, a S_o r_d}{S_o g} \tag{11.1}$$

where t = average peak shear stress, S_c = initial vertical effective stress, a = maximum acceleration at the ground surface, S_o = total overburden stress at the depth considered, r_d = stress reduction factor which decreases from 1.0 at the ground surface to 0.9 at a depth of 35 ft (10.7 m), g = acceleration due to gravity. The quantity $[t/S_c]$ is called the *cyclic stress ratio* and its magnitude relative to the strength (or density)

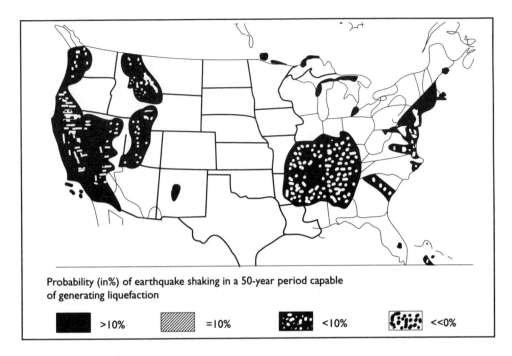

Figure 11.5 A crude liquefaction opportunity map of the contiguous United States.

Source: Youd, T. L., Recognizing liquefaction hazard, *Proceedings of the Symposium on Seismic Design and Construction of Complex Engineering Systems*, ASCE, Saint Louis, pp. 16–29. © 1988 ASCE. Reproduced by persmission of the American Society of Civil Engineers.

of fine sands determines whether or not liquefaction is likely. The cyclic stress ratio must be sufficiently high to induce liquefaction. On the basis of the field study, Figure 11.6 was developed by Seed *et al.* (1983). US Environmental Protection Agency (EPA) (Inyang, 1992) used Equation (11.1) and Figure 11.6 for evaluation of waste containment system vulnerability.

11.4 Liquefaction phenomena and characteristics for clay-like soil

As noted previously, liquefaction is generally a concern for loose, cohesionless (granular) soil. One might inquire as to why cohesive and/or fine-grained soils are not liquefiable. Theoretically speaking, clay-like soils can also experience liquefaction phenomenon, however, the bonding energies between particles for cohesive (clay) soil are much higher and more complex in comparison with granular soil. As a result, clay soils are generally considered non-liquefiable.

11.4.1 Example laboratory dynamic shear test for clay-like soil

1 Laboratory dynamic test: A laboratory cyclic simple shear apparatus was used for evaluation of the dynamic properties of Shanghai silt, as described by

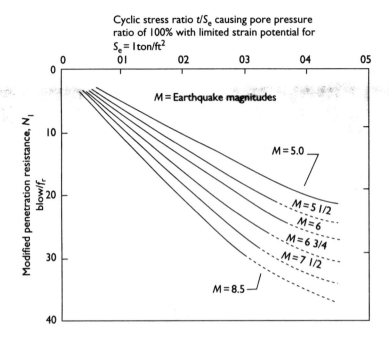

Cyclic stress ratio t/S_e causing pore pressure ratio of 100% with limited strain potential for $S_e = 1 ton/ft^2$

Figure 11.6 A chart for evaluation of liquefaction potential of sands for earthquake of different magnitude.

Source: Seed, H. B., Idriss, I. M., and Arango, I. Evaluation of liquefaction potential using field performance data, *Journal of the Geotechnical Engineering Divisons, Proceedings of the ASCE*, v. 111, no. 12, pp. 458–482. © 1983 ASCE. Reproduced by permission of the American Society of Civil Engineers.

Fang *et al.* (1981). The samples were taken from depths between 6 and 10 m (19.7–32.8 ft), and the soil was classified as a soft silt. Dynamic property tests using a staged procedure were performed on undrained samples. The staged procedure consisted of applying a known low cyclic load and observing the resultant shearing strain level achieved after three cycles of load. The process was then performed again on the same sample by increasing the cyclic load and observing the increased shearing strain level. This process was performed for six separate stages on each sample. The resultant shearing strain levels ranged from approximately $1.8 \times 10^{-2}\%$ to $2.2 \times 10^{-1}\%$. This cyclic load was applied to the sample at a frequency of 2 Hz. A summary of test results is presented in Table 11.4 and plotted graphically in Figure 11.7 as shear stress, shear modulus, G, and damping ratio versus strain, γ_r.

2 *Reference strain and normalized stress–strain curve:* For normally consolidated soils, the use of the "reference strain" approach to normalize stress–strain data provides a curve which is independent of the effective stress path and sample disturbance (Drnevich, 1979). Hence, the normalized stress–strain curve produced by this technique should apply to in situ conditions. Using data from Table 11.4 to illustrate, the computation procedure will be presented in Example 11.1. The normalization of test data presented in Figure 11.7 requires values of τ_{max} and G_{max}, at low shearing strain levels ($\leq 10^{-4}\%$). The maximum shear modulus (G_{max}) may be computed by

Table 11.4 Summary of G/G_{max} and γ/γ_r ratios with various shear modulus, shearing strain, and effective consolidation pressure for Shanghai soft silt

Sample number	σ_v (kg/cm²)	G (kg/cm²)	G/G_{max}	γ (cm)	γ/γ_r
1	0.47	95.0	0.940	0.00022	0.100
		85.0	0.842	0.00049	0.245
		75.0	0.743	0.00096	0.409
		67.5	0.668	0.00116	0.527
		65.0	0.643	0.00150	0.682
		50.0	0.495	0.00216	0.980
		$G_{max} = 101.0$ kg/cm²; $\gamma_r = 0.0022$ cm			
2	0.67	100.0	0.880	0.00022	0.100
		100.0	0.880	0.00049	0.245
		85.0	0.748	0.00096	0.436
		70.0	0.616	0.00136	0.618
		65.0	0.572	0.00162	0.736
		50.0	0.440	0.00205	0.932
		$G_{max} = 113.6$ kg/cm²; $\gamma_r = 0.0022$ cm			
3	0.80	130.0	0.975	0.00018	0.090
		115.0	0.860	0.00040	0.200
		105.0	0.790	0.00066	0.330
		95.0	0.710	0.00096	0.480
		90.0	0.675	0.00116	0.580
		85.0	0.637	0.00134	0.670
		$G_{max} = 133.3$ kg/cm²; $\gamma_r = 0.0020$ cm			

two procedures. The first procedure was presented by Hardin and Drnevich (1972) where the shear modulus, G, was shown to be equal to the following expression:

$$G = \frac{\tau}{\gamma} = \frac{1}{a + br} \tag{11.2}$$

where G = dynamic shear modulus, τ = shear stress, γ = shear strain, and **a**, **b** = constants. It was also shown that the ratio of shear modulus, G, to its maximum value, G_{max}, evaluated at low shearing strains could be given by

$$\frac{G}{G_{max}} = \frac{1}{1 + \gamma/\gamma_r} \tag{11.3}$$

where $\gamma_r = \tau_{max}/G_{max}$ = reference strain, and τ_{max} = shear stress at failure. Combining Equations 11.2 and 11.3 obtains the constants **a** and **b** as follows.

$$a = 1/G_{max}$$

$$a = 1/G_{max} \tag{11.4}$$

Rearranging yields

$$\gamma_r = \frac{\tau_{max}}{G_{max}} = \frac{a}{b} \tag{11.5}$$

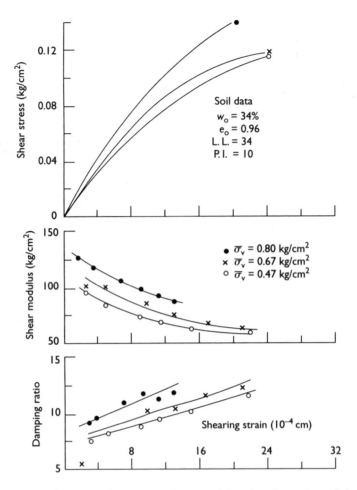

Figure 11.7 Correlation of dynamic shear stress, shear modulus, damping ratio, and shearing strain of Shanghai soft silt.

Source: Fang, Chaney, and Pandit (1981). Reprinted with permission of the University of Missouri-Rolla.

The constants **a** and **b** are determined by plotting $[\gamma/\tau]$ versus γ. The maximum shear modulus, G_{max}, can be estimated by the following empirical equation proposed by Hardin and Black (1968) as follows:

$$G_{max} = \frac{A(B-e)^2}{1+e}(OCR)^k(\sigma'_o)^{1/2} \tag{11.6}$$

where A = constant = 2630 for round-grained sand and 1230 for fine-grained material, B = constant = 2.170 for round-grained sand and 2.973 for fine-grained material, e = void ratio, OCR = overconsolidation ratio, $k \cong 0$ (for low plasticity index soil), $\sigma'_o = \sigma'_v (1 + 2 k_o/3)$, σ'_v = effective vertical stress, and k_o = coefficient of earth pressure at rest. The above equation may be checked with the values provided in Table 11.4 and Figure 11.7.

11.4.2 Code versus dynamic analysis

The code approach attempts to substitute a dynamic problem with an equivalent static problem. The parameters and variables involved in earthquake resistant design and the uncertainties involved have been divided into three major categories as reported by Castro (1975): (a) nature of ground motion including acceleration level and frequency content; (b) response of the structure to ground motion including natural frequency and mode shapes, and damping characteristics; and (c) level of safety desired including damage that can be tolerated, and balancing cost against risk.

Codes generally were developed by experts, who have given these uncertainties considerable thought. The earthquake force levels are defined for broad geographical zones. Codes establish levels of safety with a minimum of engineering time and expense. A dynamic analysis does provide a number of advantages, which a code approach lacks. With the advent of computer programs, a complicated dynamic analysis of virtually any structure is possible. Further discussions for earthquake design are given by numerous investigators such as earthquake effects on soil-foundation systems by Seed *et al.* (1991), and for building structures by Ambrose and Vergun (1995) and Hamburger (2003).

11.5 Dynamic shear characteristics of contaminated fine-grained soil

11.5.1 General discussion

In recent years, many hydraulic structures, reservoirs, and riverbanks have required consideration of seismic resistance during the design process. However, these considerations are based on soil parameters under normal conditions where the contact pore fluid is not polluted. Unfortunately, in actual cases the contact water, in general, has some degree of pollutants caused by various liquid and solid wastes, and acid rain. Therefore, the dynamic behavior of soil under normal conditions may be somewhat different in comparison to polluted water. Because of the complex nature of soil pollution, the terms "pollution" or "pollutant" are used loosely as discussed in Section 1.3. These terms will be reflected in terms of the pH or ion characteristics of the pore fluid (Sec. 4.6). Water of very low or very high pH (Sec. 4.1) is considered polluted and could be hazardous for living organisms.

11.5.2 Laboratory dynamic tests on contaminated fine-grained soil

1 Development of apparatus: The apparatus used for studying contaminated fine-grained soil is based on the Zeevaert type of dynamic torsion device (Zeevaert, 1983, 1996). According to Zeevaert (1983) the shear modulus can be obtained from

$$G = \left(\frac{\omega^2}{1 - D^2}\right)G_J \tag{11.7}$$

where ω = undamped circular frequency (rad.), D = damping ratio, $G_J = J_u h/I_p$, where J_u = polar moment of inertia of apparatus, I_p = polar second moment of specimen, and $I_p = \pi d^4/32$. The polar moment of inertia of the whole apparatus consists of the sum of moment of inertia of arm, specimen, and rod. However, the polar moment of inertia of the rod and specimen are about 10^{-12} and 10^{-8} times, respectively, smaller than the polar moment of inertia of the arm and therefore can be neglected. Also, for this testing the removable weight weights were not used. Thus the polar moment of inertia of the apparatus is approximately equal to

$$J_u = \frac{mL^2}{12} \tag{11.8}$$

where m = the mass of the arm, and L = the length of the arm. The damping ratio, D, and the circular frequency, w, can readily be evaluated from the record. Since the damping ratio is always less than 0.2, we can neglect D^2. Also the damped frequency, w_D, is

$$w_D = w(1 - D^2)^{1/2} - w \tag{11.9}$$

Thus, since $D < 0.2$, the damped and undamped natural periods are approximately equal, and instead of w we can measure the damped natural period, T, from the record and evaluate w through the relationship

$$w = \frac{2\pi}{T} \tag{11.10}$$

Thus, the formula to calculate the dynamic shear modulus directly from the record is

$$G = \frac{G_{oJ}}{T^2} \tag{11.11}$$

where $G_{oJ} = 4\pi^2 G_J = 32\pi L^2 h/3d^4$.

   2   *Damping ratio:* According to Zeevaert (1983) the damping ratio can be calculated from

$$D^2 = \Delta^2/(4\pi^2 + \Delta^2) \tag{11.12}$$

where the value of Δ is known as the logarithmic decrement and may be determined from successive amplitudes of the damped vibration. The value of D calculated from the test using Equation (11.12) has no relationship with the real damping for the soil (Du *et al.*, 1986), because this value includes the effect of frictional forces of the instrument during vibration. Although the exact value of the damping ratio for the specimen cannot be found precisely, the relative values of damping ratios between different specimens allow for a comparison of the effects of different pore fluids. Figure 11.8 shows the relationship between pore fluid pH and dynamic shear modulus.

11.6   Earthquake effects on structures and design considerations

Among the dynamic loading effects on various geotechnical structures or facilities, an earthquake is the most critical and damaging. Some of these problems are (a) structures above the ground surface including buildings and housing, bridges, retaining structures and walls, roadways, towers water tanks, and silos; (b) structures below the ground surface (shallow foundations) including footings, mat foundation, drainage pipe, and oil pipe; and (c) structures below the ground surface (deep foundations)

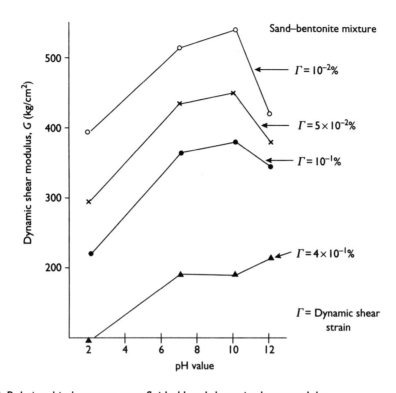

Figure 11.8 Relationship between pore fluid pH and dynamic shear modulus.

Source: Du, Mikroudis, and Fang, 1986. Copyright ASTM International. Reprinted with permission.

including caissons and pile foundations. A notable example of earthquake effects on above-ground structures was observed in the June 1964 earthquake in Niigata, Japan. This magnitude 7.5 earthquake induced liquefaction in the sandy soils underneath apartment buildings at Kawsgishi-Cho. The loss of subsurface strength resulted in a rotational failure of the structures; that is, some of buildings were tilted and partially sunk into the liquefied soils. An example of shallow foundation failures was observed in the Guatemala Earthquake of 1976. A large separation between the ground floor and subsurface soil was created by liquefaction. In terms of subsurface structures, Zeevaert (1983, 1991) and Margason (1977) observed a shear failure of concrete piles in response to horizontal earthquake forces at Mexico City. Figure 11.9 presents the interaction of batter piles and caps with earthquake loading, as reported by Margason (1977). Seed *et al.* (1991) shows a sewage purifier tank that was uplifted, because its bulk unit weight was less than the liquefied sand stratum in which it had been buried. A large collection of images of earthquake induced geostructural failures may be found through the University of California – Berkeley's National Information Service for Earthquake Engineering.

11.6.1   *Underwater and waterfront facilities*

Hydrodynamic pressure due to horizontal earthquake shock   In the design of dams, levees, and sea walls, which are located in seismic regions and exposed to water with a free surface on a wide frontal area, hydrodynamic pressure receives considerable

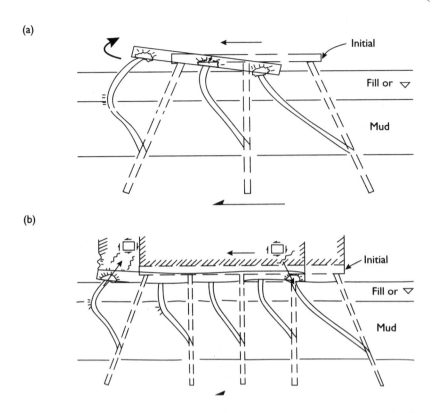

Figure 11.9 Interaction of batter piles and caps during an earthquake: (a) light dock: derck sways and rotates; and (b) heavy building platform.

Source: Margason, 1977. Reprinted with permission, Associated Pile and Fitting, LLC, Clifton, NJ, USA, Tel. 973-773-8400.

attention. Computation of hydrodynamic pressure due to a horizontal earthquake shock is more difficult in comparison with a vertical one because of the complicated mathematics involved. Westergaard (1933) has made a theoretical study and proposed an equation for hydrodynamic pressure, at a point perpendicular to the face of a structure exposed to water, due to horizontal earthquake acceleration, which may be computed by the Westergaard equation as:

$$P = \alpha\beta\gamma h^{1/2} \tag{11.13}$$

where P = hydrodynamic pressure (psf), h = vertical depth of point below water surface (ft), α = ratio of maximum horizontal earthquake acceleration to acceleration due to gravity, β = factor depending on (1) the slope, $1/S$ of the line joining the intersecting point of the water surface and the face of the structure to the point in question; and (2) the period of horizontal earthquake vibration, γ = a value depending on the total depth of water, H, from the bottom of the structure (Fig. 11.10) and the period of horizontal earthquake vibration. From design practice, the value, α, in Equation (11.13) is generally taken as 0.1; value, β, the vibration period is usually

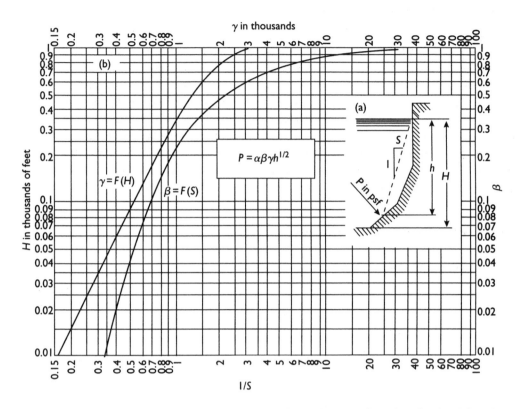

Figure 11.10 Hydrodynamic pressure on a structure due to horizontal earthquake shock based on Westergaard equation; (a) cross section of a hydraulic structure; and (b) curves for determination of hydrodynamic pressure. Applies for design vibration period of 1 s.

Source: Chow, V. T., Hydrodynamic pressure due to horizontal earthquake shock computed by curves, *Civil Engineering, ASCE*, September, pp. 52–53. © 1951 ASCE. Reproduced by permission of the American Society of Civil Engineers.

taken as 1 s, the corresponding relation between γ factor and the slope, $1/S$, has been developed by Chow (1951) as shown in Figure 11.10. After the pressure distribution along the face of the structure exposed to water is computed, the total force and its point of application can be found through integration and summing moments about a fixed point. Equation (11.13) is also used for computing the hydraulic pressure for the flood wall and has been adopted by the ASCE (1994).

EXAMPLE 11.1   (After Chow, 1951)
Given: $\alpha = 0.1$, $H = 200$ ft (61.0 m), $1/S = 1/0.08$ or 12.5, and $h = 160$ ft (48.8 m)
    Find the hydrodynamic pressure, P.

SOLUTION
From the curves (Fig. 11.10), $\beta = 0.9$ and $\gamma = 730$. Then from Equation (11.13):

$$P = (0.1)(0.9)(730)(160)^{1/2} = 831 \text{ psf } (39.8 \text{ kPa}).$$

11.6.2   Earthquake design criteria

The effects of earthquakes on various geotechnical facilities have been outlined in the previous section. The actual earthquake design criteria must follow some basic criteria including (a) probability of occurrence of strong ground shaking, (b) the characteristics of the ground motion, (c) the nature of structural behavior (deformations), (d) the behavior of materials when subjected to earthquake loading, (e) the nature of the damage, and (f) the cost of remediation. As discussed in Sections 11.3 and 11.4, there are several approaches for field and laboratory correlation studies on lique-faction. In general, techniques that measure P- and S-waves and other measurable dynamic parameters are used for design purpose. These field techniques include cross-hole, impulse, resonant footing, and surface vibration tests.

11.7   Wind and rain dynamics

11.7.1   General discussion

Wind load acts on the exposed surfaces of a superstructure, but it also affects lateral earth pressure and indirectly affects the bearing capacity as well as slope stability. There are various types of wind such as (a) wind alone without rain; (b) wind and rain mix; and (c) violent wind and rain and others such as hurricane, tornado, and El Nino effect. There are several types of wind and certain winds will seriously affect various types of civil engineering facilities.

11.7.2   Wind velocity, wind force and wind–structure interaction

1   *Wind velocity and wind force*: The wind force produced from wind velocity as proposed by API (1969) is presented by the following equation:

$$F = 0.00256 \, (V)^2 \, C_s \, A \tag{11.14}$$

where F = wind force (lb), V = sustained wind velocity (mph) at elev. +30 ft, C_s = shape coefficient, A = projected area of object. The shape coefficient, C_s, indicated in Equation (11.14) is recommended by API (1969) and are presented as Beam = 1.5, Slides of buildings = 1.5, and Cylindrical sections = 1.0.

2   *Wind–structure interaction*: Consider the situation where wind comes in contact with above-ground tanks, buildings, towers, and other structures in rigid contact with the foundation. The wind force (Eq. (11.14)) applies a turning moment to the structure and foundation. Moreover, since the wind force is likely to vary with time, the loading may result in a rocking or vibrational movement. One may evaluate the extent to which such loadings will result in appreciable movement or deformation. An average dynamic deformability, a_d^N, may be defined for each succes-sive soil layer beneath the surface and related to the applied stress and subsequent deformation, given as (Zeevaert, 1983)

$$\Delta \delta_N = a_d^N \cdot \Delta \sigma_N \tag{11.15}$$

where $\Delta\delta_N$ = the average deformation of soil layer N in response to an applied stress $\Delta\sigma_N$. The value of dynamic deformability may in turn be related to soil properties as

$$a_d^N = \frac{d_N}{2(1 + v)G^N} \qquad (11.16)$$

where d_N = the thickness of soil layer N, v = poissons ratio (often ranges between 0.3 and 0.5, for rapid loading of low-permeability soils, $v = 0.5$), G^N = dynamic shear modulus of elasticity (Eq. (11.2)).

11.7.3 Rain and rain–wind combinations

1 *Hurricane (typhoon)*: Most regions are cooled by trade winds, blowing throughout the year, and hurricanes are a feature of this kind of climate. A hurricane in the North Pacific is called a *typhoon* or *tropical cyclone*. This is a violent whirling storm in which wind blow spirally inwards toward a center of low pressure. The wind speed ranges from 65 to 174 knots/h (118–317 km/h). It may be up to 483 km across causing tremendous damage to land, nearshore, and offshore structures and leaves a wide trail of destruction behind them.

2 *Tornado (twister)*: Tornadoes and thunderstorms are frequently found imbedded within the much larger hurricane vortex. A tornado is a small but intense storm, rarely more than 600 ft (183 m) across, with winds whirling at over 300 miles/h (469 km/h) around a center of very low pressure. Strong upward air currents at the center can lift whole buildings into the air. A tornado occurs generally in specific regions (mostly inland region) at certain times of the year.

3 *Thunderstorm*: Thunderstorms form in similar conditions, where large pockets of moist air rise through cooler air. They usually developed in the summer, when a mass of warm air lies above the heated ground. The heated air rises at the same time expanding and cooling.

4 *El Nino and La Nina effects*: El Nino, a Spanish-language term referring to the Christ Child, is also the name for an unusual warming of the surface waters of large parts of the tropical Pacific Ocean. El Nino occurs rather erratically, every few years. Over the last 45 years or so, the El Nino events have started in 1952, and more recently in 1998. La Nina is the opposite of El Nino, referring to a period of cold surface waters in the Pacific. In general, these effects include intensive rainfall, flood, and high winds especially along the coastal region.

11.8 Wave and current dynamics

11.8.1 Fetch, wave height, and wavelength

Waves are generated by wind and/or earthquakes, tides, as well as anthropogenic disturbances. However, most waves are caused by wind; their characteristics can be determined by the velocity of the wind, its duration, and the fetch length as illustrated in Figure 11.11.

1 *Fetch length and wavelength*: The fetch length is the horizontal length of the generating area in the direction of the wind, over which the wind blows. When

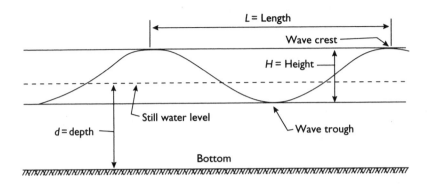

Figure 11.11 Wave characteristics.

waves approach a straight beach at an angle they tend to swing in parallel to the shore due to the retarding effect of the shallow water. Wavelength, L, is defined as the distance between consequative crests or troughs.

2   *Wave height*: The wave height indicated in Figure 11.11 is the vertical distance between the crest of a wave and the preceding trough. In deepwater waves the wave height, H, is equal to the diameter of the surficial water-particle orbit. The formation of waves in the deeper offshore areas is essentially a process involving the transfer of energy from winds to the water surface. The waves acquire a total energy, which consists of equal parts of kinetic and potential energy (Sec. 1.6).

11.8.2   Wave velocity and wave forces

1   *Wave velocity*: Wave velocity is defined as the ratio of wavelength, L, divided by wave period, T, as indicated in Equation (11.17). The wavelength is the horizontal distance between wave crests as indicated in Figure 11.11. The wave period, T, is the time in seconds required for the wave crest to move a distance of one wavelength.

$$V = L/T \tag{11.17}$$

where V = wave velocity of a deep water, L = wavelength, and T = wave period.

2   *Wave force*: The design wave for nearshore/offshore structures includes wave height, period, length, and still water depth. According to the American Petroleum Institute (API, 1969), the wave force on a structural member can be computed as

$$F = F_D + F_I \tag{11.18}$$

where F = wave force per foot of length acting perpendicular to the structural member, (lb/ft, N/m), F_D = drag force per foot of length, (lb/ft, N/m), and F_I = inertial force per foot of length, (lb/ft, N/m).

$$\text{Then } F = C_D \frac{W}{2g} DU|U| + C_m \frac{W}{g} \frac{\pi}{4} D^2 \frac{dU}{dt} \tag{11.19}$$

where C_D = drag coefficient, W = weight density of water, (lb/ft^3, kN/m^3), g = gravitational acceleration, ft/s^2, m/s^2), D = diameter of cylindrical member, (ft, m), U = horizontal particle velocity of the water, (ft/s, m/s), $|U|$ = absolute value of U, C_m = mass coefficient, dU/dt = derivative of U with respect to time, that is, horizontal acceleration of the water particle, (ft/s). The water particle velocity, U, is a function of water height, wave period, water depth, distance above bottom, and time.

11.9 Dynamics of water surface current

Water surface currents are caused mainly by winds and the rotation of the Earth. There are two general types of currents including (a) current variation with depth for shallow water channels and (b) The current force alone without waves. The current variation with depth for shallow channels can be estimated from the following equation proposed by API (1969).

$$V_x = V_s \left[\frac{X}{d} \right]^{1/7} \tag{11.20}$$

where V_x = current velocity (fps, m/s) at distance x (ft, m) above mud-line, V_s = current velocity (fps, m/s) at water surface, X = distance (ft, m) above mud-line, and d = distance (ft, m) from water surface to mud-line.

1   *Currents without waves*: The current force acting alone acting without a wave can be estimated from the following equations proposed by API (1969).

$$F_L = 0.5 C_L P V^2 A \tag{11.21}$$

$$F_D = 0.5 C_D P V^2 A \tag{11.22}$$

where F_L = lift force per unit length, C_L = lift coefficient, P = mass density, V = current velocity, A = projected area per foot of length, F_D = drag force per unit length, C_D = drag coefficient. C_D and C_L are based on the best information available for the current velocity and type of structural member involved.

2   *Current associated with waves*: Due consideration should be given to the possible superposition of current and waves. In these cases, where this superposition is deemed necessary, the current is a factor with waves. Other wave forces including wave dynamic excitation and interaction forces, wave fatigue forces, and extreme condition wave and current forces.

11.10 Machine vibration

Industrialization and automation has led to the development of a wide array of equipment and machinery. Invariably, the operation of such equipment is in relatively close contact with the ground surface. Typically, machine operation results in some unbalanced force which rotates with some degree of eccentricity and at a certain frequency. Consequently, subsurface soil layers experience both shear (S) and dilatational (P) waves. Foundation design and protection of sensitive structures require consideration of dynamic loads and engineering criteria.

11.10.1   Dynamic loads

1   Modes of vibration: In general, there are six modes of vibration relevant to foundations. Translational modes include vertical, longitudinal, and lateral, while rotational modes are noted by rocking, pitching, and yawing. These six types of vibration are illustrated in Figure 11.12. Subsequent oscillation of the foundation will depend on the deformability of the underlying soil, geometry and inertia of the foundation, and overlying structure as well as the type of excitation (Gazetas, 1991). In terms of soil deformability, the shear (G) and constrained (M_c) moduli may be used to define the passage of S- and P-waves. The shear wave velocity may be given as

$$V_s = \sqrt{\frac{G}{\rho}} \tag{11.23}$$

While the dilatational wave velocity may be written as

$$V_p = \sqrt{\frac{M_c}{\rho}} \tag{11.24}$$

where V_s and V_p represent the velocity of the shear and dilatational waves, respectively, G = the dynamic shear modulus, M_c = the constrained modulus, and ρ = the soil mass density. The shear and dilatational wave velocities may be related through poisson's ratio such that

$$V_p = V_s \sqrt{\frac{2(1 - v)}{1 - 2v}} \tag{11.25}$$

where v = Poisson's ratio and all others are defined as before.

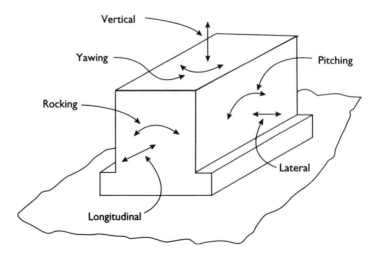

Figure 11.12 Modes of vibration for a foundation.

Soruce: Richart (1997), used with permission of Taylor & Francis.

11.10.2   Engineering criteria

The general procedure for evaluating and designing for vibrational problems involves trial and error calculations, as well as charts based on a variety of field and laboratory data. General limits for displacement amplitude are provided in Figure 11.13, while criteria for vibrations induced by rotating machinery is given in Figure 11.14. The general methodology involves the following steps (Gazetas, 1991):

- Estimate the magnitude and behavior of the vibrational loads (rotating versus reciprocating, etc.)
- Define the soil profile under consideration, noting the shear (G) and constrained (M_c) moduli as well as damping (β)
- Propose preliminary foundation type and geometry, and establish criteria (i.e. maximum deformation, etc.)

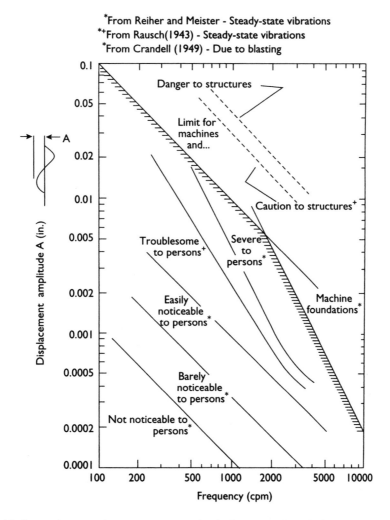

Figure 11.13 General limits of displacement amplitude for particular of vibration.

Source: Richart (1977), Used with permission of Taylor & Francis.

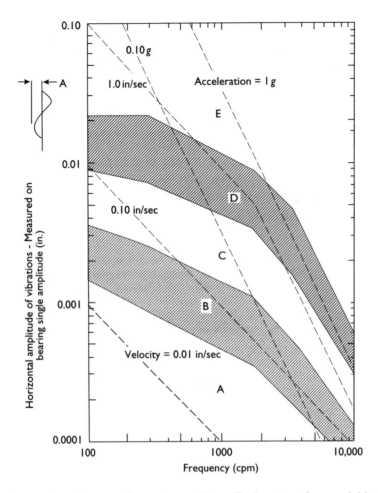

Figure 11.14 Criteria for vibrations of rotating machinery. Explanation of cases: A-No faults: typical of new equipement. B-Minor faults; correction not cost-effective. C-Faulty; correction within 10 days recommended. D-Failure is near; correct within 2 days to avoid break-down. E-Dangerous; shut down immediately avoid danger.

Source: Blake (1964), Copyright Gulf Publishing, Houston, Texas, USA. Reprinted with Permission.

- Estimate the dynamic response, and verify the extent to which criteria are met
- Revise design as necessary until criteria are met
- Monitor actual field performance, and make in-situ adjustments if necessary.

More details and calculation procedures are found in Gazetas (1991).

11.11 Other dynamic loadings

11.11.1 Construction operation

During the construction process, many additional dynamic forces are produced such as construction blasting, pile driven, densification (compaction) process and moving vehicle, etc. These are discussed as follows.

11.11.2 Blasting dynamics effects on soil and rock

Blasting is used in geotechnical engineering in three basic ways: (1) Used for quarry blasting for which it produces aggregates, crushed stone, gravel to be used as bituminous and concrete mix, base and subbase materials; (2) For removal of some undesired objectives to open up a tunnel for roadways, ditches; and (3) Used as blasting expansion pile foundation as part of densification (Ch. 7) in the ground improvement system. Regardless of the uses, subsurface considerations relative to blasting include (a) geology, (b) moisture content, (c) nonhomogeneous of ground layers, (d) amount of explosive charges, and (e) safety to adjacent structures.

1 Characteristics of blasting wave: Fundamentally, the properties of blasting wave motion that affects foundation structures is the energy which the blasting wave delivers. This energy may be represented by the amplitude of the motion it produces, the frequency of the motion, the acceleration which results from combining the amplitude and frequency, the force with which it moves a subject, or the energy itself, defined in terms of the velocity of the motion it produces. All these quantities are subject to direct measurement and various combinations of them have been used.

2 Blasting wave measurements and safe limitations: A considerable effort has been expended by the US Bureau of Mines on safety limits for blasting that will or will not cause structure damage. Most of their work is in correlating displacement, frequency, peak particle velocities, and safe distance with the blasting energy. There is a correlation with the safe distance and ground vibration. For the safe distance, Rockwell in 1927 stated that structures that are farther than 200–300 ft (61–91 m) from a blast would not be damaged. He pointed out the need for measuring vibrations from blasting in order to establish the level of vibration as a function of charge size and distance. Based on the characteristics of ground vibration one must include consideration for displacement, frequency, particle velocity. These parameters will be discussed in Section 11.12.

11.11.3 Moving vehicle and vehicle-pavement interaction

The effect of vehicular traffic on pavement and subsurface soils is a function of many variables, including the type, weight, and speed of a vehicle. Moving vehicle affects on pavement–soil interaction as reflected on pavement surface cracking due to fatigue and rutting are the two most important mechanisms leading to the deterioration of asphalt (flexible) highway pavements. Cracking is usually caused by the fatigue of repeated loading. The effects of vehicular traffic on embankment soil were studied at the AASHO Road Test. Results (HRB, 1962) are presented in Figure 11.15.

11.12 Measurement of the safe-limits under dynamic loading

11.12.1 Measurement of ground motion characteristics

1 Acceleration: In 1942, the Bureau of Mines reported particle acceleration as the best criterion for estimating damage to structures. For example, a particle acceleration

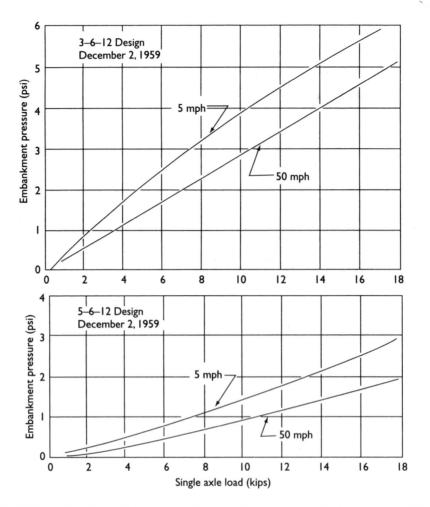

Figure 11.15 Effect of moving vehicle on embankment soil as a function of axle load and vehicle speed. Source: Data from AASHO Road Test.

of 0.1 g or less in the structure was labeled no damage, 0.1–1 g as caution, and a particle acceleration above 1 g was listed as possible damage.

2 *Energy Ratio:* Crandell (1949) presented a criterion based on vibration levels in the ground in the vicinity of the structure. This criterion called for an *energy ratio,* which is defined as

$$R_E = a^2/f^2 \qquad (11.26)$$

where R_E = energy ratio, a = acceleration (ft/sec²), and f = frequency. An energy ratio below 3 was safe and above 6 was denoted dangerous. Numerous studies were performed between 1949 and 1960 in which various criteria were suggested. During this period particle displacement of 0.076 cm (0.03 in.) was adopted by several states as a safe blasting limit.

3 *Particle velocity:* Devine (1966) proposed a criterion based on particle velocity in the ground near a structure. The unit of particle velocity is inch per second (ips). In this report, 2.9 ips (7.4 cm/s) was listed as producing no damage, but increasing velocities would then begin to produce damage. Edwards and Northwood (1960) also set forth particle velocity as the criterion for damage control; 2 ips (5.1 cm/sec) was considered safe.

$$V = H(D/W^{1/2})^{-\beta} \tag{11.27}$$

where V = particle velocity, inch per second (ips), H = constant for a particular site, W = maximum charge weight per delay, (lb.), D = distance from blast to measurement point, (ft), and β = constant. The exponent β and the constant, H, must be determined for each site considered.

11.12.2 Cracking measurement based on structural response

1 *Cracking intensity:* Measurement of cracking characteristics based on the structural response is proposed by Fang and Koerner (1977, cited by Fang, 1997). It measures the structural response as reflected by structural cracking intensity defined by Equation (11.28). Figure 11.16 illustrates the types of cracking and generalized dynamic response curve.

$$I_C = CW_{max}t \tag{11.28}$$

where I_C = cracking intensity (in.s), C = constant (depending primarily on material type, it varies from 0 to 1, with brick or concrete block construction being from 0.3 to 0.6), W_{max} = maximum cracking opening (in.) and t = duration (sec). The maximum crack opening, W_{max} and time duration, t as indicated in Figure 11.16 and Equation (11.28) must be determined experimentally.

11.12.3 Acoustic emission and impact echo

1 *Acoustic emission:* Acoustic emissions are the internally generated sounds which a material produces when it is placed under certain stress conditions. Sometimes these

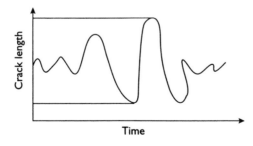

Figure 11.16 Generalized crack opening characteristics as related to loading conditions. The dashed lines define the maximum crack length, C.

sounds are audible such as wood cracking, ice expanding, soil and rock particles abrading against one another, etc. but more often they are not, due to either their low amplitude or high frequency, or both. Acoustic emission monitoring began in the 1940s by Obert and Duval (Obern, 1941) and by Hodgson in 1942, both of whom were interested in predicting rock bursts in mines, which were generated during the excavation process. Not only did these early works formulate the basic ideas of acoustic emission monitoring they also attempted to use triangulation to locate the zone of maximum stress intensity and performed laboratory tests to substantiate their field studies. Kaiser in the early 1950s (Kaiser, 1953) studied acoustic emissions in metals including aluminum, copper, lead, steel, and zinc, and this work had significant impact. Emissions were found in all materials and presumed to be from grain boundary motion induced by the applied stresses. Subsequent work in metals has brought acoustic emission technology into its current status. For soil–water system, extensive work was done by Koerner-Lord and Associates since the early 1970s (Koerner *et al.*, 1976).

2 Acoustic emission Mechanisms: Normally a piezoelectric sensor (an accelerometer or transducer) is used as a "pickup" to detect the acoustic emissions. These sensors, when mechanically stimulated, produce an electrical signal. The signal is then amplified, filtered, counted, and displayed or recorded. The counts, or recordings, of the emissions are then related to the basic material characteristics to determine the relative stability of the material being tested (a) if no acoustic emission is present, the material is in equilibrium and thus stable under that condition; and (b) if, however, emissions are observed, a nonequilibrium situation is present, which if continued could ultimately lead to failure.

3 Impact echo: Echo is also a part of sound dynamics. At present, the echo method is used to determine stability of concrete structures. However, limited data are also available for soil. The principle of the echo method is based on introducing a transient stress pulse into an object by mechanical impact on the surface of the object and sometimes it is called *impact-echo method.* The majority of the stress pulse propagates through the material along spherical wave fronts called P- and S-waves. The rest of the stress pulse generates a R-wave, which travels along the surface away from the point of impact. The P- and S- waves are reflected by internal and external boundaries. The impact-echo method is based on monitoring reflections of the P-wave, which are most dominant under the point of impact. For further discussion on basic concept see Sansalone and Carino (1986).

11.13 Summary

Soils are subjected to a variety of static and dynamic loads. Dynamic loads are derived from natural (earthquakes, waves, wind) or anthropogenic (traffic, blasting, machine vibrations) sources. In either case it is necessary to assess the nature of the applied stresses and the soil properties. In particular, the dynamic shear modulus, G, is often necessary for design. Perhaps the most extreme soil dynamics problem is the occurrence of earthquakes and potential liquefaction. Liquefaction occurs in loose, cohesionless soils, and there are a variety of empirical methods available to predict susceptibility. The dynamics of sound were also discussed, including acoustic emission and impact-echo testing methods for evaluation of soil rock behavior. These techniques have the added advantage of being a nondestructive means of evaluation, contrasted with boring holes or otherwise retrieving physical soil/rock samples.

PROBLEMS

11.1   Why is the dynamic behavior of soil important for many geotechnical problems?

11.2   Granular soil exhibits liquefaction character, but why is it of less concern in clay-like soils?

11.3   What is tsunamis action? Why is tsunami action more critical in comparison with common earthquake action?

11.4   Explain why the dynamic shear modulus of contaminated fine-grained soil is different with non-contaminated fine-grained soil.

11.5   What are the mechanics of landslides and how does an earthquake trigger landslide action?

11.6   Discuss and compare soil–structure interactions under both static and dynamic loads. Also discuss the factors affecting shear strength of cohesive soils under both static and dynamic loads.

11.7   List and discuss the methods for determining dynamic shear modulus of cohesive and cohesionless soils.

11.8   Discuss characteristics of wave action in air and under the water.

11.9   Why can dynamic loading be more critical than static loading for design from a safety viewpoint?

11.10   What is the general concept of acoustic emission and impact-echo method for determination of a material? Make comparisons, in general, between acoustic emission and impact-echo methods.

Chapter 12

Bearing capacity of shallow foundations

12.1 Introduction

12.1.1 General discussion

The bearing capacity of soil is its ability to withstand an applied loading condition. If the supporting power of ground soil is greater than the structural load, then the condition is said to be safe. This supporting power used in geotechnical engineering is referred to as the *bearing capacity* or the *ultimate bearing capacity* of ground soil. The characteristics of bearing capacity of soil belongs to the multimedia energy field, because it is not only controlled by the load applied, but also is influenced by local environmental conditions such as soil types, location of groundwater table, frost penetration depth, and characteristics of pollution intrusion. Current practice assesses the bearing capacity in terms of the load or pressure which develops under a foundation, relative to the maximum allowed. Bearing capacity may be considered in terms of either shallow or deep foundations. The *shallow foundation* is defined as the foundation width, **B**, being larger than the depth of the foundation, **D**, such as footings and mat foundations. Deep foundations may be defined as elements in which the depth is larger than the width, such as caissons and pile foundations. In this chapter, only the bearing capacity of shallow foundations is discussed.

12.1.2 Basic requirements for analysis of shallow foundations

To ensure satisfactory performance of a shallow foundation, it is necessary to provide adequate safety against shear failure of the foundation soil and to prevent excessive settlement of the foundation. These general requirements and related information are (a) subsurface conditions including soil type, depth of the groundwater table, frost penetration depth, and topographical features; and (b) footing characteristics including footing width, footing depth, footing shape, and footing base condition. There are numerous methods for determination of the bearing capacity of shallow foundations that can be further grouped into four basic approaches as theoretical approach, in situ measurement, correlation with other soil parameters, and building codes. Before discussing these various methods for determining the bearing capacity of ground soil, it is necessary to review the general ground stability analysis methods available.

12.2   Ground stability analysis

12.2.1   Methods of analyses

At the present time, analyses of shallow foundations can be made by employing one of the following five widely used methods, namely: (a) slip-line method, (b) limit equilibrium method, (c) limit analysis method, (d) finite difference method, and (e) finite element method. The first three methods are generally used in association with stability problems where the bearing capacity is sought. If instead, a settlement of foundation and stress distribution within the soil mass is of prime interest, then the fourth method must be used. A brief description of each method is presented herein as introductory information.

1   Slip-line method: This method involves construction of a family of shear slip-lines in the vicinity of the structure (footing) loads. These slip-lines, which represent the directions of the maximum shearing stresses, form a network known as a *slip-line field*. The plastic slip-line field is bounded by regions, which are rigid. For plane strain problems, there are two differential equations of plastic equilibrium and one yield a condition available for solving the three unknown stresses. These equations are written with respect to curvilinear coordinates, which coincide with the slip-lines. The boundary conditions are given only in terms of stresses, thus these equations are sufficient to give the stress distribution without any reference to the stress–strain relationship. However, if displacements or velocities are specified over part of the boundary, then a constitutive relation must be used to relate the stresses to the strains and the problem becomes much more complicated. Although solutions may be obtained analytically, numerical and graphical methods are often found necessary as presented by Brinch Hansen (1961) and Sokolovskii (1965).

2   Limit equilibrium methods: Familiar methods described in various geotechnical engineering textbooks on the subjects of bearing capacity, earth pressures, and slope stability by Terzaghi (1943), Taylor (1948), and Meyerhof (1955) are all classified as methods of limit equilibrium. They can be best described as approximate approaches to the construction of slip-line fields. Assumptions must be made regarding the shape of the failure surface and the normal stress distribution along such surface. The stress distribution usually satisfies the yield condition and the equations of static equilibrium in an overall sense. By trial and error, it is possible to find the most critical location of the assumed slip surface from which the capacity of the footing can be calculated.

3   Limit analysis method: This method uses the concept of a yield criterion and its associated flow rule, which considers the stress–strain relationship. The method is based on two theorems for any body or assemblage of bodies of elastic-perfectly plastic material, namely lower-bound theorem and upper-bound theorem. This method is relatively new in comparison with commonly used limit equilibrium method stated in Case (2), therefore, more detailed discussions are presented in the separated section (Sec. 12.2.2).

4   Finite difference method: The basic concept of the finite difference methods is the discretization procedure, which is based on replacing continuous derivatives in equations governing the physical problem by the ratio of changes in the variable over a small but finite increment. As a result of these replacing procedures, a differential equation is transformed into a difference equation. The differential equations generally involve first, second, third, and fourth derivatives.

5 Finite element method: The finite element method is a process through which a continuum with infinite degrees of freedom is approximated by an assemblage of sub-regions called *finite elements*, each with a specified but finite number of degrees of freedom. The basic property underlying the finite element method is that typical sub-regions can be studied for their behavior independent of the other elements. Therefore, once the behavior of a typical element is defined in terms of the value at the nodes of the element, the complete model is then obtained by appropriate assembly of the complete system elements.

12.2.2 Fundamentals of limit analysis

1 Basic concept: In contrast to the slip-line and limit equilibrium approaches, the limit analysis method considers the stress–strain relationship of soil in an idealized manner. In this approach, soil is assumed to have an elastic-perfectly plastic behavior satisfying the Coulomb yield criterion and its associated flow rule. This idealization establishes the plastic limit theorems on which the limit analysis is based. The two main limit theorems (Drucker and Prager, 1952) for any body or assemblage of bodies of elastic-perfectly plastic material may be stated, in terminology appropriate to soil mechanics, as follows:

2 Theorem 1 (Lower-bound theorem): The collapse load calculated from a statically admissible stress field, which satisfies all stress boundary conditions in equilibrium without violating the Coulomb yield criterion, is always lower or at most equal to the actual collapse load. If an equilibrium distribution of stress can be found which balances the applied load without violating the yield criterion, which include c, the cohesion, and ϕ, the angle of internal friction, the soil mass will not fail or will be just at the point of failure.

3 Theorem 2 (Upper-bound theorem): The collapse load, calculated from a kinematically admissible velocity field of which the rate of external work done exceeds the rate of internal dissipation, is always greater that the actual collapse load. The soil mass will collapse if there is any compatible pattern of plastic deformation for which the rate of work of the external loads exceeds the part of internal dissipation. According to the statement of the theorems, in order to properly bound the "true" solution, it is necessary to find a compatible failure mechanism (velocity field or flow pattern) in order to obtain upper-bound solutions. A stress field satisfying all conditions of lower-bound theorem will be required for a lower-bound solution. If the upper- and lower-bounds provided by the velocity field and stress field coincide, the exact value of the collapse, or limit, load is determined. Further discussions on applications to geotechnology are given by Chen and McCarron (1991), chapters 13 and 14.

12.2.3 Comments and discussions on stability analysis methods

1   Five methods described in Section 12.2.1 are related to each other in a certain way. Most slip-line solutions give kinematically admissible velocity fields and are

thus considered as upper-bound solutions, provided that the velocity boundary conditions are satisfied.

2   If the stress field within the plastic zone can be extended into the rigid region so that the equilibrium and yield conditions are satisfied, then this solution constitutes a lower-bound.

3   In view of limit analysis, each of the limit equilibrium methods utilizes the basic concept of the upper-bound rule, that is, a failure surface is assumed and the least answer is sought. However, it gives no consideration to soil kinematics, and the equilibrium conditions are satisfied only in a limited sense. Therefore, limit equilibrium solution is not necessarily an upper- or a lower-bound. However, any upper-bound solution from limit analysis will obviously be a limit equilibrium solution. Nevertheless, the method has been most widely used due to its simplicity and reasonable accuracy.

4   By means of the finite element method, it is possible to calculate the complete states of stress and strain within the soil beneath the structure (footing). The method has been proved useful for studying the bearing capacity and other soil related problems. It can locate areas of local failure and give a clear answer to the overall stability. However, the results reported so far on the value of bearing capacity of footings are not significantly better than those obtained from some accurate limit equilibrium methods. Nevertheless, the finite element method is undoubtedly of practical value since there are virtually no other methods capable of predicting the movement, the states of stress and strain, and the localized failure zones around the footing.

5   All above mentioned stability analysis methods are based on the mechanical energy concept; however, some additional environmental factors relating to stability will also be discussed in following chapters when the proper situation is encountered.

12.2.4   Planning for the foundation stability analysis

The type of ground stability analysis procedure needed to be performed depends on the ground soil properties, upper structure conditions, and its surrounding environmental situations. As illustrated in Figure 12.1, there are three general cases based on geographical conditions. A brief discussion of each case is presented as follows; (a) If the structure rests on flat firm ground (when $\beta = 0$), then, in general, the bearing capacity and settlement analyses (Ch. 9) are sufficient; (b) If the structure is to be built on a sloping hillside (when $\beta > \phi$), and/or on weak soft ground soil deposits, then the slope stability analysis (Ch. 14) must be carried out; and (c) For a vertical cut or designing of a retaining wall (when $\beta \to 90°$), an additional lateral earth pressure analysis (Ch. 13) is required.

Other conditions such as for offshore or waterfront structures, wave action needs to be considered. In seismic or problematic soil–rock regions, such as earthquakes or dynamic forces (Sec. 11.5), special attention to problematic soil behavior (Sec. 2.10) is required. From the design point of view, the duty of the geotechnical engineer is not only to analyze and design the foundation structure just beneath the main structure, but must also consider all possible environmental conditions surrounding the area. Some environmental conditions include topography, conditions of the right-of-the-way, and surface and subsurface drainage patterns.

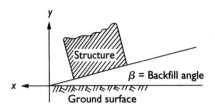

Case I: when $\beta = 0$ (on flat ground)
  Stability analysis:
    Bearing capacity and settlement analysis are sufficient. If results do not meet the proper requirements, one or a combination of the following ground improvements must be taken.
  Ground improvement techniques:
    Soil replacement
    Pre-loading
    Densification
    Grouting and stabilization
    Footings
    Deep foundation (piles, caisson, etc.)

Case II: when $\beta > \phi$ (on sloping hillside)
  Slope stability analysis is needed
Case III: when $\beta \Rightarrow 90°$ (vertical cut)
  Additional earth analysis is required. If results do not meet the proper requirements, one or a combination of the following ground improvements must be applied.
Ground improvement techniques: (Cases II and III)
  Retaining structures
  Bracing system
  Anchors
  Reinforced earth
  Grouting and stabilization
  Piles

Figure 12.1 Ground stability analysis planning and its interaction.

12.3 Loads and allowable loads

12.3.1 Loads

Loads acting on a structure include static, dynamic, and environmental loads. Static load (dead load) includes the weight of the structure and all material permanently attached to it. Permanent and fixed service equipment is usually considered as part of the dead load. Dynamic load includes live load and impact load. Live load includes all loads that are not permanently a part of the structure but are expected to be superimposed on the structure during a part or all of its useful life. Vertical loads due to wind or snow are not considered as live load. Human occupancy, partition walls, furniture, warehouse goods, and mechanical equipment are major live loads. The magnitude of live loads to be used in the design of various buildings is usually stipulated in local building codes. Railroad and highway bridges subjected to traffic loading, reaction from industrial cranes, and elevators sometimes constitute a large portion of the live load. Total loads acting on the ground soil are calculated in three categories: normal load, maximum and minimum loads, and horizontal load.

1 *Normal load*: Normal load is a vertical load, which includes static (dead load) as illustrated in Equation (12.1).

$$P_N = P_D + P_S + P_L + P_V - P_B \tag{12.1}$$

where P_N = normal load, P_D = dead load, P_S = snow load, P_L = live load, P_V = vertical reaction due to lateral earth pressure, and P_B = buoyancy load.

2 *Maximum and minimum loads*: The maximum load includes the dead load, live load, and vertical components of lateral earth pressure as shown in Equation (12.2).

To determine the *minimum load*, assume that live load is equal to zero as shown in Equation (12.3).

$$P_{max} = P_D + P_L + P_V \qquad (12.2)$$

$$P_{min} = P_D + P_V \qquad (12.3)$$

All notations are the same as previously stated.

3 *Horizontal loads*: Horizontal loads include wind load, horizontal components of a traveling crane, and unbalanced lateral earth pressure (Ch. 13) and can be represented by Equation (12.4).

$$P_H = P_W + P_h + P_u \qquad (12.4)$$

where P_H = horizontal load, P_W = wind load, P_h = horizontal force from traveling crane, and P_u = unbalance lateral earth pressure.

4 *Environmental loads*: Environmental loads relating to bearing capacity include wind load, machine vibration, earthquake load, etc. These dynamic loads have been discussed in Chapter 11. Some additional environmental loads relating to the bearing capacity problems include swelling pressure of soil, thermal expansion, and traction force.

12.3.2 Allowable load

Allowable load also called *design load* is the ultimate load divided by the factor of safety. The principle and procedure for selecting the factor of safety will be discussed in the following section.

12.4 Factor of safety

12.4.1 Fundamentals of factor of safety

A factor of safety (F_s) against shear failure of the foundation soil is always specified due to the uncertainties involved in the determination of the ultimate bearing capacity (Q_u) as well as in the analysis of the anticipated bearing pressure. The selection of an appropriate value for the factor of safety is dependent upon many factors such as the adequacy of subsurface investigations, nature of the soil formation, reliability of the shear strength parameters (c and ϕ) assigned to the foundation soil, location of groundwater table, and the accuracy of estimating the applied load. The degree of complexity as well as practical experience and judgment is used to develop an appropriate factor of safety Commonly used values for the factor of safety range from 1.5 to 3.0. There are numerous types factor of safety proposed. Some of these commonly used procedures are presented as follows:

12.4.2 Minimum factor of safety

The selection of factors of safety for design cannot be made properly without assessing the degree of reliability of all other parameters that enter into the design, such as

Table 12.1 Minimum factors of safety for design of shallow foundations

Category	Typical structures	Characteristics of the category	Soil exploration	
			Complete	Limited
A	Railway bridges Warehouses Blast furnaces Hydraulic Retaining walls Silos	Maximum design load likely to occur often; consequences of failure disastrous	3.0	4.0
B	Highway bridges Light industrial and public buildings	Maximum design load may occur occasionally; consequences of failure serious	2.5	3.5
C	Apartment and office building	Maximum design load unlikely to occur	2.0	3.0

Source: Vesic, 1975. In *Foundation Engineering Handbook*, 2nd edition, H. Y. Fang, ed., Copyright (1991) by Van Nostrand Reinhold. With kind permission of Springer Science and Business Media.

applied loads and strength characteristics of ground soil. The minimum factor of safety for the design of shallow foundations are proposed by Vesic (1975) as a guide for permanent structures in reasonably homogeneous ground soil conditions. Vesic has suggested the total factors of safety F_s in Table 12.1 on the basis of classification of structure, foundation conditions, and the consequences of failure. Lower factors of safety may be applied to temporary structures. Meyerhof (1984) discussed the total factors of safety F_s given in Table 12.2(a) and the use of the load and resistance factors (partial factors) given in Table 12.2(b).

12.4.3 Partial factors of safety

An alternative approach to the evaluation of safety of a structure and the uncertainties involved with different variables such as foundation loads or soil strength is given by the term *partial factor of safety* as introduced by Brinch Hansen (1961). A nominal state of failure is considered, in which the applied loads are multiplied by certain partial factors, while the soil strength parameters, c and ϕ, are reduced by other partial factors. A list of recommended partial factor of safety is given in Table 12.3. This procedure is more commonly referred to as the load and resistance factor design method (LRFD).

12.4.4 Localized factors of safety

Equation (12.5) presents a general form for computation of the localized factor of safety. This factor is to compensate the additional risk caused by the environmental effects. In other words, the localized factor of safety expands the conventional factor of safety.

$$F_L = \lambda \, F_s \tag{12.5}$$

Table 12.2 Values of minimum factors of safety: (a) Minimum total factors of safety; and (b) Minimum partial factors of safety

Failure type	Item	Safety factor, F_s	
(a) Values of minimum total safety factors			
Shearing	Earthworks	1.3–1.5	
	Earth-retaining structures, excavations	1.5–2	
	Foundations	2–3	
Seepages	Uplift, heave	1.5–2	
	Exit gradient, piping	2–3	
Category	Item	Load factor	Resistance factor
(b) Values of minimum partial safety factors			
Shearing	Dead loads	(f_d) 1.25 (0.85)	
	Life loads, wind or earthquake	(f_1) 1.5	
	Water pressures	(f_u) 1.25 (0.85)	
Seepages	Cohesion (c) (stability; earth pressure)		(f_c) 0.65
	Cohesion (c) (foundations)		(f_c) 0.5
	Friction (tan θ)		(f_θ) 0.8

Source: Meyerhof, G. G. (1984), Safety factors and limit state analysis in geotechnical engineering. *Geotechnique*, vol. 21, pp. 1–7. Copyright Thomas Telford Ltd. Reprinted with permission.

Note
Load factors given in parentheses apply to dead loads and water pressures when their effects are beneficial, as for deal loads resisting instability by sliding, overturning or uplift.

Table 12.3 Partial factor of safety for shallow foundations

1 *Load factors*	
(a) Dead load	1.00
(b) Steady water pressure	1.00
(c) Fluctuating water pressure	1.20 (1.10)
(d) Live load (general)	1.50 (1.25)
(e) Wind load	1.50 (1.25)
(f) Earth and grain pressure in silos	1.20 (1.10)
2 *Soil strength factors*	
(a) Cohesion, c	2.00 (1.80)
(b) Coefficient of internal friction angle, ϕ	1.20 (1.10)

Source: Based on Chen and McCarron (1991) and Brinch Hansen (1961).

where F_L = localized factor of safety, λ = correction factor, and F_s = conventional factor of safety. The localized factor of safety is a special type of factor of safety, which deals with certain types of soil or sites that frequently appear as problematic with higher risk of potential failure. In such case, the conventional factor of safety must be adjusted to account for the additional risk. For the hazardous/toxic waste sites such as an abandoned landfill, a localized factor of safety also should be used to compensate for environmental concerns. Table 12.4 summarizes some common environmental geotechnical problems.

Table 12.4 Suggested localized factor of safety for problematic soil deposits and hazardous/toxic waste sites

Types	Description	Correction factor, λ
I	Soft clay; High groundwater table	1.75
II	Residual soil; Expansive clay; and Dispersive clay region	1.50
III	Abandoned landfill sites (Clean landfill)	1.50
IV	Abandoned landfill sites (Highly nonuniform landfill)	1.75

Note
Correction factor, λ, is only a guide to engineers in predicting stability of ground soil in problematic sites and should be used with caution and judgment.

12.5 Ultimate and allowable bearing capacity

12.5.1 Ultimate bearing capacity

The determination of the ultimate bearing capacity is an essential step toward the objective of providing adequate safety against shear failure of the foundation soil. Ultimate bearing capacity evaluations presented in general are based on the measured or estimated shear strength of the foundation soil. The use of plate-load tests for ultimate bearing capacity determination is discussed in Section 12.9. Methods for ultimate bearing capacity analysis are presented in three parts, namely (a) foundations on soils in general, (b) foundations on cohesionless soils, and (c) foundations on highly cohesive soils.

For practical design purposes, soils with little cohesion are sometimes treated as cohesionless soil ($c = 0$). Similarly, many types of soils with appreciable cohesion but having a relatively small angle of internal friction (or subjected to undrained conditions) are frequently treated as highly cohesive soils with $\phi = 0$. A summary of commonly used methods for determining the ultimate bearing capacity of foundations on soil in general is presented in Table 12.5. Typically, three failure modes are considered, namely general shear, local shear, and punching shear failures. General shear failure is perhaps the more common mode, occurring in soils and rock of moderate to high strength. Failure is relatively fast (i.e. the displacement of soil and/or associated structure), and the failure surface is relatively clear. Localized shear failure may occur in somewhat weaker soils, and the failure surface is clear beneath the surface but less distinct elsewhere. Displacements associated with local shear are less dramatic but may continue with considerable time as the soil continues to yield to the overlying surcharge. Punching shear failure is generally relegated to the weakest of clays or loosest of sands wherein the foundation is pushed directly into the subsurface. Punching shear failure usually results in large settlements with a poorly defined failure surface.

12.5.2 Allowable bearing capacity

The allowable load or design load is equal to the ultimate load divided by the factor of safety as indicated in Equation (12.5). A factor of safety of 2–3 is commonly used

Table 12.5 Outline of general requirements and related information concerning the design of shallow foundations

Requirements	General procedures	Basic factors affecting foundation design
Safe against shear failure of foundation soil	1 Determine the ultimate bearing capacity (q_{ult}) 2 Select an appropriate factor of safety (FS). The allowable bearing pressure $= q_{ult}$ f	Ultimate bearing capacity (see Sec. 12.5) and desired safety factor are dependent upon: 1 Subsurface materials and site conditions (soil and environment) 2 Depth, width, shape, and other factors related to the foundation (affecting primarily q_{ult}); type and nature of the associated structure
Safe against excessive settlement of foundation	1 Estimate the settlement of foundation 2 Adjust foundation design, if necessary, to meet the requirement that the estimated settlement does not exceed the permissible amount	Foundation settlement is dependent upon factors similar to those indicated above. For design analysis, the settlement of foundations is considered to be due to on or a combination of the following causes: 1 Immediate settlement (Ch. 9) 2 Consolidation settlement (Ch. 9). In conventional methods of analysis, these two types of settlement are estimated by separate procedures

Source: Based on Chu and Humphries (1969) and Chen and McCarron (1991).

as discussed in the previous section. According to the US Navy NAVDOCKS Design Manual DM-7 (1962), the following may be used as a guide in the selection of a proper factor of safety: (a) to obtain allowable bearing pressures, apply a factor of safety between 2 and 3 for the dead load plus normal live load depending on the nature of the structure and the reliability with which subsoil conditions have been determined; and (b) the required factor of safety ranges from 1.5 to 2 for dead load plus maximum live load which includes transient loads.

$$Q_a = Q_u/F_s \tag{12.6}$$

where Q_a = allowable bearing capacity, Q_u = ultimate bearing capacity, and F_s = factor of safety. In practice, the design of foundations is made not only in accordance with engineering principles but also to meet specific requirements set forth in an applicable building code. If it is desired to use an allowable bearing pressure higher than the maximum bearing value specified in the building code, load tests are usually required to provide evidence to justify the modification of the code restrictions.

12.6 Bearing capacity determination by limit equilibrium method

12.6.1 Soil governing parameters

1   *General discussion:* The analytical approach for determination of the ultimate bearing capacity in general can be determined by limit equilibrium or limit analysis methods. The limit equilibrium techniques are commonly used. The bearing capacity of a footing depends not only on the physical properties of the soil such as cohesion and internal frictional angle but also on the geometrical characteristics of the footing (width B and depth D). There are several methods proposed to determine the bearing capacity of ground soil by the limit equilibrium approach.

2   *Superposition method:* In general, the bearing capacity of footings on ground soils have been calculated by a superposition method suggested by Buisman (1940) and Terzaghi (1943) in which contributions to the bearing capacity from different soil and loading parameters are summed. These contributions are represented by the expression

$$q_0 = c \, N_c + q \, N_q + \gamma \, B/2 \, N_\gamma \tag{12.7}$$

where q_0 = bearing capacity of ground soil, c = unit cohesion, q = surface loading, γ = unit weight of soil, B = width of footing, and N_c, N_q, N_γ = bearing capacity factors (Figures 12.2 and 12.3, Table 12.6). Equation (12.7) covers three basic terms including cohesion, surcharge weight, and friction. Terms N_c, N_q, and N_γ are theoretical values mainly dependent on the assumed footing failure shape. The parameters N listed in Equation (12.7) are all functions of the internal friction angle of soil, ϕ. For determination of the N value, Terzaghi used a quasi-empirical method and assumed that these effects are additive, whereas the soil behavior in the plastic range is nonlinear and thus superposition does not hold for general soil bearing capacities.

Table 12.6 Bearing capacity factors

ϕ	N_q	N_c	N_γ	N_q/N_c	$\tan\phi$
0	1.000	5.142	0.000	0.194	0.000
2	1.197	5.632	0.156	0.212	0.035
4	1.432	6.185	0.350	0.232	0.070
6	1.716	6.813	0.595	0.252	0.105
8	2.058	7.527	0.909	0.273	0.141
10	2.471	8.345	1.313	0.296	0.176
12	2.973	9.285	1.837	0.320	0.213
14	3.586	10.370	2.522	0.346	0.249
16	4.335	11.631	3.422	0.373	0.287
18	5.258	13.104	4.612	0.401	0.325
20	6.399	14.835	6.196	0.431	0.364
22	7.821	16.833	8.316	0.463	0.404
24	9.603	19.323	11.173	0.497	0.445
26	11.854	22.254	15.049	0.533	0.488
28	14.720	25.803	20.351	0.570	0.532
30	18.401	30.139	27.665	0.611	0.577
32	23.177	35.490	37.849	0.653	0.625
34	29.440	42.163	52.182	0.698	0.675
36	37.752	50.585	72.594	0.746	0.727
38	48.933	61.351	102.050	0.798	0.781
40	64.195	75.312	145.191	0.852	0.839
42	85.373	93.706	209.435	0.911	0.900
44	115.307	118.368	306.920	0.974	0.966
46	158.500	152.096	458.018	1.042	1.036
48	222.297	199.257	697.926	1.116	1.111
50	319.053	266.878	1089.456	1.195	1.192

Source: Chen and McCarron (1991) In *Foundation Engineering Handbook*, 2nd Edition, Fang, H. Y. ed., Copyright (1991) by Van Nostrand Reinhold. With kind permission of Springer Science and Business Media.

12.6.2   Terzaghi and Meyerhof methods – for general soil condition

Meyerhof (1951, 1955) used Terzaghi's concept (1943) and presented extensive numerical results for shallow and deep footings by assuming failure mechanisms for the footing and by presenting results in the form of bearing capacity factors, *N*. It has been generally assumed that the bearing capacities obtained by Terzaghi's method are conservative, and experiments on model and full-scale footings seem to substantiate this for cohesionless soils. Terzaghi's method as modified by Meyerhof is shown in Figure 12.2. The figure includes ultimate bearing capacity equations for footings of various shapes and a graph for determining the bearing capacity factors N_c, N_q, and N_γ contained in these equations. It will be noted that the N_γ values are given for two conditions, namely, rough base and smooth base. For footings constructed by pouring concrete directly on foundation soils, the condition is similar to the case of the rough base.

1   Local shear and reduction factor: In Terzaghi's analysis of the ultimate bearing capacity of shallow foundations, it was recognized that local shear might occur in

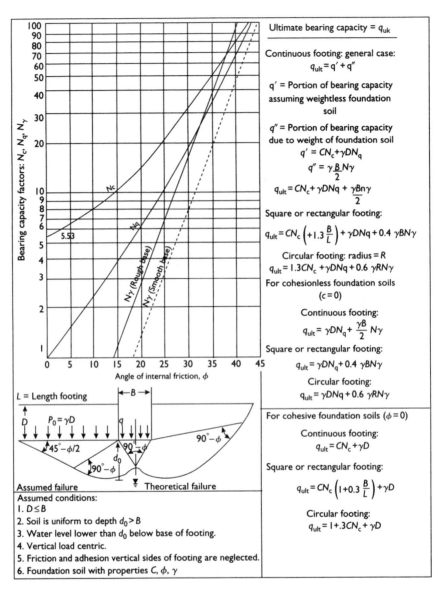

Figure 12.2 Ultimate bearing capacity of footings.

Source: US Navy, NAVDOCKS DM-7, 1962.

relatively loose soft soils below foundations. In such cases, the ultimate bearing capacity would be less than that determined by the method shown in Figure 12.2. According to an approximate method developed by Terzaghi, the ultimate bearing capacity for the case of local shear can be estimated by assuming a reduced cohesion, c, and reduced angle of internal friction from the following relationship:

$$\tan \phi' = 2/3 \tan \phi \qquad (12.8)$$

$$c' = 2/3 \, c \qquad (12.9)$$

where c and ϕ are the measured values of the angle of internal friction and cohesion, ϕ' and c' are the reduced values to be used for estimating the ultimate bearing capacity. Terzaghi's approximate method for determining the ultimate bearing capacity in the case of local shear implies that ϕ' and c' are to be used in conjunction with the bearing capacity factors, N_c, N_q, and N_γ in his original analysis. The difference between the original Terzaghi bearing capacity factors and those shown in Figure 12.2 is however relatively small for reduced values of cohesion and angle of internal friction. Sowers (1962) suggested that the reduced value of the angle of internal friction be applied to sands having a relative density (Ch. 7) less than 20% and the reduced value of cohesion be used for clays with a sensitivity (Ch. 10) of 10 or more.

12.6.3   Bearing capacity on cohesionless soils (sand)

1   General discussion: The general behavior of cohesionless soils is governed largely by the relative density (D_r). In particular, Coduto (2001) notes that the typical failure modes for dense $(D_r > 67\%)$, medium $(30\% < D_r < 67\%)$, and loose $(D_r < 30\%)$ sands is through general shear, local shear, and punching failure, respectively. The bearing capacity factors, N_q and N_γ for cohesionless foundation soils $(c = 0)$ may be obtained from Figure 12.3. The typical corresponding SPT blow counts are also indicated. The effect of local shear on ultimate bearing capacities, as discussed in the previous section, was accounted for in formulating the relationship between N_q, N_γ, and ϕ in Figure 12.3.

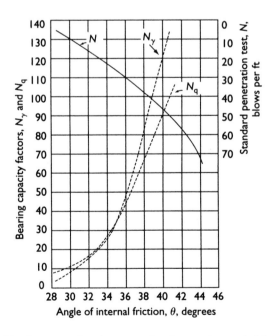

Figure 12.3 Relationship between bearing capacity factor, internal friction angel, and standard penetration test of sand.

Source: Peck, R. B., Hansen, W. E., and Thornburn, T. H. *Foundation Engineering, 2nd Edition,* © 1974 Wiley. This material is used by permission of John Wiley & Sons, Inc.

2 *Peck–Hanson–Thornburn method*: Peck *et al.* (1974) proposed the following equation for adjusting the standard penetration test (SPT) when the above condition is encountered.

$$N = 15 + 1/2 \, (N' - 15) \tag{12.10}$$

where N = SPT for estimating ϕ, from Figure 12.4, N' = SPT obtained from very fine sands below groundwater table. Adjustment of the SPT, according to Equation (12.10), is not necessary if the actual value does not exceed 15. Otherwise, a correction must be made. Further discussion on limitations and correction procedures will be presented in the next section.

12.7   Bearing capacity for cohesive soils (clay)

12.7.1   General discussion

In making broad comparisons, the behavior of clays (cohesive) and sands (cohesionless) is similar with respect to how dense the material is compacted. For example, soft normally consolidated clays may behave similar to loose sands whereby punching or local shear failure is more typical. Likewise, overconsolidated clays behave similar to dense sands where general shear failure tends to describe the failure mechanism. A particularly important issue with regard to cohesive soils is the rate of loading relative to the rate at which excess pore pressures may dissipate. If the loading rate is relatively fast, then pore water pressure develops as discussed in Chapter 10 and the condition is characterized as undrained. The friction angle, ϕ, may be neglected in these instances and the undrained shear strength is considered instead.

12.7.2   Skempton method

The ultimate bearing capacity in this case may be determined by the equations for cohesive foundation soil ($\phi = 0$) given in Figure 12.2. Another method for evaluating the bearing capacity of foundations on highly cohesive soils was developed by Skempton (1951). In Skempton's method, the ultimate bearing capacity, q, is computed by the following equations.

1 *Square or circular footings:*

$$q_{ult} = c \, N_{CS} + \gamma D \tag{12.11}$$

2 *Rectangular footing:*

$$q_{ult} = c \, N_{CS} \, (1 + 0.2 \, B/L) + \gamma D \tag{12.12}$$

3 *Continuous footing:*

$$q_{ult} = c \, N_{CC} + \gamma D \tag{12.13}$$

where q_{ult} = ultimate bearing capacity, c = cohesion, N_{CC}, N_{CS} = bearing capacity factors shown in Figure 12.4, γ = unit weight of soil within depth D, L = length of footing, B = width of footing, and D = depth of footing. The term γD in Equations

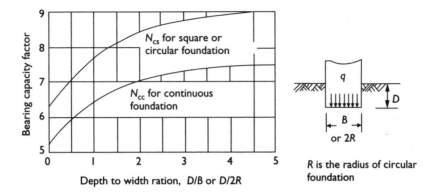

Figure 12.4 Bearing capacity factors on highly cohesive soils.

Source: Skempton, A. W. and Bjerrum, L. (1957), A contribution to the settlement analysis of foundations on clay, *Geotechnique*, vol. 7, no. 4, pp. 168–178. Copyright Thomas Telford Ltd. Reprinted with permission.

(12.12) and (12.13) represents the effect on the ultimate bearing capacity due to the weight of soil above the base of the footing. If the groundwater table is at or below the base of footing, the wet unit weight of the soil is to be used in computation. Additional groundwater considerations are presented in Section 12.12.

12.7.3 Net ultimate bearing capacity

The net ultimate bearing capacity can be computed from gross and net loading intensities as follows:

1 *Gross loading intensity*: The ultimate bearing capacity computed by Equations (12.11), (12.12), and (12.13) as well as that determined by the equations shown in Figure 12.2 refers to *gross loading intensity*. This loading intensity is based on the gross load also called the *total load*.

2 *Net loading intensity*: Frequent excavations are required for foundation construction and, as a result, the footings are placed at a certain depth below the ground surface. In such cases, the gross load reduced by the weight of soil excavated from the zone directly above the base of the footing is called the *net load*. This net load can then be used for computing the *net loading intensity*. The ultimate bearing capacity expressed in terms of net intensity can be determined from the following equation:

$$\text{Net } q_{\text{ult}} = q_{\text{ult}} - \gamma D \tag{12.14}$$

For continuous footings, the net ultimate bearing capacity can be obtained by combining Equations (12.12) and (12.13).

$$\text{Net } q_{\text{ult}} = q_{\text{ult}} - \gamma D = c\, N_{\text{CC}} \tag{12.15}$$

Similarly γD in Equations (12.12) and (12.13) will be eliminated if the ultimate bearing capacity is expressed in terms of net intensity. In comparing Skempton's method with that shown in Figure 12.2 for the case of cohesive soils ($\phi = 0$), it is noted that for footings having relatively high depth to width (D/B) ratios, the ultimate bearing capacity determined by Skempton's methods is higher than that obtained from the other existing methods. This is due to the fact that the additional shear resistance of the clay above the base of the footing is included in Skempton's analysis. The effect of variations in the width and depth of footings on the ultimate bearing capacity on highly cohesive soils is much less than that it is in the case of footings on cohesionless soils. Another characteristic of footings on highly cohesive soils is that the unit weight of the soil below the footing has no direct effect on the ultimate bearing capacity. The unit weight, however, may influence the ultimate bearing capacity indirectly because it is one of the factors that affect the shear strength of soil as discussed in Chapter 10.

3 *Accuracy of shear strength*: An important factor affecting the degree of accuracy of ultimate bearing capacity evaluations is the determination of shear strength of the soil. Although methods for estimating the shear strength of highly cohesive soils according to SPT data are available (Ch. 10), it will be noted that more reliable methods for shear strength determinations are often required for foundation designs. The specific method most suitable for this purpose is dependent upon the type of soil at the project site as well as the nature and extent of the project.

12.8 Bearing capacity determination by limit analysis method

12.8.1 General discussion

In the solution of a solid mechanics problem, three basic conditions must be satisfied, namely (a) the stress equilibrium equations, (b) the stress–strain relations, and (c) the compatibility equations. The stress–strain (constitutive) relations have been discussed in Chapter 10. In an elastic-plastic material (Fig. 10.1), however, there is as a rule a three-stage development in a solution when the applied loads are gradually increased from zero to some magnitude. The complete solution by this approach is cumbersome for all but the simplest problems, and methods are needed to furnish the load-carrying capacity in a more direct manner (Chen and Davidson, 1973). Limit analysis is a method that enables a definite statement to be made about the collapse load without carrying out the step-by-step elastic-plastic analysis.

The limit analysis method considers the stress–strain relationship of a soil, but in an idealized manner. This idealization, termed normality, establishes limit theorems on which limit analysis is based. Within the framework of this assumption, the approach is rigorous and the techniques are competitive with those of limit equilibrium and in some instances are simpler. The two main limit theorems for a body or an assemblage of bodies of an elastic-perfectly plastic material were discussed in Section 12.2.2.

12.8.2 Cohesive soil on a layered anisotropic foundation

Soil deposits frequently consist of multiple layers. Consider the situation where a foundation is placed in sand; however, at some distance below the footing is a clay

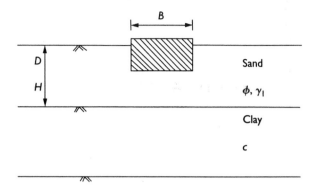

Figure 12.5 Foundation over a two-layer system for Meyerhof and Hanna solution.

layer, as given in Figure 12.5. The bearing capacity in this case may be given as (Meyerhof and Hanna, 1978; Hanna and Meyerhof, 1980)

$$q_{uv} = q_{bv} + \gamma_1 H^2 \left[1 + \frac{2D(\cos \alpha D)}{H} \right] K_s i \frac{\tan\phi}{B} - \gamma_1 H \le q_{tv} \tag{12.16}$$

where q_{uv} is the vertical component of the ultimate bearing capacity, q_{ult}, q_{bv} and q_{tv} are the vertical components of the bearing capacities found by considering only the lower and upper layers individually, α is the load inclination factor (Fig. 12.6), and K_s is the coefficient of punching shear (Fig. 12.6), α is the inclination of the load relative to the vertical. Other properties are as given in Figure 12.5 and listed previously.

12.9   In situ measurements of bearing capacity of ground soil

There are several in situ measuring techniques for estimation of bearing capacities of ground soils. The SPT is the most common method for estimating the bearing capacity and has been discussed in Chapter 2. The CBR, plate-load tests, and others are presented as follows:

12.9.1   California bearing ratio method

1   General discussion: The California Bearing Ratio (CBR) is defined as the ratio of the stress required to penetrate a soil mass with a 3 in.^{2} (19.4 cm^2) circular piston (approximately 2 in. (5 cm) diameter) to that required for corresponding penetration of a standard material. The ratio is usually determined at a 0.1 in. penetration, although other penetrations are sometimes used. Originally California procedures required determination of the ratio at 0.1 in. intervals to 0.5 in. The US Army Corps of Engineers' procedures required determination at a ratio of 0.1 in. and 0.2 in. When the ratio at 0.2 in. is consistently higher than at 0.1 in., the ratio at 0.2 in. is used. Detailed test procedures of the CBR method is given by ASTM (D1883).

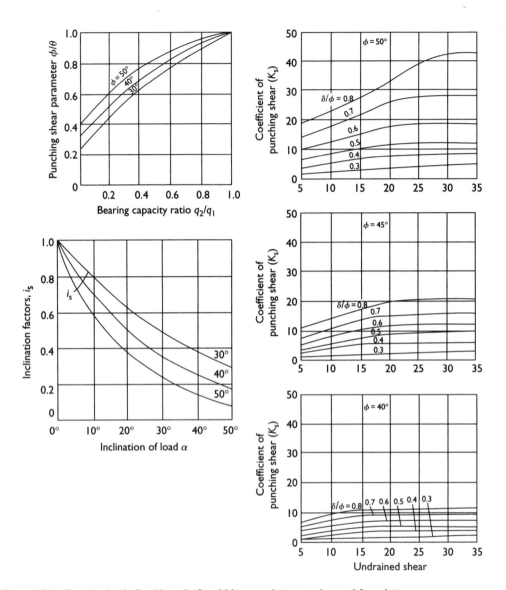

Figure 12.6 Graphical aids for Meyerhof and Hanna solution to layered foundations.

Source: Chen and McCarron (1991). In *Foundation Engineering Handbook*, 2nd Edition, H. Y. Fang, ed., Copyright (1991) by Van Nostrand Reinhold. With kind permission of Springer Science and Business Media.

2   *CBR computation procedure:*

$$\text{CBR} = 100 \, x/y \tag{12.17}$$

$$x = aD/3$$

where x = soil resistance or unit load on the piston (psi) (for 0.1 in. of penetration interval), y = standard unit load (psi) (at 0.1 in. penetration, the standard unit load is 1000 psi), a = value of one dial division (lbs), and D = actual dial reading. The standard load of 1000 psi (6.9 MPa) is approximately the load required to penetrate a mass of crushed rock by 0.10 in., and as such the CBR essentially compares the strength of a given soil to that of rock. Typical test results for fine-grained embankment soil and sand–gravel subbase and crushed stone base course materials are presented in Chapter 7. Effects of temperature and other factors on the CBR results are presented in Chapter 6.

12.9.2 Plate-load test

The *plate-load test*, also called *plate-bearing test, field load test,* or simply *load test,* is conducted by placing a square or circular rigid plate(s) directly on the foundation soil and measuring the settlement of the plate when it is subjected to increments of applied load. A standard methodology is presented in ASTM (D 1194). Information obtained from these tests may be utilized for estimating the ultimate bearing capacity and immediate settlement of the shallow foundations. Due to the relatively short duration of time in conducting plate-load tests, the observed deflection of the test plate is primarily the immediate settlement. Load test data are therefore not suitable for design analysis concerning consolidation settlement (Sec. 9.9).

1 Test equipment: The basic equipment for the plate-load test consists of (a) reaction trailer, (b) hydraulic ram and jack, (c) various sizes of steel spacers for use in trenches of different depths, (d) A 12 in. (30.5 cm) diameter cylindrical steel loading frame cut out on two sides to allow the use of a center deflection dial, (e) spherical bearing block, (f) 1 in. (2.5 cm) thick steel plates of 12 in. (30.5 cm), 18 in. (55.7 cm), 24 in. (61.0 cm), and 30 in. (76.2 cm) in diameter, and (g) 16 ft (4.9 m) long aluminum reference beam. A maximum reaction of about 12,000 pounds (53.4 kN) could be obtained with a 17,000 pound (75.6 kN) loaded rear axle of the reaction trailer. A standard hydraulic ram and hand-operated jack is used to apply the load.

2 Test procedure: A repeating type of plate-load test is commonly used. The test procedure is provided for the application of three repetitions of three different increments of loads, and the measurement of gross and elastic rebound deflections. Loads are applied for approximately 15 s without provision for the deformation to come to equilibrium. The applied loads for various plate sizes are typically selected to approximate what might be expected under loaded trucks and to prevent excessive strain in the layer.

3 Application of results: The use of load test results derived from small plates to larger full size footings is not standard practice, as noted in Bowles (1988). However, in the case of clay soils, it has been noted that the bearing capacity may be taken as independent of footing size. In that case, the bearing capacity determined from a load test may be used as an estimate of the bearing capacity for the proposed foundation. For granular soils, all three of the bearing capacity factors given by Equation (12.7) apply and Bowles (1988) indicates that

$$q_{ult,f} = M + N_\gamma \frac{B_{foundation}}{B_{loadtest}} \tag{12.18}$$

where $q_{ult,f}$ = the ultimate bearing capacity of the foundation, M = a factor that includes the values of the N_c and N_q terms, N_γ is as given previously and B is given as the width for the foundation and load test, as noted by the subscripts. The use of several size plates, as noted earlier, allows for a graphical solution. In the case of sands where the N_q term is negligible, the $q_{ult,f}$ may be estimated as

$$q_{ult,f} = q_{loadtest} \frac{B_{foundation}}{B_{loadtest}} \tag{12.19}$$

This equation is not recommended if the ratio of the widths of the foundation to the plate used in the load test is greater than about 3. However, the foregoing approach is limited and should be treated with caution for several reasons. When load tests are conducted with any size less than the full size footing, the depth of influence (pressure-bulb, Ch. 9) is less. As such, the influence of stratification may be significant. Consider the situation depicted in Figure 12.5 for a two-layered system. If a load test were conducted with small plates, the underlying clay layer might not be detected. Another difference that may emerge between a load test and a full size foundation is the effect of overburden pressure, which tends to stiffen the soils at greater depth. The ultimate bearing capacity of the foundation will reflect this additional stiffness while it may not be captured by smaller load tests.

12.9.3 Bearing capacity from standard penetration test

1   *General discussion*: The SPT (ASTM D1586) can be used for estimation of the bearing capacity of cohesionless soil such as sand, sand–gravel, and sandy silts. Terzaghi and Peck in 1948 had developed a series of design curves for which the allowable bearing capacity, q_a, is a function of the foundation width, B, and standard penetration resistance number N. Teng (1962) modified Terzaghi and Peck's work and proposed the following equation:

$$q_a = 0.72(N - 3)\left(\frac{B + 1}{2B}\right)^2 \tag{12.20}$$

Later, Meyerhof (1955, 1963) reevaluated Terzaghi and Peck's data and published the following modified equations:

$$q_a = 0.25(N) \quad \text{when } B < 4 \tag{12.21}$$

$$q_a = \frac{N}{6}\left(\frac{B + 1}{B}\right)^2 \quad \text{when } B > 4 \tag{12.22}$$

where q_a = allowable bearing capacity of ground soil, N = penetration resistance, corrected for overburden pressure, and B = width of foundation.

2   *Precautions for using penetration resistance, N value*: The SPT N value is an important parameter and used for estimation of shear strength, liquefaction potential, and bearing capacity correction. However, the angle of internal friction of soil, ϕ, estimated on the basis of SPT data by Figure 12.3 or similar figures should be used with caution. For SPT tests conducted at shallow depths, the predicted friction angle is too low.

12.9.4 Bearing capacity determination based on CBR and plate-load results

1 *Bearing capacity determination based on CBR Test*: An empirical relationship between the CBR value and the allowable bearing capacity may be given by:

$$q_a = 25 \text{ CBR} \tag{12.23}$$

where q_a = allowable bearing capacity (kPa), and CBR = California bearing ratio.

2 *Bearing capacity of embankment soil based on pavement surface deflection data*: The bearing capacity may also be estimated from surface deflection data obtained from a plate-load test. In particular, the bearing capacity of embankment soil as reflected by a modulus term from a plate-load test from the deflection of the flexible pavement surface has been studied. The surface deflection is obtained from the Benkelman-beam deflection indicator (Benkelman *et al.* 1962). A simple experimental equation based on Boussinesq elastic theory (Sec. 9.4) proposed by Fang and Schaub (1967) as

$$K_E = 0.027 \frac{Q\beta}{\Delta} \tag{12.24}$$

where K_E = elastic modulus of embankment soil determined from plate-load test, Q = wheel load, β = coefficient of pavement surface deflection, which is a function of thickness of pavement components, and Δ = pavement surface deflection (flexible pavement).

12.9.5 Bearing capacity determination based on cone penetrometer test

1 *Static cone penetrometer (SCP)*: Meyerhof (1955) proposed that soil bearing capacity can be estimated from static cone penetration test (CPT) results as

$$q_a = \frac{q_c}{15} \quad \text{where } B < 4 \tag{12.25}$$

$$q_a = \frac{q_c}{25} \left(\frac{B+1}{B} \right)^2 \quad \text{where } B > 4 \tag{12.26}$$

where q_a = allowable bearing capacity of ground soil (ksf), q_c = static cone resistance (ksf), and B = width of foundation (footing) (ft). Equations (12.25) and (12.26) are based on 1 in. (25 mm) settlement. For the bearing capacity of a raft foundation, Meyerhof suggested using $2q_a$, where q_a is computed from Equation (12.26). Sanglerat (1972) gives the following equation between allowable bearing capacity and static cone penetration results:

$$q_a = \frac{q_c}{10} \tag{12.27}$$

2   *Dynamic cone penetrometer (DCP)*: According to the European Symposium on Penetration Testing (1974) it was shown that for cohesionless materials the DCP can be related through CBR results to SCP tests or to standard penetration resistance values from SPT. All above equations are for shallow foundations and no account is taken of the possible influence of depth or overburden load. The field test can be performed rapidly by one person during a short duration.

12.9.6   Bearing capacity chart

Figure 12.7 presents the interrelationships of soil classification and some selected in situ measurement methods including CBR, CPT, SPT, and self-boring pressuremeter test (SBPMT). This chart can be used to make preliminary estimations of bearing capacity values of subgrade soil as well as its classification with given in situ parameters. It can also be used to derive an approximate correlation between different

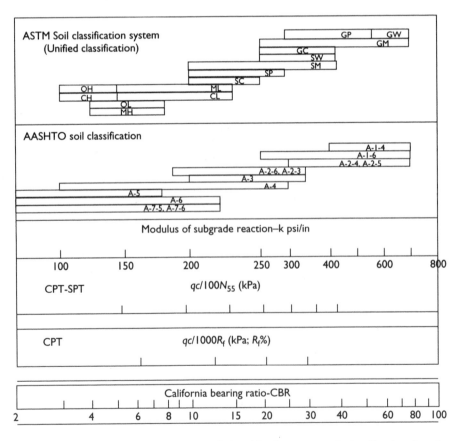

Figure 12.7 Chart for the approximate interrelationships between soil classification, bearing values and some in situ strength parameters.

Source: Pamukcu, S. and Fang H. Y., Development of a chart for preliminary assessments in pavement design using some in-situ soil parameters. In *Transportation Research Record* No. 1235, Transportation Research Board. National Research Council, Washington DC, 1989, pp. 38–44. Reproduced with permission of the Transportation Research Board.

in situ parameters, verify test results, or identify areas where a more detailed database is needed.

12.10 Building codes and special soils and rocks

12.10.1 Bearing capacity estimated from building code

1 General discussion: In practice, the design of foundations is made not only in accordance with engineering principles but also to meet specific requirements set forth in an applicable building code. If it is desired to use an allowable bearing pressure higher than the maximum bearing value specified in the building code, load tests are usually required to provide evidence to justify modification of the code restrictions. In many large cities around the world, they provide their own building code for foundation designs. These values are generally based on years of experience in their own particular zone or region. These values are simply based on the local experience and are often referred to as *presumptive pressure*. Presumptive pressures are often based on a visual soil classification such as soft clay, stiff clay, dense sand, and loose sand.

2 Typical building code data: Certain localities provide estimated allowable bearing pressures in building codes or similar documents. These values are useful for preliminary analysis and design of foundation systems. The bearing capacity values obtained from such sources do not reflect the groundwater table, depth and size of footing, or other environmental conditions. Building codes are limited to the local condition for which they were developed and reflect particular geological, environmental, and topographical considerations. Codes are specific to their zone or region.

12.10.2 Some special soil and rocks

1 *Bearing capacity of loess:* Loess is a type of wind blown silt with fine uniform grain size distribution. Figure 12.8 presents the bearing capacity of loess. Figure 12.8(a) shows the bearing capacity for the general types of loess, and Figure 12.8(b) shows the bearing capacity of newly deposited loess.

2 *Bearing capacity of rocks:* The bearing capacity of rocks can be roughly estimated from the following relationship as proposed by Broms (1966) and others.

$$q_o = 24 \, \sigma_t \tag{12.28}$$

$$q_o = 4-6 \, q_u \tag{12.29}$$

where q_o = ultimate bearing capacity (psf or kPa) σ_t = tensile strength (psf or kPa), (Ch. 8), and q_u = compressive strength (Sec. 10.7) (psf or kPa). Equations (12.28) and (12.29) are based on the linear fracture mechanics concept (Ch. 8) and are used for estimating the pile capacity (Ch. 15) when piles are driven into rock. Also, the bearing capacity of rock or stabilizing construction materials such as cement treated, bituminous treated highway base, or subbase materials can be estimated as describe in Section 8.9.

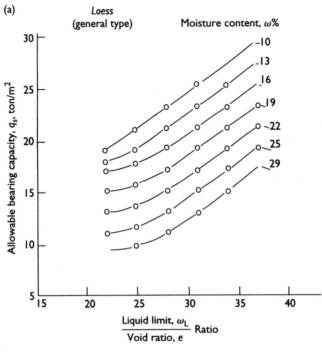

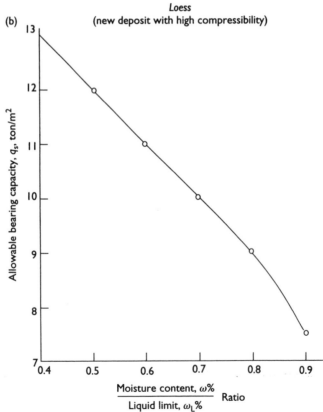

Figure 12.8 (a) Allowable bearing capacity of general types of loess. (b) Allowable bearing capacity of new loess deposits.

Source: Data from Chinese Building Code for Loess Region 1979.

12.11 Inclined and eccentric loads

A foundation may experience inclined or eccentric loadings by original design or after subsequent modification. This condition occurs whenever a foundation is subjected to both horizontal and vertical loads, as might be the case where wind and gravity loads are applied. Foundations supporting industrial facilities and/or structures with rotating machinery or moving equipment may experience eccentric loading. An example of inclined loading is provided in Figure 12.9, which shows the second longest single steel arch bridge in the world (New River Bridge, West Virginia). Notice the bridge abutments on the far side of the mountainside which support both vertical and horizontal components of load. The bearing surface is also inclined. Incidentally, the longest steel arch bridge is the Lupu Bridge constructed in 2002 in Shanghai, China. To compute the bearing capacity in these situations, a reduced area is often computed. In the case of rectangular footings, the modified length and width become

$$L' = L - 2e_1 \tag{12.30}$$

$$B' = B - 2e_2 \tag{12.31}$$

where L' and B' represent the modified length and width, L and B are the original length and width, and e_1 and e_2 are the eccentricity with respect to axis 1 and 2. Similar reductions are made for circular footings. Figure 12.10 shows the reduced areas associated with rectangular or circular footings while Figure 12.11 provides the

Bridge Abutments, *New River Bridge,* *West Virginia, USA*

Figure 12.9 Bridge abutments that are subjected to both horizontal and vertical load components.

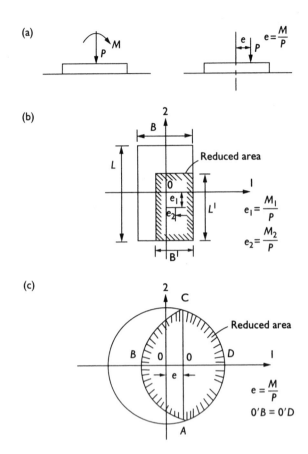

Figure 12.10 Reduced footing area for eccentric loads: (a) Equivalent loadings. (b) Reduced area – rectangular footing. (c) Reduced area – circular footing.

Source: Chen and McCarron (1991), as modified from API (1987). In *Foundation Engineering Handbook*, 2nd Edition, Fang, H.Y. ed., Copyright (1991) by Van Nostrand Reinhold. With kind permission of springer Science and Business Media.

reduced area factor as a function of a dimensionless eccentricity. Highter and Anders (1985) have provided additional graphical solutions to determine the equivalent areas of rectangular and circular footings subjected to eccentric loads. The bearing capacity equation used for inclined and eccentric loads has the following equation:

$$q_o = N_c s_c i_c d_c c + N_q s_q i_q d_q q + \gamma B/2 \, N_\gamma s_\gamma i_\gamma d_\gamma \qquad (12.32)$$

where q_o = vertical component of the footing load, the factors s, i, and d refer to the footing shape, load inclination, and footing depth, as provided in Table 12.7 (Meyerhof) and Table 12.8 (Brinch Hansen). Other parameters are as defined previously. An iterative solution is required in using the factor i as both vertical and horizontal components of the load are included for an inclined load, as shown in Figure 12.12.

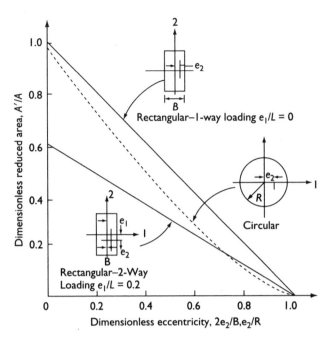

Figure 12.11 Area reduction factors for eccentrically loaded footings.

Source: Chen and McCarron (1991), as modified from API (1987). In *Foundation Engineering Handbook*, 2nd Edition, Fang H. Y. ed., Copyright (1991) by Van Nostrand Reinhold. With kind permission of Springer Science and Business Media.

Table 12.7 Meyerhof footing depth and load inclination bearing capacity modifiers

$$q_0 = N_c s_c i_c d_c c + N_q s_q i_q d_q q + \frac{\gamma B}{2} N_\gamma s_\gamma i_\gamma d_\gamma$$

For $D < B$:

$$d_c = 1 + 0.2 \sqrt{N\varphi} \frac{D}{B}$$

$$d_q = d_\gamma = 1 \qquad\qquad (\phi = 0°)$$

$$d_q = d_\gamma = 1 + 0.1 \sqrt{N\varphi} \frac{D}{B} \qquad (\phi = 10°)$$

$$i_c = i_q = (1 - \alpha/90°)^2$$

$$i_\gamma = (1 - \alpha/\theta)^2$$

$$N_\theta = \tan^2 \left(\frac{1}{4} \pi + \frac{1}{2} \theta \right)$$

Source: Chen and McCarron (1991). In *Foundation Engineering Handbook*, 2nd Edition, Fang, H. Y. ed., Copyright (1991) by Van Nostrand Reinhold. With kind permission of Springer Science and Business Media.

The bearing capacity of a foundation is also influenced by the shape, depth, and/or geometric factors which define the system. Brinch Hansen (1970) and Vesic (1975), as given by Chen and McCarron (1991), note that the bearing capacity may be computed as

$$q_o = N_c s_c d_c g_c b_c c + N_q s_q d_q g_q b_q q + \gamma B/2 \; N_\gamma s_\gamma d_\gamma g_\gamma b_\gamma \tag{12.33}$$

Table 12.8 Brinch Hansen footing depth and load inclination bearing capacity modifiers[a]

$$q_0 = N_c s_c i_c d_c c + N_q s_q i_q d_q q + \frac{\gamma B}{2} N_\gamma s_\gamma i_\gamma d_\gamma$$

For $D < B$:

$$d_c = 1 + 0.4 \frac{D}{B} \qquad (\phi = 0°)$$

$$d_c = d_q = -\frac{1 - d_q}{N_c \tan \varphi} \quad (\phi > 10°)$$

$$d_q = 1 + 2 \tan \phi (1 - \sin \phi)^2 \frac{D}{B}$$

$$d\gamma = 1$$

$$i_c = 1 - \frac{mH}{A_c(\pi + 2)}$$

$$i_c = i_q - \frac{1 - i_q}{N_c \tan \varphi}$$

$$i_q = \left[1 - \frac{H}{V + A_c \cot \varphi} \right]^m$$

$$i_\gamma = \left[1 - \frac{H}{V + A_c \cot \varphi} \right]^{m+1}$$

$$m = m_L \cos^2 \theta_n + m_B \sin^2 \theta_n$$

$$m_B = \frac{2 + B/L}{1 + B/L} \qquad m_L = \frac{2 + L/B}{1 + L/B}$$

Source: Chen and McCarron (1991). In *Foundation Engineering Handbook*, 2nd Edition, Fang, H. Y. ed., Copyright (1991) by Van Nostrand Reinhold. With kind permission of Springer Science and Business Media.

Notes
θ_n is the projected direction of load in the plane of the footing, measured from the side of the length *L*.
a As modified by Vesić (1975).

where the factors *s*, *d*, *g*, and *b* are factors that account for the footing shape, depth, ground inclination, and base inclination, respectively, and all other variables are as defined previously. Tables 12.9 and 12.10 provide the factors necessary in Equation (12.33). Further discussions are given by Meyerhof (1953), Vesic (1975), and Chen and McCarron (1991).

12.12 Effect of environmental conditions on bearing capacity

12.12.1 Effects of groundwater table on bearing capacity

The elevation of the groundwater table relative to the foundation will affect the unit weight of the soil. In addition, the shear strength of ground soil may be influenced by the elevation of the groundwater table. A procedure developed by Meyerhof (1955) for evaluating the groundwater effect on soil unit weight is shown in Figure 12.13.

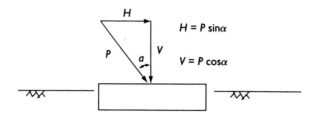

Figure 12.12 Inclined footing load.

Table 12.9 Meyerhof and Brinch Hansen footing shape bearing capacity modifiers

$$q_0 = N_c s_c i_c d_c c + N_q s_q i_q d_q q + \frac{\gamma B}{2} N_\gamma s_\gamma i_\gamma d_\gamma$$

Meyerhof (1963)

$$s_c = 1 + 0.2N_\phi \frac{B}{L}$$

$$s_q = s_\gamma = 1.0 \qquad\qquad (\phi = 0°)$$

$$s_q = b_\gamma = 1 + 0.1N_\phi \frac{B}{L} \qquad (\phi > 10°)$$

$$N_\phi = \tan^2\left(\frac{1}{4}\pi + \frac{1}{2}\varphi\right)$$

Brinch Hansen (After Vesić, 1975)

$$s_c = 1 + \frac{B N_q}{L N_c}$$

$$s_q = 1 + \frac{B}{L}\tan\varphi$$

$$s_\gamma = b_\gamma = 1 - 0.4\frac{B}{L}$$

For circular footing use $B/L = 1$

Source: Chen and McCarron (1991). In *Foundation Engineering Handbook*, 2nd Edition, Fang, H.Y. ed., Copyright (1991) by Van Nostrand Reinhold. With kind permission of Springer Science and Business Media.

The figure also shows the ultimate bearing capacity equations for various groundwater levels.

12.12.2   Bearing capacity on frozen and thawing ground

1   *Bearing capacity at frozen ground soil*: Major factors affecting the bearing capacity of ground soil are freezing–thawing and wetting–drying cycles. The mechanism of these factors including swelling–shrinkage behavior, cracking patterns, etc. has been discussed in Chapter 6. The particular influence of temperature on bearing capacity for a given study is given in Figure 12.14. There

Table 12.10 Brinch Hansen footing and ground inclination bearing capacity modifiers

$$q_0 = N_c s_c d_c g_c b_c c + N_q s_q d_q g_q b_q q + \frac{\gamma B}{2} N_\gamma s_\gamma d_\gamma g_\gamma b_\gamma$$

Footing inclination factors

$$b_q = b_\gamma = (1 - \alpha \tan \phi)^2$$

$$b_c = 1 - \frac{2\alpha}{\pi + 2} \qquad (\phi = 0°, \alpha \text{ in radians})$$

$$b_c = b_q - \frac{1 - b_q}{N_c \tan \varphi} \qquad (\phi > 0°)$$

Ground inclination factors

$$g_q = g_\gamma = (1 - \tan \omega)^2 \qquad (\phi > 0°)$$

$$g_c = 1 - \frac{2\omega}{\pi + 2} \qquad (\phi = 0°, \omega \text{ in radians})$$

$$g_c = g_q - \frac{1 - g_q}{N_c \tan \omega} \qquad (\phi > 0°)$$

Restrictions: $\alpha < 45°, \omega < 45°, \omega < \phi$.
For ground inclination use $N_\gamma = -2 \sin \omega$

Source: Chen and McCarron (1991). In *Foundation Engineering Handbook*, 2nd Edition, Fang, H. Y. ed., Copyright (1991) by Van Nostrand Reinhold. With kind permission of Springer Science and Business Media.

is a dramatic increase in strength once the temperature falls below freezing (32°F, 0°C) as the pore water begins to freeze, serving to bind soil particles and strengthen the overall mass. Notice that increasing temperatures beyond freezing have a negligible effect.

2 *Bearing capacity at spring thawing period*: While frozen soils may exhibit greater bearing capacity, thawed soils tend to be relatively weak. During periods of freezing, a suction is created whereby moisture is pulled through capillary forces into the freezing zone, as discussed in Chapter 6. As such, the net moisture content is typically increased. Upon thawing, this excess moisture serves to reduce strength. This is in part why the potholes and other signs of weather-related distress typically manifest in the spring season.

12.12.3 *Bearing capacity on wetland and compressible fill areas*

1 *Bearing capacity at a high groundwater table site*: To improve the bearing capacity of soils with a high groundwater table, a surface and subsurface drainage network and/or dewatering process (Ch. 5) must be investigated. Other ground improvement techniques are discussed in Chapter 15.

2 *Bearing capacity at compressible fill areas*: Many problems exist in connection with foundations on fill and the prospect of weak bearing capacity is one of the major problems. These problems include improper placement of the fill, inadequate compaction, use of unsuitable materials in the fill, and the presence of a compressible soil stratum below the fill. The construction of foundations on

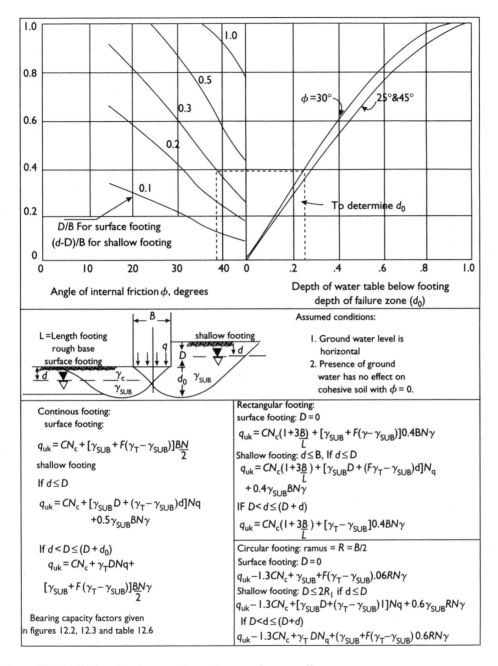

Figure 12.13 Ultimate bearing capacity with ground-water effect.

Source: US Navy, NAVDOCKS DM-7, 1962.

fill is an acceptable practice provided that the material, placement, and compaction of the fill are under engineering control and that no compressible soils exist under the fill. The presence of a soft compressible soil stratum at a certain depth below a structure is sometimes unnoticed during the design and construction stages. After completion of the structure, long-term settlement of the

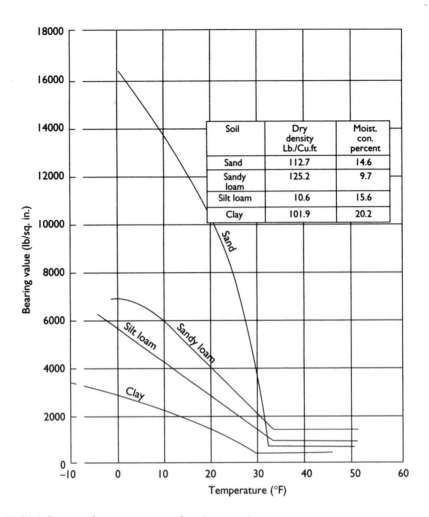

Figure 12.14 Influence of temperature on bearing capacity.

Source: Kersten, M. S. and Cox, A. E., (1951) the effect of temperature on the bearing value of frozen soils. In *Highway Research Board Bulletin* 40, Highway Research Board. National Research Council, Washington DC, 1951, pp. 32–38. Reproduced with permission of the Transportation Research Board.

foundation will occur due to the slow consolidation process of the compressible layer. In such cases, consolidation data from the soft soil must be evaluated.

3   *Bearing capacity at the landfill site*: To redevelop a former landfill site, the bearing capacity is often required. A relatively simple procedure was developed by Fang, Slutter, and Koerner cited by Fang (1997). A similar procedure was developed as described in Chapter 10 for obtaining laboratory shear parameters of compacted natural waste disposal blocks (bales) for cohesion, c, and friction angle, ϕ. Once knowing these strength parameters, one then uses the conventional bearing capacity equation (Eq. 12.7). However, from a design perspective, it is usually the settlement criteria which govern a proposed design.

4   *Pollution intrusion*: Pollution intrusion affects bearing capacity as reflected on soil compaction properties, as discussed in Section 7.5, compressibility characteristics (Sec. 9.5), and shear strength (Sec. 10.11) along with several factors including temperature, pore fluid as reflected by pH values, types and, concentrations of exchangeable ions. In all cases, pollution intrusion significantly affects all engineering behavior of all types of soil especially for the fine-grained soil.

12.13   Techniques for improvement of weak bearing capacity ground soil

The main purpose of the pre-loading and surcharging methods is to eliminate part or all of the post-construction primary consolidation and/or the secondary compression. This method is suitable for large projects such as improvement for wetlands, dredged fills, reclaimed areas, and landfill sites. Because these methods are considered a low-cost and time consuming, a year or more is required for the project in order to obtain significant results. Surcharging is the effective approach for the pre-loading method. Surcharging is used either prior to, or simultaneously with, the construction of the permanent structure. Ground improvement techniques for weak bearing capacity ground soils include (a) loading methods such as pre-loading, pre-compression, and surcharging; (b) removal of weak material; (c) in situ stabilization; and (d) structural supporting system such as pile foundations (Sec. 15.12) and caisson (Sec. 15.13). The pre-loading method improves the density and increases the bearing capacity. In the case of foundations other than earth structures, surcharging is placed on the proposed construction site before any planned construction. Construction cannot begin until the surcharging period is over and the load is removed. The surcharging period and load intensity should be estimated prior to the project's start, and it depends on the shear strength or bearing capacity of the underlying soil to be loaded. The load intensity and the resulting induced stresses should not exceed the strength of the underlying soil. The amount of preconsolidation desired is the tolerable post-construction settlement for the foundations. The material used for surcharging will depend on availability and utility. For embankment construction, a static surcharge or a rolling surcharge is used. A static surcharge is normally a layer of soil superimposed on the full length of the embankment.

12.14   Summary

The design of structures on shallow foundations requires an assessment of both settlement and bearing capacity. While settlement was discussed in Chapter 9, this chapter focused on bearing capacity which may be defined as the soil's ability to withstand a given loading condition. There are various methods of stability analysis available, including slip-line, limit equilibrium, limit analysis, finite difference, and finite element technique. Limit equilibrium and limit analysis are among the most commonly employed approaches. The bearing capacity determined by the limit equilibrium method was discussed in detail for both the Terzaghi and the Meyerhof methods. The limit analysis method can also be successfully used to analyze stability problems in geotechnical engineering. Within its framework, the method is soundly logical and gives an insight to the physical problem. The solution offers a closed-form mathematical solution. The upper-bound limit analysis can predict bearing capacities of cohesive soils with internal friction within a reasonable degree of accuracy. It can be said that the results compare favorably with existing limit equilibrium solutions. The bearing capacity can also be estimated by in situ measurements including the CBR method, plate-load test, and cone penetration method. A bearing capacity chart based on the in situ measurements was developed and presented for preliminary evaluation purposes. Special consideration of bearing capacity may be necessary when dealing with difficult subsurface and problematic soils such as loess, landfill sites, and frozen soil.

PROBLEMS

12.1   A load test was made on a bearing plate 1 ft² (0.09 m²) on the surface of a cohesionless deposit of sand having a unit weight of 110 pcf (17.2 kN/m³). The load–settlement curve approached a vertical tangent at a load of 4000 lb (1800 kg). What was the ultimate bearing capacity for the sand?

12.2   In each of the following cases, determine the allowable vertical load that can be put on an 8 ft = 20 ft (2.44 m × 6.10 m) pier footing at 0 ft, 4 ft (1.22 m), and 8 ft (2.44 m) beneath the ground surface. Use a factor of safety of 3. Plot on a single sheet the depth against the allowable footing load for each case.

12.3   a   For the case of a 5 ft × 5 ft (1.53 m × 1.53 m) footing plot, the bearing capacity as a function of depth of an embankment from 0 to 5 ft (1.525 m) for two soils is

$$c = 1200 \text{ psf } (57.5 \text{ kPa}); \ \phi = 0; \ \gamma = 110 \text{ pcf } (17.2 \text{ kN/m}^3)$$

$$c = 0; \ \phi = 30^0; \ \gamma = 105 \text{ pcf } (16.5 \text{ kN/m}^3).$$

   b   For the same two soils, plot the effect of the square footing size on bearing capacity at the surface for $B = 1$ to $B = 10$ ft (3.05 m).

12.4   A structure was built on a mat foundation at 100 ft² (9.3 m²). The mat rested at the ground surface on a stratum of uniform soft clay which extended to a depth of 150 ft (45.8 m). If failure occurred at a uniformly distributed load of 4500 psf (215 kPa), what was the average value of cohesion for the clay? (Because of the great depth of the zone of plastic equilibrium, the consolidation of the clay prior to failure can be disregarded, and it can be assumed that $\phi = 0$.)

12.5   Explain the relative merits and limitations of a plate-load tests versus the use of a rational formula for determining the allowable bearing capacity of a shallow footing. Of what use will the local building codes be in this situation? If the groundwater table is lowered will the effects be beneficial or detrimental for the footing on (a) sand (b) clay? Explain.

12.6   (a) In a laboratory CBR test on subgrade materials, a surcharge weight is usually placed on the sample during soaking. Why and how is this weight selected? (b) Briefly explain how the test data from a CRB test is converted into a CBR figure for design.

12.7   Why is the bearing capacity at a landfill site unpredictable? Is there any effective method for estimating bearing capacity at a landfill site?

12.8   Why is bearing capacity used only for a short-term prediction of ground soil supporting power? Why can it not be used for long-term prediction?

Chapter 13

Lateral earth pressure

13.1 Introduction

13.1.1 General discussion

The magnitude and distribution of the lateral pressure acting on retaining structures or foundations are important for the design of excavation bracing, retaining walls, waterfront, and near shore structures. There are two general types of lateral pressure or force, the earth pressure and environmental force. Earth pressure is the major contributor to overall lateral pressure, and it can be divided into the three scenarios as follows: active earth pressure, passive earth pressure, and earth pressure at rest. The stages of earth pressure and their applications are (a) *active earth pressure* including all types of retaining walls; (b) *active–passive earth pressures* including sheet piling and deep excavation; (c) *passive earth pressure* including basement wall, underwater down slope, and roadway sign; and (d) *earth pressure at rest* including pile foundation and bridge pier. Environmental forces such as wind, wave, current, and earthquakes are also closely related to the stability of geotechnical engineering structures.

13.1.2 Characteristics of lateral earth pressures

Figures 13.1(a) and (b) illustrate the interrelationship of lateral earth pressures at active, passive, and at rest stages. Figure 13.l(a) shows the section of a movable vertical rigid wall and Figure 13.1(b) is the load–displacement relationship. The mechanism of lateral earth pressure of these three stages can be explained as follows.

1   *Earth pressure at rest*: In examining Figure 13.1(a), it is apparent that the rigid wall may have two directions of motion, into the bank or away from the bank. For a cohesionless soil mass, it is apparent that if the wall is initially at rest and held by a force, $P = P_o$, that as the force, P, is reduced, the wall will be forced outward due to the weight of the soil. Also, as P is gradually reduced, the soil undergoes first elastic deformation, then elastic-plastic deformation, and finally uncontained plastic flow. In simple terms, if no wall movement occurs, the lateral earth pressure is referred as *earth pressure at rest*.

2   *Active and passive earth pressures*: Figure 13.1(b) shows a load–displacement curve depicting the behavior of the soil under active and passive earth pressures. The limiting force $P = P_a$ is usually defined as *active earth pressure*, p_a. The force

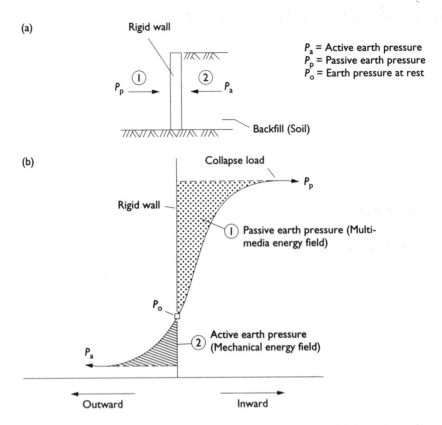

Figure 13.1 Lateral earth pressure at active, passive, and at rest stages. (a) A section of movable vertical rigid wall; and (b) Load–displacement relationships.

P_p is referred to as *passive earth pressure* which is directed toward the soil mass, p_p, and P_o referred as *earth pressure at rest*, p_o, where the wall is not moving in either outward or inward directions. Passive earth pressure is also indicated in Figure 13.1(b).

13.1.3   Coefficient of earth pressures

The *coefficient of earth pressure* is the principal stress ratio at a point in a soil mass. The characteristics of the coefficient of earth pressure are given in terms of three types: active, passive, and at rest.

1   *Coefficient of active earth pressure* (K_a): The minimum ratio of the minor principal stress to the major principal stress is the coefficient of active earth pressure. Active earth pressure is developed when a retaining wall has moved or tilted slightly in response to earth pressure from an adjacent soil mass. This is applicable where the soil has yielded sufficiently to develop a lower limiting value of the minor principal stress. K_a values are computed as indicated in Equation (13.1), based on either Coulomb's or Rankine's theories as discussed in Section 13.3.

2 *Coefficient of passive earth pressure* (K_p): The maximum ratio of the major principal stress to the minor principal stress is applicable where the soil has been compressed sufficiently to develop an upper limiting value of the major principal stress. Passive earth pressure is developed when a retaining wall is pushed against or into an adjacent soil mass. K_p values are computed as indicated in Equation (13.2), based on either Coulomb's or Rankine's theories as discussed in Section 13.3.

3 *Coefficient of earth pressure at rest* (K_o): The coefficient of earth pressure at rest is the ratio of the minor principal stress to the major principal stress. This is applicable where the soil mass is in its natural state without having been permitted to yield or without having been compressed. K_o values are measured either in laboratory or in situ conditions. Some empirical equations proposed by various researchers are discussed in Sections 13.3.4 and 13.11.

13.2 Methods for analysis of lateral earth pressure

There are two basic approaches for estimating the lateral earth pressures, the theoretical and empirical approaches which are discussed as follows:

1 *Theoretical approaches*: Theoretical approaches for estimation of lateral earth pressure include (a) Coulomb's earth pressure theory, (b) Rankine's earth pressure theory, (c) Limit analysis, and (d) Elasticity theory.

2 *Empirical approaches*: Empirical approaches are for the design of indeterminate foundation structures such as bracing excavation and anchored bulkhead. Some semi-empirical approaches for the determination of lateral earth pressure will be discussed in Sections 13.9 and 13.13.

13.3 Coulomb earth pressure theory (Wedge theory)

13.3.1 Principles and assumptions

The earliest analytical solution was the Coulomb method developed in 1776. It is also called the *wedge theory* Assumptions of the wedge theory include (a) backfill material is cohesionless soil, (b) the failure surface and applied pressure surface are planes, (c) friction exists between the wall and the soil, (d) the failure wedge may be treated as a rigid body, (e) failure is two-dimensional, and (f) the soil is isotropic and homogeneous.

13.3.2 Active earth pressures

Figure 13.2 shows the condition of Coulomb's active earth pressure complete as represented by the failure wedge and the force polygon. Figure 13.2(a) shows the failure wedge for a retaining wall against an inclined surface and Figure 13.2(b) presents the corresponding force polygon. Based on Figure 13.2, the active earth pressure is shown in Equation (13.1).

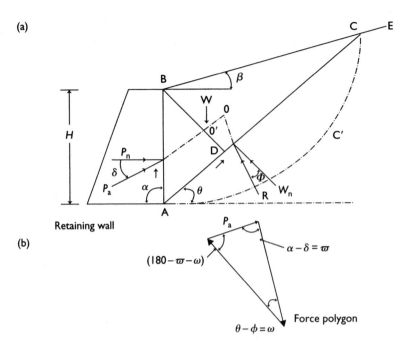

Figure 13.2 Coulomb's active earth pressure. (a) Failure wedge; and, (b) Force polygon.

$$P_a = \tfrac{1}{2} K_a \gamma H^2 \qquad\qquad (13.1)$$

where K_a = Coulomb active earth pressure coefficient and is given by

$$K_a = \frac{\sin^2(\alpha + \phi)}{\sin^2\alpha \sin(\alpha - \delta)\left[1 + \sqrt{\dfrac{\sin(\phi + \delta)\sin(\phi - \beta)}{\sin(\alpha - \delta)\sin(\alpha + \beta)}}\right]^2} \qquad\qquad (13.2)$$

and the angular terms are as defined in Figure 13.2. Note that δ refers to the angle of friction at the wall–soil interface. Equation (13.2) can be simplified if the retaining wall is vertical, smooth, and the backfill has no inclination as follows:

$$K_a = \frac{1 - \sin\phi}{1 + \sin\phi} \qquad\qquad (13.3)$$

13.3.3 Passive earth pressures

Coulomb's passive earth pressure assumed failure wedge and force polygon and are illustrated in Figures 13.3(a) and (b). Figure 13.3(a) represents the failure wedge, and Figure 13.3(b) shows the force polygon of the failure wedge. Based on Figure 13.3, the passive earth pressure is shown in Equation (13.4).

$$P_p = \tfrac{1}{2} K_p \gamma H^2 \qquad\qquad (13.4)$$

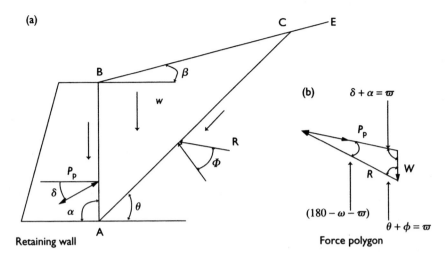

Figure 13.3 Coulomb's passive earth pressure. (a) Failure wedge; and (b) Force polygon.

where K_p = Coulomb's passive earth pressure coefficient and is given by

$$K_p = \frac{\sin^2(\alpha + \phi)}{\sin^2\alpha \sin(\alpha - \delta)\left[1 - \sqrt{\dfrac{\sin(\phi + \delta)\sin(\phi - \beta)}{\sin(\alpha - \delta)\sin(\alpha + \beta)}}\right]^2} \qquad (13.5)$$

and the angular terms are as described in Figure 13.3. As for the active condition, Equation (13.5) may be simplified for smooth, vertical walls retaining backfill with no inclination as

$$K_p = \frac{1 + \sin\phi}{1 - \sin\phi} = \frac{1}{K_a} \qquad (13.6)$$

Graphic solutions of Coulomb's method for both active (Sec. 13.2) and passive earth pressure (Sec. 13.3) cases developed by Culmann in 1875 will be presented in Section 13.6.

13.3.4 Earth pressure at rest (K_o)

The earth pressure at rest is defined and illustrated in Figure 13.1. The coefficient of earth pressure at rest, K_o, is determined experimentally. Table 13.1 presents the typical values for coefficient of earth pressure at rest. A common relationship used to determine K_o is given by (Mayne and Kulhawy, 1982)

$$K_o = (1 - \sin\phi')OCR^{\sin\phi'} \qquad (13.7)$$

Where ϕ' denotes the effective friction angle and OCR equals the overconsolidation ratio (Ch. 9). Further discussions on laboratory and in situ measurement techniques of earth pressure at rest, K_o, for both cohesive and cohesionless soils are presented in Section 13.13.

13.4 Rankine earth pressure theory

13.4.1 Principles and assumptions

Rankine in 1857 proposed a procedure for cohesionless soils based on the Coulomb method shown in Equations (13.1) and (13.4) for horizontal ground surfaces, dry cohesionless soils, and smooth walls. The assumptions for Rankine's theory are virtually the same as those noted in Coulomb's theory with the exception that wall friction is neglected. As such, the resultant of the normal and shear forces are assumed to act parallel to the ground surface.

13.4.2 Active earth pressures

One assumption of the Rankine theory is that pressure distribution is a hydraulic static pressure distribution as shown in Figure 13.4. Where p_a is the active lateral earth pressure and P_a is the resultant of the lateral earth pressure. Since the pressure

Table 13.1 Typical values for coefficient of earth pressure at rest

Soil types	Coefficient of earth pressure at rest, K_0
Sand–gravel	0.35–0.60
Sand, dense	0.40
Sand, loose	0.60
Silt–clay	0.45–0.75
Clay, stiff	0.50
Clay, soft	0.60

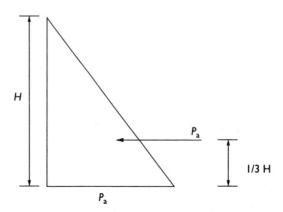

Figure 13.4 Hydraulic static pressure distribution.

distribution is assumed to be in a hydraulic state pressure, therefore, the P_a is the area of the triangle as shown in Figure 13.4.

1   *Unit active earth pressure*:

$$p_a = \gamma h K_a \tag{13.8}$$

2   *Resultant active earth pressure*:

$$P_a = \tfrac{1}{2} \gamma h^2 K_a \tag{13.9}$$

where p_a = unit active earth pressure, P_a = resultant active earth pressure, h = height of the wall, γ = unit weight of the backfill soil, and K_a = coefficient of active earth pressure. For a simple condition where the backfill is level, K_a may be computed similar to that shown with Coulomb's theory:

$$K_a = \tan^2 \left(45° - \frac{\phi}{2} \right) = \frac{1 - \sin \phi}{1 + \sin \phi} \tag{13.10}$$

where ϕ = angle of internal friction of the backfill soil.

13.4.3   Passive earth pressures

1   *Unit passive earth pressure*:

$$p_p = \gamma h K_p \tag{13.11}$$

2   *Resultant passive earth pressure*:

$$P_p = \tfrac{1}{2} \gamma h^2 K_p \tag{13.12}$$

where p_p = unit passive earth pressure, P_p = resultant passive earth pressure, h = height of wall, and K_p = coefficient of passive earth pressure. For a simple condition where the backfill is level

$$K_p = \tan^2 \left(45° - \frac{\phi}{2} \right) = \frac{1 + \sin \phi}{1 - \sin \phi} \tag{13.13}$$

As noted, Rankine analysis neglects the influence of wall friction. Figures 13.5(a) and 13.5(b) show the effect of wall friction or roughness on values of K_a and K_p, respectively.

13.5   Earth pressure for cohesive soil – the modified Rankine theory

13.5.1   General discussion

Originally, Rankine's theory as discussed in Section 13.4 deals with cohesionless soil, that is the cohesion, c, is not considered as indicated by its absence in the foregoing equations. Bell (1951) developed active and passive earth pressures for cohesive clay

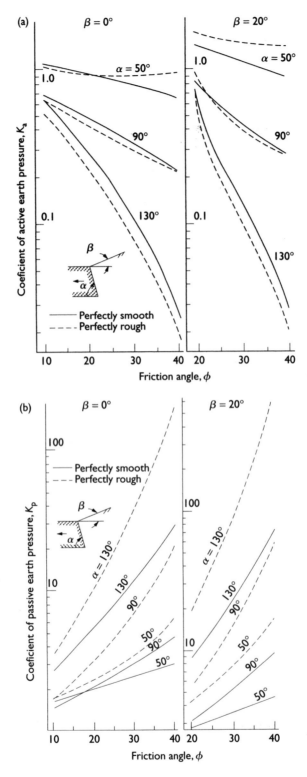

Figure 13.5 Effect of wall roughness on the coefficient of earth pressure. (a) Active; (b) Passive. (After Chen and Rosenfarb, 1973.)

based on the Rankine's and Coulomb's equations. He recognized that Mohr's circle (Sec. 10.3.2) could be used to obtain equations that incorporate cohesion. These are given as follows:

13.5.2 Active earth pressure cohesive soils

1 Unit active earth pressure:

$$p_a = \gamma h K_a - 2c\sqrt{k_a} \tag{13.14}$$

2 Resultant active earth pressure:

$$P_a = \frac{1}{2}\gamma h^2 K_a - 2ch\sqrt{K_a} + \frac{2c^2}{\gamma} \tag{13.15}$$

where p_a = unit active earth pressure, P_a = resultant active earth pressure, h = height of the wall, c = unit cohesion, and K_a is the coefficient of active earth pressure as defined previously.

13.5.3 Negative earth pressure and depth of tension crack

As the preceding equations would imply, the introduction of a cohesion term allows for the development of tension in the soil mass. Soil is an assemblage of particles and generally does not have the ability to withstand tension. Instead, cracks develop to relieve the stress. The depth of a tension crack along a wall and backfill can be estimated by the zone in which tensile forces are mobilized. In practice, backfill soil will separate from the wall and vertical cracks may form in the upper part of the soil layer. This vertical depth is measured from top of the wall to the depth when negative earth pressure is developed. This is given by

$$z_c = \frac{2c}{\gamma} K_a \tag{13.16}$$

where z_c = critical crack depth, c = unit cohesion, and K_a = coefficient of active earth pressure. Note that in soft cohesive soils where ϕ is taken as zero, the formula reduces to $2c/\gamma$.

13.5.4 Passive earth pressure for cohesive soils

1 Unit passive earth pressure:

$$P_p = \gamma h K_p - 2c\sqrt{k_p} \tag{13.17}$$

2 Resultant passive earth pressure:

$$P_p = \frac{1}{2}\gamma h^2 K_p + 2ch\sqrt{K_p} \tag{13.18}$$

where p_p = unit passive earth pressure, P_p = resultant passive earth pressure, h = height of wall, c = unit cohesion of soil, and K_p is the coefficient of passive earth pressure as defined previously.

13.6 Culmann graphical procedures based on Coulomb theory

Culmann (1875) proposed a graphical procedure, which is based on the Coulomb theory as discussed in Section 13.3. Culmann's graphical procedures can be used for determination of both active and passive earth pressures. With the active earth pressures, it also can be used for four different loading conditions including (a) uniform backfill soil, (b) surcharge load, (c) point load, and (d) combination of all three loads. For this reason, the Culmann graphical procedure is particularly useful. The technique as applied to active pressure is presented as follows. In the case of a uniform backfill, the procedure as outlined by Murthy (2002) is appropriate and shown in Figure 13.6:

1 Construct the wall and the backfill surface profile to a convenient scale;
2 Draw the line AE at an angle of ϕ above the horizontal and AD at an angle of $(\alpha - \delta)$ from AE. AE is the ϕ-line while AD is the pressure line;
3 Select and draw wedges ABV, AB1, AB2, etc. On line AE, mark off the points AV, A1, A2, etc. representing the weights of the individual wedges ABV, AB1, AB2, etc. and plotted to a reasonable scale, such that all points can be plotted on AE. The weight per unit length of wall is simply the soil unit weight multiplied by the area of the wedge.
4 Draw lines from each point V, 1, 2, etc. parallel to AD so that the wedges defined by sides AV, A1, A2, etc. are intersected at points V', 1', 2' etc.
5 Connect the points V', 1', 2', etc. This smooth curve is referred to as the Culmann Line or Pressure locus.
6 Find the point on the curve for which a tangent at the point is parallel to the line AE, this point is denoted as C'.
7 Draw a line from C' parallel to the line AD to connect with line AE at point C''. *The magnitude of the line C'C'' is equal to the active earth pressure, P_a.*
8 Draw a line from A to C' and continue to C at the surface of the backfill. *Line AC defines the rupture surface or critical sliding surface.*

For passive earth pressure, the same scheme may be used with the shape of the Culmann Line inverted from the shape of the line for the active case. Alternatively, other graphical procedures include the logarithmic-spiral method or friction circle (Ch. 14) method. Details on these applications are beyond the scope of this chapter and can be found in Terzaghi (1943).

13.7 Lateral earth pressure determined by elasticity theory

Utilization of principles of the elasticity theory for estimation of lateral earth pressures on retaining structure caused by various types of loading conditions have

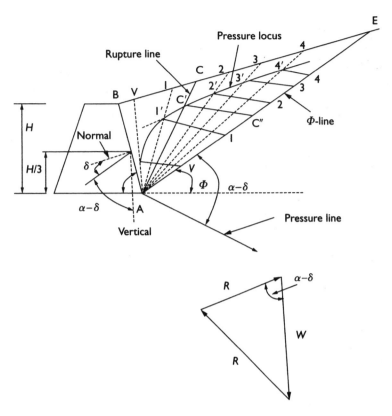

Figure 13.6 Illustration of Culmann's graphical procedure for active earth pressure.

been studied including (a) strip load, (b) point load, and (c) line load. The principles of the elasticity theory related to the lateral earth pressures have been examined by Timoshenko and Goodier (1951). In terms of strip loads, Terzaghi (1954) developed a method for assessing their effect on earth pressure when the loads are parallel to the retaining structure. Examples of this loading scenario include continuous wall footings, highways, and railroads. The actual lateral pressure against a rigid wall is twice the value determined by theory of elasticity as reported by Terzaghi (1954). Emphasis is given here to point and line loads, discussed as follows:

13.7.1   Point surcharge load

Any load concentrated on a small contact area may be treated as a point load as proposed by Terzaghi (1954). The intensity of lateral pressure varies not only with the depth but also with the horizontal distance from load to the wall. Subsequently, US Navy modifications (1962) were made to this theory to reflect experimental data and coupled with the Boussinesq equation (Sec. 9.6) as shown in Figure 13.7.

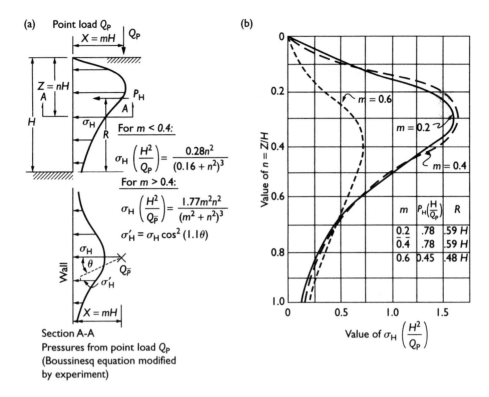

Figure 13.7 Lateral earth pressure influence diagrams due to a surface point load. (a) Force diagram; (b) Influence diagram.

Source: U.S. Navy, NAVDOCKS DM-7, 1962.

Teitgen and Fiedler method: The effect of point surcharge on retaining structures can be determined by semi-empirical equations. The equations for the simplified chart method proposed by Teitgen and Fiedler (1973) is presented as follows:

a   *Resultant force*

$$P_h = \frac{Q_p X C_1 C_2}{H}$$ (13.19)

b   *Location of resultant force*

$$h = Hn$$ (13.20)

where P_h = the resultant induced horizontal force, Q_p = the point surcharge load, X = the perpendicular distance from the point surcharge load to the face of the retaining wall, C_1 = value found from Figure 13.8, C_2 = value found from Figure 13.8, h = the vertical distance to the resultant from the ground surface, H = the retaining wall height, and n = the desired fraction $(0 < n < 1.0)$ of the total depth, H, see Figure 13.8(b).

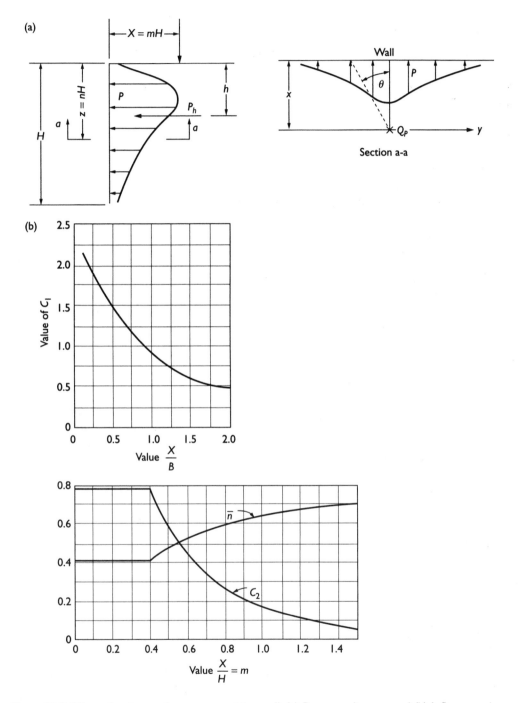

Figure 13.8 Effect of point surcharge on retaining wall. (a) Pressure diagram; and (b) Influence values.

Source: Teitgen, F. C. and Fiedler, D. R., (1973), Effect of point surcharge on retaining walls, *ASCE Civil Engineering*, Engineer's Notebook, Nov. p. 82. © 1973 ASCE Reproduced by permission of the American Society of Civil Engineers.

EXAMPLE 13.1   (*After* Teitgen and Fiedler, 1973)

Given: Q_p = 10 kips (44.5 kN);   B = 12 ft (3.66 m);

H = 10 ft (3.05 m);   X = 8 ft (2.44 m).

Compute the resulting horizontal force and its location using the technique of Teitgen and Fiedler.

SOLUTION

X/B = 0.67, so from Figure 13.8(b)
C_1 = 1.25, X/H = 0.80, so from Figure 13.8(b)
C_2 = 0.275, n = 0.595. By Equation (13.19):

$$P_h = \frac{(10)(8)}{10}(1.25)(0.275) = 2.75 \text{ kips (12.24 kN). By Equation (13.20):}$$

h = 10 (0.595) = 5.95 ft (1.81 m) from top.

Note: A design moment could then be estimated by applying the resultant force, P_h, conservatively as a point load.

13.7.2   Line surcharge load

A continuous wall footing of narrow width may be taken as a line load when located parallel to the retaining wall as proposed by Terzaghi (1954). The procedure is based on the theory of elasticity. The stress at any depth, z, on a retaining structure caused by a line load of intensity, q, per unit length may be calculated. Based on Timoshenko and Goodier (1951) and Terzaghi (1954), the US Navy (1962) developed a modified version of lateral earth pressure design chart as shown in Figure 13.9.

13.8   Lateral earth pressure determined by semi-empirical method

Semi-empirical methods may be used to assist in the selection of potential retaining wall sections used to withstand lateral earth pressures. Preliminary estimates of retaining wall sections are needed for (a) the preparation of estimates; (b) determining the most economic shape of the wall; and (c) the analysis of the capability of the selected wall to resist the forces that will act on it. In making the tentative selection of the section for analysis of forces, the designer is guided by experience and by various published charts or tables. If the analysis shows that the wall is unsatisfactory, the dimensions of the selected section should be revised, and a new analysis will be made. A set of simple design charts was developed for determination of preliminary cross-section of retaining wall based on a semi-empirical earth pressure method (Terzaghi and Peck, 1967). Use of these charts first requires consideration of the relevant type of retaining wall and distribution of forces. As illustrated in

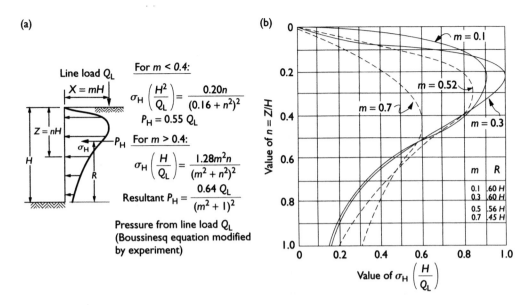

Figure 13.9 Lateral earth pressure influence diagrams due to a surface line load. (a) Force diagram; and (b) Influence diagram.

Source: US Navy, NAVDOCKS DM-7, 1962.

Figure 13.10, the weight of the wall and pressure of the backfill are related to the geometric dimensions of the slope, height, and width of the wall. It is assumed that the resultant force of the backfill pressure and the weight of the wall act on one-third of the bottom base.

Estimation of the cross-section proceeds by classifying the backfill material for a given problem according to five types as suggested by Terzaghi and Peck (1967). These types are given as:

Type I: Coarse-grained soil without an admixture of fine soil particles, very free-draining (clean sand, gravel or broken stone);

Type II: Coarse-grained soil of low permeability due to an admixture of particles of silt size;

Type III: Fine silty sand, granular materials with conspicuous clay content, or residual soil with stones;

Type IV: Soft or very soft clay, organic silt, or soft silty clay;

Type V: Medium or stiff clay that may be placed in such a way that a negligible amount of water will enter the spaces between the chunks during floods or heavy rains.

Each of these backfill types has its own design chart as shown in Figure 13.11. The charts can be used to determine the necessary width and height of a retaining wall (as denoted by b_o/h on the x-axis) as a function of wall slope (y-axis) and given backfill slope.

The design of retaining walls consists essentially of the successive repetition of two steps: the tentative selection of the dimensions of the wall, and the analysis of the

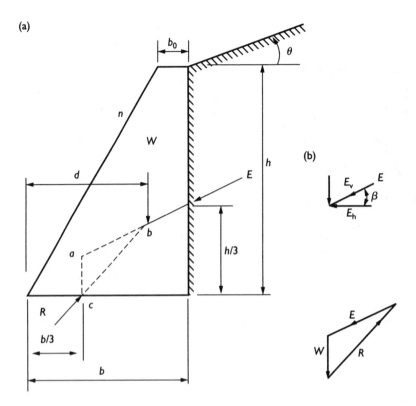

Figure 13.10 Cross-section and force diagram of a gravity retaining wall. (a) Cross-section; and (b) Force diagrams.

ability of the selected structure to resist the forces that will act on it. If the analysis indicates that the structure is unsatisfactory, the dimensions are altered and a new analysis is made. In order to make the analysis, some basic steps are listed as follows: (a) estimating the magnitude of the forces that act above the base of the wall, including the pressure exerted by the backfill and the weight of the wall itself; (b) investigating the stability of the wall with respect to overturning; and (c) estimating the adequacy of the underlying soil to prevent failure of the wall by sliding along a plane at or below the base to withstand the pressure beneath the toe of the foundation without failure and allowing the wall to overturn and to support all the vertical forces, including the weight of the backfill, without excessive settlement, tilting, or outward movement.

13.9   Wall stability and lateral environmental pressures

13.9.1   Wall stability due to earth pressure and surcharge loading

To check the stability of a retaining wall, the following steps are necessary. The stability analysis includes (a) overturning, (b) sliding failure, (c) bearing failure, and (d) settlement analysis.

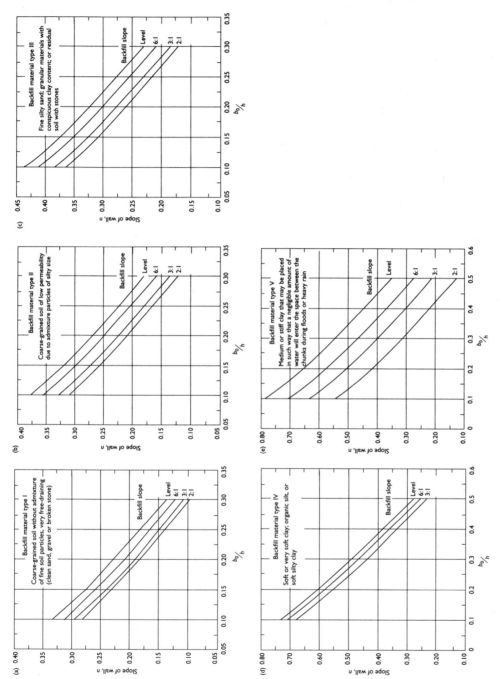

Figure 13.11 Design charts developed according to various backfill materials, height, width, and slopeface of wall. (a) Backfill material type I; (b) Backfill material type II; (c) Backfill material type III; (d) Backfill material type IV; and (e) Backfill material type V.

1   *Overturning analysis*: factor of safety against overturning

$$F_s = \frac{\text{Stabilizing moment}}{\text{Overturning moment}}$$ (13.21)

2   *Sliding failure analysis*:

$$F_s = \frac{\text{Horizontal resistance}}{\text{Horizontal force}}$$ (13.22)

3   *Bearing failure analysis*: factor of safety, F_s, against bearing failure

$$F_s = \frac{\text{Ultimate bearing capacity}}{\text{Bearing pressure}}$$ (13.23)

4   *Settlement analysis*: Settlement analysis proceeds as discussed in Chapter 9.
5   *Other stability analysis*: Slope stability analysis as discussed in Chapter 14.

13.9.2   Wall stability due to earthquake loading

1   *General Discussion:* The analysis of wall stability due to earthquake loading may be performed with Mononobe–Okabe's active earth pressure equation (Okabe, 1924; Mononobe and Matuo, 1929) to compute the active earth pressure coefficient with earthquake effect. The modified equation is based on Coulomb's active pressure equation (Eq. 13.1) with modifications to take into account the vertical and horizontal coefficients of acceleration induced by an earthquake as illustrated in Figure 13.12.

In examining Figure 13.12, H = height of wall, i = slope of the backfill with respect to the horizontal, β = slope of the back of the wall with respect to the vertical, δ = angle of friction between the wall and the soil, α = angle between failure plane and horizontal line, ϕ = friction angle of soil, F = resultant of shear and normal forces along the failure plane, BC, P_{AE} = active force, W = weight of wedge, S = shear force, N = normal force, $k_h W$ and $k_v W$ = the inertia forces in the horizontal and vertical directions.

2   *Assumptions of Mononobe–Okabe's equation:* The Mononobe–Okabe equation is based on the following assumptions: (a) backfill material is assumed to be cohesionless soil; (b) the movement of the wall is sufficient to produce minimum active pressure; (c) the shear strength of the dry cohesionless soil can be given by $s = \sigma' \tan \phi$, where σ' is the effective stress and s is shear strength; (d) at failure, full shear is mobilized along the failure plane, BC; and (e) the backfill soil behind the retaining wall behaves as a rigid body.

3   *Solution of Mononobe–Okabe's equation:* Figure 13.12(b) shows the forces considered in the Mononobe–Okabe solution. The forces on the failure wedge ABC per unit length of the wall are weight of wedge, W, and active force, P_{AE}. The active force, P_{AE}, is the resultant of shear and normal forces along the failure plane, F, and $k_h W$ and $k_v W$ are the inertia forces in the horizontal and vertical directions respectively,

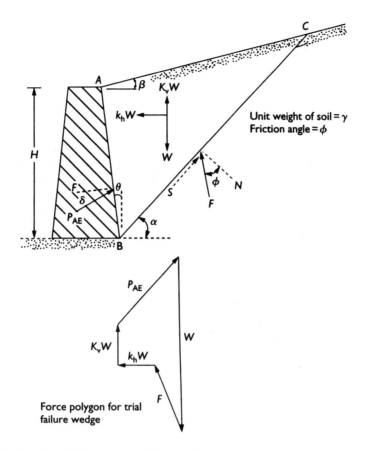

Figure 13.12 Derivation of Monobode–Okabe equation.

where k_h and k_v are the horizontal and vertical components of earthquake acceleration divided by g, the acceleration due to gravity, respectively. The combined effect on the active earth pressure may be given as

$$P_{AE} = \tfrac{1}{2}\gamma H^2 (1 - k_v) K_{AE} \tag{13.24}$$

Equation (13.24) is referred to as the *Mononobe–Okabe active earth pressure equation*. For the active force condition (P_{AE}), the soil wedge ABC located behind the retaining wall exists at an angle α from the horizontal (Fig. 13.12). The value of K_{AE} in Equation (13.24) is the active earth pressure coefficient with the earthquake effect and can be obtained from Equations (13.25) and (13.26). Various types of charts or tables are available (Das, 1992) to simplify the computation procedures.

$$K_{AE} = \frac{\cos^2(\phi - \eta - \phi)}{\cos\eta\,\cos^2\theta\,\cos(\delta + \theta + \eta)\left[1 + \sqrt{\dfrac{\sin(\phi + \delta)\sin(\phi - \eta - \beta)}{\cos(\delta + \theta + \eta)\cos(\beta - \theta)}}\right]^2} \tag{13.25}$$

Where all other variables have been defined previously and in Figure 13.12. The value of η may be found by:

$$\eta = \tan^{-1}\left[\frac{k_h}{1 - k_v}\right] \tag{13.26}$$

13.9.3   Lateral environmental loading

In addition to the lateral earth pressures as discussed in previous sections, environmental lateral forces act on structures such as buildings, waterfront structures, nearshore and offshore structures including (a) *land structures* which include water pressure and seepage forces, earthquake loads, wind loads, and traction forces, and (b) *marine structures* which include: earthquakes and tsunamis, wind loads, wave forces, currents, hydrodynamic pressures, ice forces, and mooring pulley forces. The nature of these environmental forces was discussed in Chapter 11.

13.9.4   Water pressure and seepage force

Water pressure may act laterally against a foundation structure. Considering the foundation structure as a whole, the lateral hydrostatic pressure is always balanced, but the hydrostatic buoyancy or uplift force must be counteracted by the dead load of the foundation structure. If not, some provisions must be made to anchor the foundation. If the backfill soil contains a large amount of water and if no proper drainage system is provided (or blocked), the water seeps through the backfill in a downward direction. Seepage water increases overall earth pressure by increasing the total unit weight of the backfill soil.

13.9.5   Wind load, wave, and other environmental loading

1   *Wind load:* Wind loads act on all exposed surfaces of a structure. The design pressure is usually stipulated in local building codes or design manuals. In most cases, wind loads affect structures above the ground surface such as buildings, bridges, TV towers, etc. Wind loads also affect foundation structures below the ground surface such as highway signposts and foundations of tall buildings. Some of these effects will be discussed in Section 13.12. Also, as discussed in Chapter 11, typhoons, hurricanes, and tornados are special types of wind load, as they are particularly violent. These types of wind loads have specific locations and seasons. Typhoons and hurricanes occur generally along the coastal areas while the tornado occurs in land such as the south and mid-west United States and occur in late summer and early fall.

2   *Surface wave force and currents:* Waves are generated by wind and/or by earthquakes, tides, etc. Most of the time waves are caused by wind. The characteristics of the wave are determined by the velocity of the wind, the duration of the wind, and the fetch length (Fig. 11.5). Designs for nearshore/offshore structures involve wave height, period, length, and still water depth (Sec. 11.8). Also as discussed in Section 11.8, currents are the driving forces of the oceans. Surface currents are caused mainly by winds and the rotation of the Earth.

3 Ice force in the water: Intermittent freezing and thawing of rivers and lakes often leads to detached masses of ice that are moved about by wind and current. These broken ice masses float on the water and cause lateral forces, sometimes undermining the stability of nearshore and offshore structures and foundations. This type of force can be estimated by following equation proposed by Teng (1962):

$$F = C f_c A \tag{13.27}$$

where F = ice force (lb), C = coefficient, f_c = compressive strength of ice (psi), and A = area struck by ice (in^2). Ice strength, f_c, varies with temperature, salt content, load rate, etc.

4 Mooring pull and ship impact: Various types of nearshore, offshore structures as well as dock structures are provided with mooring posts for anchoring boats. The magnitude of the mooring pull may be assumed to be equal to the capacity of the wind used on the boat. In most codes it is suggested to use a ship impact of 25 ton (22.7 Mg) to greater than 100 ton (90.7 Mg) for design purposes.

5 Traction force: Traction forces are due to moving railway and highway traffic and due to hoist and crane wheels. The lateral components of these forces are transmitted to the soil layer and foundations and must also be considered for certain projects. The American Railroad Engineering Association (AREA) and the American Association of State Highway and Transportation Officials' (AASHTO), specifications contain information on the magnitude of such traction forces.

13.10 Coefficient of earth pressure at rest (K_o) and other friction forces

Many naturally occurring sediments as well as man-made fills are deposited and compacted in horizontal layers where no lateral yielding occurs. Under such conditions the ratio of lateral to vertical stresses is known as the coefficient of earth pressure at rest or just called K_o. Al-Hussaini (1981) made a study on K_o and comparison of various measuring techniques for determining this parameter. Field measurements of K_o may also be obtained with Pressuremeter test (PMT) or Dilatometer test (DMT) as discussed in Chapter 10.

13.10.1 K_o for clay-like soil

Figure 13.13 shows the relationship of K_o versus soil types as reflected by plasticity index, I_p. A linear relationship is found such that as I_p increases, so does K_o. Note, however, that there is considerable scatter in the data.

13.10.2 K_o for sand

Laboratory measurement of K_o for sand has been made with the assistance of instruments including linear variable differential transducer (LVDT) and strain gauges as reported by Al-Hussaini (1981). Results obtained between theoretical and experimental studies for fine sand are presented in Figure 13.14. Significant variation for these results is observed.

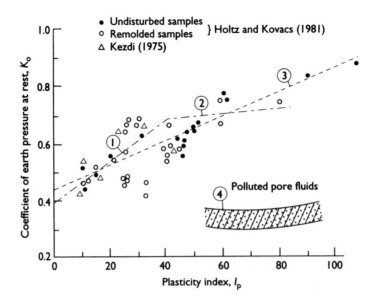

Figure 13.13 Variations of K_o for various types of soil as reflected on the plasticity index, I_p.

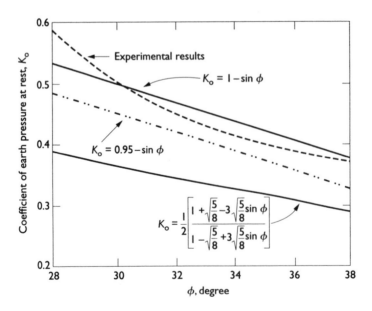

Figure 13.14 Comparisons between theoretical and experimental tests results on K_o of sand.

Source: Al-Hussaini (1981). Copyright ASTM INTERNATIONAL. Reprinted with permission.

13.10.3 Friction force and contact angles

The frictional force between soil and soil is defined by the friction angle, given by ϕ. The concepts and mechanism of frictional force between soil and soil have been discussed in Chapter 10. When describing the friction force between soil and some other material, such as a retaining wall, the term *contact angle* is used instead. The frictional force between soil and walls or other structures such as bulkheads and pile foundations (Sec. 15.12) may also be referred to as skin friction.

13.11 In situ measurements of lateral earth pressures

13.11.1 In situ earth pressure measurements

Instruments for obtaining in situ lateral earth pressure measurements include the pressure cell, LVDT, slope indicator (inclinometer), as well as conventional surveying equipment. Pressure cells and slope inclinometers are discussed as follows:

1 *Pressure cells*: Pressure cells are used to measure the free-field stresses within soils or the soil pressures acting against structures. There are three general types: (a) acoustic (vibrating wire), (b) electric pressure cells, and (c) hydraulic pressure cells. All these pressure cells are commercially available.

2 *Slope inclinometer*: The slope inclinometer measures the direction and magnitude of horizontal movement of soil. It consists of a probe with two sets of wheels. The probe is inserted into a cased-in borehole and measurements are taken as a function of depth to assess the extent of tilting. There are several types of slope inclinometers available. It has been used for determining the profile of a wide variety of nearly vertical surfaces. These devices are mainly used in connection with earth-fill and rock-fill dams, retaining structures, landslide, piling and sheet piling, and ground subsidence. Use of pressure cells in conjunction with inclinometers was conducted to develop the results presented in Figure 13.15.

13.11.2 Comparison of earth pressures between theoretical and experimental results

The loads or pressures on a wall system are a function of both design and local environmental factors. Figures 13.15(a) and (b) show a comparison between theoretical and experimental lateral earth pressure results from Bethlehem Steel Corporation and Bank of California excavation. In examining Figure 13.15(a), the slope indicator is used in comparison with Coulomb's earth pressure theory. Significant differences are found between theoretical and experimental results. Figure 13.15(b) shows a comparison of apparent earth pressure as calculated from the observed tie loads to those assumed in design at sections A and B. In each case the observed and calculated earth pressures are very similar.

13.11.3 Factors affecting lateral earth pressures

The loads on a wall system are a function of many factors including both design and random environmental variables. In this discussion, a distinction between types of loading will be made on the basis of whether the system is braced or tieback because of the following factors:

1 *Loading types and types of wall systems*: Loading types and types of wall systems affect the lateral earth pressure. Using a tieback wall system as an

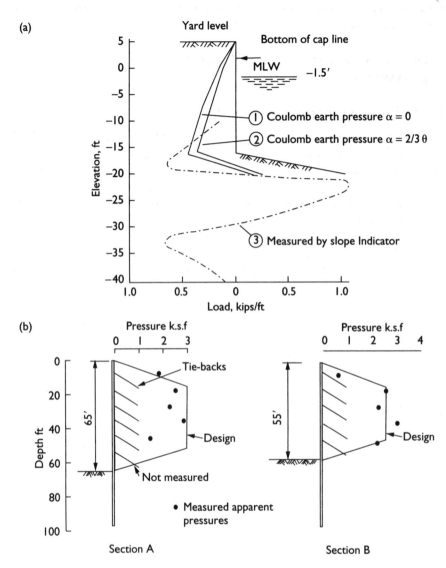

Figure 13.15 Comparison between theoretical and experimental lateral earth pressure results. (a) Bethlehem Steel Corporation. Sparrow Points, MD; (b) Bank of California excavation.

Source: Dismuke (1970), Clough (1976).

example, the supports of a tieback wall are generally significantly more flexible than those of a braced wall, leading to a different distribution of earth pressures. The tieback wall is commonly pre-stressed with resultant loads equal to or above those of active pressure conditions while the braced wall is rarely subjected to pre-stress levels of this magnitude.

2 *Environmental factors*: Environmental factors include weathering, floods, seasonal variations, as well as moving vehicles around the structural sites.

13.12 Earth pressures around excavations and other special cases

Excavations of more than 20 ft (6.1 m) are classified as a *deep excavation*. The pressure variation shown for land cofferdams has had some verification from field studies and can vary widely with field installation practice and soil characteristics. In addition to braced and tieback walls, other special cases are also presented in this section as (a) underwater slopes, (b) lower part of foundation structures, and (c) geosynthetic-reinforced soil (GRS) walls (Ch. 15).

13.12.1 Braced or tieback walls

Lateral earth pressures distributed along the braced or tieback walls are summarized and discussed by Dismuke (1991). A condensed version of such a case is presented here. It is significant to note that Rankine or Coulomb's method are typically not used in the case of braced excavation and/or tieback walls, primarily because even slight wall movement is generally not tolerated in these cases and also because of the staged nature of the construction sequence. Recall that development of active or passive earth pressure conditions is predicated on wall movement away or into the soil mass. Tieback walls come in a variety of forms, although the general configuration involves a steel tendon or similar element that is secured to the retaining wall on one end and grouted in the soil on the other end. Pressures acting on braced walls rarely assume the familiar triangular distribution with depth, and in many cases it is nonuniform. A trapezoidal or rectangular pressure distribution is typically assumed. The actual pressure distribution is a function of both the type of braced system and the soil properties. Several equations have been used to predict unit pressures in various soils for these situations, the most prominent of which have been developed by either Terzaghi and Peck (1967) or Tschebotarioff (1973). Equations from each of these investigators are presented as follows for sand, soft to medium clay, and stiff clay:

1 *Cohesionless soil (sand)*:

   a Terzaghi's and Peck (1967)

$$p = 0.65\,K_a\gamma H \tag{13.28}$$

where K_a = coefficient of active earth pressure as defined previously, γ = the unit weight of soil, H = height of wall.

   b Tschebotarioff (1973)

$$p = 0.8\,K_a\gamma H \cos\delta \tag{13.29}$$

where δ = the wall to soil interface friction angle and the other parameters are as defined before.

2 *Soft to medium clay*

   a Terzaghi and Peck (1967)

$$p = 1.0\,K_a\gamma H \tag{13.30}$$

where K_a is defined by

$$K_a = 1 - m \frac{2q_u}{\gamma H} \tag{13.31}$$

where q_u = the undrained strength of the clay, m = a reduction factor depending on the value of N. N is a stability number and is defined by

$$N = \frac{\gamma H}{c} \tag{13.32}$$

where c = the cohesion and the other parameters are as noted before. If $N > 3$–4, and the clay has been preloaded, then $m = 1$, otherwise a value of $m < 1$ should be selected. As its name implies, the value of N can also be used to assess the performance of clay in an excavation. Specifically, if the value of $N > 3$–4, then movement at the base of the excavation is likely, while values < 6 suggest that base failure is likely.

b   Tschebotarioff (1973)

$$p = 0.375\,\gamma H \tag{13.33}$$

3   *Stiff fissured clay*

a   Terzaghi's and Peck (1967)

$$p = (0.2\text{–}0.4)\,\gamma H \tag{13.34}$$

b   Tschebotarioff (1973)

$$p = 0.3\,\gamma H \tag{13.35}$$

Reviewing various methods of determining pressures in excavations and soil movements and comparing the results with data from field excavations show that movements of the soil outside of the excavation and strut loads cannot be adequately predicted in most field conditions. As such, design conservatism is particularly warranted in these situations.

13.12.2   Heave and piping

1   *Heave:* Heave is the upward movement of soil caused by expansion or displacement resulting from phenomena such as moisture absorption, removal of overburden, frost action (Sec. 6.6), and driving of piles (Sec. 15.12). Heave and piping are common failure modes of retained excavations. Bottom heave in excavations in clay soil is influenced by shear strength and loading history of the clay.

2   *Piping:* Piping is the movement of soil particles by percolating water leading to the development of channels. Sometimes, it is called subsurface erosion. In excavation, it is referred to as blowing, blowout, or boiling. It is an upward movement of soil material in the base of an excavation, cofferdam, or basement because of groundwater pressure normally associated with insufficient toe penetration of sheeting. An equation proposed by Terzaghi and Peck (1967) cited by Dismuke (1991a) estimates

whether or not the excavation is safe against the piping. A factor of safety of 1.5 is recommended for determining the resistance to heave.

$$F_s = 2N\gamma \left(\frac{\gamma_1}{\gamma_2}\right) K_a \tan\phi \tag{13.36}$$

where F_s = factor of safety, $N\gamma$ = bearing capacity factor of the soil below the excavation, γ_1 = unit weight of soil above the bottom of the excavation, and γ_2 = unit weight of soil below the excavation. Piping occurs if the water head is sufficient to produce critical velocities in cohesionless soils. This results in a "quick" condition at the bottom of the excavation.

13.12.3   Passive earth pressure on underwater walls and bulkheads

Passive pressure in underwater soil that slopes downward away from sheet pile bulkheads is difficult to calculate with Coulomb's equation (Sec. 13.3) because of the uncertainty in the angle of slope and friction. Also Culmann's graphical solution (Sec. 13.6) is rather lengthy. For this type of problem, however, the graphical vector solution as proposed by Bigler (1953) simplifies computations considerably. Numerical illustrations are also presented as follows: Figure 13.16 illustrates vector solutions of passive earth pressures on walls or bulkheads. In Figure 13.16, the angle of internal friction in the soil is assumed to be 35°, the weight of the soil, w, (pcf), and the angle of friction of soil against the wall is 16°. This angle could be zero, but many engineers assume that friction against the wall increases like passive resistance. It is included here and it increases the resulting resistant force:

1   On the ground slope, **OE** (Fig. 13.16), points **A, B, C, D** are marked off at convenient distances horizontally from the wall **OQ**;
2   The areas of triangle **AOQ, BOQ**, etc. are computed;
3   These areas are measures of the weight of soil per foot of width and are laid off vertically on the stress diagram as **O'A'**, **O'B'**, etc. in Figure 13.16;
4   Surfaces **AQ, BQ**, etc. in Figure 13.16 are possible failure planes;
5   The resisting forces along these failure planes are at an angle of 35° with the normal to the planes;
6   The vectors representing these forces are drawn in Figure 13.16 from **A', B', C'**, etc;
7   Vector **O'X'** drawn at the assumed angle of friction (16°) with the wall will then give the earth resistance per foot of length;
8   The minimum value of this vector, **OP**, is the minimum resistance offered by the earth, in this case 162 w.

Coulomb's theory (Sec. 13.3) is based on the assumption of plane failure for passive resistance, and the critical plane for failure is that one for which the passive thrust is a minimum. Therefore, enough vectors must be drawn to locate the minimum length of the vector **O'P** in order to find the minimum possible earth resistance. Then, from Coulomb's equation (Eq. 13.1)

$$P_p = \tfrac{1}{2} p_p H^2 = 162\, w \tag{13.37}$$

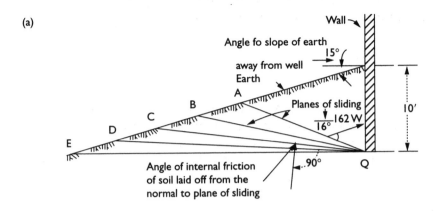

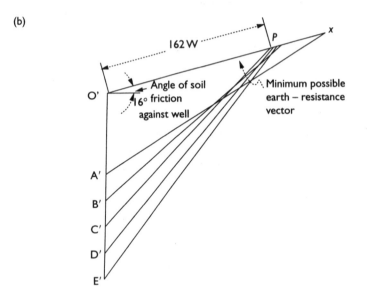

Figure 13.16 Vector solution of passive earth pressure on walls and bulk heads. (a) Schematic sketch of wall and backfill; (b) Vector stress diagram.

Source: Based on Bigler (1953).

From Figure 13.16, $H = 10$ ft (3.05 m), solve for p_p as

$p_p = 3.2 w$, the equivalent hydrostatic passive pressure per sq ft per ft of depth.

13.12.4 Passive earth pressure at lower part of foundation structures

1   Piers supported by passive earth pressure: A short-cut method has been developed by Robbins (1957) to save time in the oft-repeated operation of designing a concrete pier to resist a horizontal force. It takes into account the passive earth pressure. Figure 13.17 was prepared based on Equation (13.38).

$$P_P - P_A = \frac{4W_P(2 + 3F)}{X^2} \tag{13.38}$$

where P_P = total passive earth pressure, P_A = total active earth pressure, W_P = total horizontal load applied, $F = H/X$ (Fig. 13.17(a)), and X = distance pier extends below grade. These are the values on the curve lines in Figure 13.17(b). To find P_P use the Equation (13.39):

$$P_P = W_e \tan^2 (45° + \tfrac{1}{2} \phi) \tag{13.39}$$

where W_e = weight of earth and ϕ = friction angle of earth material.

A nomograph (Fig. 13.17(b)) gives the depth of a pier required below grade (X, curved lines) when the applied load and height above grade are known.

EXAMPLE 13.2 (After Robbins, 1957)
A horizontal load of 300 lb is applied to a single post stanchion at a point 15 ft above grade. Determine the depth and width of a concrete pier required to resist this force. The pier is to be earth formed, using the passive earth pressure as a resisting force.

SOLUTION
Enter the graph, Figure 13.17, at H = 15 ft, and move to the right along the dashed line. The curves represent the depth of excavation. The dashed line intersects the 6 ft curve at the 300 lb point (see horizontal scale). This means that a pier 12 in. wide and 6 ft deep will resist a 300 lb point (see horizontal force applied to it 15 ft above grade). Since the 5 ft curve is intersected at 180 lb, if the pier extended below grade only 5 ft, it would have to be 20 in. wide. The equation would be

300/180 × 12 in. = 20 in.

From Equation (13.34) where W_e = weight of earth material = 100 lb/ft³, ϕ = 33°, then

$$P_p = W_e \tan^2 (45 + \tfrac{1}{2} \phi) = 340 \text{ lb } (1.51 \text{ kN}).$$

2 Pole embedment to resist lateral loads: The passive earth pressure varies widely for different soil types and environmental conditions as discussed in Section 13.2. A simple approach for such problems was proposed by Patterson (1958) and based on Rutledge's work. The Rutledge chart for determination of required depth of embedment of a post is presented in Figure 13.18.

3 Other passive earth pressure problems: The passive earth pressure and lateral resistance of a subsurface structure such as tall building, highway sign posts, TV tower, and lower part of group pile also requires consideration during the design process. De Simore (1972) pointed out that the passive earth pressure affects the lower part of foundation structures of tall buildings, and the interested reader is referred to the original work for more details.

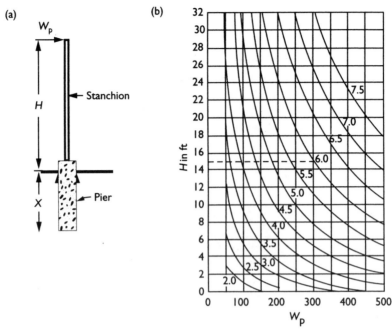

(a)

W_p

Stanchion

H

Pier

X

(b)

No monograph gives depth of pier required below grade (X, curved lines) when load applied (W_p, horizontal scale) is pounds per sq.ft. of projected width of pier normal to applied load, is known, and when height above grade of application of applied load (H, vertical scale) is known also known.

Figure 13.17 Piers supported by passive earth pressure. (a) Diagram defines terms W_p, H and X; and (b) Nomograph gives depth of pier required below grade when applied load.

Source: Robbins, N. G., Piers supported by passive earth pressure, *Civil Engineering*, ASCE, April, p. 276. © 1957 ASCE. Reproduced by Permission of the American Society of Civil Engineers.

13.12.5   Lateral earth pressure on geosynthetic reinforced soil (GRS) wall systems

Detailed description of GRS wall systems will be presented in Chapter 15. The lateral earth pressure used in the GRS wall systems is based on Rankine's or Coulomb's theories similar to those discussed in horizontal force acting on a rigid wall in the beginning of this chapter. However, some modifications are proposed by various investigators for specified applications. Additional information may be found in Wu (1994) and Koerner (1998).

13.13   Summary

The focus of this chapter has been on earth pressure; that is, the pressure that is exerted in the horizontal direction. In the case of water, pressure is the same in all directions, however in soil the pressure in the horizontal direction is generally different than it is in the vertical direction. Three types of earth pressures, active, passive,

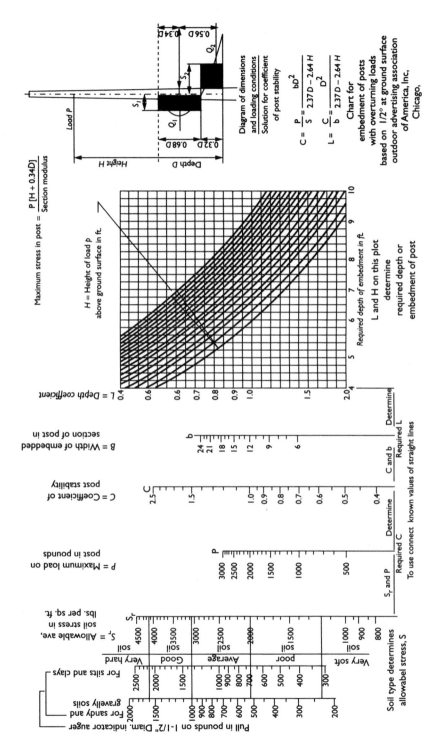

Figure 13.18 Rutledge chart for embedment of posts with overturning loads. (a) Diagram of dimensions and loading conditions, and (b) Chart for determining required depth of embedment of post.

and at rest, were identified and discussed. The extent to which any of these conditions exists depends on whether movement occurs away from soil (active), into soil (passive), or not at all (at rest). These pressures are needed to design a variety of structures, although a retaining wall is the most common example. The horizontal earth pressure is needed to analyze the structure for failure by either sliding or overturning. Although many researchers have expanded and modified equations for use in specific situations, earth pressures are basically determined from either the Coulomb or Rankine method of analysis. Differences of major characteristics between Rankine and Coulomb, and limit equilibrium and limit analysis methods for determination of lateral earth pressures are identified and discussed including assumptions, computation procedures, together with numerical examples. Colmann's graphical solution of Coulomb's method is discussed in detail. Additionally, lateral environmental forces or pressures also exist as a consequence of such activities as wind, water, and seismic events, and have to be accounted for in the design process when relevant.

PROBLEMS

13.1 What assumptions were made in (a) Rankine's and (b) Coulomb's earth pressure theories? Under what conditions will Rankine's and Coulomb's yield identical results?

13.2 Develop an expression for the resultant earth pressure exerted by a cohesionless backfill with a horizontal surface against a vertical retaining wall by (a) Rankine's method and (b) Coulomb's method, assuming the angle of wall friction to be zero.

13.3 A gravity retaining wall 15 ft (4.575 m) high, whose inside face is inclined at an angle of 10° to the vertical (away from the backfill), restrains a deposit of graded sand and gravel with $\phi = 35°$, mass unit weight = 120 pcf (18.8 kN/m³), and angle of wall friction = 10°. The surface of the backfill is inclined at 20° above the horizontal and extends at this slope for considerable distance from the face of the wall. Which theory will require a heavier wall: Rankine's or Coulomb's: (Give numerical values in support of your reasoning.)

13.4 From Problem 13.2, if the wall had a vertical face and a horizontal backfill, at what distance from the top of the wall does the total force produced by a 5 kips per foot (22.2 kN) line load begin to decrease in magnitude? At what distance above the base will the force due to the surcharge act under these conditions? How far from the top of the wall must the surcharge be placed so that it causes no increases in the earth pressure against the wall?

13.5 A concrete wall is 12 ft (4 m) high, 5 ft (1.5 m) thick at the base, and 2 ft (0.6 m) thick at the top. One face is vertical. What are the maximum and minimum unit pressures under the base of the wall due to its weight?

13.6 A vertical retaining wall 12 ft (4 m) high supports a medium coarse sand and gravel backfill whose surface is horizontal and carries a uniform distributed load of 80 psf (3.8 kPa). The soil properties are as follows: Friction angle (soil–soil) = 32°, Friction angle (soil–wall) = 22°, Void ratio (e) = 0.58, Specific gravity of solid = 2.70. The free water level is 3.5 ft (1.07 m) above the base of the wall, and the capillary rise may be considered negligible. Determine the magnitude (per ft of wall), direction, and point of application of the resultant force acting on the wall.

Chapter 14

Earth slope stability and landslides

14.1 Introduction

Slope stability and *landslides* belong to one system. Landslides are the result of slope instability. It occurs in many parts of the world, especially in those areas with problematic soils/rocks and/or adverse environmental conditions. They are usually caused by excavation, undercutting the foot of an existing slope, improper surface and subsurface drainage systems, tunnel collapse of underground caverns, surface and subsurface erosion, or by a shock caused by earthquake or blasting, which liquefies the soil.

In analyzing the landslide problem, engineers and geologists often look at it from different points of view. The geologist regards a landslide as one of many natural processes acting as part of the geological cycle. They are interested only in the ground movement with respect to the geological and hydrological features. On the other hand, the geotechnical engineer is interested in the soil types, their engineering behavior, the maximum height of the slope, and maximum slope angle in terms of a safety factor. In most cases, they do not understand the geological formation and environmental factors that cause a landslide. Even within the engineering group there are different perspectives: the practitioner is interested in the measurements of soil–rock properties, ground movements, and local environmental conditions to design a solution, while the theoretician is interested in idealizing the failure surface in order to fit it into a mathematical description for use in subsequent efforts to model the system.

Since the landslide problem is not a simple matter, it requires knowledge from other disciplines. Therefore, a joint effort from geologists, geotechnical engineers, and seismologists is required to tackle this problem. There are numerous state-of-the-art publications concerning slope stability and landslides with these various aspects emphasized. In this chapter, a general review of landslides and slope stability is given with emphasis on environmental aspects and controls.

14.2 Factors affecting slope instability

Factors affecting earth slope instability are (a) External loading conditions including surcharge loading, earthquake actions, blasting vibration, moving vehicle, and construction operation; and (b) environmental factors including rainstorms hurricane (typhoon), flash flood, El Nino and La Nina effects, dry–wet and freeze–thaw cycles, acid rain and acid drainage, pollution intrusion, tree/vegetation roots, and animal, insect, and microbiological attack.

The external loading conditions can be divided into two groups: dead load and environmental load. The dead load, in general, is also called a surcharge load, which is also called a *static load*. The environmental loads are mostly dynamic in nature. They can be violent, such as earthquake and blasting vibrations as discussed in Chapter 11. The internal factors include volume changes, shrinkage and swelling, and surface and internal cracking of soil mass, which consequently changes bond stress between soil particles and loss shear strength, bearing capacity, etc.

14.3 Slope failure phenomena and mechanisms

In most cases, while a landslide or slope failure may sometimes seem to occur suddenly, the underlying processes actually occur gradually or progressively. The associated phenomena include ground cracking, shrinking, erosion, surface creep, which then leads to surface slip and excessive settlement at the prefailure stage. When the slope soil reaches a certain level, such as from points **a** to **b** in Figure 14.1, the soil's internal resistance is no longer able to hold together due to the external loads and at that point the landslide or ground failure begins as shown from points **b** to **c**.

14.3.1 Phenomena of slope failure at prefailure stage

Prefailure phenomena of an earth slope as shown in Figure 14.1 includes surface erosion, creep, cracks and slip, etc. These phenomena are generally referred to as *progressive failures* or types of failure phenomena in which the ultimate shearing resistance is progressively mobilized along the failure surface. Progressive failures related to landslides and surface soil erosion have been recognized by geologists and agricultural scientists since the early days and for geotechnical engineering which is considered for the design of various earthen structures. Since surface movement is

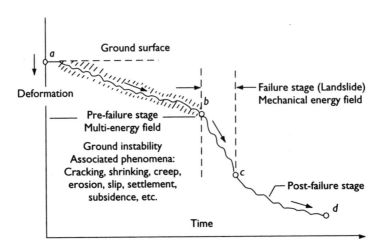

Figure 14.1 Prefailure and failure conditions of an earth slope.

always related with surface creep and landslides, many researchers have attempted to measure in situ ground creep rates. The rate of these movements varies during the seasons of the year, and movements are often confined to the shallow ground surface soil layer. The rate increases as failure approaches and the actual time of a landslide can frequently be predicted by monitoring the ground surface movements such as (a) *surface erosion and creep*: surface erosion and creep are the major part of progressive failure phenomena; (b) *cracks and slip*: earth slope cracks due to wet–dry and freeze–thaw cycles; and (c) *settlement and subsidence*: when large cracks appear on the slope surface, settlement (Sec. 9.7) and/or ground surface subsidence (Sec. 16.6) occurs.

14.3.2 Mechanisms of slope failure

There are numerous mechanisms of earth slope failure that have been suggested. Figure 14.2(a) and (b) illustrate the mechanics of slope failure in general. Figure 14.2(a) presents the forces acting on a wedge section, and Figure 14.2(b) is the force diagram of the wedge section. In examining Figure 14.2(a), the slope failure plane **a-b** may be a straight line, circular arc, logarithmic-spiral, or irregular pattern. Regardless of the type of possible failure surface it may be assumed that the general concept of how the slope will fail is virtually the same. **W** is the weight of soil of the wedge, **S** is the shear strength of soil along the failure surface **a-b**, and **R** is the resultant with angle ϕ, the angle of internal friction. During the structure's lifetime, the weight of soil, **W**, may change slightly according to variations in the degree of saturation as influenced by the weather. However, the shear strength, S, of soil can change dramatically as discussed in Section 10.4. Following are some possible failure mechanisms proposed to explain the slope failure mechanism along the failure plane.

1 *Mechanical–physical concept*: The slope failure mechanism has been explained with a mechanical–physical concept by Culmann in 1866, Resal in 1910, and many others. This approach considers the applied stress (i.e. from self-weight and surcharge) relative to the strength along some assumed failure plane. Terzaghi

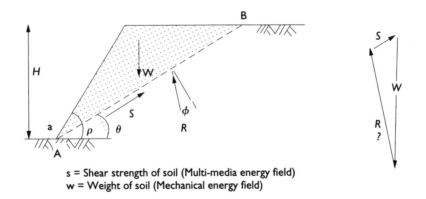

s = Shear strength of soil (Multi-media energy field)
w = Weight of soil (Mechanical energy field)

Figure 14.2 Slope failure mechanism. (a) Forces acting on earth slope; and (b) Force diagram.

(1943) considered slope failure to be similar to a slaking process and explained the process in terms of mechanical energy considerations.

2   *Physicochemical concept*: This approach considers slope failure in terms of physico-chemical concepts including (a) the mechanism of water attack on cohesive soil system, (b) electrical causes (Sec. 6.8), and (c) ion exchange effects. These explanations are discussed in previous sections. Others, as discussed in Section 4.7 are the ion exchange effect reported by Seifert *et al.* (1935) and Matsuo (1957).

3   *Linear elastic fracture mechanics concept*: Fang (1994) used the concept of linear elastic fracture mechanics (LEFM) to explain the mechanism of slope failure reflected from the cracking and fracture behavior of soil mass as discussed in Section 8.8.

4   *Particle-energy-field theory*: Fang (1997) have used the particle-energy-field theory to explain slope failures and the underlying mechanism as a function of various energy fields. There are three types of energies or mechanisms involved in a single landslide action, namely potential, kinetic, and mass transport phenomena, as shown in Figure 14.3. Potential energy is manifested by the weight of soil and moisture prior to movement. Once movement occurs (as reflected by rotation/translation of the entire soil mass or by percolation/infiltration of moisture through the soil), the process is characterized by kinetic energy. Mass transport phenomena describe the movement of dissolved ions that move within the pore fluid (moisture). Depending on local variations of the type and concentration of ions, part of the soil matrix may become more or less susceptible to a slope failure. This is because strength in soil is a function of the extent to which forces may be distributed through soil particles, and ionic composition influences the nature and orientation of particle to particle interaction, as discussed in Chapter 3.

14.4   Slope stability analysis methods

14.4.1   General discussion

The first major contribution on the stability of earth slopes was made by Collin in 1846. There are numerous methods currently available for performing the slope stability analysis. The majority of these methods may be classified as limit equilibrium

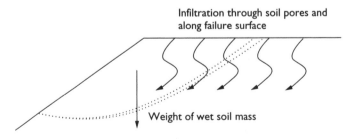

Figure 14.3 Slope failure considerations in terms of potential energy (before movement), kinetic energy (after movement), and mass transport phenomena (dissolved ions within pore fluid).

and limit analysis methods. The limit equilibrium method is widely used at the present time due to its simplicity. There are numerous state-of-the-art publications concerning slope stability and landslides with these various aspects emphasized (Turner and Schuster, 1996).

14.4.2   Types and classification of slope stability analysis

1   *Characteristics of slope stability classifications*: There are several ways to classify slope stability analysis methods, including (a) classification based on fundamental concepts such as limit equilibrium and limit analysis; (b) classification based on types of failure surface such as straight line, circular arc, logarithmic-spiral, or irregular; and (c) classification based on energy field such as single or multimedia energy field analysis.

2   *Types of classification*: Slope stability analysis are primarily categorized according to either limit equilibrium or limit analysis approaches (a) The limit equilibrium approach covers straight-line failure plane including the Culmann method; circular arc failure surface which includes the Swedish circle method, Taylor ϕ-circle method, Bishop method, Paterson method and Haung method; and non-circular failure surface including logarithmic-spiral failure surface, and irregular failure surface; and (b) the limit analysis approach covers the straight-line failure surface (simple cut), and logarithmic-spiral failure surface.

14.4.3   Selection of strength parameters

There are various slope stability analysis procedure requirements for various strength parameters. For example, for short-term stability analysis, the total strength (Sec. 10.3) is needed. However, for long-term stability analysis, the effective strength (Sections 5.5 and 10.3) is required. For stability analysis on overconsolidated clay deposit, the residual shear strength (Sec. 10.13) is suggested. Therefore, the selection of strength parameter is an important part of slope stability analysis procedure as discussed in Ch. 14.

14.4.4   Factor of safety

As discussed in Section 12.4, the factor of safety or degree of safety is used by engineers to indicate the extent to which the resisting forces exceed the driving forces for failure, or the ratio of available strength to required strength. This can be expressed in terms of shear strength, the components of shear strength (c, ϕ), moments, and heights. The present concept for determining the factor of safety for a slope is based on Coulomb's law (Sec. 10.3), and the factor of safety is the ratio of available shear strength to the required shear strength.

$$F_s = \frac{S}{\tau} \tag{14.1}$$

or

$$F_s = \frac{\tan \phi}{\tan \phi_c} \quad \text{and} \quad \frac{H_c}{H} \tag{14.2}$$

Table 14.1 Recommended factors of safety for slope stability analysis in residual region[a]

Class	Cutting type	Factor of safety	
		(A) Comprehensive site investigation[a]	(B) Cursory site Investigation[b]
1	Road cutting or cutting in remote area where probability of life at risk, owing to failure, is small	1.1	1.2
2	Road cutting on main arterial route where main line communications can be cut and risk to life is possible	1.2	1.3
3	Areas adjacent to buildings where failure would affect stability of building, e.g. car park. Risk to life significant	1.2	1.4
4	Cuts adjacent to buildings where failure could result in collapse of building. Risk to life very great	1.4	Not applicable

Source: Binnie and Partners (1971); Chiang (1979).

Notes
a Such a site investigation would, in addition to normal boring and drilling, include a program of laboratory testing to determine shear strength parameters for both soils and rock failures. Joint system surveys would be carried out and likely effects of heavy rainfall on the slopes would also be considered. These effects would be included in the soils and rock stability analyses.
b Site investigation under such a classification would be limited to determination of the boundaries of the various grades of material, the type of rock, and also predominant joint patterns in the case of rock stabilibty problems. Shear strength parameters would be derived from back-analysis of failures.

where F_s = factor of safety, S = available shear strength of soil, τ = required shear strength of soil, ϕ = internal friction angle of soil, ϕ_c = critical internal friction angle of soil, H = height of slope, and H_c = critical height of slope. Note the critical height of the slope is the maximum height at which a slope remains stable, while the critical friction angle refers to required friction angle. The factor of safety can also be obtained from practical experience as illustrated in Table 14.1, which provides some guidance in selecting the appropriate factor of safety for slope stability analysis.

14.5 Culmann method – straight line failure plane

14.5.1 General discussion

The Culmann method developed in 1866 represents a typical limit equilibrium solution. It assumes the whole wedge section as a free body. The method assumes that failure occurs on a plane (straight line) passing through the toe of the earth slope. The Culmann failure mechanism is as shown in Figure 14.2. In examining Figure 14.2, W = weight of soil in the wedge, S = total cohesion along the failure plane AB, β = slope angles, R = result force necessary to hold wedge in equilibrium, H = height of the earth slope, and ϕ = friction angle of the soil.

14.5.2 Failure mechanism and stability factor

From the geometrical relationships shown in Figure 14.2, the weight of soil in the wedge is

$$W = \tfrac{1}{2}\, \gamma L\, H \csc \rho \sin (\rho - \theta) \tag{14.3}$$

where γ = unit weight of the soil, L = length of the failure plane AB; H = height of embankment; and ρ = slope angle. If c is the unit cohesion, then the total cohesion, C, is

$$C = c\, L \tag{14.4}$$

where L = length of failure plane AB as indicated in Figure 14.2. Substitution of Equation (14.3) and (14.4) into the *Law of Sines* expressed for the force diagram in Figure 14.4(b) yields

$$\frac{\gamma H}{c} = \frac{2 \sin \rho \, \cos \phi}{\sin^2[(\rho - \phi)/2]} \tag{14.5}$$

The term $\gamma H/c$ in Equation (14.5) is a dimensionless expression called the *stability factor*, or *stability number* (N_s). The critical stability factor (most dangerous plane) may be obtained by minimizing the first derivative of the stability factor with respect to θ. This yields

$$N_s\,(\text{critical}) = \frac{4 \sin \theta \, \cos \phi}{1 - \cos (\theta - \phi)} \tag{14.6}$$

where N_s = stability factor, θ = slope angle, and ϕ = friction angle. Stability numbers larger than the critical stability number are likely to fail. The stability factor is also sometimes defined in reverse; that is, c divided by γH, in which case numbers smaller than the critical value are likely to fail.

14.5.3 Stability factor for a vertical cut

Many excavations involve the creation of a vertical cut. In such situations where $\rho = 90° = \pi/2$, Equation (14.6) becomes (using radians)

$$N_{s(\text{critical})} = \frac{\gamma H}{c} = \frac{4 \cos \phi}{1 - \sin \phi} = 4 \tan \left(\frac{\pi}{4} + \frac{\phi}{2} \right) \tag{14.7}$$

It is often useful to compute the height of a cut, beyond which failure may occur. This *critical height* of an earth slope, H_c, may be given as (using degrees)

$$H_c = \frac{N_s c}{\gamma} = \frac{4c}{\gamma} \tan \left(45 + \frac{\phi}{2} \right) \tag{14.8}$$

In the case of soft clay where ϕ may be taken as zero, the critical height may be taken as

$$H_c = N_s \frac{c}{\gamma} \tag{14.9}$$

Notwithstanding assumptions that often conflict with field conditions, the Culmann method has been widely and successfully used for slope stability analyses because of its simplicity.

EXAMPLE 14.1
A vertical trench 18 ft (5.50 m) deep is to be constructed in soft clay having a shear strength of 500 psf (24 kPa) and a unit weight of 112 pcf (17.6 kN/m³). What is the maximum safe depth of cut that can be made without bracing?

SOLUTION
From Equation (14.7),

$$N_{s(critical)} = \frac{4\cos\phi}{1 - \sin\phi} = \frac{4\cos(0)}{1 - \sin(0)} = \frac{4}{1} = 4$$

$$H_c = N_s \frac{c}{\gamma} = 4 \cdot \frac{500\ \text{lb/ft}^2}{112\ \text{lb/ft}^3} = 17.9\ \text{ft}$$

So, theoretically the maximum depth is about 18 ft (5.5 m), however, a factor of safety of at least 1.5 would ordinarily be applied, reducing this depth by 17.9 / 1.5 to about 11.9 ft (3.6 m).

14.6 Limit equilibrium method – circular arc failure surface

14.6.1 Swedish circle method (method of slices)

The *Swedish circle method*, also called the *method of slices*, was developed by Fellenius of the Swedish Geological Institute in 1927. It considers an earth slope with a failure surface defined by an arc of a larger circle which is then divided into equal slices as shown in Figure 14.4. Each slice is then analyzed for equilibrium. This approach was inspired by an assessment of many slope failures in Sweden, where the failure plane assumed an arc shape. To simplify an otherwise statically indeterminate problem, it is assumed that the forces acting on the sides of each slice (from adjoining slices) have zero resultant force in the direction perpendicular to the failure arc. The resulting equation for the factor of safety is

$$F = \frac{c'L + \tan\phi'\sum_{i=1}^{n}(W_i\cos\alpha_i - u_i\Delta L_i)}{\sum_{i=1}^{n}W_i\sin\alpha_i} \tag{14.10}$$

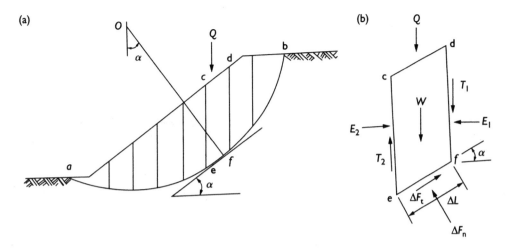

Figure 14.4 Circular failure surface and method of slices. (a) Cross-section of failure circle; (b) Force diagram.

where c' = effective cohesion, L = length of the entire failure arc, ϕ' = effective friction angle of soil, α_i = the angle between slice i and the horizontal, u_i = the pore pressure for a given slice, ΔL_i = the length of slice i, and W = weight of slice i. While used extensively because of its simplicity and history, use of Equation (14.10) may result in factors of safety that are 10–60% less than reported by other methods (Lambe and Whitman, 1979).

14.6.2   Bishop method of slices

1   General discussion: The Bishop method is similar to the Swedish circle method and differs primarily according to the direction over which forces are considered. In particular, the resultant of side forces is assumed to act in the horizontal direction with zero magnitude in the vertical direction. If a slope consists of several types of material with different values of c and ϕ, and if the pore pressures, u, in the slope are known or can be estimated, the Bishop method of slices (Bishop, 1955) is useful. From Figure 14.4, the mass of soil, acdbfe, is divided into vertical slices. The forces acting on each slice are evaluated individually on the basis of limit equilibrium. As before, the equilibrium of the entire mass is determined by summation of the forces on all the slices. Consider the forces on an individual slice cdfe, as shown in Figure 14.4. They consist of the weight of the slice, W, the surface load acting on the slice, Q, the normal and shear forces, F_n and F_t, acting on the failure surface, ef, and the normal and shear forces, E_1, T_1, E_2, and T_2, on the vertical faces, cdfe. The system is again statically indeterminate, and it is necessary to make certain assumptions regarding the magnitudes and points of application of the forces, E and T.

2   Bishop short-hand procedure: The Bishop short-hand procedure is commonly used to determine the factor of safety. Consideration of the above noted forces results in the following equation for the factor of safety:

$$F = \frac{\sum_{i=1}^{n} c' \Delta L_i + \left[W_i \cos \alpha_i - u_i \Delta L_i \right] \tan \phi'}{\sum_{i=1}^{n} W_i \sin \alpha_i} \tag{14.11}$$

where all variables are as defined previously. In Figure 14.4, an additional surcharge load Q is shown, and the analysis is similar except that it is added to the W term, that is $(W + Q)$.

3 *Bishop long-hand procedure:* The accuracy of the analysis may be improved by taking forces, E and T, as shown in Figure 14.4(b) into consideration. For the slice in Figure 14.4(b), the summation of forces in the vertical direction gives

$$\Delta F_n \cos \alpha = (W + Q) + (T_1 - T_2) - u \Delta L \cos \alpha - \Delta F_t \sin \alpha \tag{14.12}$$

The factor of safety, F, is then found through the following equation:

$$F = \sum_{i=1}^{n} c' \Delta L_i \cos \alpha_i + \left[(W_i - u_i \Delta L_i \cos \alpha_i) + (T_1 - T_2) \right] \tan \phi'$$
$$\times \left[\cos \alpha_i + \left(\tan \phi' \sin \frac{\alpha_i}{F} \right) \right]^{-1} \Bigg/ \left[\sum_{i=1}^{n} W_i \sin \alpha_i \right] \tag{14.13}$$

The factor of safety, F, from Equation (14.13) is found through successive approximation of the quantity $T_1 - T_2$. Trial values of E_1 and T_1 to maintain equilibrium of each slice, and the conditions $\Sigma(E_1 - E_2) = 0$, $\Sigma(T_1 - T_2) = 0$ are used. The calculation is reduced if the term $\Sigma(T_1 - T_2) \tan \phi$ is assumed to be 0. Next, an arbitrarily selected value of F is used to start the iteration procedure. This assumed value of F is placed where it first appears in the numerator of Equation (14.13) and the equation is solved for a new value of F, together with the soil properties c, ϕ, u, and the slope geometry α. If the calculated value differs appreciably from the assumed value, a second approximation is made and the computation is repeated. A chart developed by Janbu *et al.* (1956) helps to simplify the computation procedure.

Bishop (1955) claimed that the above approximation taking $\Sigma(T_1 - T_2) \tan \phi$ as 0 results in an error of only about 1%. The error introduced by using Equation (14.11) is about 15%. Thus Equation (14.13) is recommended for use. The calculations outlined above refer to only a one trial circle. Several circles must be analyzed until the minimum value of factor of safety is determined. Hand calculations, graphical methods, and computer programs may be used.

14.6.3 Taylor method (friction circle method)

The Taylor method is based on the friction circle method (Taylor, 1937, 1948) which is illustrated by the diagram shown in Figure 14.5. The radius of the circular failure surface is designated by R. The radius of the friction circle is equal to $R \sin \phi'$. Any line tangent to the friction circle must intersect the circular failure arc at an oblique

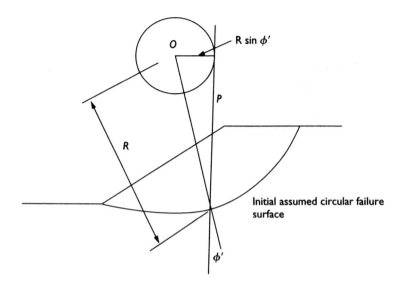

Figure 14.5 Taylor's friction circle method.

angle, ϕ'. Therefore, any vector representing an intergranular pressure at the angle ϕ' to an element of the failure surface must be a tangent to the friction circle. Similar to the previous methods, the failure surface is divided into segments and a trial and error solution is used to find the factor of safety. The details of this method may be found elsewhere (Lambe and Whitman, 1979; Murthy, 2002). Another development by Taylor is the *stability factor*, N_s, a pure number, depending only on the slope angle, β, and friction angle of soil, ϕ. This has been defined previously for vertical cuts and given as Equation (14.9). The relationships between N_s, β, and ϕ are shown in Figure 14.6. This method is based on total stresses and assumes that the cohesion, c, is constant with depth. Use of Figure 14.6 extends the applicability beyond vertical cuts and allows for slopes of varying angles to be analyzed.

EXAMPLE 14.2
An embankment has a height of 30 ft (9.1 m). The soil properties are cohesion equals 800 psf (38.3 kN/m^2), friction angle equals 25°, and the unit weight of soil is 122 pcf (19.2 kN/m^3). Find the slope angle, which corresponds to a factor of safety of 2.0.

SOLUTION
The given height, $H = 30$ ft. For a factor of safety $= 2.0$, the critical height is given by Equation (14.2):

$$H_c = F \cdot H = 2.0 \cdot 30 \text{ ft} = 60 \text{ ft}.$$

The stability number, from Equation (14.9), is given by:

$$N_s = \frac{\gamma H_c}{c} = \frac{122 \text{ lb/ft}^3 \, 60 \text{ ft}}{800 \text{ lb/ft}^2} = 9.15 \approx 9$$

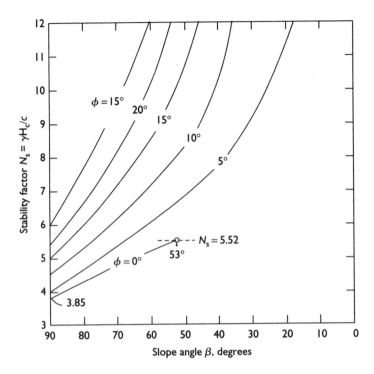

Figure 14.6 Stability factor with Taylor's method.

Then, by consulting Figure 14.6 with $N_s = 9$ and $\phi = 25°$, a slope angle of approximately 70° or a 1:3 horizontal to vertical slope is obtained. Note that lowering the required factor of safety results in steeper slopes up to a maximum 90° vertical cut.

14.6.4   Huang's method

Huang's method (1980) is based on the Swedish (Fellenius, 1927) and Bishop's (1955) method and also assumes a circular failure plane. Huang's method has the advantage of considering other factors such as (a) locating the most dangerous failure circles (failure planes), (b) analyzing multiple soil layers, (c) porewater pressure effect in each layer, and (d) seismic effects.

1   *Locating the potential center of the failure circle*: The procedure for locating the most dangerous failure surface is shown in Figure 14.7. In the figure, the height of slope, H, and a slope $S:1$ (horizontal : vertical) are given. Let $ab = a'b' = 0.1$ SH. The empirical assumption that $ab = 0.1\ SH$ is based on field data but yields good results. Point o is the intersecting point of lines of aa' and bb'. The center of potential failure circle must be along the line oo, and the circle must pass through points a and a'. By trial, a most dangerous failure circle is drawn. The vertical distance YH can be measured graphically. The value of **Y** is needed for computing other parameters as shown in Figure 14.7.

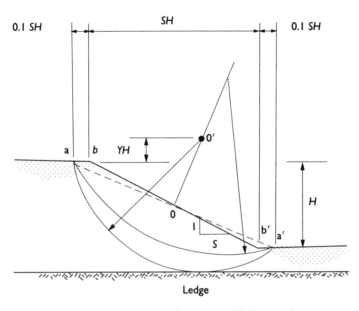

Figure 14.7 Procedures for locating the center of a potential failure circle in a typical earth slope.

Source: Huang, Y. H., Stability charts for effective stress analysis of nonhomogeneous embankments. In *Transportation Research Record* No. 749, Transportation Research Board. National Research Council, Washington DC, 1980, pp. 72–74. Reproduced with permission of the Transportation Research Board.

2 *Computing of the factor of safety with seismic effects*: Slopes can be particularly sensitive to earthquakes and other sources of ground movement. Huang's method may be used in such cases as follows. When a failure circle is determined (Fig 14.7), the average shear stress developed along the failure surface can be calculated by consideration of the moments. The moment at the center of a circle due to both the weight of the sliding mass and the corresponding seismic force is equated to that due to the average shear stress distributed uniformly over the failure arc. The amount of shear stress which develops is proportional to the unit weight of the soil and the height of the slope. The average shear strength along the failure surface is also a function of the unit weight of soil and slope height. The factor of safety is the ratio between the shear strength and the shear stress and is given by

$$F = \frac{c'/\gamma H + (1 - r_u) \tan \phi'/N_f}{1/N_s + C_s/N_e} \tag{14.14}$$

where F_s = factor of safety, c' = effective cohesion, γ = unit weight of soil, H = height of slope, r_u = pore pressure ratio (ratio of porewater pressure to overburden pressure), ϕ' = effective angle of internal friction, N_f = friction number (Fig. 14.8), N_s = stability number (Fig. 14.8), C_s = seismic coefficient (ratio of seismic force to the weight of structure), and N_e = earthquake number (Fig. 14.8). Equation (14.14) shows that the factor of safety depends on four geometric parameters (H, N_s, N_f, and N_e) and three soil parameters (r_u, c', and ϕ').

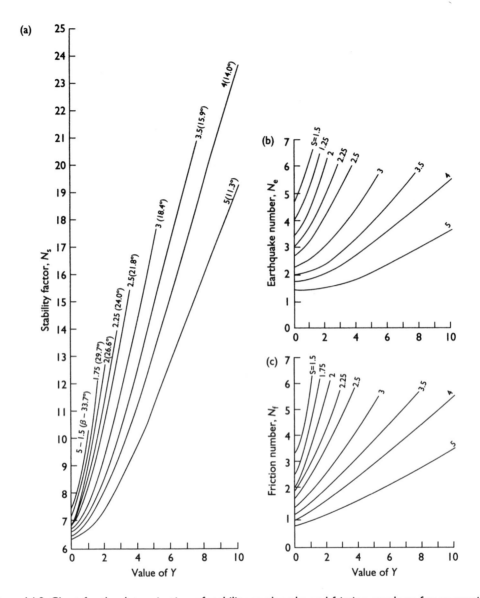

Figure 14.8 Chart for the determination of stability, earthquake and friction numbers for computing the factor of safety in slope stability analysis. Parameter, Y, used in these charts is determined from Figure 14.7 (a) Stability factor; (b) Earthquake factor; and (c) Friction factor.

Source: Huang, Y. H., Stability charts for effective stress analysis of nonhomogeneous embankments. In *Transportation Research Record* No. 749, Transportation Research Board. National Research Council, Washington, DC, 1980, pp. 72–74. Reproduced with permission of the Transportation Research Board.

EXAMPLE 14.3   (After Huang, 1980)
Figure 14.9 shows a 2.5:1 slope, 20 m (65.6 ft) high, composed of three different soil layers. The soil data including c', ϕ', and γ, as well as the location of the groundwater table, are given. Assuming a seismic coefficient of 0.1, determine both the static and the seismic factors of safety.

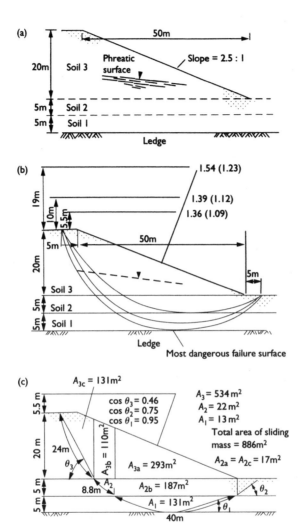

Figure 14.9 Example problem for Huang's method of slope stability analysis.

SOLUTION

1   First locate the most dangerous failure surface as shown in Figure 14.9(b) and determine the distance YH with Figure 14.9(c)

YH = 5.5 m,

$$Y = \frac{5.5 \text{ m}}{20 \text{ m}} = 0.275$$

2   Determine N_s, N_f, and N_e from Figure 14.8 with $S = 2.5$ (given) and $Y = 0.275$:

$N_s = 7.0,$ $N_f = 2.0,$ $N_e = 2.8$

3   Determine the average unit weight of soil:

$$\gamma = \frac{(131)(18) + (221)(19) + (534)(20)}{131 + 221 + 534} = 19.5 \text{ kN/m}^3 \text{ (124.2 pcf)}$$

4   Determine the average effective cohesion, c':
Measure the length of the failure arc through soils 1, 2, and 3 and note the lengths to be 40 m (12.2 ft), 17.6 m (5.4 ft), and 24 m (7.3 ft), respectively. Then the average effective cohesion, c' is given by

$$c' = \frac{(40)(5) + (176)(7.5) + (24)(10)}{40 + 176 + 24} = 7.0 \text{ kPa (1.0 psi)}$$

5   Determine the average coefficient of friction, $\tan \phi'$:

Friction is developed from the component of weight of overlying soil that is normal to the failure surface. The necessary $\cos \theta$ values are given in Figure 14.9(c). A weight may be computed for each layer as

$W_1 = (131 \cdot 18 + 187 \cdot 19 + 293 \cdot 20) \cdot 0.95 = 11{,}182 \text{ kN/m}$
$W_2 = (2 \cdot 17 \cdot 19 + 110 \cdot 20) \cdot 0.75 = 2{,}135 \text{ kN/m}$
$W_3 = (131 \cdot 20) \cdot 0.46 = 1{,}205 \text{ kN/m}$

The average value for $\tan\phi'$ may then be given as

$$\tan \phi' = \frac{(11{,}182)(\tan 25) + (2135)(\tan 30) + (1205)(\tan 35)}{11{,}182 + 2135 + 1205} = 0.502$$

6   Determine the average pore pressure, r_u:

$$r_u = \frac{A_{sw}\gamma_w}{A_t\gamma}$$

where A_{sw} and A_t = the area of sliding mass under water and total area of sliding mass, respectively, and γ_w and γ = the unit weight of the water and the average unit weight of the soil, respectively. If A_{sw} was measured to be 527 m², then r_u is given by

$$r_u = \frac{527 \cdot 9.8}{886 \cdot 19.5} = 0.299$$

7   With all of the above values now calculated, the factor of safety, F, may be computed from Equation (14.14)

(a) Static factor of safety

$$F = \frac{(7/19.5 \cdot 20) + (1 - 0.299) \cdot 0.502)/2.0}{(1/7) + (0/2.8)} = 1.36$$

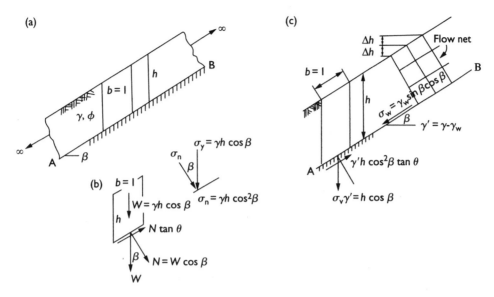

Figure 14.10 Cross-sections and free-body diagrams of infinite earth slope of a cohesionless soil. (a) Infinite slope; (b) General free-body of slope element; (C) Infinite slope with seepage forces present.

(b) Seismic factor of safety

$$F = \frac{(7/19.5 \cdot 20) + (1 - 0.299) \cdot 0.502/2.0}{(1/7) + (0.1/2.8)} = 1.09$$

As such, the influence of seismic activity is to reduce the factor of safety.

14.7   Infinite earth slopes

The infinite slope is a constant slope that, while in reality has some finite length, may be considered infinite if the failure plane is parallel to the slope (contrast with the circular planes described above) and if the depth to the failure plane is small relative to the slope height. Typically, uniform soil properties under constant environmental conditions are assumed. A typical cross-section of an infinite slope is shown in Figures 14.10 and 14.11. The subsurface soil may be homogeneous but may consist of variable strata of different soils as long as all strata boundaries are parallel to the surface of the slope. The concept was proposed by Taylor (1948) for stability analysis of natural earth slopes. It has been extended for the analysis of seafloor slope stability of marine deposits. There are two general cases: cohesionless and cohesive soils, each of which may be considered with and without seepage forces.

1   Cohesionless soils: The cross-section and free-body diagram of an infinite earth slope without seepage force is illustrated in Figure 14.10(a). The driving force for failure is the weight of the soil, while the resistance is derived from the shear strength at the

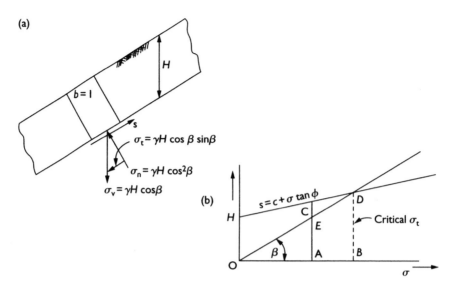

Figure 14.11 Cross-sections and free-body diagrams of finite slopes in cohesive soil. (a) Infinite cohesive slope; (b) Mohr's circle for stress conditions of (a). For σ_t = AE, slope has F > I; for σ_t = BD, slope has F = I, σ_t = tangential stress.

failure plane. Considering the geometry shown, the normal stress imparted by the weight may be written as

$$\sigma_n' = \gamma h \cos \beta \cos \beta = \gamma h \cos^2 \beta \tag{14.15}$$

Similarly, the shear stress may given as

$$\tau = \gamma h \cos \beta \sin \beta \tag{14.16}$$

At failure, the above noted shear stress (Eq. 14.16) will be equal to the shear strength as defined by Coulomb's law (Ch. 10). By equating the shear stress to the shear strength and incorporating the equation for the normal stress (Eq. 14.15), we have

$$\tau \equiv \sigma_n' \tan \phi' = \gamma h \cos \beta \sin \beta = \gamma h \cos^2 \beta \tan \phi' \tag{14.17}$$

From the above equation, the γh terms divide out, a $\cos \beta$ term is remains on the right hand side and a $\sin \beta$ term remains on the left hand side. The equation reduces to

$$\tan \beta = \tan \phi' \tag{14.18}$$

As may be observed, the stability of infinite slopes of cohesionless soils may be assessed through comparing the effective friction angle of the soil to the slope angle and the factor of safety may be simply stated as

$$F = \frac{\tan \phi'}{\tan \beta} \tag{14.19}$$

The above equation applies regardless of whether the soil is dry or totally submerged (as in the slope of a lake or ocean). If however water is flowing through the slope, then the relationship must be modified to consider seepage-induced pore pressures as shown in Figure 14.10(b). The buoyant unit weight γ_b (Ch. 3) may be introduced to account for these pressures relative to the total unit weight γ_t with the factor of safety given by

$$F = \frac{\gamma_b/\gamma_t \tan \phi'}{\tan \beta} \tag{14.20}$$

Considering Equation (14.20) and the fact that buoyant unit weight is approximately half that of total unit weight, slopes without seepage may be about twice as steep as those with seepage for the same factor of safety.

2 *Cohesive soil:* By definition, cohesive soils have the added benefit of cohesion to resist the shear stresses imposed by the weight of slopes. In examining Figure 14.11(a), we have

$$\sigma_n = \gamma\, h\, \cos^2\beta \tag{14.21}$$

$$\tau = \gamma\, h\, \sin \beta \cos \beta \tag{14.22}$$

As before, Coulomb's failure criteria indicates that the shear strength may be given as

$$s = c_d + \sigma \tan \phi_d \tag{14.23}$$

where s = resisting shear strength, c_d = required (design) cohesion, and ϕ_d = required (design) friction angle.

Substitution of Equations (14.21) and (14.22) into Equation (14.23), we obtain

$$\gamma\, h\, \sin \beta \cos \beta = c_d + \gamma\, h\, \cos^2\beta \tan \phi_d \tag{14.24}$$

or the design cohesion, c_d, is

$$c_d = h \cos^2 \beta\, (\tan \beta - \tan \phi_d) \tag{14.25}$$

The critical value of clay thickness (height) is

$$H = \frac{c_d}{\gamma} \frac{\sec^2 \beta}{\tan \beta - \tan \phi_d} \tag{14.26}$$

As discussed in Section 14.5.2, the stability factor can be used to describe overall stability in terms of the cohesion, unit weight and slope height. In this particular case, the stability factor is given as

$$N_s = \frac{C_d}{\gamma H} = \cos^2 \beta\, (\tan b - \tan \phi_d \tag{14.27}$$

Note that stability factors are sometimes written with the numerators and denominators being switched, as above. To avoid confusion in calculating factors of safety, remember that cohesion is the "resisting" force while the unit weight and slope height are the "driving" forces for failure. When seepage forces are involved, the stability numbers are modified for the buoyant unit weight, as shown by Equation (14.28):

$$N_s = \cos^2\beta \left(\tan\beta - \frac{\gamma_b}{\gamma_t}\tan\phi_d\right) \tag{14.28}$$

14.8 Earthquake loading effects – limit equilibrium solutions

Earthquakes trigger the failure of earth slopes frequently. There are several approaches proposed by various investigators. Among these, Huang's method was discussed in Section 14.6.4. Koppula's method (1984) applies to the stability analysis of slopes in cohesive soils. This method assumes some nonzero value of shear strength at the ground surface as well as a linear increase with depth. The effect of an earthquake is analyzed by treating the earthquake loading as an equivalent horizontal force. The factor of safety, F, is defined as the ratio of the resisting moment to the driving moment and is given by Equation (14.29):

$$F = N_1\frac{a_0}{\gamma} + N_2\frac{c_0}{\gamma H} \tag{14.29}$$

where F = factor of safety, N_1 and N_2 = stability factor, γ = unit weight of soil, H = height of slope, and a_0, c_0 = constants used to express the relationship between the strength of the soil with depth, given by

$$C = c_0 + a_0 z \tag{14.30}$$

where C = shear strength of soil at depth z below the ground surface, c_0 = shear strength of soil at ground surface, and a_0 = gradient at which the soil strength varies with depth. The above equations are used in conjunction with charts developed to relate the stability numbers N_1 and N_2 to the slope angle as a function of an earthquake's horizontal acceleration, A, as shown in Figures 14.12, 14.13, and 14.14. Further explanation is given in Examples 14.4 and 14.5.

EXAMPLE 14.4   (*After* Koppula, 1984)
Consider a cohesive slope of height, H inclined at 60° to the horizontal. Let the shear strength of the soil be given by $a_0/\gamma = 0.02$ and $c_0/\gamma H = 0.3$. Determine the factor of safety, F_s, when the seismic coefficient $A = 0$ (no earthquake) and $A = 0.4\,g$ (strong earthquake), where g = the acceleration due to gravity (9.81 m/s²).

SOLUTION
When $\beta = 60°$ and the seismic coefficient $A = 0$, from Figure 14.12 we obtain the stability factor $N_1 = 3.2$, and from Figure 14.13 we obtain the stability factor $N_2 = 5.3$. The factor of safety F, can be computed from Equation (14.29) as

$$F = (3.2)(0.02) + (5.3)(0.3) = 1.65$$

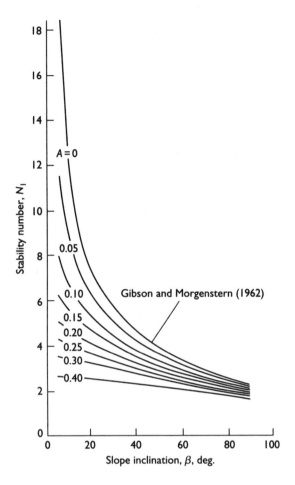

Figure 14.12 Relationship between stability number N_1, slope inclination β and seismic coefficient A.
Source: Koppula (1984). Copyright ASTM International. Reprinted with permission.

When the seismic activity is increased to 0.4 g, the values of N_1 and N_2 are found to be 2.0 and 2.75, respectively, and F is calculated as

$$F = (2.0)(0.02) + (2.75)(0.3) = 0.87$$

14.9 Slope stability problems solved by limit analysis methods

The emphasis of this chapter has been placed on limit equilibrium based methods, which tend to be the most commonly used. However, the limit equilibrium approach neglects the relationship between stress and strain in soil (Ch. 10). The limit *analysis* method includes direct consideration of the stress–strain relationship. The method was first introduced to the earth slope stability problem by Drucker and

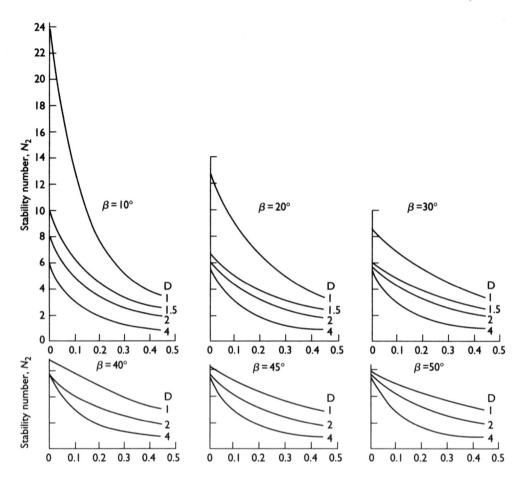

Figure 14.13 Relationship between stability number N_2 and seismic coefficient A for various slope inclinations β.

Source: Koppula (1984). Copyringht ASTM INTERNATIONAL. Reprinted with permission.

Prager in 1952. Many additions and refinements have been made since the initial introduction by Snitbahn *et al.* (1975) and others. The major advantage of this method is that it generally provides a closed-form mathematical solution and a clear picture of the failure mechanism. Two general types of failure planes exist: the straight-line and the logarithmic-spiral type. These may be considered in connection with various environmental conditions, seismic conditions and with soil heterogeneity. Figure 14.15 shows a straight-line failure plane while Figure 14.15 represents a logarithmic-spiral failure plane as analyzed by limit analysis. A comparison of the factors of safety computed on the basis of limit equilibrium and limit analysis is given in Table 14.2. Note that the values are quite similar and virtually the same in many cases. Limit analysis is beyond the scope of this introductory text and the interested reader is referred to Fang and Mikroudis (1991) for more details and examples.

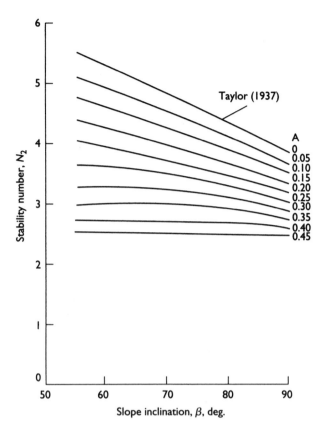

Figure 14.14 Relationship between stabililty number N_2, slope inclination β ($\gtrsim 55°$) and seismic coefficient A.

Source: Koppula (1984). Copyright ASTM INTERNATIONAL. Reprinted with permissions.

14.10 Environmental effects on slope failures and landslides

14.10.1 Rainfall and rainstorm

Soil erosion is caused by the drag action, especially rainfall on the surface of bare or unprotected soil surfaces. It involves a process of both particle detachment and transport. It has been found that the amount, intensity, and distribution of rain upon the soil, and the amount and velocity of runoff are related to soil erosion. If the intensity of rain is low, the total rainfall may not cause excessive erosion. Intense rain of extremely short duration may not cause much soil loss, but a combination of long duration and high intensity in a given rainfall will seriously affect runoff, surface erosion, and slope instability. Further discussion on soil erosion caused by water and rainfall will be presented in Section 16.5. In tropical regions, rainstorms with long duration and high intensity are a general occurrence. In urban environments, tree and vegetation cover is scarce, so the likelihood of a landslide or slip during a heavy rainstorm is great. Many case studies from Rio de Janeiro, Brazil, Bonaventura and

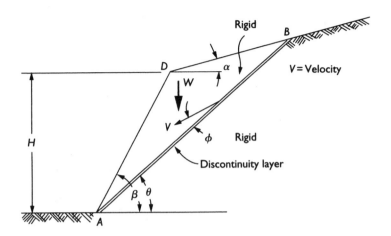

Figure 14.15 Straight-line plasticity failure mechanism – velocity field (upper-bound solution).

Table 14.2 Comparison of stability factor by limit equilibrium and limit analysis methods.

Slope angle β, degrees	Friction angle ϕ, degrees	Limit equilibrium				Limit analysis	
		Culmann	Fellenius slices	ϕ Circle	Log-spiral	Log-spiral	Straight line
90	0	4.00	3.83	3.83	3.83	3.83	4.00
	5	4.36	4.19	4.19	4.19	4.19	4.36
	15	5.20	5.02	5.02		5.02	5.20
	25	6.30	6.06	6.06	6.06	6.06	6.30
75	0	5.22	4.57	4.57	4.57	4.57	5.22
	5	5.85	5.13	5.13	5.14	5.13	5.85
	15	7.45	6.49	6.52		6.57	7.45
	25	9.80	8.48	8.54		8.59	9.80
60	0	6.95	5.24	5.24	5.24	5.25	6.95
	5	8.06	6.06	6.18	6.18	6.17	8.06
	15	11.30	8.33	8.63	8.63	8.64	11.30
	25	17.30	12.20	12.65	12.82	12.75	17.30
45	0	9.60	5.88	5.88	5.88	5.86	9.60
	5	12.00	7.09	7.36		7.33	12.00
	15	20.20	11.77	12.04		12.05	20.20
	25	43.50	20.83	22.73		22.95	43.50
30	0	14.90	6.41	6.41	6.41	6.51	14.90
	5	21.20	8.77	9.09		9.17	21.20
	15	55.20	20.84	21.74		21.71	55.50
	25	500.00	83.34	111.1	125.0	120.0	500.00
15	0	30.40	6.90	6.90	6.90	7.35	30.40
	5	66.60	14.71	14.71	14.71	14.80	66.60

Bucaramanga, Colombia, Hong Kong Island, etc. have been reported. Morganstern and de Matos (1975) reported a landslide pattern in a residual soil region. They classified the landslide pattern into three types of failure caused by rainfall as shown in Table 14.3.

Table 14.3 Landslide pattern in residual soil regions caused by rainfall

Homogeneous slopes	Normally intense rainfalls	Planar slide
	Very intense rainfalls	Rotational slide
	Normally intense rainfalls	Avalanches (rapid flow)
Heterogeneous slopes	Very intense rainfalls	Rolling
		Complex slides

Source: Morgenstern and de Matos (1975).

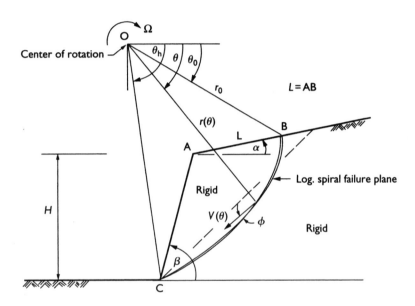

Figure 14.16 Failure mechanism for the stability of an embankment limit analysis method.

In general, landslide occurrences are limited to the upper zones approximately 3–7 m below the ground surface. The depth of a saturated zone or thickness of the wetting band can be estimated from the limiting rate of infiltration as shown in Equation (5.1) in Section 5.2. In order to understand how rainfall affects slope instability, the following example is used to illustrate the failure mechanism (Fig. 14.16). Figure 14.3 shows that the slope failure due to rainfall can be divided into two environmental zones: the mechanical energy and multimedia energy fields. For example, for the initial stage of rainfall, the flow movement in the soil mass is due to potential and kinetic energies and will follow Darcy's Law. Due to the complex soil–water interaction for fine-grained soil, the flow movement becomes a mass transport phenomenon as discussed in Chapter 5. In such a case, the flow movement should follow Fourier's Law or Ohm's Law as discussed in Chapter 6. Also, as discussed in Chapter 6 and Chapter 11, El Nino's effect causes heavy rain especially along the coastal regions. Heavy rainstorms trigger the landslides. Flood, wind, tornado (twister), and thunderstorm also cause landslides.

14.10.2 Pollution intrusion effects on slope stability

As discussed in Section 1.3, the characteristics of the acid rain and acid drainage water are such that slope stability problems may be exacerbated. In general, acid pore fluid will affect the following soil properties: (a) speeds up the ion exchange activity (Ch. 4), (b) causes decomposition processes, (c) increases the corrosion process, and (d) increases the geomorphic (aging) process (Ch. 4) of soil.

14.10.3 Seasonal effects, wet–dry and freeze–thaw cycles

1   *Seasonal effects include wet–dry and freeze–thaw cycles*: The mechanism of wet–dry and freeze–thaw cycles have been discussed in Chapter 6. These affects include bearing capacity, lateral earth pressure, settlement, shear strength, etc. For a slope stability analysis, the wet–dry and freeze–thaw cycles are more critical because of (a) loosening of the soil particle assemblage (increase in void ratio), (b) reduced bonding strength between soil particles, (c) surface water intrusion, (d) pollution intrusion, and (e) development of cracks.

2   *Wet–dry cycle cause mudflow*: As indicated in Figure 14.17(a), the potential failure surface is related with the depth of the wet–dry zone. This depth can be either measured in situ or estimated from Equation (5.1) as described in Chapter 5.

3   *Solifluction caused by freeze–thaw cycles*: Solifluction is a special form of creep caused in regions where the ground freezes (Sec. 6.6). In warm seasons the upper part of the mantle thaws (saturated), while the lower part remains solidly frozen. The saturated part of soil flows sluggishly under its own weight; this type of movement (landslide) is known as *solifluction* or called *soil-flowage* as illustrated in Figure 14.17(b). Commonly, it is the mixing of soil with coarse rock fragments to form a mass of debris referred to as a *debris flow*. The potential failure surface is related with the depth of the freeze–thaw zone. This depth can be either in situ measured or can be estimated by Equations (6.7) and (6.8) as described in Section 6.7.

14.10.4 Tree roots and wind on earth slope stability

1   *Vegetation and tree roots*: Vegetation and tree roots are used for earth reinforcement. However, there are some disadvantages as some tree roots affect the stability of earth slopes. The presence of vegetation and tree roots creates many channels for the conduction of free water in a soil mass. Patterns of subsurface flow are disturbed and boundary layers do not form in the soil. The type or pattern of tree branches and the patterns of tree roots are interrelated. Essentially, all trees or vegetation can have a deep or large distribution of roots. An approximate relationship between them has been proposed by Fang (1997) as follows:

$$(W_T) (H_T) (\gamma_T) (\sigma_T) \approx (W_R) (D_R) (\gamma_R) (\sigma_R) \tag{14.30}$$

where W_T, W_R = width of tree branches or roots, H_T = height of tree, γ_T, γ_R = density of tree branches or roots, and σ_T, σ_R = tensile strength of tree branches or roots. Tree roots can be used for the stabilizing of ground soil, but can also damage the soil structure and undermine the stability of earth slopes.

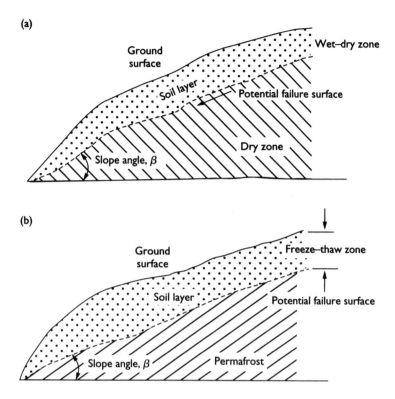

Figure 14.17 Seasonal effects on earth slope stability. (a) Landslide: wetting–drying cycle; and (b) Solidification: freezing–thawing cycle.

2 *Tree–wind interaction relating to the stability of earth slope:* The tree–wind system also affects the stability of earth slopes as illustrated in Figure 14.18. In examining Figure 14.18, the locations (a) through (h) are as follows: (a) earth slope, (b) potential failure plane which can be estimated from standard slope stability analysis as discussed in Section 14.4; (c) potential failure zone; (d) tree and associated branches above the ground surface, (e) roots below the ground surface, (f) the wind load, when the wind load acts on a tree, and a certain portion of soil layer will be disturbed through the tree roots as shown in (g) and (h). In such cases, the potential slope failure surface of the earth slope changes. In other words, tree roots can serve to enhance or worsen the stability of earth slopes, depending on the interplay between the tree, slope and failure plane as indicated in Figure 14.18.

14.10.5 Landslide on problematic soils and rocks

1 *Landslides on problematic soils:* Landslides occur in many parts of the world, especially in those areas with problematic soils and adverse environmental conditions. These problematic soils include residual soils, dispersive clay, and expansive clay and are extremely sensitive to water, exhibiting low shear strength and large volume change susceptibility as discussed in Chapter 2.

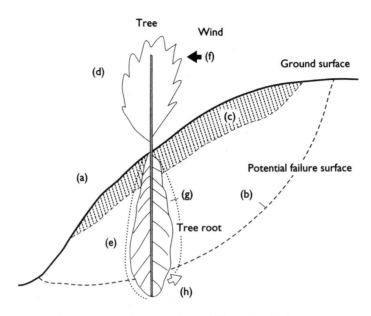

Figure 14.18 Tree–wind interaction relating to the stability of earth slopes.

2   *Landslides on overconsolidated clay deposits*: Overconsolidation is primarily
caused by mechanical loading although environmental factors such as acid rain,
acid drainage, and hazardous/toxic intrusion may also contribute. To solve landslide problems in overconsolidated clay deposits, one must start from the slope
failure causes and mechanisms. The overconsolidated pressure caused by loading
is greater than that caused by the environmental factors. The latter one will take
a longer time to produce an overconsolidated pressure. On the other hand, the
time required for the first-time slope failure indicates that the environmental
causes take longer. There are numerous studies on various aspects of overconsolidated clays including landslide and slope instability problems (TRB, 1995).

14.11   Mudflow and debris flow

14.11.1   Mudflow and mudslide

1   *General discussion: Mudflow*, also referred to as a *mudslide*, is a part of an ava-
lanche action and also is one of a special case of landslides or slope failure. Among
all slope failure types, the mechanism of mudflow is the most complex and least
understood, because it deals with fine-grained cohesive soil that is very sensitive to
the local environment, especially with water. *Mudslide* is a general term when mud is
in a unsaturated natural condition. When mud is saturated, this is referred as
mudflow. The mudflow is not necessarily composed exclusively of soil, as it is mixed
with some fine rock debris or gravel. In arid and semiarid regions, fine rock debris
becomes water-soaked on steep slopes after heavy rains, ultimately moving downward
as a mudflow. Table 14.4 presents major factors affecting or causing mudflow.
   Hurricanes', tornados', El Nino's, and La Nina's (Ch. 11) effects generally
trigger mudflows most of the time. El Nino's effects range from extreme droughts to

Table 14.4 Major factors affecting or causing mudslides

A	*Adverse weathering conditions* Rainfall, torrential rain, hurricane (typhoon), tornado, flood, flash flood, El Nino, and La Nina effects
B	*Problematic soils and rocks* Weathering rocks and residual soils, dispersive clays, expansive clays, and overconsolidated clay deposits
C	*Topographical locations and features* Flat hillside bare soil surface poor surface, and subsurface drainage systems

record-setting floods. For example, the extensive damages in 1999 from the series of California and North and South Carolina floods are related to El Nino's effects.

2   Causes and failure mechanisms of mudflow: Mudflows occur only under certain conditions. Commonly, it happens to certain types of fine-grained soil deposits such as residual soils and weathering rocks, dispersive clays, expansive soils, and overconsolidated clays. These soil deposits are very sensitive to the local environment and especially when combined with water. It occurs at specific topographic features such as relatively flat sloping hillsides and also under certain specific weather conditions such as heavy and intensive rainstorms. Rain affects mudflows in two ways. The first is to saturate the soil and reduce the adhesive ability and bonding strength between soil particles, and second, raindrops serve to relocate soil particles, moving them into a failure pattern. In addition, soil–rainwater interaction involves considerations of flow through porous media and mass transport phenomena, as discussed in Chapter 5. Therefore, the failure patterns and mechanisms relating to rainfall intensity of a mudflow are more complicated than simple circular arc or logarithmic-spiral failure surfaces commonly assumed and cannot be solved by the conventional limit equilibrium or limit analysis techniques. More sophisticated techniques, including finite element analysis and detailed in situ measurements, may provide more answers, although continued research is needed.

14.11.2   Debris flow and volcanic mudflow

Debris flow is a general term which covers not only mud but mud–rock mixtures as well. It occurs in general at a high altitude. There are three basic factors causing debris flow: (a) steep slopes at high altitude, (b) sufficient water resources surrounding the site, and (c) relatively loose rock pieces and soil. For evaluating a debris flow, climatic and geological conditions as well as surface vegetation and tree root type and distribution are particularly relevant. A volcanic mudflow is a special type of mudflow. This type of mudflow is characterized predominantly by fine-grained tephra. *Tephra* is a collective term for material that has been ejected from volcano, irrespective of size, shape, or composition.

14.11.3   Stability on landfill slopes

Stability analyses are a major part of the design procedure for waste containment facilities and may be accomplished by the techniques presented in this chapter, providing the strength parameters of the waste and/or other relevant lining/cover materials that are available. The importance of stability analysis was not always recognized, as historically stability problems were seen as "operational" problems to be solved by the responsible landfill owner or operator. However, concerns for groundwater quality, as potentially impacted by leachate production and escape, have led to considerable regulatory scrutiny over the design and monitoring of waste containment systems. Landfills have evolved into highly engineered structures, characterized by multiple layers of soil and geosynthetic material. These facilities are susceptible to a number of different rotational, translational, and sliding failure modes, including base failure of native soils as well as failure through the waste matrix itself. Given the multicomponent nature of virtually all landfill covers and liners, sliding failure is a particularly common concern. Sliding failure may occur in covers or liners, the most notable of which occurred at the Kettleman Hills Landfill as described by Seed *et al.* (1990) and Mitchell *et al.* (1990). This particular failure was developed by sliding along interfaces within a composite geosynthetic/compacted clay liner system and resulted in lateral and vertical displacements of up to 35 and 14 ft, respectively. The emerging trend of operating landfills as "bioreactors" wherein additional liquids are added to the waste matrix to accelerate decomposition places even more concern on landfill stability. The addition of liquids (i.e. water and/or leachate) results in increased unit weight and may reduce the effective stress as liquids fill the pore spaces within the waste. In particular, Isenberg *et al.* (2001) have demonstrated that the factors of safety for a landfill may be reduced by 75–80% by current inaccuracies in estimating shear strength alone. Given typical shear strength parameters, the authors note that this may result in unacceptable factors of safety in most designs, let alone any further reductions imposed by additional moisture. These concerns proved catastrophic for the Dona Juana Landfill in Bogota, Columbia, where excessive liquid injection is believed to have caused massive slope failure (Hendron *et al.* 1999).

14.12   Prevention, control, and remedial action on landslides

For a particular landslide or potential landslide, there is seldom one and only one method of treatment. In general, the most economically effective means of prevention consists of a combination of two or more methods. Some recommendations for preventing and controlling landslides include (a) minimizing the cutting of a hillside (reduce slope angle) in order to reduce the risk of creating instability of the slope. Use a series of terraces instead of one long slope; (b) using retaining structures, the minimum depth of which should be deeper than the possible failure surface; (c) both surface and subsurface drainage systems should be properly installed. Divert all surface water away from potential failures areas. Inspect the drainage system regularly. Install internal drainage, such as horizontal drain (Sec. 5.7) to release the porewater pressure in the soil mass; and (d) in the case of particularly sensitive zones where human life is in jeopardy, patrol on a 24 h basis all potential landslide areas during intensive rainfall and advise the immediate evacuation of the area in danger if necessary. Remedial actions may include (a) geometric methods including flattening slopes and pressure berms; (b) hydrological methods including surface drains, vertical sand drains, horizontal drains, and lowering of the groundwater level; (c) physicochemical methods including chemical grouting, soil stabilization, and thermal treatment; and (d) mechanical methods including compaction (Ch. 7), rock-bolts, piles (Sec. 15.12), toe walls, retaining walls (Ch. 13), sheet piling (Sec. 15.8), and reinforced earth (Sec. 15.9).

14.13   Summary

Slope stability and landslides belong to one system and landslides are the result of slope instability. While a slope failure may appear suddenly and without obvious warning, the underlying mechanisms tend to occur gradually or progressively. Slope stability problems are generally assessed in terms of a factor of safety. The driving force for failure is a function of the weight of the soil while the resisting forces are derived from the friction angle and/or cohesion of the soil along an assumed failure plane. Emphasis has been placed on limit equilibrium methods where stability is considered through a force balance. Limit analysis methods, however, may be used to capture the behavior of the stress–strain relationships. Both methods yield similar factors of safety in many cases. Special considerations and calculations are necessary when designing for slope stability in seismic zones where the factor of safety would otherwise be reduced. Slope stability is also a critical design element for landfills and other waste containment systems where facility integrity is critical. There are various remedies for slope failures which have been discussed and summarized.

PROBLEMS

14.1   What is the relationship between slope stability and landslides? Describe some phenomena at the prefailure stage during landslides.

14.2   Explain why acid rain and/or acid drainage will cause landslides more than just rainfall. How is rainfall intensity related to a landslide?

14.3   Why does slope failure occur without warning in overconsolidated soil deposits?

14.4   A cut 30 ft (9.15 m) deep is to be made in a deposit of highly cohesive soil that is 60 ft (18.3 m) thick and is underlain by basalt. The shear strength of the soil is constant at 500 psf (24 kN/m²). The unit weight of the soil is 120 pcf (18.8 kN/m³). The factor of safety of the slope must be 1.25. Estimate the slope at which the cut should be made.

14.5   A cut is to be made in a soil having total unit weight = 105 pcf (16.5 kN/m³), cohesion = 600 psf (28.7 kN/m²), and frictional angle $\phi = 15°$. The side of the cut slope will make an angle of 45° with the horizontal. What should be the depth of the cut slope for a factor of safety of 3?

14.6   A slope of 2 (horizontal) to 1 (vertical) is cut in homogeneous saturated clay, $S = 2000$ psf (95.8 kN/m²). The slope is 40 ft (12.3 m) high, the mass unit weight of soil is 120 pcf (18.8 kN/m³). Determine the factor of safety of the slope assuming a plane surface of sliding.

14.7   Consider a cohesive slope with $a_o/\gamma = 0.1$ and $c_o/\gamma H = 0.1$ as soil strength parameters, situated in a seismic zone with seismic coefficient, $A = 0.1g$ intensity. Determine the factor of safety, F.

Chapter 15

Fundamentals of ground improvement systems

15.1 Introduction

Natural or man-made soil deposits are not always in a stable condition and in many cases they need modification or improvement. Moreover, as the population grows, sites which may have been considered deficient in some way are receiving renewed interest. The purpose of ground modification or improvement is generally to increase the strength, reduce the settlement or to change the permeability of existing soils. With regard to strength, typical scenarios include strengthening ground soil (a) before failure occurs, (b) during soil's useful life period, and (c) after premature or unexpected failure. This chapter presents the fundamental considerations and basic requirements of ground modification and the following chapter presents typical geotechnical problems with special focus on environmental aspects.

15.1.1 Characteristics of ground improvement systems

Ground improvement or ground modification engineering is the collective term for any mechanical, hydrological, physicochemical, biological methods or any combination of such methods employed to improve certain properties of natural or man-made soil deposits. The purposes of this improvement are

1 *Strengthen ground soil before failure occurs*: This type of ground improvement generally happens where the soil is weak with low bearing capacity, and groundwater table is high.
2 *Strengthen ground soil during soil's useful life period*: This type of ground improvement is generally necessary for proper maintenance or to repair certain potential failure areas to prolong soil's useful life.
3 *Strengthen ground soil after premature or unexpected failure*: In many cases, ground failure is unexpected. However, it is required to examine the causes of failure before the ground improvement start.
4 *Temporary ground improvement systems*: This type of ground improvement system is used in certain conditions and certain locations, such as underwater repair, or where the permanent structure is under construction.

15.1.2   Basic considerations of ground improvement systems

To design an effective ground improvement system, some basic factors must be considered and the interactions among soil, ambient environment, and the improvement system itself must be evaluated. Some of these factors are listed as follows: (a) *sensitivity of soil to environment*: Soil is more sensitive to local environment than any other construction material used today. Each soil type responds to the environment differently as discussed in Chapter 4; (b) *dealing with vast amounts of materials*: In general, ground improvement systems deal with a vast amount of earth material. Presently, the annual figures of the volume of earth materials used in the construction field are in the billions of tons and the highest among all other construction materials; (c) *ground soil pollution*: Due to population growth, a progressive living standard and industrial progress, much of the air, water, and land has become exposed to varying amounts of pollution; (d) *problematic and/or adverse ground conditions*: In addition, more land is needed and many soil deposits previously claimed to be unfit for residential housing or other construction projects are now being used. Such areas include: wetlands, collapsible soil regions, mining subsidence areas, landfills sites, etc. To overcome these natural or man-made problematic soil deposits for use either as foundation material or as borrow material, additional improvement is required for conventional construction purposes; and (e) *selection of material*: Other needs to be considered recently include the need for energy conservation and potential material shortages. These issues represent a challenge to the engineering profession in searching for alternative or low-cost materials to be used in ground improvement systems.

15.2   Load factor and environmental-load factor design criteria

Conventional soil mechanics, if unamended with an environmental perspective, is ineffective at analyzing soil behavior under true field conditions as discussed in Ch. 1. The difficulty arises when complicated soil–water–environment reactions become significant. To account for these interactions, the environmental-load factor design approach is recommended. Before discussing environmental-load factor design criteria, it is necessary to review the load and resistance factor design method, commonly employed in steel design.

15.2.1   Load and resistance factor design criteria

Load and resistance factor design (LRFD) (AISC, 1993) is a method used to design steel members with greater efficiency than achieved by using an overall, bulk safety factor. The design for ultimate load is obtained by applying factors to the different service loads. Load factor design involves, first of all, a "loading function." This loading function involves a consideration of types of loads and the factors to be applied to each case. Load factors are numerically greater than one, that is, the anticipated load is increased for purposes of design. Second, there is a "resistance function" or "limits" for the structural member. These values are numerically less than one, that is, the

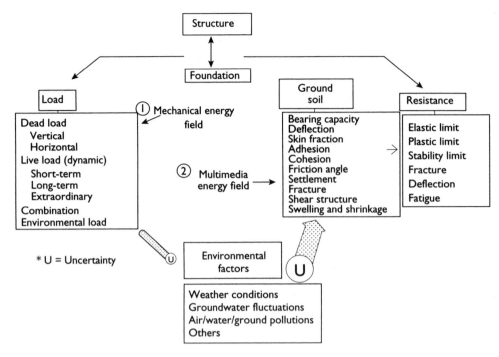

Figure 15.1 Environmental-load factor design criteria in geotechnology.

expected strength of the member is reduced for purposes of design. One of the unique features of load factor design is the use of multiple load factors. Dead loads are subject to less variation and uncertainty than live loads, and on this basis it is not unreasonable to assign a lower load factor to the dead loads than to the live loads. From a structural engineering viewpoint, this approach shows some advantages such as: encouraging the use of probability in design, and designs result in a more economical structure. The criteria involves two functions as: (a) *load function*: This function involves various types of loads, and the factors to be applied to each. There is dead load, live, short-term, long-term, extraordinary loads, and combinations as shown in Figure 15.1; and (b) *resistance function*: There is a resistance function or limits applied to the structural usefulness. The design process equates the two through analytical techniques. Figure 15.1 also shows the environmental factors affecting the overall structural system and will be discussed in the following section.

15.2.2 Environmental-load factor design criteria

The conventional approach for analysis and design of most foundations or other geostructures is based on allowable or working stress conditions. Regardless of loading types and environmental conditions, this approach uses the ultimate or failure load divided by a factor of safety as discussed in Section 12.4. The relevant factor of safety is usually provided by building codes, specifications, standard textbooks, and handbooks. Unfortunately, ground soil is very sensitive to the local environment such as water content, temperature and pore fluid composition, which will significantly change soil behavior. A list of loading conditions, environmental conditions, and corresponding parameters to be considered for environmental-load factor design criteria are presented in Figure 15.1. Essentially, design criteria which neglect environmental factors are susceptible to a greater likelihood of performance failures. As discussed in Ch. 4, fine-grained soils are more sensitive to the environment than large soil particles because smaller soil particles have a greater surface area per unit volume or mass.

15.3 Structure–soil and soil–structure interactions

15.3.1 Characteristics of load–soil interactions

When a load is applied to a soil mass, deformation may result. This deformation will depend upon load types, loading sources, ground soil properties, drainage conditions, stress history, and environmental conditions as discussed in Chapter 9. However, the most important factor is the nature of the load–soil interaction. There are two general types of load–soil interactions: the structure–soil and soil–structure interactions. Load derived from the superstructure to the ground soil is called structure–soil interaction, and load derived from the subsurface soil represents soil–structure interaction. In addition to the structural load, the structure–soil interaction also includes wind load (Sec. 11.7). The soil–structure interaction includes blasting, machine vibration, moving vehicle, pile driving during construction, and seismic loads which all relate to local environments as shown in Figure 15.2. Local environmental factors

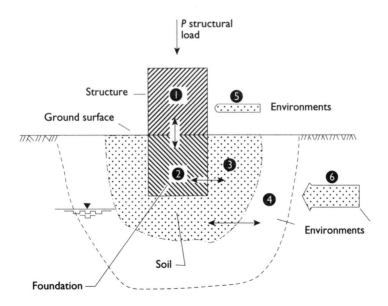

Figure 15.2 Structure–foundation–soil–environment interactions.

include weather conditions, groundwater fluctuations, degree of ground-water-air pollution, etc.

15.3.2 Structure–soil and soil–structure interactions

1 General discussion: If the structural load acts on the foundation first, the foundation transfers the load into the ground soil, then the soil responds to the combination of structure and foundation loads. This load transfer mechanism is referred to as structure–foundation–soil interaction. The response is not only caused by structural load but by foundation type as well, because the type of foundation plays an important role for soil mass response to the structural loads. Wind loads acting on a building is a typical structure–foundation–soil interaction problem. In some cases the load acts on the soil mass first, then the soil will transfer the load into the foundation and structures as in earthquake and blasting loading. This type of load–soil response is referred as soil–foundation–structure interaction. Soil–structure interaction caused by blasting loading as discussed in Chapter 11.

2 Complete analysis of soil–structure interaction: Using the earthquake effects on soil–structure interaction with respect to design of nuclear power plants as an example, ASCE (1979) presents the methods for analyzing soil–structure interaction effects. The problem of accounting for soil–structure interaction definition is illustrated in Figure 15.3. A complete analysis must (a) Account for the variation of soil properties with depth, (b) Give appropriate consideration to the material nonlinear behavior of soil, (c) Consider the three-dimensional nature of the problem, (d) Consider the complex nature of wave propagation which produced the ground motion, and (e) Consider possible interaction with neighboring structures.

3 Idealized interaction analysis: Idealized interaction analysis includes: (a) Kinematic interaction analysis and (b) Inertial interaction analysis. In examining Figure 15.3(b) vertical wave propagation is used to replace the actual complex ground motion pattern while retaining a specified motion of control point.

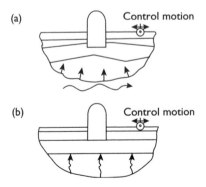

Figure 15.3 Complete and idealized complete analyses of soil–structure interaction effects for design of nuclear power plant. (a) Complete solution; and (b) Idealized complete solution.

15.4 Ground instability causes, failure modes, and classifications

15.4.1 Natural causes of ground instability

1   *General discussion*: Natural causes of ground instability include tectonic movements, earthquakes, geothermal events, floods, wetting–drying, and freezing–thawing cycles, flora–fauna as well as other geological hazards. Soil responds to these causes in various ways, according to the type, mineralogy, local environment, and so on. Earthquakes, for example, affect the behavior of granular soils like sand and gravel dramatically. These soils provide adequate bearing capacity under ordinary circumstances, but may liquefy and have larger settlement during an earthquake.

2   *Intensive rainfall*: Intensive rainfall will reduce the strength of soil, especially in residual soils and dispersive clay deposits which lead to erosion, progressive failure, subsidence, and landslides.

3   *Geographical location*: Geographical locations and weather conditions can also affect material behavior. Construction in cold regions is different from construction in desert areas. In extremely hot weather, rapid evaporation of moisture content in the concrete mass affects the concrete strength during the curing period, thus the concrete will not reach its proper design strength. Also for embankment compaction, rapid loss of moisture causes shrinkage cracks and reduces the maximum unit weight of compacted soil.

4   *Flood*: Ground failure due to flooding is based on the effect of water content changes in the soil–water system and is of concern in both partially and fully saturated soils. Flooding is also a major problem for river bank failure and contamination of both surface and ground water systems and accelerates the corrosion process on various foundations and waterfront structures, bridges as well as pavement components.

5   *El Nino and La Nina effects*: As discussed in Section 11.8, these effects generally include intensive rainfall, flood, high wind, and tide wave especially along the coastal regions.

15.4.2 Man-made causes of ground instability

1   *Caused by construction operations and pollution intrusion*: Ground instability caused by construction operations include: dewatering (Sec. 5.7), blasting (Sec. 11.11), deep excavation (Sec. 13.12), moving vehicles (Sec. 11.11), and pile driving (Sec. 15.12). Pollution intrusion routes and processes have been discussed in Section 1.3. Acid rain, acid snow and acid drainage covers large areas

2   *Human errors and unexpected factors*: Human errors and unexpected factors in design and construction deficiencies such as error in assumed loads, changes in use of the superstructures as well as tree roots and insects can all cause ground failure. Soil is generally subjected to the corrosive power of the carbon dioxide cycle, to acids produced during the decomposition of successive vegetation, and to enzymes secreted by microorganisms (Sec. 4.11). Wet–dry and freeze–thaw processes (Sec. 6.7) that have changed subsurface behavior will in turn change

Table 15.1 Summary of major causes and reasons leading to ground instability

I	*Problematic natural soil deposits and rocks*
	Weathering rocks and residual soils
	Clay shales
	Karst region and sinkholes
	Expansive clays
	Dispersive clays
	Collapsible silts and loess
	Organic soils
II	*Natural causes which weaken or damage soil – structure systems*
	Tectonic movement, earthquakes
	Geothermal
	Flora and fauna
	Flood, dry spells, hot and humid, wet and dry cycles
	Freezing–thawing cycles
	Tornado or hurricane
III	*Subsidences*
	Dewatering, mining
	Oil, coal and gas removal
IV	*Air–water–land pollution*
	Industries wastes
	Chemical wastes
	Nuclear wastes
	Acid rains, acid mine drainage
V	*Design and construction deficiencies*
	Error in assumed loads
	Changes in use of upper structures
	Construction operations
	Dewatering during the construction
VI	*Other unexpected factors*
	Human error
	Material properties
	Construction methods and equipment deficiencies

strength, settlement, and bearing capacity. A summary of major causes and reasons leading to ground instability is presented in Table 15.1.

15.4.3 *Ground improvement systems and classification*

1 *Ground improvement models and phenomena*: In most cases, the causes of ground instability involve more than one reason, and the improvement techniques also involve more than one method. Figure 15.2 shows a schematic diagram illustrating the ground failure modes and phenomena before and after ground failure and its related processes. While massive failures seem to occur suddenly, the process actually happens gradually or progressively. Prefailure phenomena include cracking, shrinking, surface creep, etc., which then leads to surface slip and excessive settlement as discussed in Ch. 14.

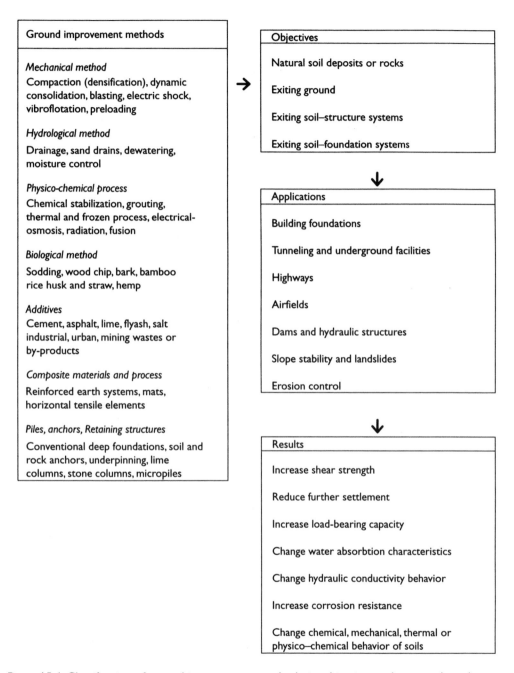

Figure 15.4 Classification of ground improvement methods, its objectives and expected results.

2 *Ground improvement system and classification*: For practical purposes, the ground improvement methods can be grouped into seven types as shown in Figure 15.4, together with the objectives, applications and expected results from ground improvement. In general, the result is increased bearing capacity and shearing resistance, and decreased compressibility and hydraulic conductivity.

15.5   Ground improvement techniques

There are an impressive array of ground improvement techniques available to the engineer. Many of techniques have been perfected and some cases patented by specialty contractors, however in general they include (a) Mechanical energy techniques such as compaction (Sec. 7.3), dynamic consolidation (Sec. 7.9), blasting (Sections 7.9 and 11.11), dewatering (Sections 5.7 and 15.7), and drainage systems (Sec. 5.7); (b) Thermal energy techniques (Sec. 6.5): heat treatment (Sec. 6.5), fusion process, and ground freezing techniques; (c) Electric energy techniques: such as electroosmotic process (Sec. 6.12) and electrochemical techniques (Sec. 6.11); (d) Multimedia energy techniques (composite material): such as admixture, stabilization, solidification, and vitrification. The relevance of a given technique depends on a host of factors, such as cost, site geometry and constraints, objectives, etc., however the general applicability of some may be inferred by the grain size of the candidate soil. Figure 15.5 presents a chart applicable grain size ranges for ground soil improvement methods.

15.5.1   Ground improvement by single energy action

Ground improvement by single action or interaction means that there is no chemical reaction involved or considered. Such action occurs in most mechanical energy

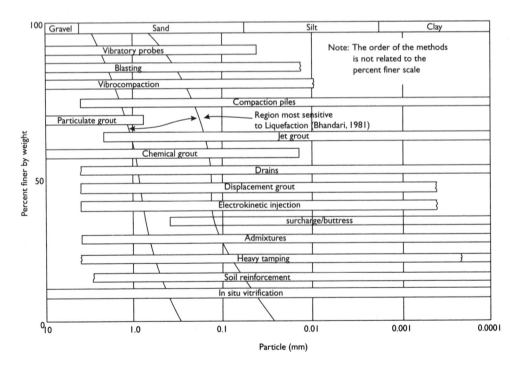

Figure 15.5 Applicable grain size ranges for soil improvement methods.

Source: Ledbetter, 1985, U.S. Army Corps of Engineers.

Note
Jagged lines at the ends of a bar indicate the uncertainty of applicability of the method.

problems such as compaction, dynamic consolidation, dewatering as listed in Table 15.1. Another example is the blending of granular materials to obtain a desired grain size distribution, as might be the case for filter or drainage applications.

15.5.2 Ground improvement by multimedia energy actions

Ground improvement with multimedia energy action, includes chemical, physico-chemical, and/or biological reactions in addition to the application of some physical technique. Some examples include

1 *Admixture*: One or two materials mixed into one mixture consider physico-chemical interactions between them. Such actions might involve, soil–lime, soil–flyash, soil–cement mixtures. Admixtures relating to concrete technology are not included.

2 *Fixation*: similar to admixture as stated in Case (1) as one or two materials mixed into one mixture, although fixation refers primarily to the immobilization of a particular contaminant. For example, naturally occurring but toxic metals such as arsenic or chromium may be fixated by combining cementitious materials together with soil.

3 *Solidification*: Solidification uses admixtures but the mixture sooner or later will become a hardened material such as soil cement. Regardless of the specific process, the main objectives of solidification are: (a) improvement of physical properties (mechanical stabilization), (b) encapsulation of pollutants (immobilization by fixation), and (c) reduction of solubility and mobility of the toxic substances (immobilization by isolation).

4 *Vitrification*: The vitrification technology originated in the 1950s when researchers began studying ways of locking radioactive waste in glass. Studies with vitrified waste show that glass can be ten thousands times more durable than other waste forms. The process of vitrification originated from rapidly solidified magma. Because of the rapid cooling rate and high liquid viscosity of oxide and silicate, molecules cannot move sufficiently to form a crystalline structure. Hence the amorphous (glass-like) structure is formed.

15.6 Ground improvement structural systems

Composite structural systems consist of more than one type of structural member. All foundations or geostructural members are buried completely or partially under the ground surface. If steel fibers or reinforced structural members are added into a soil it may be called a *composite material*. Composites in structural engineering in general include steel–concrete composites, fiber concrete composites, lime–bamboo composites, etc. Foundation structures such as footings, piles, caissons, and their components may be used as: (a) Retaining structures: including retaining walls (Ch. 13), sheet piling, flexible bulkhead, geosynthetic-reinforced soil (GRS) wall; (b) Anchors: soil anchors (anchor used in clay deposit), sand anchors (anchor used in cohesionless soil deposit), rock anchors (anchor used in rock mass); (c) Nailing, pins, mini-piles; (d) Reinforced earth system; (e) Pile foundations, Caissons, injection footings and (f) Drainage systems.

15.7   Geosynthetics

15.7.1   General discussion

Geosynthetics are fabric-like materials made from polymers such as polyester, polyethylene, polypropylene, polyvinyl chloride, nylon, chlorinated polyethylene, and others. The term geosynthetics includes geotextiles, geomembranes, geogrids, geonets, and geocomposites. Each type of geosynthetic performs one or more of the following five major functions: (a) drainage, (b) filtration, (c) moisture barrier, (d) reinforcement, and (e) separation. Many geosynthetics serve more than one of these functions. A discussion of geosynthetic types, namely geotextiles, geomembranes, geogrids, and geonets is provided as follows.

1   *Geotextiles:* Geotextiles are flexible, porous, polymeric fabrics used primarily for separation, drainage, reinforcement, and filtration. They are typically made from polypropylene or polyester, but other types have also been used (Koerner, 1991). A typical example involves the use of geotextiles in separating a stone base aggregate material from the underlying soil subgrade.

2   *Geomembranes:* Geomembranes are sheets of plastic (polyvinyl chloride, PVC and high-density polyethylene, HDPE are common) with extremely low permeability, usually in the range of 10^{-11}–10^{-14} cm/s (Koerner, 1998). The low permeability feature of these materials is used primarily as a moisture barrier, as is needed for example in waste containment applications.

3   *Geogrids:* Geogrids are characterized by their large opening size. Some geogrids are made from punched sheets that are drawn to align the polymer molecules. Other geogrid constructions are formed by welding together oriented strands or by weaving or knitting yarns and coating them to form a grid configuration. Geogrids are typically used in mechanically stabilized earth (MSE) walls, wherein wall support is derived from the soil to geogrid shear strength.

4   *Geonets:* A Geonet is an abbreviation of geosynthetic drainage nets. Geonets are formed by the continuous extrusion of polymeric ribs at acute angles to each other. They have large openings in a netlike configuration and the primary function of geonets is drainage.

15.7.2   Geosynthetics used for drainage system

One of the critical parameters in the design of composite systems with geosynthetics is the permeability. When the composite system is intended for the function of filtration or drainage, the geotextile component must be permeable enough to allow the flow of water, yet its openings small enough to prevent movement of soil particles. To satisfy these conditions, the geosynthetic and the soil components need to be compatible, that is the soil particles should not clog or wash through the geosynthetic material. A number of approaches have been developed to help select the appropriate geosynthetics that are compatible with soil found at a project site. These

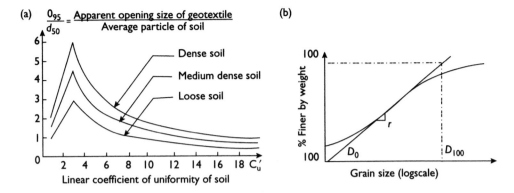

Figure 15.6 Retention criteria for geotextile filter. (a) Retention criteria based on C_u' and (b) determination of C_u'.

Source: Giroud, 1982, reprinted with permission.

approaches utilize the apparent opening size (AOS) or 95% "retained on" a US standard sieve opening size referred to as the O_{95} of the geosynthetic. The soil properties that are used in the compatibility analysis are either D_{50} (Ch. 2), or D_{85}, and C_u, ($C_u = D_{60}/D_{10}$) the coefficient of uniformity. Giroud (1982) recommended a *linear coefficient of uniformity*, C_u', be derived from the central linear part of a gradation curve. This retention criterion for geotextile filters is shown in Figure 15.4. C_u' used in Figure 15.6 equals:

$$C_u' = D_{100}/D_0 \qquad (15.1)$$

where C_u' = linear coefficient of uniformity of soil, D_{100} = particle size corresponding to 100% finer, and D_0 = particle size corresponding to 0% finer.

Carroll (1983) recommended that for selection of appropriate geosynthetics for drainage system as

$$O_{95} < (2 \text{ or } 3) \, D_{85} \qquad (15.2)$$

where O_{95} = 95% retained on a US standard sieve, and D_{85} = particle size corresponding to 85% finer.

15.7.3 Geosynthetic-composite systems and structures

1 General discussion: Geosynthetic-composite systems and structures include many varieties of geosynthetic-composite liners and geosynthetic-composite walls. These composite systems or walls engineered for drainage, filtration, erosion control, or liquid barrier functions are expected to perform under adverse effects of changing physical and chemical environment.

2 Earth pressure computations: Principles for the lateral earth pressure methods are discussed in Chapter 13. However, some modifications for the geotextile structural systems by various investigators are summarized in Figure 15.7.

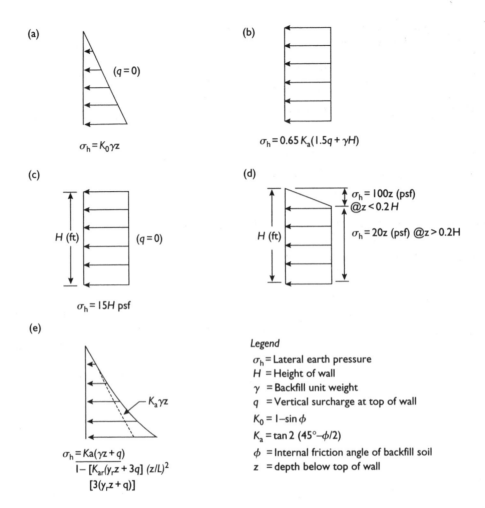

Figure 15.7 Lateral earth pressure for designing geotextile structural systems. (a) Forest service, (b) Broms, (c) Collin (geotextile), (d) Collin (geogrid), (e) Bonaparte *et al.*

Source: Reprinted from *Geotextiles and Geomembranes*, vol. 12, Claybourn, A. F. and Wu, J. T. H., Geosynthetic-reinforced soil wall design, pp. 707–724 © (1993), with permission from Elsevier.

3   Comments on environmental aspects of geosynthetic-composite systems: Some of the common factors in design of composite systems, such as a soil/geotextile system that serves filtration or drainage functions are the permeabilities of the soil and the geotextile and the retention of soil by the geotextile. There are well established methods to measure or predict these characteristics and design composite systems with components that are compatible over time. There have been a number of case studies that show good performance of existing engineered facilities. However, there is much room for research when longer-term performance is to be estimated under possible changes in the soil environment such as changes in the pore fluid chemistry, availability of water, biological activity, etc. These types of activities have been shown to influence soil's physical and chemical parameters significantly over time, especially in the fine-grained size ranges. Physical or chemical changes in the soil component of a composite system will alter the designed compatibility of soil with adjacent geotextile and may ultimately impair the functioning of the system. Physical and chemical changes may also take place in the geotextile components as well. The foregoing serves as an introduction to geosynthetics, with particular emphasis on those materials used in drainage and filtration. Those interested in a more complete treatment of geosynthetics are directed to Koerner (1998).

15.8 Sheet piling and other types of walls

15.8.1 Sheet piling and bulkhead structures

1 *General discussion:* Sheet piling is a thin metal element tied together to make a vertical wall, and sometimes it is called a *flexible retaining wall*. There are several types of sheet piling walls. Major uses are as a retaining wall against soil or water or both soil and water. There are three distinct types of sheet piling structures, simple sheet piling structure, bulkhead, and anchored bulkhead.

2 *High strength interlock sheet piling:* Due to construction demand, the sizes and depths of sheet piling structures that are available have increased significantly. In some cases the standard sheet piling sections are not satisfactory. The steel industry has developed a high strength steel sheet piling. The relatively new sheet piling is designated as PSX32 and PSX35 with an interlock-strength of 28,000 psi (193 MPa), 75% higher than the previous strength of similar flat-web sections. The steel from which the new sheet piling has been manufactured has a minimum yield strength of 45,000 psi (310 MPa). This value, which is above the current standard specification requirements of ASTM A-328, ensures high-load performance of both the web and the interlock. The high strength steel is generally provided automatically by the steel company when a new section is specified under the ASTM A-328 specification.

3 *Bulkhead and anchored bulkhead:* A bulkhead, sometimes is referred to as a *seawall*, is a structure constructed along a shore line of loose mounds or heaps of rubble, or masonry walls supplemented with treated timber, steel or reinforced concrete sheet piling driven into the beach and strengthened by a wale, guide and brace pile. A bulkhead serves the same general purpose as a retaining wall. The bulkhead itself consists of a single row of sheet piles of which the lower ends are driven into the ground surface. The wall without tie-rod is referred to as *bulkhead*. If tie-rod is used, it is called *anchored bulkhead*. The lateral earth pressure is taken up partially by anchor rods, which are tied to the sheet piles

15.8.2 Special types of walls

The following are various types of walls used in geotechnical engineering. These walls are used in different field conditions with a specific purposes. The design concepts are similar to the conventional retaining wall structures as discussed in previous sections.

1 *Bearing wall* and *breast wall:* A wall that supports vertical load, as a floor or roof is called *bearing wall*. A wall built against a bank of earth or rock to prevent it from falling is called *breast wall*.

2   *Cut-off wall* (*curtain wall*): A structure constructed, underground, to impede the flow of water as: (a) under stream beds in arid regions to extend to the surface to form a reservoir, (b) under earth dams to prevent trickles from developing into dangerous channels, (c) under concrete dams to prevent under-scour, and (d) under earth or concrete levees. These walls may be made of steel sheet piling, concrete, puddle clay, injected grout or other material.

3   *Mud wall*: An earth diaphragm or impervious cut-off-wall in a dam or a wall above the beam seats of a bridge abutment designed to support the approach slab and retain the earth behind the abutment.

4   *Slurry trench wall* (*diaphragm wall*): A watertight concrete cut-off wall or a combination concrete structural cut-off wall poured in an excavated and fluid (bentonite slurry) filled trench. Also called *diaphragm wall*.

5   *Training wall*: A structure constructed along a river of loose mounds or heaps of rubble, with or without a surrounding masonry wall, timber, close timber piling, wood sheet piling, steel sheet piling or reinforced concrete to direct the flow of the river into a more favorable, fixed channel.

15.8.3 Cofferdam and cellular structures

1   *Cofferdam*: Cofferdams are structures built to exclude earth and water from an area in order that work may be performed there under reasonably dry conditions. A cofferdam does not have to be entirely watertight to be successful. It may be cheaper to permit some flow into the working area; water is then removed with pumps (Ch. 5).

2   *Cellular structures* (*cellular cofferdam*): A structure of interlocking steel sheet piling to make a self-sustaining cofferdam with separate inside and outside walls. There are two general types of cellular structures namely, circular-type and diaphragm-type: (a) *Circular-type cellular structure*: A structure constructed of interlocking steel sheet piling consisting of circular cells joined with connecting arcs. The arcs are installed after the cells are completed; the cells and arcs are filled with granular soils; and (b) *Diaphragm-type cellular structure*: A structure made of steel sheet piles with each of the inner and outer walls consisting of a series of arc segments, which are connected at their intersections with diaphragms that extend through the cofferdam to form a series of cells. The cells are filled with earth, sand, gravel or rock.

There are two major methods used for analysis of stability of cellular structures used in many countries, namely the Terzaghi (1943) and Cummings (1960) methods. Other methods such as TVA, Bureau of Yards and Docks and Corps of Engineers methods are generally derived from the Terzaghi method with some minor modifications. Brinch Hansen (1953) proposed an alternate design method to evaluate the stability of cellular structures on rock or soil. This method has been widely used in Europe. In addition, a alternate design method for analyzing the stability of cellular structures has been proposed by Kurata and Kitajima (1967). The method is based on the model study of thin-walled steel tubes filled with sand. The modes of failure included sliding, overturning, tilting, and deformation. The design procedure indicated that the effective width of cellular structure should be determined by consideration of sliding, overturning, deformation, and reaction and the thickness of the wall.

15.9 Reinforced earth systems

15.9.1 Characteristics of reinforced earth systems

The use of *reinforced earth* as an engineered composite material system was developed by Vidal in 1969. It is formed by the association of a frictional noncohesive soil with thin plate metallic reinforcements. A structural system constructed with this material behaves as a coherent gravity mass, which avoids stress concentrations in the ground soils, distributes forces evenly within the whole mass and withstands differential ground settlement. The term "reinforced soil" refers to a soil strengthened by a material capable of resisting tensile stresses and interacting with the soil through friction and/or adhesion. With no practical height or length limitations, reinforced earth provides the necessary design flexibility to meet requirements for vertical earth retention structures for various highway, railroad, and embankments. A typical reinforced earth system is shown in Figure 15.8.

15.9.2 Modified reinforced earth systems

Reinforced earth systems have many uses such as (a) Mitigating slope instability, (b) Increasing weak bearing capacity of ground soil, and (c) reducing settlement and ground surface subsidence. The original reinforced earth system is made of thin metal strip as indicated in Figure 15.10, although geosynthetic materials have become more standard (15.10). In order to reduce the cost, several low-cost reinforced earth systems have been proposed.

1   *Sandwich type of reinforced earth mat*: Sandwich type of reinforced earth mat consisting of a layer of another material such as quicklime between two sheets of material such as wicked cardboard. This type of mat is applicable to high water content soft clay area. The sandwich layer is placed on each clay layer.

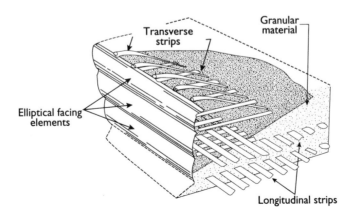

Figure 15.8 Typical reinforced earth system.

Source: Reinforced Earth Co., Reprinted with permission.

2 *Bamboo–lime reinforced earth mat*: This low-cost system has particular value in developing countries where the high tensile strength of the bamboo may be used to an advantage.

15.10 Geosynthetic-reinforced soil (GRS) systems

15.10.1 General discussion

Geosynthetic-reinforced soil (GRS) walls, also known as mechanically stabilized earth walls (MSE), derive their support from multiple layers of geosynthetic sheets or strip embedded in the backfill behind the face of the wall. Use of geosynthetics (geotextile and geogrid) as reinforcement has many advantages over other reinforcement materials such as increased resistance to corrosion and bacterial action, compared with metallic reinforcement. Geotextitle-reinforced soil walls with wrapped-face were first constructed in Siskiyou National Forest in Oregon in 1974 and Olympic National Forest in Shelton, Washington in 1975 by the US Forest Service. A typical configuration of the US Forest Service (USFS) wrapped-faced GRS wall is shown in Figure 15.9.

15.10.2 Failure modes of GRS retaining walls

Failure modes of GRS retaining walls can be divided into external and internal failure modes as illustrated in Figures 15.10(a) and (b). The external stability is generally evaluated by considering the reinforced soil mass as a rigid retaining wall with earth pressure acting behind the wall. The wall is checked, using methods similar to those for conventional stability analysis of rigid earth retaining structures (Sec. 13.2). The internal stability of GRS walls requires that the wall be sufficiently stable against failure within the reinforced soil mass, that is, the reinforcement is not over-stressed and its length is adequately embedded. Internal failure modes include tensile rupture failure of reinforcement and pullout failure of reinforcement.

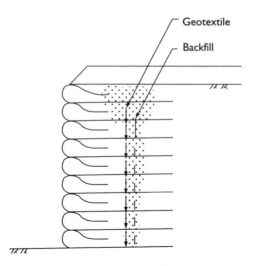

Figure 15.9 Typical configuration of a USFS wrapped-faced GRS wall.

Source: Wu (1994). Reprinted with permission.

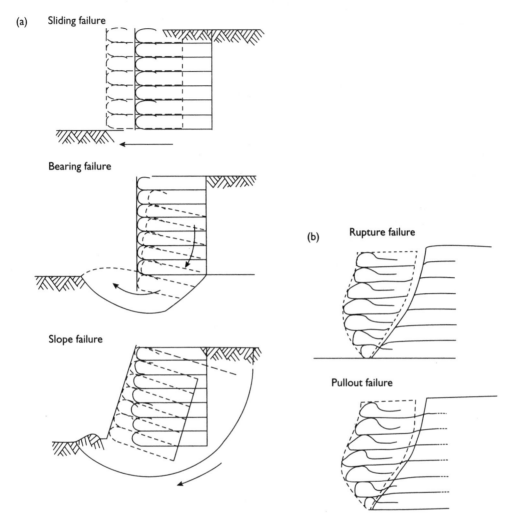

Figure 15.10 Failure modes of GRS Walls. (a) External failure modes and (b) internal failure modes.
Source: Wu (1994). Reprinted with permission.

15.10.3 Design concept of GRS retaining walls

Design concepts of GRS retaining walls can be categorized into three groups: ultimate-strength method, service-load method, and performance-limit method. Brief discussions of each method are presented as (a) Ultimate-strength method is based on the method of limit equilibrium (Ch. 12). To provide adequate safety margins, the ultimate-strength design method applies safety factors to the ultimate strength of the soil, reinforcement and facing, to the calculated forces and moments; (b) Service-load method is similar to the ultimate-strength method in that it is also based on the method of limit equilibrium. However, the design is primarily for the service load at which the wall movement and required reinforcement stiffness and strength are determined; and (c) The performance-limit method, on the other hand, allows direct determination of the wall movement and other performance characteristics of the wall. The design is obtained by limiting the wall deformation and/or other wall performance characteristics to ensure satisfactory performance of the wall. Details of design procedures for the GRS retaining walls and case studies are given by Wu (1994).

15.11 Anchors, nailing, and pins

15.11.1 Anchor systems

1 General Discussion: An anchor is a mechanical system designed to resist a lateral or upward force. It is used to resist hydrostatic uplift forces, or to support various retaining structures and excavation bracing. Anchors are used to resist a force in any direction. The most commonly used anchor unit is the grouted bar or tendon, which develops resistance to the applied load by the mobilization of shear forces along the soil–anchor wall interface.

2 Anchor types and classifications: Anchor types and classifications are presented as (a) classification based on uses: such as soil anchors, sand anchors, rock anchor, and composite anchor; (b) classification based on geometry of anchor and construction procedures: such as spread anchors, helical anchors, and grouted anchors; and (c) classification based on applications and required bearing capacity: such as short bar anchor, long bar anchor, and cable anchor.

3 Selection of a suitable type of anchor: The selection of a suitable type of anchor for securing generally depends on the soil type, groundwater conditions, project constraints, and cost. Juran and Elias (1991) and others note the following common anchor types: (a) *low pressure grouted straight shafted ground anchors* are used in many soil types. Hollow-stem augers (Ch. 1) are typically used to bore the hole and apply tremie grouting under low pressure conditions; (b) *low pressure grouted anchors* are usually tremie grouted (at pressures < 150 psi) in holes that are cased if cohesionless and possibly uncased in cohesive soil and rock; (c) *pressure injected anchors* are used for sandy and gravelly soils wherein the grout is injected at pressures higher than 150 psi; (d) *cable anchors* are useful for the transfer of considerable tensile forces from the structure to the deeper zones of the bedrock. With regard to manipulation, short anchoring bars are the simplest in terms of preparation, placing, and prestressing. Longer bars are rather more difficult to handle, and for long anchors, cables are preferable. Bar anchors are made of reinforcement steel. Cables are composed of patented, cold drawn wires. Cold drawn wire is manufactured at the iron and wire mills of high quality carbon steel melted in the furnaces, and (e) *composite anchor*: Most composite anchor is made from steel and concrete, steel-concrete, and steel fiber. The low-cost composite anchor is made from bamboo–lime composite, and bamboo–lime and biological fiber.

15.11.2 Soil nailing and pins

Soil nailing is an in situ soil reinforcement technique. The basic concept of soil nailing consists of reinforcing the ground by passive inclusions closely spaced, to create a cohesive gravity structure and thereby to increase the overall shear strength of the in situ soil and restrain its displacement (Juran and Elias, 1991). The steel reinforcing elements used for soil nailing can be classified as (a) driven nails, (b) grouted nails, (c) jet-grouted nails, and (d) corrosion-protected nails.

1 *Driven nails:* Driven nails are small diameter (15–46 mm) rods, or bars, or metallic sections. They are closely spaced and create a rather homogeneous composite reinforced soil mass.

2 *Grouted nails:* Grouted nails are generally steel bars. They are placed in boreholes with a vertical and horizontal spacing varying typically from 1 to 3 m. The nails are usually cement-grouted by gravity or under low pressure.

3 *Jet-grouted nails:* are composite inclusions made of a grouted soil with a central steel rod, which can be as thick as 30–40 cm. The jet-grouting installation technique provides recompaction and improvement of the surrounding ground and increases the pullout resistance of the composite inclusion.

4 *Corrosion-protected nails:* Corrosion-protected nails use double protection schemes similar to those commonly used in ground anchor practice.

The use of *Soil pins* is similar in principle to soil nailing. Short steel pins used commonly to protect the rockslides and failure of unstable rock slopes.

15.12 Pile foundations

Pile foundations are a major part of geotechnology. It involves complex structure–soil–foundation–environment interactions as discussed in Section 15.3. Therefore, discussions of various aspects of pile foundations are necessary and will be covered in considerable detail in this section.

15.12.1 Characteristics of pile foundations

1 *Function of pile foundations:* Piles are structural members used to transmit structural loads through a material or stratum of poor bearing capacity to one of adequate bearing capacity material. This load transfer may be by friction, end-bearing, or both, depending on whether the load is resisted by friction along the surface of the pile, or whether the pile end (point) rests on a soil stratum strong enough to carry the load. The load carried by friction is called *friction pile*, and the load carried by hard soil stratum is called *end-bearing pile*. The pile may utilize both friction and end-bearing to carry the imposed structural load. The structural load may be static or time-dependent, vertically or laterally transmitted to the soil stratum from single piles or pile groups. The typical use of pile for ground improvement and foundation engineering is: use for distribution of load, transfer load to firm soil stratum, resist the uplift pressure, and resist inclined or lateral loads. Other indirect uses of piles are listed as to eliminate objectionable settlement, compact granular soil stratums in order to reduce their compressibility, anchor structures subjected to hydrostatic uplift or overturning, and protect waterfront structures from wear caused by floating objects.

2 *Pile driving process:* Piles are usually inserted by driving with a steady succession of compaction blows by means of a hammer on the top of the pile. The common types of pile driving hammers include single-acting hammer, double-acting hammer, differential-acting hammer and diesel hammer. Hammer power sources include steam, air, and hydraulic. The diesel hammer is a single piece of equipment, which contains power source and hammer. At present, it seems that the diesel hammer is

widely used over other types. However, for selection of pile driving hammer, noise pollution in residential area must be considered. The selection of a pile type and its appurtenances is mainly dependent on environmental conditions. If piles are driven into salt water, the environmental conditions to be considered are wave action (Sections 11.8 and 13.12), moving debris, ice, and marine borer attack. If concrete pile is used, strong chemicals in water or in alkali soils could cause serious deterioration. If a steel pile is used in an environment with high dissolved solids and close proximity to electrical currents, then electrolysis deterioration may result. Types of soil also affect the selection of pile types. Piles to be driven through obstructions to bedrock with the least driving effort and soil displacement would favor a steel H-pile or open-end pipe pile.

15.12.2 Soil–pile interaction

15.12.2.1 Soil–pile interaction explained by mechanical energy field concept

1   *Load transfer characteristics*: The mechanism of load transfer in a soil–pile system is complex, and it involves unknown variables. Figure 15.11(a) shows a simple load transfer from a simple pile. Where, P_u is the ultimate pile capacity, P_p is the load carried in point bearing, f is friction, and P_f is the load carried by friction along the pile (Fig. 15.11(b)), the typical load friction distribution diagram's skin surface (Fig. 15.11(c)). Vesic (1970) suggested the use of the finite element method for analysis of load transfer of piles. This method allows the introduction of a complete stress history of the pile-soil system along with nonlinear and stress-dependent response of adjacent soils.

2   *Friction resistance between soil and pile*: Skin friction or adhesion is the friction force or resistance between the soil and pile. The coefficients of skin friction with

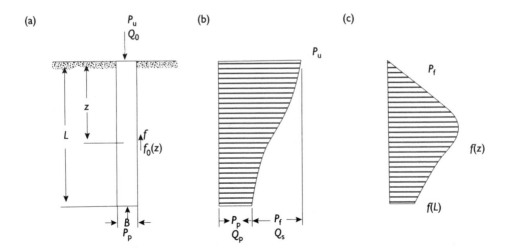

Figure 15.11 Load transfer from a single pile. (a) Vertical load on a single pile; (b) load distribution along a pile shaft; and (c) friction along a pile shaft.

Source: Vesic (1970).

various pile materials and soil types were developed by Potyondy (1961) and others. When a pile is loaded, the resistance available at the pile–soil interface is gradually mobilized from ground surface downwards. In general, a pile is loaded, the resistance available at pile–soil interface is gradually mobilized from ground surface downwards. The skin resistance of a friction pile is computed using either a combination of total and effective, or only effective stresses. There are several methods presently available to obtain unit frictional or skin resistance of pile in clay. Tomlinson (1971) and Vijayvergiya and Focht (1972) methods are commonly used and are briefly described as follows.

a   *Tomlinson method (1971):*

$$f_s = c + q \, K \tan \phi \tag{15.3}$$

where f_s = skin resistance or adhesive factor, c = average cohesion for the soil stratum of interest, q = effective vertical stress on the pile element, K = coefficient of lateral earth pressure, and ϕ = the effective friction angle between soil and pile.

b   *Vijayvergiya and Focht method (1972):* This method assumes that the displacement of soil caused by pile driving results in a passive lateral pressure at any depth, and the average unit frictional or skin resistance can be given as:

$$f_s = \lambda \, (q + 2c) \tag{15.4}$$

where f_s = average unit frictional or skin resistance, λ = the value of λ will change with depth, varied from 0.1 to 0.5, q = average effective vertical stress, and c = undrained cohesion. This method is also called the λ method.

3   *Phenomena of soil during pile driven process:* When a pile is driven into ground soil, the surrounding soil along the pile is compressed and remolded. Based on Broms' (1966) findings, the compressed zone extends from 1 to 3 diameters

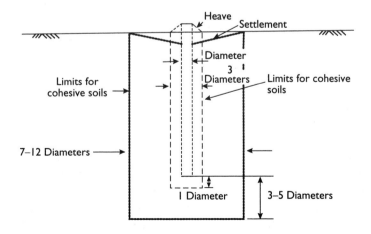

Figure 15.12 Zones of compaction and remolding due to pile driving.

laterally and about 1 pile diameter below the pile point as shown in Figure 15.12. The compressed zone becomes larger when soil gets stiffer. Driving of a pile displaces soil laterally and thus increases the horizontal stress acting on the pile. Test results (Lambe and Whitman, 1979) of the horizontal stress acting on piles in sand indicated a wide variation between the vertical and horizontal stresses on a pile driven in sand.

15.12.2.2 Soil–pile interaction explained by energy field concept

Soil–pile interaction is explained by the energy field concept as follows: On a molecular scale, a compression wave is started whenever a solid particle is struck. When a drop hammer strikes the pile cap, the molecules of the pile material (say a steel pile) at the top surface are subjected to a net force caused by the hammer. According to Newton's Second Law, this force causes acceleration, and the molecules start to move downward. At this point, they push on neighboring molecules and a pulse is transmitted to the tip of the pile. When sea shock wave or impact load travels along the pile, a pressure is momentarily built up where ever the molecules are closer together than is normal (before the pile was driven). The behavior for particles (molecules) around the pile and soil will depend upon the type of pile, soil types, and local environmental conditions. Distortion phenomena during the pile driving process is mainly due to that which gives rise to an elastic force that pushes the next molecules along.

15.12.3 Pile design concept and criteria

1 *Spacing and length of pile*: When several piles are clustered, it is reasonable to expect that the soil pressures developed in the soil mass as resistance will overlap. With sufficient overlap, either the soil will fail or the pile group will settle excessively. To avoid the overlap, the spacing of the piles could be increased. But large spacings are impractical, since a massive and heavy pile cap is required to be cast over a group of piles for the column base and to transmit applied loads to all the piles. Both theory and practice have shown that the total bearing capacity of a group of friction piles, particularly in clay, may be less than the product of the bearing capacity of an individual pile multiplied by the number of piles in the group. The reduction in value per pile depends on the size and shape of the pile group, spacing, and length of the piles. The length of pile is a factor in selecting pile spacing. A spacing of at least 10% of the length was required to avoid group action. Figure 15.13 developed by Gupta (1970), which shows a relationship between relative density of soil (Sec. 7.5), spacing and size of pile. It is readily seen that the same relative density can be obtained at different spacing, depending upon pile diameter.

2 *Negative skin friction (NSF)*: It is a downward drag acting on the piles due to relative movement between the piles and the surrounding soil. The drag force may occur when piles are driven through compressible soils or at the newly placed fill. As the soil consolidates, the fill moves downward, which develops friction forces on the perimeter of the pile and to carry the pile farther into the ground. Lowering of ground water level in such compressible soils may also bring about negative skin friction. The pile capacity under these conditions should be reduced to compensate the drag due to NSF. Figure 15.14 presents the comparison of the mechanisms of positive and negative

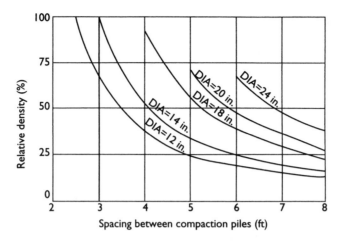

Figure 15.13 Relationship between relative density, spacing, and diameter of piles.

Source: Gupta, S. N. (1970), Discussion of sand densification by piles and vibroflotation, by C. E. Basore, and J. D. Boitano, *Journal of the Soil Mechanics Foundation Division Proceedings of the ASCE*, v. 96, no. SM4, pp. 1473–1475. © 1970 ASCE. Reproduced by permission of the American Society of Civil Engineers.

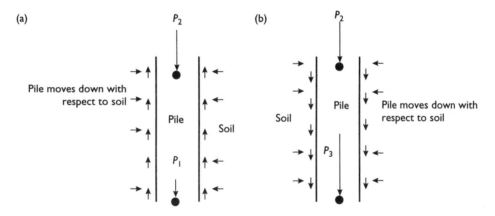

Figure 15.14 Mechanism of skin friction of pile foundations. (a) Positive skin friction and (b) negative skin friction.

skin friction. In examining Figure 15.14, for positive skin friction, the pile moves down with respect to soil; however, for NSF, the soil moves down with respect to the pile.

Field observations of NSF have been made on a pipe pile driven into the compressible silt. It was found that the NSF developed along a portion of the pile shaft extending to about 70–75% of the pile length. Reported on long pipe pile in marine clay, it was found that the greatest unit skin friction values developed in the upper part of the soil profile where the excess porewater pressures (Sec. 5.5) had dissipated. It is also indicated that the NSF is related to the effective stresses.

3 Uplift pressure: Uplift forces on a pile may be caused by hydrostatic pressure, wind force, earthquake, ice, frost action, and lateral forces. The type of pile with the largest perimeter is generally chosen to resist uplift if a friction pile is used for this purpose. When piles are required to resist uplift force in excess of the dead load of the structure, the following steps are commonly suggested: (a) the piles must be anchored sufficiently into the cap, the cap tied to the column (pile shaft), and the cap designed for the uplift stresses; and (b) concrete piles must be reinforced with longitudinal steel for the full net uplift. Splices in all types of piles should be designed to the full uplift.

15.12.4 Pile capacity determination

There are four basic ways to estimate pile capacity, and these are: static formula, dynamic formula, correlation with other simple in situ devices, and pile load test. More recently, a hybrid static-dynamic method (aka STATNAMIC) has been developed and become quite popular. Static and dynamic formulas of great variety have been used in the past and still new ones are being proposed. Several common methods for estimation of pile capacities are presented as follows: (a) static formula, (b) dynamic formula includes dynamic hammer and wave equation, (c) correlation with other in situ measurement devices include static cone and Standard Penetration Test (SPT), and (d) pile load test.

1 *Static formula:* The static formulas are based simply on adding the estimated tip resistance and skin friction. The pile tip point or tip resistance is calculated by using conventional bearing capacity formulas that are discussed in Chapter 12, and skin resistance is calculated by assuming either a constant friction value for entire depth of penetration or friction which increase linearly with depth.

$$P_u = P_p + P_f \qquad (15.5)$$

where P_u = ultimate pile capacity, P_p = load carried in point bearing (tip resistance), and P_f = load carried by friction along perimeter of pile (shaft resistance). The point resistance of pile end can be estimated by analogy to shallow foundation behavior (Ch. 12)

$$q_o = c\,N_c + q\,N_q \qquad (15.6)$$

where q_o = ultimate unit resistance at the pile tip, c = undrained shear strength of soil below the pile tip, N_c, N_q = bearing capacity parameters (Sec. 12.5), and q = mean normal stress at the pile tip.

2 *Pile capacity determined from dynamic hammer:* The basic assumption of dynamic formulas is that the energy of the hammer is related to the ultimate resistance of the pile multiplied by the average set of the pile for the last few blows of the hammer. The simpler formulas attempt to account for energy losses by large factors of safety including the weight of pile and hammer. The more elaborate formulas attempt to evaluate losses on the basis of Newtonian impact and elastic strain energy of pile cap, pile itself, and ground soil as illustrated in Equation (15.7).

$$RS = WH \quad \text{or} \quad R = WH/S \qquad (15.7)$$

where R = dynamic resistance of the soil or the ultimate capacity of the pile, S = penetration per blow, W = weight of the ram (hammer), and H = the height of the ram fall. Equation (15.7) does not account for various energy losses and other uncertainties. There are numerous dynamic pile driving formulas that are available and compressive review of these formulas is given by PCA (1951) and Chellis (1961).

Two dynamic pile driving formulas, the Engineering News and Terzaghi formulas are presented with numerical examples.

a   *Engineering News formula*: A. M. Wellington of the *Engineering News* (1888), introduced an additional factor, C to allow for losses of energy.

$$R = \frac{2WH}{S + C} \qquad (15.8)$$

where R, W, H, S terms are previously defined in Equation (15.7).

b   *Modified Engineering News formula*:

$$R = \frac{2WH}{S + 0.1(P/W)} \qquad (15.9)$$

where R, W, H, and S terms are as previously defined, and P = pile weight.

c   *Terzaghi formula*:

$$R = \frac{AE}{L}\left[-0.15 \pm \sqrt{0.15^2 + \frac{2WH(W + Pe^2)/(W + P)}{AE/L}}\right] \qquad (15.10)$$

A comparison of the load-carrying capacity of a identical piles in the same soil as determined by the *Engineering News* (Eq. (15.8)), *Modified Engineering News* (Eq. (15.9)) and Terzaghi formulas (Eq. (15.10))

EXAMPLE 15.1   (After PCA, 1951)
Compute pile capacity of identical piles in the same soil
A = cross-sectional of 20-in. square pile = 400 sq. in. (2580 cm^2)
L = length of the pile = 40 ft (12.2 m)
W = weight of ram = 3.75 tons (33.4 kN)
P = weight of pile = 8.25 tons (73.4 kN)
h = fall of ram = 4.0 ft (1.22 m)
H = 12 h, the fall of ramin. = 48 in. (121.92 cm)
S = penetration per blow = 0.15 in. (0.381 cm)
R = resistance of pile according to various formulas
E = modulus of elasticity of concrete (2,000,000 psi) 2 × 10^6 psi (13,789 MPa)
e = coefficient of restitution
k = hammer blow efficiency from Equation

SOLUTION
1   *By the Engineering News formula* (Eq. (15.8))

$$R = \frac{2 \times 3.75 \times 4.0}{0.15 + 0.1} = 120 \text{ tons (1067 kN)}$$

2   *By the Modified Engineering News formula* (Eq. (15.9))

$$R = \frac{2 \times 3.75 \times 4.0}{0.15 + 0.1\,(16,500/7500)} = 81 \text{ tons (720 kN)}$$

3   *By the Terzaghi formula* (Eq. (15.10))

$$\frac{AE}{L} = 400 \times \frac{2,000,000}{40 \times 12} = 1,660,000 \text{ lb/in. (2907 kN/cm)}$$

$$WH\left(\frac{W + Pe^2}{W + P}\right) = 7500 \times 48\left(\frac{7500 + 16,500 \times 0.5^2}{7500 + 16,500}\right) = 174,000$$

Then substituting in Equation (15.10):

$$R = 1,660,000\left[-0.15 \pm \sqrt{0.15^2 + \frac{174,000 \times 2}{1,660,000}}\right]$$

$$= 1,660,000 \, (-0.15 \pm 0.48).$$

Since the resistance cannot be a negative quantity,
$R = 0.33 \times 1,660,000 = 550,000$ lb $= 275$ tons (2446 kN)
Comparison of computed pile capacities:

1   *Engineering News formula* 120 tons (1067 kN)
2   *Modified Engineering News formula* 81 tons (720 kN)
3   *Terzaghi formula* 275 tons (2446 kN).

Based on aforementioned results, there are significant variations, therefore, it should noted that pile driving formulas are only a guide to the engineer in predicting safe pile bearing capacities and should be used with caution and judgment.

   3   *Wave Equation*: Smith (1962) presented a practical means for calculating the response of a pile to the impact of a hammer by means of a finite difference equation known as the wave equation. The wave equation provides some significant parameters during pile driving processes. These parameters include hammer type, size, cap, helmet, and soil conditions. The equation attempts to describe the travel of an impulse or wave down a pile as it is being driven. Values of solution depend on the reliability of the input data, therefore, only computer solution is practical.
   The analysis of wave equation is carried out by considering the hammer ram striking an elastic cushion with an initial velocity, v_o (Fig. 15.15(a)). Resulting forces on the drive head and pile cause the pile to penetrate the soil. Soil resistance is provided in the analysis in the form of skin friction and point bearing; both are considered as elastic-plastic with plastic behavior occurring at deflection, Q. Strain rate effects in the soil are accounts for by using a viscous damping factor. Simulation of the hammer-pile-soil system is illustrated in Figure 15.15(b) where the pile is considered to be a series of springs and masses. Digital computer programs have been developed to treat this problem; they provide the following information: (a) stresses and deflections at any point in the pile as a function of time; and (b) ultimate dynamic pile load capacity at the time of driving versus resistance (blows per inch.). Using wave equation for estimating the pile capacity and other related parameters computer programs and proper knowledge for designer to use (Hirsch *et. al.* 1970).
   4   *Estimation from static cone data*: The use of static cone penetrometer data to estimate the pile capacity was outlined by Sanglerat (1972). The method consists of extrapolating the cone bearing pressure to a bearing pressure that corresponds to the

(a)

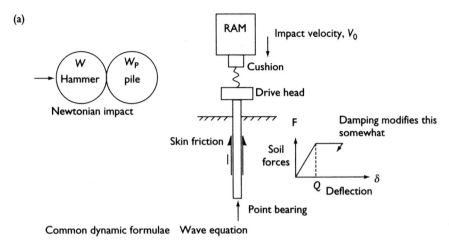

(b)

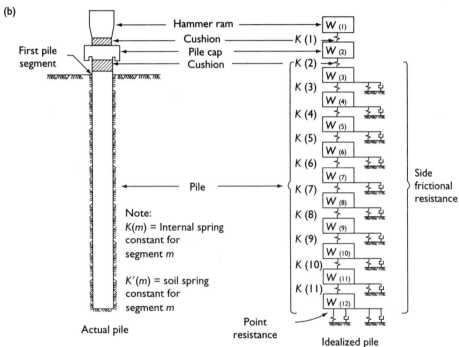

Figure 15.15 Characteristics of wave equation for determination of pile capacity. (a) Characteristics of wave equation; and (b) Springs and masses.

Source: Hirsch, T. J., Lowery, L. L., Coyle, H. M. and Samson, C. H., Pie-driving analysis by one-dimensional wave theory, state-of-the-art. In *Highway Research Record No. 333*, Transportation Research Board. National Research Council, Washington, DC, 1970, pp. 33–54. Reproduced with permission of the Transportation Research Board.

selected pile diameter. The cone pressure is an ultimate or failure pressure. Hence the extrapolated pressure is also a failure pressure and must be reduced by a safety factor for design. The extrapolation is as follows:

$$q_{cd} = 1/2\ (q_{c1} + q_{c2}) \tag{15.11}$$

where q_{cd} = ultimate pressure for the diameter (d) of the pile, q_{c1} = average q_c = 3.5 d (below the pile base) q_{c2} = average q_c = M × d (above the base), d = diameter of the pile, M = 8 (for sand) M = 1 (for very stiff saturated clays). Based on static cone penetrometer data for estimation of pile capacity is suited for a well-defined end-bearing stratum where skin friction above the pile base is negligible. A safety factor of 2 to 3 is normally applied (2–2.5 when skin friction is neglected, 2.5–3 when skin friction is considered). The local sleeve resistance is used as the pile adhesion; hence, allowable pile loads are determined as:

$$Q_a = 1/F_s \, (q_{cd} \, A_p + L \, A_s) \tag{15.12}$$

where Q_a = allowable pile load (capacity), F_s = safety factor, q_{cd} = determined from Equation (15.11) A_p = pile tip area, L = pile length, and A_s = gross pile skin area.

5   *Pile capacity computed from SPT N-values*: Calculation of pile capacity in cohesive soils on the basis of the SPT N-value has been proposed by Friels (1979) and others. The procedure for computing the ultimate pile capacity is as follows:

$$Q_u = Q_s + Q_p \tag{15.13}$$

$$\text{or } Q_u = f_s A_s + q_p A_p \tag{15.14}$$

$$Q_u = \alpha \, c \, L \, p + c \, N_c \, A_p \tag{15.15}$$

where Q_u = ultimate pile capacity, Q_s = ultimate pile capacity at shaft, Q_p = ultimate pile capacity at pile tip, f_s = shaft friction or soil-pile adhesion, A_s = surface area of shaft embedded in the soil, q_p = bearing capacity of soil at pile tip, A_p = end area of pile. The bearing capacity (q_p) can be computed from conventional bearing capacity procedure (Sec. 12.5) from $c \, N_c$ suggested by Vesic (1975), where the bearing capacity factor, N_c is 9 for a deep foundation, and c is the cohesion.

The shaft friction (f_s) can be estimated from [$\alpha \, c$], where c is the cohesion and α is an adhesive coefficient generally varying from 0.5 for stiff clays to 1.0 for soft clays. The cohesion c can be estimated from the SPT N-value by the following:

$$c = N/7.5 \text{ (ksf)} \quad \text{or} \quad N/15 \text{ (tsf)} \tag{15.16}$$

where L = embedded pile length, p = perimeter of pile, and N = average STP N value.

6   *Pile loading test*: When conducting an in situ pile loading test, two test methods can be employed. The load can be applied by weights such as iron ingots or concrete blocks. The other method is to use two or more reaction piles connected by a beam, the load being applied to the pile by jacking against the beam. For a detailed procedure see ASTM standard (D1195–71). O'Neill *et al.* (1997) presented a load testing of deep foundations including pile foundations. The loading systems and linear inertial mass vibrators have permitted testing of much larger foundations to failure than was possible in the past and have allowed for tests to be performed at a rate of loading that closely replicate the loading events being modeled.

15.12.5   *Pile capacity for group piles*

There are two commonly used empirical methods to compute pile group bearing capacity. One is based on group efficiency and the other on block failure. The group

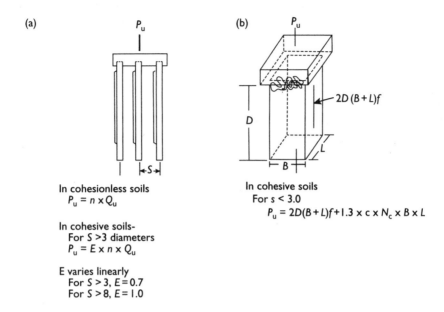

Figure 15.16 Bearing capacity of pile group. (a) Efficiency formula, and (b) Block failure.

efficiency may be defined as the ratio of group capacity to the sum of the individual capacities. The group efficiency of friction piles in cohesive soil is normally less than one. For cohesionless soils, Vesic (1969) reported the maximum group efficiency is equal to 1.7 at spacing of 3–4 pile diameters. The efficiency reduces with an increase in pile spacing. The reason is that the efficiency of a pile group in sand is generally greater than one. Vesic explained that the driving of adjacent piles increases the horizontal effective stress, and the driving of adjacent piles tends to increase the relative density of sand, thereby causing an increase in the friction angle of the sand. Figure 15.16(a) shows the bearing capacity of pile group based on group efficiency, and Figure 15.17(b) shows the group pile capacity based on block failure.

1 *Bearing capacity of pile group based on group efficiency (Fig. 15.16(a))*
   For cohesive soil: (for $s > 3$ diameter)

$$P_{u} = E\, n\, Q_{u} \tag{15.17}$$

For cohesionless soil:

$$P_{u} = n\, Q_{u} \tag{15.18}$$

where P_{u} = ultimate group pile capacity, E = group efficiency, values varies linearly, for $s = 3$, $E = 0.7$, for $s \geq 8$, $E = 1.0$, n = number of piles, s = spacing between piles, Q_{u} = ultimate load for each individual pile.

2 *Block failure concept* (Fig. 15.16(b)): The group pile capacity for cohesive soil can be estimated from block failure concept as follows: [Spacing < 3.0 diameter of pile]

$$P_{u} = 2D\,(B + L)\,f + 1.3\,c\,N_{c}\,B\,L \tag{15.19}$$

where P_u = ultimate group pile capacity, D = depth of pile block, B = width of pile block, L = length of pile block, f = friction resistance, c = cohesion, and N_c = bearing capacity parameter (Sec. 12.5)

EXAMPLE 15.2
Friction pile of a 24-in.^{2} (154.8 cm^2) reinforced concrete section are to be used with an embedded length of 40 ft (12.2 m) in a soft clay layer. The clay is known to have an unconfined compressive strength of 800 psf (38.3 kPa) and to be very uniform throughout the deep layer. An isolated footing load at this site will exert a concentric load to be required pile group of 250 tons (2224 kN).

a   What is the design allowable bearing capacity per one pile using a factor of safety of 2?
b   If the value obtained in (a) is used, how many piles will be required in the group in order to support the intended load?

SOLUTION

a   Allowable bearing capacity of a single pile
p = perimeter of simple pile = (4) (24 in./12) = 8 ft (2.44 m)
L = length of pile = 40 ft (12.2 m)
c = cohesion = 1/2 q_u = 800/2 = 400 psf (19.2 kPa)
Allowable bearing capacity of a single pile = pLc/F_s

$$= \frac{(8)(40)(400)}{2} = \frac{128,000}{2} = 64,000 \text{ lb } (284.7 \text{ kN})$$

b   Number of piles required, N

$$N = \frac{250 \times 2000}{64,000} = 7.8 \text{ (use 8 piles)}$$

There are numerous formulas for estimation of pile capacity in both single and group piles as discussed by Fellenius (1991), US Army (1993) and many others.

15.12.6   Factors affecting pile capacity

1   *Groundwater fluctuation*: Seasonal groundwater fluctuation frequently occurs. When groundwater decreases, the soil surrounding the pile is dried, consequently, soil mass will shrink and adhesion between the soil and pile is reduced.
2   *Frozen ground and freezing–thawing cycles*: When soil is frozen, then the bearing capacity is increased as discussed in Section 12.12. Effect of frozen ground soil relating to the pile capacity is given by Phukan (1991). In particular, uplift forces may be exerted on piles in contact with the freezing zone. However, when ground thaws, bearing capacity decreases significantly.
3   *Ground soil corrosion*: Ground corrosion affects on pile capacity as reported by Kinson *et al.* (1983), Dismuke (1991b) and many others. Figure 15.17 shows

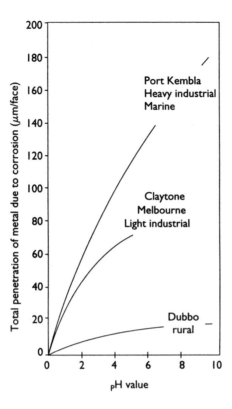

Figure 15.17 Corrosion loss of badly exposed mild steel.

Source: Based on Kinson, *et al.* (1983).

corrosion loss of boldly exposed mild steel. In all cases, pH value increase as penetration of metal due to corrosion increases. Other factors affecting pile capacity include sinkhole, ground cavity, larger boulders, larger tree roots, etc.

15.12.7 Comments on pile foundations

1   Structural loads transmit combinations of vertical and lateral, static and time-dependent loading to pile foundations. The allowable deformations or deformation tolerances of the structure at the foundation, due to structural constraints or foundation restrictions should be examined. It is recommended that the structural engineer and architect work with the geotechnical engineer prior to the start of a project. Examination of structural loads, types of loading, deformation tolerances as well as the supporting medium are necessary in order to develop a satisfactory soil-foundation system.

2   The allowable pile capacity is generally limited by building codes. This limitation frequently takes the form of an allowable stress expressed as a percentage of the yield or ultimate strength of the pile material. Often, neither the stress restriction nor the pile load tests have any relationship to the pile capacity because such

variables as actual structural loading, the manner in which the load test must be conducted and the ultimate soil–pile capacity are not properly accounted for in the codes. Pile load test criteria specified by numerous codes should be compiled, and the allowable pile capacities determined by these criteria should be compared to those determined by static or dynamic design methods.

3   The field inspector must observe and record all pertinent information. It is recommended that the engineer meet with field engineers, pile inspectors, and pile driving contractors prior to the start of driving. It is the duty of the engineer to see that only experienced and competent personnel are employed and that equipment used is adequate for the work at hand.

15.13   Drilled caissons, piers, pressure injection footings, and others

15.13.1   Drilled caissons and drilled caisson pile

A *drilled caisson* is a type of deep foundation that is constructed in place by drilling a hole into the ground soil to the required depth, which is often to bedrock or stratum. Then the inside is cleaned out and filled with concrete, or reinforced concrete to form a vertical column-type supporting member. The caisson is primarily a compression member with an axial load applied at the top, a reaction at the bottom, and lateral support along the sides. The bearing capacity of drilled caisson is similar to pile foundations supported by two major sources, the skin friction and end-bearing at the base of the caisson. The friction of the caisson is smaller than pile foundation, because caissons are not driven, they do not make tight contact with the surrounding soil as do piles.

15.13.2   Pier and drilled pier

A *pier* is a large size deep foundation. In general, they are used to support bridges and considered as a part of the bridge foundation. However, a pier also has several other meanings including (a) that which is constructed by placing concrete in a deep excavation, (b) a structure built perpendicular or oblique to the shoreline of a body of water for mooring ships, (c) a plain, detached mass of masonry, usually serving as a support, and (d) the pier of a bridge. The term pier is also used to describe to column-like foundations, similar to piles as discussed in Section 15.13. The *drilled Pier* is a larger diameter, up to 10 ft (3.05 m) or more, opening excavated to bearing strata and filled with concrete-cased or uncased. Table 15.2 presents advantages and disadvantages of piles, caissons, and footings (Ch. 12) to support structural loading.

15.13.3   Pressure injected footing

Pressure injected footings (PIF) developed by Franki Foundation Company have been used in many ground improvement systems. This special type of footing is also known as *Franki pile, displacement caisson, high-load bulb pile* or *compacted concrete pile*. The PIF is a hybrid element, and is considered as composite material system. It combines some of the properties of driven piles, drilled piers and spread

Table 15.2 Advantages and disadvantages of various foundation systems to support structural loading

Criterion	Footings	Caissons	Piles
1 Availability of construction contractors	Many	Few	Fair amount
2 Equipment required for installation	Minimal	Large amount	Small amount
3 Consolidation of lower layers	None	Some	Quite a bit
4 Type of load transfer on bearing surface	End bearing	Mostly end bearing	End and friction bearing
5 Penetration through debris	Good	Good	Fair
6 Depth restrictions	5 m	50 m	100 m
7 Resistance to horizontal loads	Good	Good	Very good with battered piles
8 Problems with groundwater	Dewatering required	Dewatering or drill casing required Segregation of concrete	None
9 Problems with solutional erosion	Bearing surface may be inspected Sinkholes require special treatment and design alterations	Footing bearing surface may be inspected	Cannot tell exactly where bearing strata is
10 Cost for medium depth foundation (<5 m)	Low	High	Medium
11 Cost for deep foundation (<7 m)	High	Highest	Lowest

footings into one system or one unit. PIF consists of a concrete shaft, which may be uncased or have a permanent steel casing as soil conditions may require, and an enlarged concrete base. The base is somewhat spherical in shape and forced into a bearing soil by driving zero slump concrete into the soil with high energy impact from a drop hammer. The main feature is the enlarged case, formed by displacement, which compacts the surrounding bearing soil both laterally and downward into a dense matrix.

15.13.4   Low-cost and energy saving piles

There are numerous types of low-cost and energy saving piles developed recently, including (a) mini concrete piles, (b) sand and stone piles, (c) lime column or lime piles, (d) bamboo reinforced lime piles; as well as a pile made from compacted municipal solid wastes (Sec. 16.9). Hu *et al.* (1981) investigated the stress–strain relationships of these alternative pile types. Many of these alternatives have application where conventional techniques (e.g. steel reinforced concrete) are cost-prohibitive. This is frequently the case in developing countries. However depending on the application, they may also provide equal or better technical performance, and as such they offer engineers more options when implementing sustainable design practices.

15.14 Summary

Failure modes, phenomena, causes, and classification of geotechnical ground improvement systems were discussed and summarized. Soil–structure and structure–soil interactions were also explored. It is indicated that for fine-grained soil, the soil-structure interaction, such as bacteria, suspended organic matter, colloids, and various ionic species in the porewater must be considered. They have a tendency to adhere and accumulate on a structure or soil-structure system. Many fine-grained soils have swelling-shrinkage potential that is largely controlled by the pore fluid chemistry.

There are many instances in which the soils at a given site need to be improved in order to adequately support anticipated loading conditions. Generally, soils are modified to increase strength, reduce settlement or alter the hydraulic conductivity (e.g. reduce seepage or enhance drainage). There is an impressive array of ground improvement techniques at the disposal of the engineer. These techniques may involve blending additives, precompaction or consolidation or the use of geosynthetics. In certain cases, subsurface soils need to be bypassed altogether with piles to reach a more competent strata. These piles derive their strength at the point of contact with the rock or firm strata and, in many cases, from the friction along the length of the pile. Methods for estimation of pile capacity by static, dynamic hammer, wave equation, and pile loading tests are presented and discussed. Factors that affect pile capacity including temperature and other environmental factors are also presented. At present time, no effective method for computing the pile capacity either for a single pile or group piles has been generally accepted by the engineering community. Most engineers are likely to agree that the field pile load test is the only method that can be used to determine a reliable pile capacity at a site. However, the static or dynamic method or both can be used for preliminary estimates of bearing capacity.

PROBLEMS

15.1 Why would soil need to be improved? List the technical and financial factors that affect ground improvement planning? Discuss the site investigator, designer, and contractor's viewpoints of a ground improvement program.

15.2 Discuss the significant differences between load factor design criteria and environmental-load- factor design criteria. Why are environmental factors important?

15.3 Explain the differences of mechanisms between soil–structure, soil–foundation–structure, structure–soil and structure–foundation-soil interactions.

15.4 How can air-water pollution affect the geotechnical behavior of ground soil? How does acid rain affect foundation structures? How does acidic water affect the embankment soils and concrete mixtures?

15.5 Explain why steel pile and sheet piling are not suitable for use in terms of polluted or a saltwater waterfront environment ?

15.6 A concrete wall is 12 ft (3.7 m) high, 5 ft (1.5 m) thick at the base and 2 ft (0.6 m) thick at the top. One face is vertical. What are the maximum and minimum unit pressures under the base of wall due to its weight?

Chapter 16

Problems in environmental geotechnology

16.1 Introduction

The concepts and fundamentals of geotechnology have been discussed from Chapters 1 to 14. Chapter 15 presents the structure–foundation–soil interactions and ground improvement systems. There are numerous environmental geotechnical problems; however, these problems require knowledge from other disciplines. Due to the limited space of this closing chapter, only a subset of relatively important and commonly encountered subjects such as wetland, marine margin land, dredging and reclaimed land, ground surface subsidence, waste control facilities (including radioactive nuclear wastes and radon gas), landfill technology, arid land, and desert regions are presented in this chapter with brief discussions on each subject with emphasis on causes, failure mechanisms, and prevention and control from an environmental geotechnology point of view. Finally, the environmental geotechnology perspective and new instruction and research areas are proposed.

16.2 Wetlands and flood plain

16.2.1 Characteristics of wetlands

Wetland is a general term, which includes marshes, swamps, flood plains, bogs, as well as rice paddies, and the man-made wetlands. The formation of wetland varies greatly in age, especially man-made ones which are of relatively recent origin, while others had their beginnings following the retreat of the glaciers. In wetland areas, most soils belong to organic soils. These soils are solid constituents consisting predominantly of vegetable matter in various stages of decomposition or preservation. They are commonly designated as bog, muskeg, and moor soils with differentiation between peat and muck soils on one hand, and coastal marshland soils on the other.

16.2.2 Definitions and classifications of wetland

1 *Definitions*: There are several definitions of wetlands frequently used in various literature sources such as (a) US Fish and Wildlife Service (USDA, 1969) which defines it as a land where water is the dominant factor determining the nature of soil development and types of plant and animal communities living at the soil surface; and (b) Mitsch and Gosselink (1993) which defines it as those areas that are

inundated or saturated by surface water or groundwater at a frequency and duration sufficient to support, and that under normal circumstances do support, a prevalence of vegetation typically adapted for life in saturated soil conditions.

2   *Wetland classification system*: Scientists recognize five major wetland systems: marine, estuarine, lacustrine, riverine, and palustrine. Marine and estuarine include coastal wetlands. The other three categories represent freshwater systems. All five types of wetlands are commonly used to designate distinct wetland types: (a) *Marshes*: marshes are characterized by soft-stemmed herbaceous plants, such as cattails and pickerweed; (b) *swamps*: swamps are dominated by woody plants namely trees and shrubs; and (c) *bogs*: bogs are peatlands, usually lacking an overlying layer of mineral soils.

3   *Flood plain*: A flood plain is also a part of wetlands. It is the lowland that borders a river, which is usually dry but is subject to flooding when the stream overflows its banks. *Rice paddies* are a man-made wetlands used for agricultural purpose. The behavior of rice paddies is similar to flooded soil but the composition is different and in particular the rice paddy contains a large amount of fertilizers.

16.2.3   *Environmental geotechnical problems of wetlands*

Utilization of wetland material for various engineering uses can create additional useful lands; however, there are many detrimental effects for such uses. For example, highway earth fills on wetlands may compromise sensitive ecological habitats and environmental systems. Research findings (De Santo and Flieger, 1995) indicate the following effects:

1   *Ecological effects*: The ecological effects of earth fills on wetlands include (a) inhibition of storm water and tidal distribution, (b) increased water turbidity and alteration of water circulation patterns, (c) removal of natural filtration systems and introduction of exotics, and (d) inhibition of movements of animals and humans. Also, they discuss wetland functions and values by descriptive approaches to visualizing and assessing wetland systems.

2   *Environmental geotechnical effects*: Geotechnical engineering aspects of wetland material include. (a) high water content, (b) low bearing capacity, (c) low hydraulic conductivity, (d) low shear strength, and (e) large settlement. All of these characteristics render the material to be undesirable from a load-carrying capacity perspective. Typical problems include hydrological, physico-chemical, water quality, erosion, and sedimentation effects. One of the primary benefits of wetlands, whether natural or constructed, is their ability to improve the water quality of stormwater runoff. Development of smart technology of dewatering systems, surface and subsurface drainage network in wetland regions are important.

16.3   Coastal margins and marine deposits

16.3.1   *General discussion*

In recent years, human activities have brought many changes to coastal environments. From an ecological point of view, as reported in 1996, the effects of loss of wildlife habitats in the states of Connecticut and California are most striking, because about 50% and 90% are lost respectively. Interception of water and sediment including waterworks for irrigation, storm protection, and power have reduced the area of wetlands on many coasts and shifted channel locations. Associated problems include increased pollution of nearshore environments including industrial and agricultural chemicals, an increasing rate of sea level rise due to the greenhouse effect (Sec. 16.4), and sea level rise.

16.3.2 Properties of sea water

Most of the dissolved constituents in the ocean are ions. The terms salinity and chlorinity are commonly used to characterize the properties of seawater. The *salinity* of seawater as defined by oceanographers is the mass in grams of the solids in one kilogram of seawater evaporated to a constant mass at 480°C (896°F). *Chlorinity* is defined as the number of grams of the chloride ion (Cl^-), bromide ion (Br^-) and iodide ion (I^-) contained in one kilogram of seawater. The salinity of seawater is directly related to the chlorinity. The relationship between temperature, salinity, and density are linear. For a given density of seawater, when temperature increases, the percent of salinity also increases.

16.3.3 Characteristics of marine sediments

1 Marine sediments: The marine sediments are predominantly depositional rather than erosional. Thus, marine sediments exhibit more uniformity than normally found on land. Typically, marine sediments are broadly classified on whether the sediments are land derived (Terrigenous) or are the result of marine activity (Pelagic). The Pelagic sediments can be further divided into inorganic and organic materials. Inorganic Pelagic materials are typically clay size material. Deposits with Pelagic clay are primarily found off areas of major deserts (Sec. 16.7). The organic materials are primarily the skeletal remains of marine organisms. These materials are either calcium carbonate ($CaCO_3$) or silica (SiO_2). Figure 16.1 presents the classification of carbonate sediments.

The presence of calcium carbonate ($CaCO_3$) in the marine sediment is controlled by the biological productivity and the calcium carbonate compensation depth (CCD). The CCD is the depth of the water columns at which $CaCO_3$ is dependent on the amount of biological material available. The marine sedimentary composition of the marine margins are affected predominantly by their close association with the continents.

2 Factors affecting the characteristics of marine sediments: The factors affecting the geotechnical properties of marine margins and sediments are as follows: (a) rapid sedimentation by terrigenous materials (sands, silts, and clay), (b) desiccation of upper sediment layers due to sea level declines, (c) porewater pressure changes due to wave action, (d) beach erosion, and (e) high levels of biological activity including bioturbation.

On the shelf itself, the segregation action of wave attacks (Sec. 11.8) on the shore material, and the resulting seaward transportation causes most marine deposits to have a relatively narrow particle size range. These relatively uniform deposits make them susceptible to liquefaction due to cyclic loading. Geometric constants (Sec. 3.3)

| | | Fine grained | | | Medium–coarse grained | | | | | |

Increasing grain size (mm)

| | | 0.002 | 0.006 | 0.02 | 0.06 | 0.2 | 0.6 | 2 | 6 | 20 | 60 |

		Fine	Medium	Coarse	Fine	Medium	Coarse	Fine	Medium	Coarse
Soil	Carbonate mud or clay	Carbonate silt			Carbonate sand clastic/bioclastic/oolitic			Carbonate gravel clastic/bioclastic		
Rock	Calcilutite (carbonate mudstone)	Calcisiltite (carbonate siltstone)			Calcarenite (carbonate sandstone) clastic/bioclastic/oolitic			Calciruidite (carbonate conglomerate or breccial) clastic/bioclastic		
	Fine grained limestone				Detrital limestone			Conglomerate limestone		
	Crystalline limestone				Crystalline limestone					

Degree of induration/cementation	Cone resistance q_c MN/m^2
Very weakly indurated/cemented	0–2
Weakly indurated/cemented	2–4
Firmly indurated/cemented	4–10
Well indurated/cemented	>10

Figure 16.1 Classification of carbonate sediments.

Source: Clark, A. B. and Walker, B. F. (1977), proposed scheme for the classification and nomenclature for use in the engineering description of Middle Eastern Sedimentation rocks, *Geotechnique*, v. 27, no. 1., pp. 1–10. Copyright Thomas Telford Ltd. Reprinted with permission.

of granular marine deposits have been obtained from more than 30 sampling locations worldwide (Chaney and Fang, 1986). The ranges of maximum and minimum values of effective sizes D_{10} are from 0.07 to 1.43 and uniformity coefficient C_u from 1.2 to 3.4. All samples are taken at a low water mark. In addition to their gradation characteristics their void ratio can exceed their critical void ratio and therefore are in a potential liquid state. They may be changed into actual macromeritic liquids (Sec. 3.5), not only in specific shear zones but throughout the whole granular system. Relatively small energies such as machine noises, vibrations, or minor earthquakes acting on the deposits may trigger the soil movement and failure.

   3 Geotechnical properties of marine deposits: Some marine deposits also contain high percentages of sensitive clay (Sections 4.10 and 10.9), and as such sediments are in a potential macromeritic liquid state becoming actual liquids after destruction of their inter-particle bonds. Of course, any man-made disturbance during the construction period will reduce the strength and bearing capacity significantly. Table 16.1 presents typical geotechnical data and their ranges of marine deposits.

Table 16.1 Typical geotechnical data and their ranges of marine deposits

Parameters	Ranges
Sizes composition (%)	
Clay 2 micron	35–60
Silt	40–60
Sand	10–20
Clay minerals (%)	
Illite	60–75
Kaolinite	10–25
Montmorillonite	5–20
Phyiscial properties	
Moisture content (in situ)	60–180
Activity (A)	0.33–1.33
Sensitivity	1.60–26
Liquid limit (LL)	72–121
Plastic limit (PL)	34–51
Field moisture equivalent	65–78
Centrifuge moisture equivalent	55–68
Shrinkage limit (SL)	7–10
Void ratio (e)	0.5–9.0
Compression index (C_c)	$\geqslant 2.0$

Source: Chaney and Fang (1986). Copyright ASTM International. Reprinted with permission.

16.4   Saltwater intrusion, estuaries, and greenhouse effects

16.4.1   Saltwater intrusion

Saltwater intrusion is a dynamic equilibrium phenomenon of groundwater movement along the coastal aquifer as illustrated in Figure 16.2. It may reach thousands of meters inland, which has a significant effect on groundwater supply. This phenomenon is also called *encroachment*, which is a shoreward movement of saltwater from oceans into coastal aquifers due to the over-pumping of groundwater. The interface between saltwater and fresh water are treated as two immiscible fluids separated by an interface with a slope. While it might appear that the two should mix quite easily, the differences in density and temperature (e.g. saltwater is more dense) do in fact lead to a stratification of fresh water and seawater that has been documented in both groundwater and surface water. Diagrams illustrate the relationship of rainfall, ground surface runoff, and groundwater movement as related to a pumping well. Depending on its proximity to saltwater as well as the nature and hydraulic conductivity of the aquifer, a pumping well for groundwater supply or remediation purposes will encourage saltwater intrusion. The slope of the interface indicated in Figure 16.2 can be estimated by mathematical approximations, a laboratory viscous-fluid model study and/or in situ measurements. Factors affecting the saltwater intrusion line include climatological factors such as rainfall, topographical factors such as surface

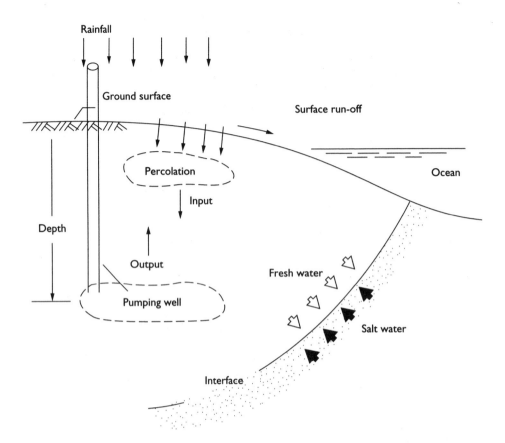

Figure 16.2 The characteristics of interface of saltwater intrusion along the coastal aquifer. Diagram also illustrates the relations of rainfall, surface runoff, and groundwater movement relating to pumping well.

runoff, subsurface soil conditions such as ground percolation ability, and pumping well characteristics such as number and the depth of pumping wells. Saltwater intrusion is a serious pollution problem along the coastal region. Careful planning is needed when pumping groundwater for industrial uses. Over pumping must be avoided and controlled.

16.4.2 Estuaries

An *Estuary* is the area where the rivers meet the sea (Fig. 16.3(a)). They are fragile and easily disrupted and are an important part of ecosystems on Earth. Figure 16.3(b) shows a simple diagram illustrating how the river and tidal currents mix in an estuary. The density of saltwater (from oceans) is heavier than fresh water (from rivers), so water near the seafloor is saltier than on the top layer. Estuaries are influenced by river flow, tidal range, and sediment distribution. Unfortunately, these factors themselves are changing continually. In general, most estuaries may never attain steady

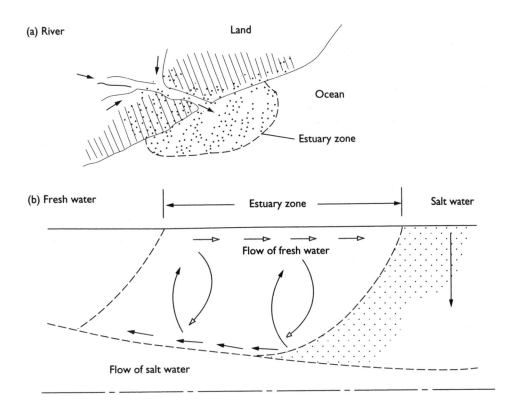

Figure 16.3 Schematic diagrams illustrating estuary areas and their interaction with environment. (a) General area of estuaries; and (b) River (fresh water) and ocean (saltwater) interaction.

state conditions and no two estuaries are alike, based on case studies. Marine sediments in estuary zones are soft, fine-grained, clay-like materials with low bearing capacities, high hydraulic conductivity, and high sensitivity.

16.4.3   Global warming and greenhouse effects

Global warming, which has been reported frequently in recent years, has been attributed primarily to increasing emissions of carbon dioxide (CO_2) and methane (CH_4) (and other gases) from both natural (e.g. volcanoes) and anthropogenic sources (e.g. automobiles, industry). The gases are transparent and allow sunlight in but retain the heat that the Earth emits with only some escaping back into space. Whether the current increasing temperature trend is a function of natural fluctuations or the result of industrialization, the average global temperature has been modeled to rise between 2°F and 6°F (−16.6°C and −14.4°C) by the year 2100. Such increases in temperature would raise the sea level several feet in the next century. From a geotechnical engineering viewpoint, this would lead to (a) more wetland formation, (b) increased coastal erosion, (d) increased saltwater intrusion, and (e) increased foundation and slope stability problems in coastal margins.

16.5   Soil erosion

16.5.1   Erosion causes and mechanisms

Soil erosion is caused by the drag action of wind, rainfall, and wave action or flood on bare or unprotected soil surfaces. It involves a process of both particle detachment and transportation. It is a serious problem particularly as it relates to construction and maintenance. Based on the US Department of Agriculture's (USDA) estimates in 1990, the total cost of sediment damage and dredging resulting from soil erosion is approximately $US 600 million yearly. It costs approximately $US 20 million annually just to remove sediment from irrigation ditches. These figures do not take into account the damages to agricultural areas, homes, roadways, bridges, and recreational areas, which also result from the loss of soil.

Soil erosion occurs primarily by two mechanisms. Either the soil is washed away by water, or the soil particles are blown away by an air-stream of sufficient velocity. Both mechanisms would predict that the stabilization of the soil would be proportional to the size of the soil particles. The surface chemistry approach to this problem involves an investigation of the soil–liquid interfacial properties with the objective of determining what the chemical additives are. The four general categories are (a) water erosion, (b) wind erosion, (c) erosion during construction, and (d) environmental causes (Flood, El Nino effect).

16.5.2   Water and wind erosions

1   *Water erosion*: An equation estimating the soil loss by water erosion can be estimated from a semi-empirical equation known as the Universal Soil Loss Equation (USLE) given by the US Agricultural Research Service. A modified equation based on the USLE for predicting soil loss due to water erosion on highway construction sites was proposed by Israelsen *et al.* (1980). Details of this method are provided by the NCHRP of Transportation Research Board report by Israelsen *et al.* (1980).

2   *Wind erosion*: Two types of wind erosion exists. Loose soil particles are picked up by moving air and carried from one place to another. This process is called *deflation*. Wind erosion potential may be estimated in a manner similar to that for water.

16.5.3   Erosion problems during construction

Uncontrolled water and wind erosion resulting from construction operations has a significant impact on the environment. The sediment that is produced pollutes surface water, restricts drainage, fills reservoirs, damages adjacent land, and upsets the natural ecology of lakes and streams. Besides harming the environment, soil erosion during construction increases costs and causes extensive delays and repairs.

Controlling erosion during construction including drainage ditches, covers, terracing, contour cultivation, fences, soil stabilization, and straw, hay, or artificial turf have been used during construction for both water and wind erosion control. Conventional soil stabilization is generally of a permanent nature. Since construction and maintenance are temporary, conventional stabilization methods may not be suitable. New materials and/or methods are available to stabilize the soil surface on

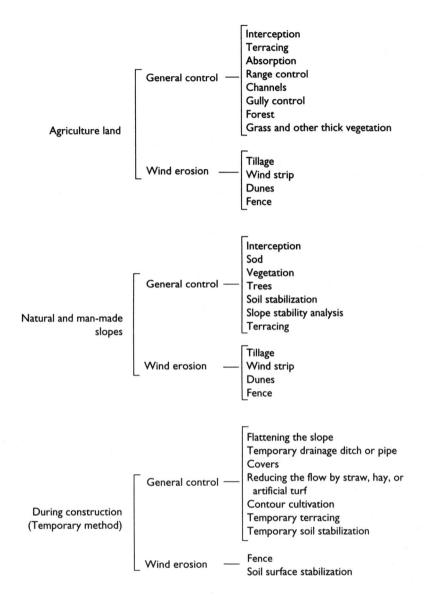

Figure 16.4 Methods for preventing and controlling soil erosion.

a temporary basis in order to minimize soil erosion from both water and wind during construction or maintenance.

16.5.4   Methods for prevention and control of soil erosion

There are a number of measures that may have a place either as temporary or as permanent installations. These techniques include divisions, grasses or paved waterways, barrier pipe outlets, bench terrace or berm, soil stabilization, retaining structures, etc. Figure 16.4 summarizes some methods for preventing or controlling soil erosion.

16.6   Ground surface subsidence

16.6.1   General discussion

In the broadest sense of the term, *subsidence* may be defined as the deformation or settlement of soil mass in any direction caused by various external loading, internal stress, and unbalanced environmental factors. In non-technical terms, it may be called *ground surface movement*. From the geotechnical engineering viewpoint, subsidence can be classified based on its origin, load-deformation mechanism, appearance of the ground surface movement, and soil particle moving velocity as proposed in Table 16.2. The classification presented in Table 16.2 indicates the general concept. If there is a vertical subsidence, it may involve traditional settlement analysis (Sec. 9.7).

Table 16.2 Classification of ground surface subsidence

A   Based on origin

Man-made	*Natural*
Fluid removal (water, gas, oil)	Earthquakes
Mining	Floods
Deep excavations, trench, quarry, borrow pit	Roots, flora, and fauna
	Surface and subsurface erosion
Construction operations (Pile driving, compaction)	Limestone sinkholes
	Tectonic activity
	Permafrost
Vibration (blasting, traffic)	
Geothermal	

B   Based on the load-deformation mechanism

Unbalanced stresses	*External loading*	*Fluid content or temperature*
Mining	Building loading	Fluid removal
Quarry and borrow pits	Earthquakes	Geothermal
Deep excavation, trench	Vibration	Shrinkage and swelling
Tunneling	Blasting	wet–dry
Limestone sinkholes	Flooding	Freezing–thawing
Tectonic movements		

C   Based on appearance of surface movement

Slow subsidence occurs	*Rapid subsidence occurs*
Fluid removal	Abandoned mining
Floods	Deep excavations, trench, quarry, and borrow pits
Tectonic activity	Limestone sinkholes
Geothermal	Tunnels

D   Based on the soil particle moving velocity

Subsidence type	*Velocity m/year*
Extremely rapid	0.6–10
Very rapid	0.4–0.6
Rapid	0.2–0.4
Moderately	0.08–0.2
Slow	0.006–0.08
Very slow	0.0004–0.006

Source: After Fang (1997).

If the particle velocity is larger than 0.06 m per year on a steep slope, or the backfill slope angle is larger than the internal friction angle, ϕ; (Sec. 10.3), it may cause a landslide (Sec. 14.3); if the particle's velocity is larger than 3 m/s, it will be classified as an earthquake (Sec. 11.2). If vertical subsidence is not uniform, it may result in a differential settlement problem.

Ground surface movement in the form of creep takes place in almost all subsidence areas. The rates of these movements vary during the year and are often confined to the surface layer. The rate increases as failure approaches, and the actual time of a subsidence can frequently be predicted by monitoring the ground movements. The subsidence in its early stage may be very small but can lead to failure if the rate and momentum increases. It has been found that the time to rupture is inversely proportional to the strain rate and is independent of the soil type.

Among these causes of ground surface subsidence, the removal of water, mining, and deep excavation are directly related to geotechnical engineering. These man-made causes will also create ground instability and can be prevented or reduced, if proper measures are carefully taken.

16.6.2 Natural and man-made causes on ground surface subsidence

16.6.2.1 Natural ground subsidence

1 *Tectonic movement and earthquakes*: Tectonic movements in the Earth's crust are a form of subsidence, which requires special precautions. Geologically active faults must constantly be monitored to determine the rate of movement in order to compensate for any future displacement in the design of a structure or foundation.
2 *Floods*: Subsidence due to flooding is based on the effects of water content changes on the soil–water system and is of concern in both partially saturated and saturated soils.
3 *Flora and fauna of the soil*: Soil is generally subjected to the corrosive power of the carbon dioxide (as carbonic acid) formed through respiration and fermentation as well as acids produced during decomposition of successive vegetation and due to enzymes secreted by microorganisms. The organisms change the surface layer of the soil by visible channels made by the roots of plants or burrows of animals and insects. The process of decomposition can cause subsidence through creation of zones of increased void volume and reduced strength.

16.6.2.2 Man-made ground subsidence

There are numerous man-made causes of ground subsidence such as (a) dewatering, (b) mining, (c) removing natural gas or petroleum, and (d) construction operations such as deep excavations. These causes are discussed in the following sections.

16.6.3 Subsidence caused by dewatering

Surface subsidence caused by dewatering has been a common problem in different parts of the world such as Bangkok, Houston, Las Vegas, London, Mexico City, Shanghai, Taipei, Tokyo, and Venice. In areas where the amount of available surface fresh water is limited, or where ever increasing industrial and municipal needs must be met, the only solution often left lies in the pumping of water-bearing aquifers. Once this becomes necessary, several problems can occur depending on the amount of pumping and the subsurface formation and geotechnical properties. In places where the ground-water table has been lowered in highly compressible soil sediments, shrinkage cracks can develop at the ground surface; due to evaporation through these initial cracks, they are able to extend deeper into the ground. Fifteen meters (45.7 ft) of subsidence has been observed in Mexico City as reported by Zeevaert (1983).

16.6.4 Mining subsidence

Geotechnical ground improvement in mining regions has become an area of extreme importance within the last two decades. Previously, mining activity took place in largely agricultural areas far from centers of population and as a result, ground surface subsidence was not of major consequence. However, due to the scarcity of land resulting from increased urban sprawl, these areas must now be considered as potential building sites. Also, the fact that large amounts of coal reserves remain under urban regions coupled with an increased demand for coal as energy and the economic necessity for maximum extraction adds to the problem. Ground surface subsidence caused by active mining effects on existing structures at surrounding areas must be considered. In general, subsidence resulting from active or abandoned mines is categorized according to 1 of 2 types, namely discontinuous and continuous surface deformations. The cause of these deformations is typically due to the failure of the arch at the mine opening.

1 *Active mining*: For active mining, control is obtained through the prediction of subsidence due to different mining configurations with approaches to prediction being either empirical or phenomenological. The maximum extraction and optimum mining configurations are then based on the existing structure's ability to accept the effects of the predicted subsidence for a trial configuration.
2 *Abandoned mines*: The stability of subsidence above abandoned mines is dependent on (a) the size and distribution of existing coal pillars, (b) condition of the rock above and immediately below the mine, and (c) the weight and thickness of the overburden. For construction over abandoned mines, minimization of subsidence effects is obtained either by support of the mine roof and overburden, or through the direct support of the proposed building, in which case the mine void is bypassed through the use of caissons or other deep foundation elements (Sec. 15.12) extending to the mine floor.

16.6.5 Construction operations

Three major construction operations which can cause ground surface movement are dewatering, pile driving, and deep excavation. Causes by dewatering have been discussed in Section 16.6.3. Pile driving and deep excavation will be presented as follows:

1 *Pile driven process*: The effect of pile driving in terms of surface movement is the occurrence of soil heave as discussed in Section 15.12. Whenever piles are driven, a certain amount of soil will be displaced both vertically and horizontally, with this amount dependent on the rate at which excess pore pressures, which build up during

driving, can dissipate. For cohesionless soils, the net soil displacement will be small due to quick drainage and resulting compression in and around the pile. Clay, however, is too impermeable to undergo volume change during the short period of driving, so the result being that a volume of clay equal to the volume of the pile is displaced laterally and upward. This effect is compounded by a close spacing of piles where zones of displacement can overlap causing a greater surface heave or settlement.

2 *Deep excavations*: Subsidence due to deep excavations such as trenches, quarries, tunneling, etc. is based on the same principle, which is applied to mining subsidence namely, the existence of unbalanced stresses in the area surrounding the open cut. Again, there is the tendency to reach a state of equilibrium and force closure of the opening. Among the factors which increase this tendency for closure include seepage forces (Sec. 12.9), which act on the sides of the cut during the dewatering phase of construction.

3 *Blasting and dynamic consolidation process*: Construction blasting (Sec. 11.11) and dynamic compaction operations (Sec. 7.8) will also cause ground surface subsidence through vibrations that serve encourage movement along failure planes. Moreover, such vibrations can lead to liquefaction (Sec. 11.3) which in turn may reduce the effective stress to zero.

16.7 Arid land and desert region

16.7.1 General discussion

A *desert* is defined loosely as a deserted, unoccupied, or uncultivated area. *Desertification* can be defined as the diminution or destruction of the biological potential of land that can lead ultimately to desert-like conditions; grazing lands cease to produce, dry-land agriculture fails, and irrigated fields are abandoned owing to salinization, water-logging, or some other form of soil deterioration. The majority of desert environments are located around the equator in areas where the temperature is high and there is a lack of rainfall. A majority of people believe that the desert is a geological cycle, which has a natural cause. However, this concept is not the whole picture, because deserts are also found in many semiarid regions. These deserts are caused by human behavior due to poor land management and a lack of soil–water conservation systems.

The general classification of tropical and desert regions is based on rainfall. A review shows that the annual rainfall in desert areas is less than 15 cm (6 in.) per year. In addition to low rainfall and high temperature, most desert regions have high winds. Low rainfall is a major feature in the desert region; however, in some places there is also heavy rain. Convection (Sec. 6.3) causes much of this precipitation on a desert. As the columns of air rising over hot places are cooled, the moisture they contain can condense and fall as localized but heavy rains.

16.7.2 Desert–environment interaction

There are three features controlling the desert climate: the high temperature, low precipitation, and high evaporation rates. In addition to these three factors, the sand–heat, sand–wind, and sand–water interactions must be examined.

1 *Sand–heat interaction*: The ground surface on a desert receives 90% of the incident solar radiation, which heats the ground and lower air layer. In contrast,

humid lands absorb 40% of the incident solar radiation. The remaining solar radiation in humid lands is dispersed in a variety of ways. Thirty percent is deflected by the water surface and land cover, 10% by dust particles, and 20% by clouds. At night, deserts turn cold because 90% of the heat generated by the solar radiation escapes back into the atmosphere. Temperatures in desert areas have been shown to vary in a somewhat regular pattern, reflecting both the annual and diurnal cycles of solar radiation. Superimposed on these regular cycles are fluctuations of variable duration and amplitude created by changing climatic conditions.

2 *Sand–wind interaction:* Sand–wind interaction has two possible mechanisms. The first is when loose particles of sand and silt are picked up by the wind and carried from one place to another, they strike against each other in the air. A second is when wind driven sand particles strike against pebbles or boulders on the ground with the result that additional particles hit other sand particles on the ground, making the sand particles on the ground jump upward. The height and moving distance depend on the size of the sand and pebbles. There are three basic patterns of sand dunes caused by sand–wind interaction: (a) *transverse dunes*: these dunes are the product of moderate, one-directional winds, which move only light or loose sand. Tumbling air eddies swirl heavier grains to the side, which tends to made ridges; (b) *longitudinal dunes*: these dunes occur when stronger one-directional winds move both fine and coarse sand particles cutting long troughs parallel with the path of the wind; and (c) *star dunes*: these dunes form in areas where the wind blows from all directions. Star dunes remain stationary.

The direction of wind and its speed are important to the design of antidesertification measures, because the direction and speed of the wind will control the patterns of sand dunes. From a geotechnical engineering point of view, when we build houses, highways, airport, railroads, dig a ditch, or plant a tree in desert areas, we must know the direction of the blowing wind in a particular desert, otherwise, the sand dune will move in and inundate everything in its path if the proper protection is not made. To prevent this occurrence, we must understand the interaction of sand and wind.

16.7.3 Desert water sources and interaction

1 *General discussion:* Water in desert regions appears in various forms. Periodically, there are heavy rains widely known as *cloudbursts*. The surface water resulting from this kind of rain disappears by a combination of surface runoff, evaporation, and infiltration. In a desert, the regolith is generally loose and dry and where bare, it is easily eroded. Basins formed by faulting and other movement of the crust dictate a larger part in determining the general sculpture of the land in an arid region than a moist one, because only rarely is water abundant enough to fill the basins. There are many lakes found in the arid lands and desert regions. The chemical character of the water in the lake and the type of precipitation that result depends on the types of rock formation underlying the lake basin. Lakes are called by various names depending on their chemical make up such as *salt lake, alkali lake,* and *bitter lake.*

2 *Water sources in desert region:* In the desert, although there is less surface water, there are many underground rivers. Certain regions have rich groundwater

reservoirs. In the Gobi and Taklamakan Deserts, large amounts of water are obtained from melted snow runoff from the Heavenly Mountain.

3   Water management: The more scarce the water, the greater the need for effective utilization of advanced technology for the acquisition and development of water supplies along with a realistic resource management program. A summary of commonly used water management techniques for arid and desert regions are (a) reducing wastewater, (b) rainwater collected from hillsides, (c) irrigation of saline water, (d) reducing evaporation from water surface, (e) reducing transpiration, and (f) utilization of solar energy from the desert to melt snow/ice from high mountains.

16.7.4   Desert soils and soil–water systems

1   Desert soil profile: A desert soil profile consists of three basic layers: the covering surface layer known as the desert varnish and desert pavement, the main part of the surface layer is sand or sand-gravel; and subsurface layer. A typical soil profile is shown in Figure 16.5. Surface soils have four distinguishing characteristics: (a) surface layer (dust layer) contains various constituents such as rock debris with which it mingles; (b) desert areas are less moist, therefore, desert soils are less chemically altered than soil in humid regions; (c) lack of water and lack of the leaching process in desert area, soils are generally saltier than humid soil (evaporation is greater than precipitation); and (d) the chemical weathering effect in desert soil is less, so the soils retain many more features of the past or parent material.

2   Desert varnish: The uppermost layer of the soil profile (Fig. 16.5) on many desert rock surfaces is a thin, dark surface patina known as *desert varnish*. Most desert varnishes are very old, but some are formed recently. The causes and formation of desert varnish are still debated. Early geomorphologists thought that desert varnishes were caused by the evaporation process, which will carry iron (Fe) and manganese (Mn) from underlying rock layers. Some scientists have found evidence that in some desert varnishes, elements are scavenged and fixed by lichens and bacteria.

3   Desert pavement: The top most surface layer of most desert soils consists of angular stones, known as the *desert pavement*. It is a dense surface layer, and most

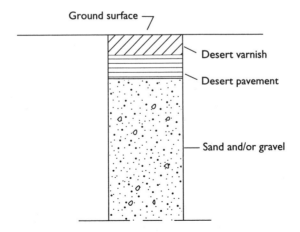

Figure 16.5 Typical desert soil profile including desert varnish and the desert pavement.

vehicles can drive on this layer without developing surface ruts. Desert pavement forms due to wetting–drying and hot–cold cycles.

4 *Desert subsurface soil layer:* The soil beneath the desert pavement in most cases has soluble salts. In humid areas, these salts are washed out entirely. However, in desert areas, there is a sequence of these accumulations. There are three distinct groups: (a) very dry desert soil has a layer of salt (sodium chloride, NaCl); (b) slightly dry desert soil, the common salt is washed out and gypsum (calcium sulfate dihydrate, $CaSO_4 \cdot 2H_2O$) is washed down to form a soil horizon; and (c) the wetter edges of the desert, the subsurface soil layer is formed by calcretes (calcium carbonate, $CaCO_3$), common salt and gypsum having been leached away. Some desert soils have other hard subsurface layers or horizons, which are due to a wetter past. These layers include *laterites* (high iron content) and *silcretes* (high silica content).

5 *Desert valley soils and soil–water system:* In general, many desert soils have a pebbly surface, created partly when the wind removes fine grains, partly when occasional storms wash grains away, and partly when pebbles rise to the surface as soil is intensely heated and cooled. In the lower parts of desert landscapes, there are two major types of soil existing in these areas: (a) vertisols and (b) salinization. In some semiarid land, where there is not too much salt, silica washed down in drainage water mixes and in combinations form a black color organic type of clay known as *vertisols*, which also forms through the breakdown of basalt rock in semiarid conditions. The second type of soil is far from fertile. In the silt deposited by rivers, salt may be brought to the surface by capillary action from a water table. When surface water evaporates, the residual salt crystals remain on the surface.

Rain patterns in the arid regions are unpredictable, because most deserts are located beneath semi-permanent anticylone, into which moist air, rain-bearing frontal systems, or tropical cyclones can only occasionally penetrate. When it happens, rapidly rising hot desert air, which cools and condenses as it rises, causes highly localized rainfall. In areas with fair amounts of vegetation and a thick layer of soil, most water generated by rainstorm can be absorbed. In arid lands, the surface layer becomes saturated at a much quicker rate.

16.7.5 Antidesertification measures

Geotechnical aspects of antidesertification measures include two general approaches: reducing wind velocity thereby decreasing sand content in the air, and consolidating surface and resistance in air–sand interaction.

1 *Reducing wind velocity:* Reducing wind velocity leads to a reduction of sand content in the air. There are two approaches: biological control methods and mechanical control methods. Biological control methods include stopping the movement by sodding, such as stopping sand movement by planting trees, planting trees between sand dunes, and surface sodding. Mechanical control methods include sand barriers, semi-visible cover up type barriers, and using gravel or clay cover on a sand layer.

2 *Consolidating sand surface:* The purpose for consolidation of the sand surface is to increase the resistant in the air–sand interaction. The use of chemical sprays on sand surface is a common technique.

16.8 Dredging technology and reclaimed land

16.8.1 Natural characteristics of dredging material

Dredging refers to the redistribution or movement of underwater deposits from one place to another in the water or out of the water. The equipment used to move such deposits is called a *dredge*. The dredge is operated by mechanical and hydraulic processes. In general, the dredging material has the following adverse geotechnical engineering characteristics: (a) fully saturated, very soft fine-grained clay; (b) low shear strength, low bearing capacity, and low permeability; (c) potentially contaminated with hazardous/toxic substances; and (d) contains large amounts of organic matter, crude oil residual, marine remains, gas, and air bubbles.

16.8.2 Reclaimed land

Reclaimed land, in general, is referred to as man-made land because of the use of material which is dredged from the bottom of harbors and ports when deepening navigation channels. The process serves two major purposes: cleaning up of the navigation channels and expansion of waterfront land. However, to use this dredged material for landfill purposes, an additional process must be added. Because dredged material itself is a fine-grained "muddy" soil, it has adverse geotechnical properties as listed in the previous section. It is also difficult to treat with conventional ground improvement techniques due to the following reasons: (a) in general, the reclaimed land covers a relatively large area; (b) dredging material is extremely sensitive to load environments such as temperature; (c) extreme difficulty in removing water or gas/air trapped in the soil mass by conventional dewatering processes; and (d) the costs associated with improving dredged material generally exceeds the cost of selecting an alternative construction site.

Ground improvement techniques in such reclaimed land conventionally include pre-loading, surcharge stabilization, and dewatering techniques for the purpose of reducing excessive water and air content and air. Consequently it increases the unit weight, bearing capacity, shear strength, etc. for use in construction. For dewatering (Sec. 5.7), current methods include wellpoint, vacuum, and electrokinetic processes (Sec. 6.12).

16.9 Municipal solid wastes and landfill technology

16.9.1 General discussion

There are numerous types of wastes and these may be broadly classified as municipal, hazardous, and radioactive. The sources for these wastes are generally domestic, industrial, agricultural, medical, or nuclear. Different types and characteristics of wastes require different methods for containment and/or treatment. Summarized types of wastes and disposal options are presented in Table 16.3. In this section, only the domestic (municipal solid waste) and landfill technology are presented and discussed, while subsequent sections are devoted to hazardous and radioactive waste.

Table 16.3 Types of wastes and disposal options

Types of wastes	Form of wastes	Re-use recycling	Immobilization technologies		
			Containment	Solidification	Vitrification
Domestic	Garbage	X	X		
	Sludge	X	X	X	
Industrial	Solid	X	X	X	X
	Liquid	X	X	X	
Mine	Mine Drainage	X	X	X	
Agricultural	Fertilizer		X	X	
Radioactive	Mixed		X		X
	Nuclear	X	X		X

Source: Meegoda, J. N., Ezeldin, A. S., Fang, H.Y., and Inyang, H. I. Waste immobilization technologies, *ASCE Practice Periodical of Hazardous, Toxic and Radioactive Waste Management*, v. 7, no. 1, pp. 46–58. © 2003 ASCE. Reproduced by permission of the American Society of Civil Engineers. Reprinted with Permission by ASCE.

16.9.2   Nature of municipal solid wastes

The nation's 100 largest urban areas generate more than a billion tons of garbage each day. Based on the 1976 estimate as reported by newspapers, it costs US$ 4.5 billion dollar just to manage the disposal of this garbage each year. Garbage sometimes is referred to as *urban refuse* or *municipal solid waste* (MSW). Waste disposal material generally referred to as garbage consists of anything that cannot be further used or recycled economically, thus its composition varies from country to country, community to community, as well as from season to season. Archaeological records indicate that land disposal of anthropogenic waste has been the method of choice for thousands of years. Wilson (1977) notes that the earliest recorded regulations for municipal solid waste were used by the Minoan civilization, which dates from 3000 to 1000 BC on the island of Crete in the Mediterranean Sea. Solid waste was disposed of in pits and covered with earthen material at regular intervals. Modern day methods are quite similar. Despite recent emphasis on source reduction as well as the growth of recycling and incineration, land disposal continues to be the dominant form of MSW disposal, accounting for 55% of the overall waste stream (US EPA, 2003). In 1998, US businesses, institutions, and residents generated, on an average, approximately 190 billion kg of municipal solid waste or about 2 kg per capita per day. This generation rate has increased by about 67% since 1960 and is also considerably higher than other nations (US EPA, 2003). The unit weight varies from 50 pcf (800 kg/m^3) to 400 pcf (6400 kg/m^3) depending on the amount of metal and debris. Since the term garbage is very loosely defined, other wastes such as sludge, agricultural, and industrial wastes may be part of garbage and end up dumped into landfills.

Preliminary classification of urban refuse consists of three groups, namely, materials that degrade relatively fast, relatively slow, and essentially not at all as shown in Table 16.4. Within the wastes there are several types such as sludge and industrial waste. A brief description of each waste is presented as follows:

1   *Sludge*: Solid, semisolid, or liquid waste material and water can be called *sludge*. Sludge results from the concentration of contaminants in water and wastewater

Table 16.4 Classification of fresh garbage

1 *Fast degradable materials*
   (a) Kitchen trash
   (b) Garden trash
   (c) Dead animals and manure
   (d) Papers and paper products
2 *Slow degradable materials*
   (a) Textiles, toys, and rugs
   (b) Glass and ceramics
   (c) Plastic, rubbers, and leathers
3 *Relative non-degradable materials*
   (a) Metals and appliances
   (b) Demolition and construction materials
   (c) Soils and rocks

treatment processes. Typical wastewater sludge contains from 0.5% to 10% solid matter.

2 *Industrial wastes*: Industrial wastes include heavy metals, rubber tires, organic fats, fatty acids, and many others. Among these wastes, the heavy metals can be recovered from a landfill and chemical waste site.

3 *Potentially incompatible wastes*: Many wastes when mixed with other wastes or other materials can produce effects that are harmful to human health and the environment. A detailed list of these items has been prepared by the US EPA (1990) as (a) heat or pressure; (b) fire or explosion; (c) violent chemical reaction; (d) toxic dust, mists, fumes, or gases; and (e) flammable fumes or gases.

16.9.3 Landfill technology

As indicated in Table 16.3, there are numerous waste types existing and various disposal methods are available. However, for each type of waste, there are several types of options, and interrelationships of waste treatment approaches are illustrated in Figure 16.6. Geotechnical engineers are frequently involved in the design and analysis of waste containment systems, which are the most common means of dealing with MSW.

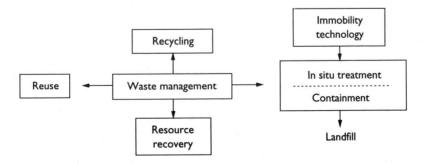

Figure 16.6 Interrelationship of waste treatment technology.

16.9.4 Landfill stability

There are several basic problems in landfill design which require geotechnical engineering knowledge. Problems involved include densification (compaction) of landfill systems, both surface and subsurface drainage systems, stability of landfill slopes, and landfill control facilities. In the following section, only the compaction and slope stability of landfills is emphasized.

1 Compaction of landfill: When waste (garbage) delivery is made by truck and dumped into the landfill sites, in some cases, this garbage is spread into a thinner layer, mixed with some locally available earth material, and compacted by conventional compaction equipment and procedures (Ch. 7) for the purpose of (a) covering up of unattractive landfill sites, (b) minimize odor and/or, (c) prevent animal and bird vandalism. There is no standard rule or regulation how garbage should be dumped or compacted at the present time. As discussed in Section 7.3, the main purpose of the compaction process is to change loose material into a denser condition by reducing further settlement, increasing bearing capacity, and increasing shear strength. From a landfill operator perspective, the interest in compaction is tied to the need to maximize the amount of waste which can be put into a permitted landfill. This need is usually expressed in terms of maximizing "air space," which refers to capacity for more waste, not air void volume. In landfill areas, when proper compaction is applied, it can also reduce potential fire hazards in the landfill. The less air trapped in the landfill, the less potential for fire hazards.

2 Process of compaction in landfills: The process of compaction plays an important role in landfill stabilization. It requires planning during the waste disposal process period. The following procedures may be considered: (a) garbage comes in all types and sizes and it cannot be perfectly distributed in the landfill. However, efforts can be made to distribute waste uniformly within a given compacted layer; (b) if not, heavier items should be dumped to the center of landfill for the purpose of controlling the stability of the fill; (c) spread the newly dumped garbage as thin as possible. Since garbage is an unstable material, it requires stabilizing in order to reduce excessive settlement. Mix locally available soil with the garbage or add fly ash, lime, or others; and (d) heavy rolling is suggested. The weight of the roller is related to the thickness of landfill layer. In some cases, use of dynamic consolidation procedures (Sec. 7.8) to control the compaction in landfill areas recently has come into practice. Alternatively, the bioreactor approach, which involves the addition of air, water, and/or leachate to accelerate the decomposition of the degradable fraction of the waste, may also be used to enhance stabilization.

3 Landfill slope stability analysis: Stability analyses proceed as discussed in Chapter 14, with the exception that properties of waste must now be known or assumed. Stability calculations are very much time-dependent, according to the decomposition of the waste itself. The critical height, H_c, (Sec. 14.4) of the slope is important, and when old or abandoned landfill sites are to be used for other purposes such as commercial parking lots, recreation parks, or as part of highway routes, etc., excavation into the landfill is sometimes required. In such case, the slope angle existing in the landfill is important. In general, landfill slope failure is similar to earth slope failure with the exception that landfill failures are caused by highly heterogeneous materials and a nonuniform decomposition process. Moreover, the types of failure also include falls, topples, slips, and slides. Falls and topples are due to the lack of cohesion between loose refuse pieces. Slope failure potential is directly related to the compaction control during the waste disposal compaction process as discussed in Step 2. The better control of the compaction, the less risk for slope failures.

4 Settlement analysis of landfill: As discussed in Section 9.10, current practice of settlement prediction is still based on the Terzaghi one-dimensional consolidation theory with modifications to overcome the environmental effects on future settlement. Due to the nature of the problem, the settlement analysis of landfill cannot be solved by mathematical equation(s). Semi-empirical methods such as the equations discussed in Section 9.10 may be the best approach at the present time.

16.10 Hazardous and radioactive waste

In addition to MSW, industrial development and military legacies have led to the production of other classifications of waste, hazardous, and radioactive waste. In terms of mass, the amount of hazardous waste generated is similar, 194 billion kg in 1995 (US EPA, 1997) while the nature and concerns of hazardous waste have made it the leading environmental issue as measured by federal and private investments (LaGrega *et al.*, 2001). The presence of radioactive waste derives mainly from that generated or regulated by the US Department of Energy (US DOE). Reported on a cumulative basis, there are approximately 380,000 m^3 of high-level, 220,000 m^3 of transuranic and 3.3 million m^3 of low-level radioactive waste (US DOE, 1997). According to the US DOE, 89% and 68% of the radioactive waste by radioactivity and volume, respectively, are from weapons production while the rest is attributed to nuclear power generation. Details on hazardous waste management may be found in LaGrega *et al.* (2001). This section provides emphasis on radioactive waste.

16.10.1 Radioactive nuclear wastes

Among all wastes generated either by nature or man-made from various sources, radioactive nuclear waste is the most complex because of its unstable and unpredictable behavior. It can change forms and properties during the decay process. It is not only a hazardous/toxic substance by itself, but it easily contaminates its surrounding areas including the soil–water–air system in the environment. At the present time, radioactive nuclear wastes are not classified as hazardous and not covered by the Resource Conservation and Recovery Act (RCRA). It is controlled directly by the US Atomic Energy Act. Because of the governmental control system and administrative policy, nuclear waste containment systems are not generally known to the public in comparison with the other waste containment facilities. Table 16.5 presents a comparison of general characteristics between radioactive and common landfill municipal solid waste.

There are approximately 128 varieties of unstable radioactive elements in existence with only about 68 that are controllable. Radioactive elements will contaminate larger territories including ground soil, rivers, streams, aquifers, as well as air. Significant amounts of radioactive waste are found in harbors and marine sediments,

Table 16.5 Comparison of general characteristics between radioactive nuclear waste and common landfill municipal solid waste

Characteristics	Radioactive waste	Common landfill
1 Stability of mineral structure	Unstable	Stable
2 Hazardous	Yes	Yes
Toxic	Yes	Maybe
3 Decay process	Complex	Relatively simple
4 Contaminated area	Large	Limited area
5 Interacts with environment	Not well understood	Fairly well documented
6 Awareness	Little	Much better
7 Current research	Little	Advanced
8 Potential effect on human health	Great	Less
9 Control methods	Little	O.K. (in general)

as well as in landfill sites. The Federal Nuclear Waste Policy Act of 1982 and the DOE in 1983 gave general guidelines for the location of an underground site to serve as the nation's first permanent storage facility for high-level radioactive waste. Yucca Mountain, Nevada has been scrutinized as the potentially best location for this site. The permanent storage of high-level radioactive waste will be in geologic repositories specifically mined for that purpose. Among the many problems with long-term confinement of high-level radioactive waste are the threat of human intrusion and the possibility of contaminant transport through groundwater migration.

16.10.2 Characteristics of radioactive material and nuclear waste

1 Radioactivity and radioactive decay: Radioactivity as a phenomenon is a natural process in which energy is released by certain substances in the form of invisible radiation. Certain nuclei are unstable and undergo radioactive decay into a more stable form. The energy released in a radioactive transformation must satisfy the Law of Energy Conservation. During the decay process, the amount of radioactive nuclei which remains decrease as time elapses. Each type of radioactive nuclei is characterized by a quantity called *half-life*, which refers to the time required for 50% of the material to decay. There are five general types of radioactive decay, namely alpha (α)-decay, beta (β)-decay, gamma (γ)-decay, electron capture, and proton emission. Among these five decay processes, the α-decay, β-decay, and γ-decay are closely related to engineering problems. The α-particles have extremely high ionizing action within their range as they are composed of two protons and two neutrons as illustrated in Equations (16.1) and (16.2). The β-particles exhibit relatively low energy levels and can travel further and faster than α-particles, but like α-particles, β-are rapidly attenuated by a thin layer of solid material. The γ- rays are often emitted by nuclei following radioactive α- or β-decay.

2 Engineering properties of radioactive elements: There are numerous radioactive elements involved in the series. However, only some elements are of general concern from an environmental geotechnology point of view such as uranium (U), radium (Ra), radon (Rn), and radon daughters. Properties of the radon family are discussed in Section 16.11. In this section only the uranium and radium information is presented. Uranium is the chief source of the elements uranium and radium. The mineral is a combination of the oxides of uranium, UO_2 and UO_3, with small amount of the other elements. Uranium itself never occurs free in nature but is found chiefly as an oxide in the mineral pitchblende where it is associated with radium. It is a hard metal but malleable and is soluble in mineral acids. Chemically, uranium has a number of isotopes, and its atomic weight varies from 234 to 239. It has the highest atomic weight of all the materials occurring normally in nature, being 92 on the periodic table of the elements. Uranium is highly unstable and can be made to disintegrate with explosive violence. Radium is a peculiar radioactive element scattered in minute quantities throughout almost all classes of rocks. It is commercially obtainable from uranium ores. The ratio of radium to uranium ore is generally about 1:3,000,000.

16.10.3 Sources and classification of radioactive wastes

1 *Sources of radioactive wastes*: Six general sources of nuclear wastes include: (a) mill tailing, the sludge generated in the extraction process from uranium mines and mills with piles of powdered rock containing large amounts of radium and radon gas; (b) residual from nuclear power plants assemblies of fuel rods stored under water near a nuclear reactor; (c) nuclear waste from commercial use, such as waste from hospital and private research laboratories; (d) sources of γ-energy for non-destructive testing; (e) nuclear explosives in construction and mining operations; and (f) nuclear waste from the manufacture of nuclear weapons. Among these resources, the wastes generated from nuclear power plants are the most critical and of major concern (Fang, 2002).

2 *Types or classifications of nuclear wastes*: Nuclear waste is classified as (a) High-level radioactive wastes (HLRW) which is generated during the reprocessing of spent reactor fuel containing thousands of curies per cubic meter (Ci/m^3). HLRW contains uranium (U^{235}), plutonium (Pu^{239}), strontium (Sr^{90}), cesium (Cs^{137}), and others; and (b) Low-level radioactive waste (LLRW): LLRW is defined loosely. In general, LLRW comes from commercial use such as hospitals and private research laboratories. Most LLRW has a low radioactivity of about 35 Ci/m^3. However, some LLRW is extremely radioactive and may contain relatively large quantities of fission products with a half-life longer than 25 years.

16.10.4 Disposal or management of radioactive nuclear wastes

There are five types of disposal or management methods for radioactive elements. They are storage, vitrification and solidification, isolation, emplacement, and elution. These five methods are discussed in the following section; however, the controlling of radioactive toxic radon gas (Ra) is presented in Section 16.11.

1 *Storage (container and dump sites)*: Radioactivity decreases in intensity by a factor of 10–100 in 50 years. In the United States, most of the agent reactor fuel has been left for 10 years or more in water-filled pools at individual plant sites

waiting for their permanent disposal which was set for 1998 by mandate in the Nuclear Waste Policy Act (1982): most are LLRW.

2 *Vitrification and solidification*: The vitrification technology originated in the 1950s when scientists began studying ways of locking radioactive waste in glass. Subsequent research on vitrified waste showed that glass could be 10,000 times more durable than other waste forms, including cement. The process of vitrification is similar to the natural occurring obsidian (silicate glass) originated from rapidly solidified magma.

3 *Isolation*: Waste can be isolated from the natural environment by storing it in remote places such as deep seafloor deposition, bottom of deep mine shafts, distant and deep places in deserts or space such as the moon.

4 *Emplacement*: For the effective emplacement of waste, the safety considerations as related to waste transportation and insertion in a repository are important. To achieve effective radioactive waste management systems, three basic criteria must be considered: (a) the waste form itself must be inert and insoluble in the repository environment; (b) the canister and over-pack material must confine any radionuclides that do leach out the primary waste form; and (c) the rock formation should be impermeable. The main interest in rock interaction with nuclear waste is attempting to determine what type of rock will interact less or be less sensitive to waste and remain in an impermeable state.

5 *Elution*: Elution means removal or extraction of radioactive element from waste as part of a recovery process. At present, only laboratory experiments have been performed.

16.11 Radon gas

16.11.1 *Radon gas and noble gas family*

Colorless, odorless, tasteless radioactive and toxic radon gas (Rn) is produced naturally in the ground by normal decay of uranium (U) and radium (Ra) and widely distributed in trace amounts in the Earth's crust. Radon gas is a noble gas, and as such it is also considered to be an inert gas, which lacks significant chemical or biological activity. It contains only one atom (monoatomic molecule) and does not mix with other elements. However, based on recent findings, radon gas can be found in the soil, water, and air and is influenced by local environments, such as temperature, pH value, ion exchange reaction, redox reaction, etc. Most radon gas is concentrated in the oxidation belt, which is at a relatively shallow depth below the ground surface. Under normal conditions, the amount of radon gas seeping into the atmosphere or entering into residential buildings is very little and will not be harmful to human health.

16.11.2 *Mechanism of radon–environment interaction*

There are four steps for radon gas (Rn^{222}) to be released from rock formations and its subsequent seepage into the soil–water–air system.

1 *Radon moving from parent rocks into host rocks*: The parent rock is igneous rock such as granite, basalt, gabbro, diorite, etc. The host rocks include

limestone, dolomite, conglomerate, breccia, etc. One common feature of the host rock is that they possess large porosity or cavities.

2   *Radon release from rock-mineral*: Radon is released from rock-minerals by alpha (α)-recoil processes (Fleischer, 1983). Fang (1990) introduced the concept of environmental stresses to the α-recoil process. The α-recoil derives its name from the process by which a radon atom recoils from a decaying parent radium atom. On decay, radium (Ra) emits a α-particle (He) to form radon (Rn^{222}) as illustrated in Equation (16.2).

3   *Radon gas interaction with water and air*: Radon gas interacts with water and air through common types of transport mediums, such as dust (Sec. 2.9) in the air and suspension in the water or in the soil–water system. Interaction between radon and dust-suspension is through physical types of adsorption action, that is, van der Waals's adsorption (Sec. 4.6).

4   *Interrelationship among uranium, radium and radon*: The interrelationship among uranium (U), radium (Ra), and radon (Rn) can be represented by Equations (16.1) and (16.2) as follows:

$$
\begin{array}{llll}
\text{Radium (Ra}^{226}) & & \text{Radon (Rn}^{222}) & \text{2 protons} \\
\text{88 protons} & -\alpha\text{-decay} \rightarrow & \text{I86 protons} + & \text{2 neutrons} \quad (16.1) \\
\text{138 neutrons} & & \text{136 neutron} & {}^{4}\text{He}
\end{array}
$$

Equation (16.2) can be rewritten into a nuclear equation as

$$^{226}\text{Ra} \xrightarrow{\alpha\text{-decay}} {}^{222}\text{Rn} + {}^{4}\text{He} \qquad (16.2)$$

In examining Equations (16.1) and (16.2), the radium nucleus (Z = 88), mass weight (A = 226), has 88 protons and 138 neutrons. When Ra^{226} emits an α-particle (^{4}He), two protons and two neutrons are carried away. Therefore, the residual nucleus has 86 protons and 136 neutrons. The product of radium decay (the daughter) is a different element. The atomic electron structure changes following the decay event. To accommodate the new nuclear particle change, it releases two of its 88 electrons. These two electrons eventually attach themselves to the emitted α-particle and form a neutral atom of helium (^{4}He). Thus, the original neutral radium atom decays and two neutral atoms are formed, one radon (Rn) and one of helium as shown in Equation (16.2). In addition to Equations (16.1) and (16.2), the interrelationship of U, Ra, and Rn elements can be considered as part of the thermal energy field.

16.11.3   Radon gas mitigation methods

There are two major methods to mitigate the radon gas at the present time: (a) The passive approach, and (b) The active approach to the problem. The passive approach is recommend by the US Environmental Protection Agency (US EPA, 1986) and Department of Environmental Resources (DER, 1985).

1   Passive mitigation method (EPA/DER method): The passive mitigation method uses special equipment to measure the amount of radon gas existing in the atmosphere for a specified period of time and sent to a laboratory for analysis. There are

Table 16.6 Radon mitigation procedures

Step	Mitigation method
1	Remove large portions of radon gas in subsurface soil layers by dewatering or/and alteration of subsurface drainage pattern channeling subsurface water (radon contaminated water) away from site in question
2	Some radon gas escaping from ground soil onto the air can be removed by alteration of surface drainage pattern by channeling surface water. Some radon gas in the atmosphere will mix with dust and float in the air and join with natural dynamic equilibrium of the ecosystem
3	Since large portions of radon gas are removed by steps 1 and 2, the remaining radon gas content will be significantly reduced. It still has the possibility that some will enter into buildings. In such cases, routine house cleaning can be used effectively

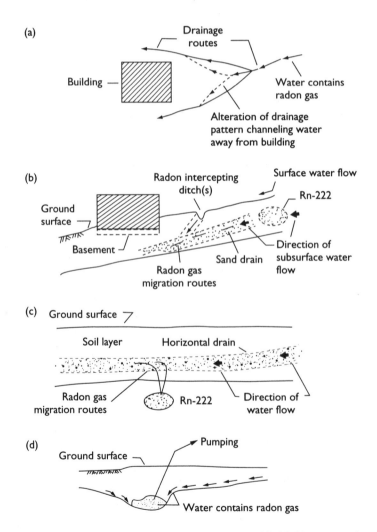

Figure 16.7 Drainage network to change radon migration route(s). (a) Alteration of surface drainage patterns, (b) Change both surface and subsurface drainage systems; (c) Installation of horizontal drainage systems; and (d) Pumping radon-water from the subsurface.

two types of measuring devices, namely the "Charcoal Canister" and the "Alpha Track Detector." The test period for Charcoal Canisters is about 3–7 days. The minimum test period for the Alpha Track Detector is 2–4 weeks. The measurement results from these two types of devices are reported as the working level (WL) or concentrations of radon gas as picocuries per liter (pci/L). WL is a measurement of energy release. The value of 1 WL is the amount of radon daughter, the decay of which will result in the emission of 1.3 billion volts of electron energy. If converted to heat, it might raise the temperature of a cup of water about a half a degree. A picocurie is 1 trillionth of a curie. One picocurie per liter of air is about two radon atoms disintegrating per minute in every liter of air in a room.

2 *Active mitigation method*: The active mitigation method was proposed by Fang (1990, 1997). The method is based on the dewatering technique by controlling the surface and subsurface drainage patterns. The purpose for dewatering of surface and subsurface waters is to dilute and/or redistribute radon gas in the environment before seeping into the atmosphere or building. Systematic approaches together with dewatering systems for radon mitigation procedures are summarized in Table 16.6 and Figure 16.7. Figure 16.7 is a drainage network to change radon migration routes. Figure 16.7(a) shows the alteration of surface drainage patterns, channeling radon contaminated surface water away from building. Figure 16.7(b) shows changes to both surface and subsurface drainage systems. Figure 16.7(c) shows the removal of subsurface radon-water by the installation of horizontal drainage systems, and Figure 16.7(d) presents the removal of subsurface radon-water by pumping techniques (Ch. 5).

16.12 Waste control facilities (containment systems)

The central objective of a landfill (whether for municipal, hazardous, or nuclear) facilities is to isolate the waste from the natural site hydrogeology as well as sensitive receptors. To that end, the prevalent approach has been to prevent the infiltration of moisture into or out of waste containment facilities through the use of earthen or synthetic barrier materials. In modern landfills, there are multiple layers and components to mitigate against such contamination. Of these components, the low-permeability barrier material in the cover and liner are the most critical. The cover prevents moisture from entering the landfill from above and generating leachate, while the liner prevents leachate from reaching groundwater supplies and the ambient environment below. Clay-based earthen materials are the most prevalent in barrier designs (Reddi and Inyang, 2000). However, landfill design, and therefore the anticipated performance, has changed in response to regulations, the principle of which is the RCRA of 1976. The trend has been to handle the aforementioned generation rates of waste with fewer, larger landfills. In 1960, there were more than 20,000 active landfills in the US while as of 1998, there were only 3091 (Tammemagi, 1999; US EPA, 2003). The thousands of landfills constructed prior to RCRA and its subsequent amendments did not necessarily incorporate barrier liners. For such facilities, the performance of the cover is of even greater importance in minimizing leachate generation and off-site migration as compared to fully lined landfills.

Currently, MSW and hazardous waste landfills are designed for 30 years of post-closure life (US EPA, 1990, 1994). For low-level radioactive waste disposal,

512 Problems in environmental geotechnology

institutional control and maintenance are anticipated to end 100 years after closure (Nyhan *et al.*, 1997). Beyond this time period, barriers are expected to provide passive resistance toward infiltration to prevent the migration of radionuclides for an additional 300–500 years (US NRC, 1982). The risks posed to human and ecological health, however, do not end after these arbitrarily defined post-closure periods. There is thus the motivation to determine and understand the performance of these materials over time. Specifically, while the low permeability and low-cost features of clayey earthen materials are attractive in landfill design, questions remain as to their long-term durability and response to life-cycle stresses such as freeze–thaw action or desiccation (Daniels *et al.*, 2003).

16.12.1 Design considerations

There are several basic techniques for waste containment purposes including pumping, capping, draining, and the use of slurry barrier walls. The selection of various types of containment systems is based on the type of waste materials and geohydrological and geoenvironmental conditions of the waste site. Often, more than one containment method is used in a given location. With time, the waste material might slowly biodegrade or chemically change to nontoxic forms, or new treatment methods may become available for the detoxifying of the waste. In other words, the containment is used to "buy time" under emergency or temporary conditions. To select proper and effective containment systems, some basic factors of the characteristics of wastes must be evaluated, including the types of wastes, form, size, location of wastes, and the extent and concentration of contamination.

There are two types of approaches for controlling the wastes, namely active and passive approaches. (a) *Active approach*: an active approach for containment is one that requires ongoing energy input. Examples of active components include disposal wells, pumping wells, and treatment plants. Disposal wells include injection wells. Hazardous waste may be pumped into deep wells to allow for percolation through porous or permeable subsurface strata and then contained within surrounding layers of impermeable rock or soil. Pumping wells include pumping to create a pumping ridge and treatment process or techniques which change the physical, chemical, or biological composition of any hazardous waste and so render it non-hazardous and safe for transport, capable of recovery, and/or storage to reduce its volume; and (b) *Passive approach*: the passive approach of a containment system includes those methods that do not require ongoing energy input. Typical examples of passive components include hydraulic barrier walls, top seal (cap), and bottom seal (liner). The structural containment components are illustrated in Figure 16.8 and a discussion of each system follows:

1 *Vertical hydraulic barrier wall*: vertical hydraulic barriers walls include (a) soil–bentonite slurry walls: they are designed to have proper density, viscosity, and filtrate loss properties. Trench depths must reach an impervious soil layer. This is important, otherwise the liquid waste may leak through this impervious layer; (b) cement–bentonite slurry walls: the trench is excavated in a similar

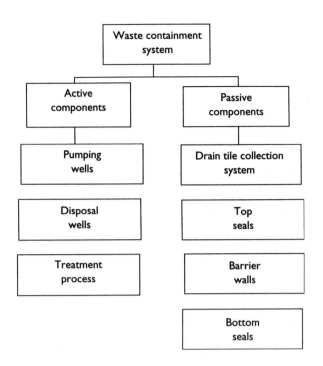

Figure 16.8 Types of containment systems.

Source: Evans and Frang (1986). Copyright ASTM INTERNATIONAL. Reprinted with permission.

manner to the soil–bentonite slurry wall. However, cement is added to the slurry. The cement–bentonite–water slurry is left in the trench and allowed to solidify; and (c) composite barriers system: the composite barrier offers several advantages over a single type of barrier. A composite barrier will provide resistance to a wider range of contaminants (Evans, 1991). Several proposed composite barrier systems currently used are presented as hybrid cutoff walls constructed with high-density polyethylene (HDPE).

2   *Top seal (cap or cover)*: the purpose of top seal is to (a) control surface water so as to maximize surface runoff and minimize infiltration into landfills, (b) protect against burrowing animals which may dig into landfill areas, (c) reduce odor, and (d) reduce leachate production and/or contaminant transport potential. There are several types of materials that can be used as top seal barrier layers including natural clay caps, bentonite clay caps, synthetic membrane caps, waste paper sludge caps, and scrap rubber tire cement mixture caps.

3   *Bottom seal (liner)*: The barrier wall and top seal can be applied for existing or newly constructed containment systems. However, the bottom seal must be constructed prior to waste placement and as such applies to new facilities. The major purpose of a liner is to prevent leachate or liquid waste from migrating into non-polluted aquifers. The barrier layer liner systems can consist of native clays, processed clays, and geosynthetic membrane as previously discussed for top seals.

16.12.2 Precautions during design, construction, and maintenance stages

The effectiveness of waste containment hinges on careful evaluations of the overall planning, analysis, and design of the entire system prior to construction. Some of these relatively important issues relating to the overall effective system may assist the designer to be aware or take precaution to design and construct safer future waste containment systems including seismic zones and dynamic loads, potential floods, locating the impervious soil layer, sinkhole or cavity nearby the landfill site, and workmanship and maintenance. Detailed discussions on these considerations are given by Reddi and Inyang (2000), and Qian et al. (2002).

1 Seismic zone and dynamic loading effects: The potential failure modes of landfills in seismic zones according to Inyang (1992) can be classified into three main categories on the basis of causes: (a) faulting through a landfill, (b) ground and component strains without liquefaction, and (c) ground and component strains due to liquefaction.

2 Shrinkage, swelling, and cracking: As discussed in Sections 4.4 and 8.3, shrinkage, swelling, and cracking in soil are natural processes that occur frequently in earthen structures due to thermal energy imbalances in the soil mass. The two common causes of energy imbalance in soil mass are nonuniform moisture and temperature distribution. These phenomena are significant around most hazardous/toxic waste sites.

3 Workmanship and maintenance: Workmanship during the construction stage and proper maintenance after the construction must be properly managed. Construction quality assurance is required to correctly transfer the design as contained in reports and drawings into the real systems as constructed in the field. Careless or incompetent construction can defeat the best design.

16.13 Environmental geotechnology perspective

16.13.1 Instruction and research trends and directions

1 Instruction: Since 1925, when K. Terzaghi introduced the concept of soil mechanics into the civil engineering field, basic concepts and theories have been established which have greatly improved modern design and construction technology in civil engineering. However, the majority of the teaching effort is placed on the physical–mechanical behavior of soil. In some institutions, geotechnical courses have become part of engineering mechanics courses, thereby ignoring the fundamental behavior of the soil itself. Since soil is very sensitive to the local environment as discussed in Chapter 4, courses such as soil science, physical chemistry, geology, microbiology, etc. are also recommended to students.

2 Research: Most research efforts are placed on loading and short-term duration to predict long-term performance. Great efforts are placed on waste control systems, but other research such as radioactive nuclear wastes, acid rain, acid mine drainage, antidesertification measures, etc. receive less attention. At present, most design approaches are based on load factors with very little consideration of environmental

factors. The problems of soil behavior cannot be solved in a vacuum, without considering all interrelated aspects. These aspects include changes in temperature, pressure, porewater composition as well as the load condition and soil properties.

16.13.2 Energy recovery, recycling, and reuse of wastes

Energy cannot be wasted. It must be recovered, recycled, and reused. They are all interrelated to each other, as indicated in Figure 16.6. Recycling and energy recovery from waste has increased over the past several decades. Aluminum cans, plastic containers, and used papers have been recycled and used again. Heavy metals such as lead (Pb), mercury (Hg), zinc (Zn) trapped in the ground soil in the landfill areas have been recovered using electrokinetic (Sec. 6.8), electromagnetic (Sec. 6.9), and other processes as part of soil decontamination and energy recovery programs. Therefore, the composition of many landfills has changed since the energy recovery and recycling program began. Reuse of wastes for conventional construction material has been studied extensively in recent years such as the utilization of scrap rubber tires used as light weight aggregates and petroleum contaminated soil used for low-volume highway construction materials. Similarly, uses of by-products of steel manufacturing and coal combustion have also been found. The extent to which materials are reused will increase as disposal sites become increasingly more difficult to site and permit.

16.13.3 New areas in geotechnology

1 *General discussion*: As discussed in Chapter 1, geotechnology requires knowledge from other disciplines in order to understand how soil responds to changing environments. Likewise, other disciplines need our knowledge to assist in solving their problems. Such areas include problematic soil regions, mining subsidence areas, sanitary landfills, cold regions, desert areas, and wetlands. Some areas have been evaluated and assisted by geotechnical groups to some extent and some are of less concern. Some relatively important areas have been briefly discussed in this chapter. Some additional new areas are further outlined in this section as follows:

2 *Geomicrobiology*: As discussed in Section 1.1, geomicrobiology is the study of the characteristics of vegetations and bacterial activities in the ground soil and their effect on the engineering behavior of soil–water system. It has only recently been discovered that the vast majority of bacteria have not been cultured. Many bacteria produce enzymes and exopolymeric substances which can serve to influence properties such as hydraulic conductivity, shear strength, and compressibility (Daniels and Cherukuri, 2005; Daniels *et al.*, 2005). The extent of this interaction has implications for natural behavior as well as for purposes of soil improvement. Brief discussions are presented in Section 4.12. Further studies are needed.

3 *Archaeology, archaeo-geotechnology, and geo-archaeology*: Archaeology is an interdisciplinary comprehensive field. Since 1960, this field has gradually shifted into science rather than an art field. The major subjects for archaeology's concern deal with chronological methods, site survey, excavation, and scientific and anthropological analysis of past cultures to gain insight into the basis of civilization. In most cases, the remaining evidence of past cultures is quite delicate and requires protective techniques to preserve the archaeological provenience for more accurate current as

well as future studies. Therefore, there is a great need for the application of the science of geotechnical engineering to assist the art and science of archaeology in order to protect the archaeological provenience, which usually lies buried to some extent in the ground soil.

Geotechnology and archaeology both deal with earth sciences and must be correlated in certain ways. The archaeological study covers five basic stages: (a) planning stage, (b) excavation stage, (c) retrofit stage, (d) stage for estimation and identification of the ages of archaeological test sites and artifacts, and (e) protection and preservation of sites and artifacts. In many archaeological excavations, numerous artifacts are damaged during the excavation process and many are ruined due to the pressure release after the removal from deep burial sites and exposure to the air. Determination of overburden pressures, porewater pressure, and swelling–shrinking characteristics of in situ conditions are important. Also research into the mechanism of the soil corrosion process, soil–water interaction, and physicochemical behavior of soil is needed.

16.14 Summary

While geotechnical engineering is typically considered in terms of stress, strain, and seepage, the multidisciplinary nature of many problems warrants consideration of other issues. These include wetlands, deserts, subsidence, and waste management. Wetlands, flood plains, and coastal margins represent a sensitive ecosystems, and special considerations have to be made when working in these areas. Soil erosion is caused by water, wind, and construction operations. Ground surface subsidence is one of the major problems in environmental geotechnology. The causes of subsidence are both natural and man-made. Among these causes, dewatering and mining are most critical. There are four major factors which lead to the development of a desert-like environment: low precipitation, high temperature, high proportion of evaporation to precipitation, and poor land management. The predominant erosive factor is the combination of sand and wind. Geotechnical aspects of antidesertification measures are summarized including biological, mechanical, and chemical control methods. Radioactive wastes are unstable and unpredictable. There are five basic approaches on disposal or management of nuclear wastes including storage, solidification, isolation, emplacement, and elution. Two major procedures for controlling radon gas are the active and passive approaches. Almost 90% of all hazardous wastes are in liquid form. An understanding of the pore fluids on liner behavior is essential to the design of various components of waste control facilities. Waste control facilities include hydraulic barriers, top seal (covers), and bottom seal (liners).

PROBLEMS

16.1 What are the so-called sensitive geological and ecological regions? Discuss two of them in detail.

16.2 What are the major environmental geotechnology problems in wetlands, estuaries, and marine margins? What causes the swamp? How would you control the swamp and convert it into useful land?

16.3 Discuss erosion types, causes, and mechanisms of soil erosion processes.

16.4 Discuss the soil profile in the desert region. Define and discuss desert pavement and desert varnish.

16.5 How many types of desertification measures are there? Discuss in detail one of the desertification measures.

16.6 What is the nature of dredged materials, and what is reclaimed land? Why it is difficult to improve reclaimed land, and why does it need improvement?

16.7 What is radioactive nuclear waste?

16.8 Why is radon gas classified as one of the six inert gases?

16.9 It is required by the design that landfill liners have a thickness of 2–3 ft (0.61–0.92 m) with a permeability of less than 1×10^{-7} cm/s. If it is assumed that only vertical flow will occur, how long will it take for a contaminated particle to travel through the clay liner? Comment on other factors that affect the travel time.

16.10 What are the basic factors for utilization of contaminated wastes to be used as conventional construction material? Identify and characterize ground soil pollution. How does ground pollution relate to ground soil color and cracking patterns?

16.11 Comment on environmental geotechnical prospectives with respect to instruction and research. What is your opinion on the conventional approach to soil mechanics and foundation engineering?

References

Acar, Y. B. and Olivieri, I. (1989), Pore fluid effects on the fabric and hydraulic conductivity of laboratory compacted clay, *Transportation Research Record* 1219, pp. 144–159.

AISC (1993), *Manual of Steel Construction-Load and Resistance Factor Design*, 2nd ed. American Institute of Steel Construction, Inc., Part 2, pp. 1–45, Chicago, Illinois.

Aldrich, H. P., Jr. (1956), Frost penetration below highway and airfield pavements, *Highway Research Board Bulletin* 135, pp. 124–149.

Algermissen, S. T. (1969), Seismic risk studies in the United States, *Proceedings of the 4th World Conference on Earthquake Engineering*, Santiago, Chile, v. 1, pp. 14–27.

Al-Hussaini, M. (1981), Comparison of various methods for determining k_o, *ASTM Special Technical Publication* (STP) 740, pp. 78–93.

Alther, G., Evans, J. C., Fang, H. Y., and Witmer, K. (1985), Influence of inorganic permeants upon the permeability of bentonite, *Hydraulic Barriers in Soil and Rock*, I. A. Johnson, R. K. Frobel, N. J. Cavalli, and C. B. N. Perterson (eds), *ASTM Special Technical Publication* (STP) 874, pp. 64–73.

Ambrose, J. and Vergun, D. (1995), *Simplified Building Design for Wind and Earthquake Forces*, 3rd edn. John Wiley & Sons, NY, 353p.

Andersland, O. B. and Anderson, D. (1978), *Geotechnical Engineering for Cold Regions*, McGraw-Hill, 566p.

Anderson, J. N. and Lade, P. V. (1981), The expansion index test, *ASTM Geotechnical Testing Journal*, v. 4, no. 2, pp. 58–67.

Andrews, R. E., Gawarkiewicz, J. J., and Winterkorn, H. F. (1967), Comparison of the interaction of three clay minerals with water, dimethyl sulfoxide and dimethyl formamide, *Highway Research Record* no. 209, pp. 66–78.

API (1969), API Recommended Practice for Planning, Designing and Constructing Fixed Offshore Platforms, API RP 2A, American Petroleum Institute, Division of Production, Dallas, TX, 16p.

Army Corps Manual (1994), *Bearing Capacity of Soils*, US Army Corps of Engineers, Manual no. 7.

Arulanandan, K., Sargunam, A., Loganathan, P., and Krone, R. B. (1973), Application of chemical and electrical parameters to prediction of erodibility, *Highway Research Board Special Report* 135, pp. 42–51.

ASCE (1958), Progress report on glossary of terms and definitions in soil mechanics, *Journal of the Soil Mechanics and Foundations Division, Proceedings of the ASCE*, v. 84, no. SM4, part 1, pp. 1–43.

ASCE (1979), *Analyses for Soil-Structure Interaction Effect, for Nuclear Power Plants*, ASCE, New York, 155p.

ASCE (1994), *Retaining and Flood Walls*, Technical Engineering and Design Guides as Adapted from the US Army Corps of Engineers, no. 4, 313p.

Aschenbrenner, B. C. (1956), A new method of expressing particle sphericity, *Journal of the Sediment Petroleum*, v. 26, pp. 5–31.

ASTM (1970), *Special Procedures for Testing Soil and Rock for Engineering Purposes*, 5th edn. *ASTM Special Technical Publication* (STP) 479, 630p.

ASTM (2003), *Annual Book of ASTM Standards*, Soil and Rock, American Society for Testing and Materials, West Conshocken, PA, v. 4.08.

Azzouz, A. S., Krizek, R. J., and Corotis, R. B. (1976), Regression analysis of soil compressibility, *Soils and Foundations*, Japanese Society of Soil Mechanics, v. 16, no. 2, pp. 19–29.

Bano, L. F. and Letey, J. (eds) (1969), Water repellent soils, *Proceedings of the Symposium on Water Repellent Soil*, University of California, Riverside, CA.

Barden, L. and Slides, G. (1971), Sample disturbance in the investigation of clay structure, *Geotechnique*, v. 21, no. 3, pp. 211–222.

Batschinski, A. (1913), Untersuchungen uber die innere reibung von Flussigkeite n. Z. *Physics and Chemistry*, v. 84, pp. 643–706.

Beattie, A. A. and Chau, E. P. Y. (1976), The assessment of landslides potential with recommendations for future research, *J. Hong Kong Inst. Eng.* Hong, Kong, v. 2, pp. 12–33.

Bell, A. L. (1951), The lateral pressure and resistance of clay and the supporting power of clay foundations, *A Century of Soil Mechanics*, Institution of Civil Engineers, London, pp. 93–134.

Benkelman, A. C., Kingham, R. I., and Fang, H. Y. (1962), Special deflection studies on flexible pavement, *Highway Research Board Special Report* 73, pp. 102–125.

Benson, C. H. (1991), Index tests for evaluating the effect of leachate on a soil liner, *Proceedings of the 2nd International Symposium on Environmental Geotechnology*, v. 2, pp. 222–228.

Berggren, W. P. (1943), Prediction of temperature distribution in frozen soil, *Transactions, American Geophysical Union* (AGU), Part III.

Bergna, H. E. (1950), Electrokinetic behavior of clay minerals, *Transactions of the 4th International Congress of Soil Society*, vol. 3, pp. 75–80.

Bertram, G. E. (1940), "Experimental Investigation of Protective Filters", Soil Mechanics Series Report no. 67, Graduate School of Engineering, Harvard University, Cambridge, MA.

Bieniawska, Z. T. (1980), Rock classification: state-of-the-art and prospects for standardization, *Transportation Research Record* 783, pp. 2–9.

Bieniawski, Z. T. (1989), *Engineering Rock Mass Classifications: A Complete Manual for Engineers* and *Geologists in Mining, Civil and Petroleum Engineering*, John Wiley & Sons, New York.

Bigler, W. D. (1953), Vector solution of soil pressures, *Engineering News-Record*, April 16, p. 16.

Biot, M. A. (1941), General theory of three-dimensional consolidation, *Journal of Applied Physics*, v. 12, no. 2, February, pp. 155–164.

Bishop, A. W. (1955), The use of the slip circle in the stability analysis of slopes, *Geotechnique*, v. 5, no. 1, pp. 7–17.

Bishop, A. W., Green, G. E., Gerga, V. K., Anresen, A., and Brown, J. D. (1971), A new ring apparatus and its application to measurement of residual strength, *Geotechnique*, v. 21, no. 4, pp. 273–328.

Bjerrum, L. (1954), Geotechnical properties of Norwegian marine clays, *Geotechnique*, v. 4, no. 2, p. 49.

Blake, M. P. (1964), New vibration standards for maintenance, Hydrocarbon Processing and Petroleum Refiner, vol. 43, no. 1, January, pp. 111–114.

Bolt, B. A. (2004), *Earthquakes*, 5th ed. W. H. Freeman and Co., NY, NY, 378p.

Boussinesq, J. (1883), Application des Potentials a L'Etude de L'Equilibre et du Mouvement des Solides Elastiques, Gauthier-Villars, Paris.

Bowles, J. E. (1988), *Foundation Analysis and Design*, 4th edn. McGraw-Hill, New York.

Brandt, H. (1955), A study of the speed of sound in porous granular media, *Journal of the Applied Mechanics*, v. 22, pp. 479–486.

Brinch Hansen, J. (1953), *Earth Pressure Calculation*, Danish Technical Press, Copenhagen, Denmark, 254p.

Brinch Hansen, J. (1961), A general formula for bearing capacity, Bulletin, no. 11, Danish Geotechnical Institute, Copenhagen.

Broms, B. B. (1966), Methods of calculating the ultimate bearing of piles, A summary, *Sols Soils*, no. 18–19, pp. 21–31.

Broms, B. B. and Wong, K. S. (1991), Landslides, chapter 11, *Foundation Engineering Handbook*, 2nd edn. Kluwer Academic Publishers, Boston, MA, pp. 410–446.

Brown, J. D. and Meyerhof, G. G. (1969), Experimental study of bearing capacity in layered clays, *Proceedings of the 7th International Conference of Soil Mechanics and Foundation Engineering*, Mexico City, v. 2, pp. 1–20.

Brown, E. (1977), Vibroflotation compaction of cohesionless soils, *Journal of the Geotechnical Engineering Division, Proceedings, ASCE*, v. 103, no. GT12, pp. 1437–1451.

Brumund, W. F., Jonas, E., and Ladd, C. C. (1976), Estimating in-situ maximum past preconsolidation pressure of saturated clays from results of laboratory consolidometer tests, *Transportation Research Board Special Report* 163, pp. 4–12.

Buckingham, E. (1907), Studies of the movement of soil moisture, U.S. Department of Agriculture, *Bureau of Soils Bulletin* no. 38, Washington, DC.

Buisman, A. S. K. (1936), Results of long duration settlement tests, Proc. 1st Intern. *Conf. Soil Mechanics Foundation Engineering*, Harvard Univ. Mass, v. 1, p. 103.

Buismann, A. S. K. (1940), *Ground mechanics*, Waltman, Delft, 190p.

Burggraf, F. (1939), Field test on shearing resistance, *Proceedings, Highway Research Board*, v. 19, pp. 484–489.

Burmister, D. M. (1949), Principles and techniques of soil identification, *Proceedings, Highway Research Board*, pp. 402–433.

Burmister, D. M. (1951), The application of controlled test methods in consolidation testing, *ASTM STP* 126, p. 83.

Byers, H. G., Anderson, M. S., and Bradfield, R. (1938a), General chemistry of the soil, *Soils and Men*, Yearbook of Agriculture, USDA, pp. 911–928.

Byers, H. G., Kellogg, C. E., Anderson, M. S., and Thorp, J. (1938b), Formation of soil, *Soils and Men*, Yearbook of Agriculture, USDA, pp. 948–978.

Caquot, A. and Kerisel, J. (1948), Tables for the calculation of passive pressure, active pressure and bearing capacity of foundations, Gauthier-Villars, Paris.

Carroll, R. G., Jr. (1983), Geotextile filter criteria, *Transportation Research Record* 916, pp. 46–53.

Casagrande, A. (1932), The structure of clay and its importance in foundation engineering, *Journal Boston Society of Civil Engineering*, v. 19, no. 4, p. 168.

Casagrande, A. (1936), The determination of the preconsolidation load and its practical significance, *Proceedings of the 1st International Conference of Soil Mechanics and Foundation Engineering*, v. 3, p. 60–64.

Casagrande, A. (1937), Seepage through dams, *Journal of the New England Water Works Association*, June, pp. 5–12.

Casagrande, A. (1948), Classification and identification of soils, *Transactions of ASCE*, v. 113, pp. 901–992.

Casagrande, L. (1952), Electro-osmotic stabilization of soil, *Journal of the Boston Society of Civil Engineers*, v. 39, January, pp. 51–83.

Casagrande, L. (1983), Stabilization of soils by means of electroosmosis, state-of-the-art, *Journal of the Boston Society of Civil Engineers*, v. 69, no. 2, pp. 225–302.

Castro, G. (1975), Liquefaction and cyclic mobility of saturated sands, *Journal of the Geotechnical Engineering Div., Proc. ASCE*, v. 101, no. GT6, pp. 551–569.

Cedergren, H. R. (1974), *Seepage, Drainage, and Flow Nets*, 3rd edn, John Wiley & Sons, New York.

Chandrasekharan, E. C., Boominathan, S., Sadayan, E., and Narayanaswamy Setty, K. R. (1969), Influence of heat treatment on the pulverization and stabilization characteristics of typical tropical soils, *Highway Research Board Special Report* 103, National Research Council, Washington, DC, pp. 161–172.

Chaney, R. C. and Fang, H. Y. (1986), Static and dynamic properties of marine sediments: a state-of-the-art, *ASTM Special Technical Publication* (STP) 923, pp. 74–111.

Chaney, R. C., Ramavjaneya, G., Hencey, G., Kanchanastit, P., and Fang, H. Y. (1983), Suggested test method for determination of thermal conductivity of soil by thermal-needle Procedure, *ASTM Geotechnical Testing Journal*, v. 6, no. 4, pp. 220–225.

Chapman, D. L. (1913), A contribution to the theory of electrocapillarity, *Philosophical Magazine and Journal of Science*, Series 6, v. 25, no. 6, p. 475.

Charles, J. A., Burford, D., and Watts, K. S. (1981), Field studies of the effectiveness of dynamic consolidation, *Proceedings of the 10th International Conference of Soil Mechanics and Foundation Engineering*, v. 1, pp. 617–622.

Chellis, R. D. (1961), *Pile Foundations*, McGraw-Hill Book Co, New York.

Chen, W. F. (1994), *Constitutive Equations for Engineering Materials*, Part B, 37, v. 2, Plasticity and Modeling, Elservier, Amsterdam, 1129p.

Chen, W. F. and Davidson, H. L. (1973), Bearing capacity determination by limit analysis, *Journal of the Soil Mechanics and Foundations Division, Proceedings, ASCE*, v. 99, no. SM6, pp. 433–449.

Chen, W. F. and Drucker, D. C. (1969), Bearing capacity of concrete blocks and rock, *Journal of the Engineering Mechanics Division, Proceedings, ASCE*, v. 95, no. EM4, pp. 955–978.

Chen, W. F. and McCarron, W. O. (1991), Bearing capacity of shallow foundations, chapter 4, *Foundation Engineering Handbook*, 2nd edn. Kluwer Academic Publishers, Boston, MA, pp. 144–165.

Chen, W. F. and Rosenfarb, J. L. (1973), Limit analysis solutions of earth pressure problems, *Soils and Foundations*. The Japanese Society of Soil Mechanics and Foundation Engineering, v. 13, no. 4, pp. 45–60.

Chen, W. F. and Saleeb, A. F. (1994), *Constitutive Equations for Engineering Materials*, Part A 37, v. 1, Elasticity and Modeling, Elsevier, Amsterdam, 580p.

Childs, E. C. and Collis-George, N. (1950), The permeability of porous materials, *Proceedings, Royal Society*, London, v. 201A, pp. 392–405.

Chow, V. T. (1951), Hydrodynamic pressure due to horizontal earthquake shock computed by curves, *Civil Engineering, ASCE*, New York, September, pp. 52–53.

Chu, T. Y. and Humphries, W. K. (1969), *Foundation Design Manual: With Special Emphasis on Applications to Subsurface Conditions in South Carolina*, University of South Carolina.

Clark, A. B. and Walker, B. F. (1977), Proposed scheme for the classification and nomenclature for use in the engineering description of Middle Eastern Sedimentation rocks, *Geotechnique*, v. 27, no. 1, pp. 1–10.

Claybourn, A. F. and Wu, J. T. H. (1993), Geosynthetic-reinforced soil wall design, *Geotextiles and Geomembranes*, v. 12, pp. 707–724.

Clough, G. W. (1976), Deep excavations and retaining structures, chapter 14, *Analysis and Design of Building Foundation*, Envo Publishing Co. Lehigh Valley, pp. 417–466.

Coduto, D. P. (2001). *Foundation Design, Principles and Practices*, Prentice-Hall, Upper Saddle River, NJ.

Collins, K. and McGown, A. (1974), The form function of microfabric features in a variety of natural soils, *Geotechnique*, v. 14, no. 2, pp. 223–254.

Collins, R. J. and Ciesielst, S. K. (1994), *Recycling and Use of Waste Materials and By-products in Highway Construction*, NCHRP Synthesis of Highway Practice 199, Transportation Research Board, Washington DC, 84p.

Conklin, A. R. (2003), Infiltration and percolation, *Contaminated Soil Sediment and Water*, January/February, pp. 50–52.

Cooling, L. F. and Smith, D. B. (1935), Hollow circular cylinders tests, *Journal of the Institution of Civil Engineers*, London, v. 3, pp. 333–343.

Cooper, H. H., Jr. and Jacob, C. E. (1946), A generalized graphical method for evaluating formation constants and summarizing well-field history. *Transactions American Geophysical Union (AGU)*, v. 27, pp. 526–534.

Cour, F. R. (1971), Inflection point method for computing c_v, *Journal of the Soil Mechanics and Foundations Division, Proceedings ASCE*, v. 97, no. SM5, pp. 827–831.

Crandell, F. J. (1949), Ground vibrations due to blasting and its effect upon structures, *Journal Boston Society of Civil Engineering*, v. 36, pp. 222–245.

Croney, D. and Coleman, J. D. (1961), Pore pressure and suction in soil, *Pore Pressure and Suction in Soils*, Butterworths, Inc. London, pp. 31–37.

CRRI (1976, 1979), *Collected Papers for Landslides*, Chinese Railroad Research Institute, Beijing (in Chinese).

Culmann, K. (1866), *Die graphische Statik*, Berlin.

Cumberledge, G., Hoffman, G. L., and Bhajandas, A. C. (1976), Curing and tensile strength characteristics of aggregate-lime-pozzolan, *Proceedings, Transportation Research Board Annual Meeting*, 25p.

Cummings, E. M. (1960), Cellular cofferdams and docks, *Transactions ASCE*, v. 125, pp. 13–34.

Czeratzki, W. and Frese, H. (1958), Importance of water in formation of soil structure, *Highway Research Board Special Report* 40, National Research Council, Washington, DC, pp. 200–211.

Dakshanamurthy, V. (1979), Stress-controlled study of swelling characteristics of compacted expansive clays, *ASTM Geotechnical Testing Journal*, v. 2, no. 1, pp. 57–60.

Daniels, J. L. and Cherukuri, R. (2005), "The influence of biofilm on barrier material performance" *ASCE Practice Periodical of Hazardous, Toxic and Radioactive Waste Management*, 9 (4), 245–252.

Daniels, J. L., Inyang, H. I., and Iskandar, I. (2003), Durability of Boston blue clay in waste containment applications, *ASCE Journal of Materials in Civil Engineering*, v. 15, no. 2, pp. 144–152.

Daniels, J. L., Cherukuri, R., Hilger, H. A., Oliver, J. D., and Bin, S. (2005), "Engineering Behavior of Biofilm Amended Earthen Barriers Used in Waste Containment" *Management of Environmental Quality, An International Journal*, 16 (6), 691–704.

D'Appolonia, D. J. and Lambe, T. W. (1970), A method for predicting initial settlement, *Journal of the Soil Mechanics and Foundations Division, Proceedings, ASCE*, v. 96, no. SM2, pp. 523–544.

Das, B. M. (1992), *Principles of Soil Dynamics*, PWS-Kent Publ. Cp., Boston, MA, 570p.

De Beer, E. E. (1967), Bearing capacity and settlement of shallow foundations on sand, *Proceedings, Symposium on Bearing Capacity and Settlement of Foundations*, Duke University, 1965 pp. 31–43.

De Beer, E. E. (1969), Experimental data concerning clay slopes, *Proceedings 7th International Conference of the Soil Mechanics and Foundation Eng.* v. 2.

Deere, D. U. (1963), Technical description of rock cores for engineering purposes, *Rock Mechanics and Engineering Geology*, v. 1, p. 18.

Deere, D. U. (1968), Geological considerations, *Rock Mechanics in Engineering Practice*, K. G. Stegg and Zienkiewicz O. C. (eds), John Wiley & Sons, London.

Deere, D. U. and Miller, R. P. (1966), Engineering classification and index properties for intack rock, US Air Force Weapons Laboratory, Kirtland Air Force Base, *Technical Report* AFNL-TR-65-116.

Demars, K. R., Richardson, G. N., Yong, R. N., and Chaney, R. C. (eds) (1994), Dredging, remediation and containment of contaminated sediments, *ASTM Special Technical Publication* (STP) 1293, Philadelphia, PA.

DER (1985), *General Remedial Action Details for Radon Gas Mitigation*, Pennsylvania Department of Environmental Resources, Harrisburg, PA, 38p.

Derjaguin, B. V. (1960), The force between molecules, *Scientific American*, v. 203, pp. 47–53.

De Santo, R. S. and Flieger, T. A. (1995), Wetland functions and values: descriptive approach to visualizing and assessing wetland systems, *Transportation Research Record* no. 1475, pp. 123–132.

De Simore, S. V. (1972), Distribution of wind loads to soil, *Proceedings, International Conference on Planning and Design of Tall Buildings*, ASCE, New York, v. 1a, pp. 35–59.

Devine, J. F. (1966), Avoiding damage to residences from blasting vibration, Highway Research Record no. 135, National Research Council, Washington, DC, pp. 35–42.

Dismuke, T. D. (1970), Stress analysis of sheet piling in cellular structures, *Proceedings, Conference on Design and Installation of Pile Foundations and Cellular Structures*, H. Y. Fang and T. D. Dismuke (eds), Envo Publishing, Lehigh Valley, pp. 339–365.

Dismuke, T. D. (1991a), Retaining structures and excavations, chapter 12, *Foundation Engineering Handbook*, 2nd edn, Kluwer Academic Publishers, Boston, MA, pp. 447–510.

Dismuke, T. D. (1991b), Durability and protection of foundations, chapter 25, *Foundation Engineering Handbook*, 2nd edn, Kluwer Academic Publishers, Boston, MA, pp. 856–867.

Dismuke, T. D., Chen, W. F., and Fang, H. Y. (1972), Tensile strength of rock by the double-punch method, *Rock Mechanics*, Springer-Verlag, v. 4, pp. 79–87.

Drnevich, V. P. (1979), Evaluation of sample disturbance on soils using the concept of reference strain, *Technical Report*, US Army Engineer Waterways Experiment Station, GL-79-13, September.

Drucker, D. C. and Prager, W. (1952), Soil mechanics and plastic analysis or limit design, *Quarterly Journal of Applied Mathematics*, v. 10, no. 2, pp. 157–165.

Drucker, D. C., Gibson, R. E., and Henkel, D. J. (1957), Soil mechanics and work-hardening theories of plasticity, *Transactions*, ASCE, v. 122, pp. 338–346.

Du, B. L., Mikroudis, G. K., and Fang, H. Y. (1986), Effect of pore fluid pH on the dynamic shear modulus of clay, *ASTM Special Technical Publication* (STP) 933, pp. 226–239.

Elton, G. A. H. (1948), Electroviscosity I: the flow of liquid between surfaces in close proximity and II: experimental demonstration of the electroviscous effect, *Proceedings of Royal Society*, London, 194A, pp. 259–274 and pp. 275–287.

Elzeftawy, A. and Dempsey, B. J. (1976), Unsaturated transient and steady state flow of moisture in subgrade soil, *Transportation Research Record* 612, pp. 56–61.

Eno, F. H. (1929), The influence of climate on the building, maintenance, and use of roads in the United States, *Proceedings of Highway Research Board*, v. 9, pp. 211–249.

Evans, J. C. (1991), Geotechnics of hazardous waste control systems, chapter 20, *Foundation Engineering Handbook*, 2nd edn, Kluwer Academic Publishers, Boston, MA, pp. 750–777.

Evans, J. C. and Fang, H. Y. (1986), Triaxial equipment for permeability testing with hazardous and toxic permeants, *ASTM Geotechnical Testing Journal*, v. 9, no. 3, pp. 126–132.

Evans, J. C. and Fang, H. Y. (1988), Triaxial permeability and strength testing of contaminated soils, *ASTM Special Technical Publication* (STP) 977, pp. 387–404.

Eyraud, C, Lareal, P., and Gielly, J. (1965), General report: physicochemical properties of water, *RILEM Bulletin*, New Series no. 27, pp. 15–18.

Fang, H. Y. (1960), Rapid determination of liquid limit of soils by flow index method, *Highway Research Board Bulletin* 254, pp. 30–35.

Fang, H. Y. (1986), Introductory remarks on environmental geotechnology, *Proceedings 1st International Symposium on Environmental Geotechnology*, v. 1, pp. 1–14, Envo Publishing, Lehigh Valley, Penna.

Fang, H. Y. (1989), Particle theory: a unified approach for analysis soil behavior, *Proc. 2nd Intern. Symposium on Environmental Geotechnology*, Envo Publ. Co., Bethlehem, PA, v. 1, pp. 167–194.

Fang, H. Y. (1990), Expert systems for assessment of radon gas, *ASCE Environmental Engineering 1989 Specialty Conference*, J. F. Malina, Jr (ed.), ASCE, New York, pp. 97–104.

Fang, H. Y. (1992), Environmental geotechnology: a perspective, *Proc. Mediterranean Conf. on Environmental Geotechnology*, ed. by M. A. Usmen and Y. Acar, A. A. Balkema, Publ., Rotterdam, pp. 11–19.

Fang, H. Y. (1994), Cracking and fracture behavior of soil, *Fracture Mechanics Applied to Geotechnical Engineering*, L. E. Vallejo and R. Y. Liang (eds), *ASCE Geotechnical Special Publication* no. 43, ASCE, New York, pp. 102–117.

Fang, H. Y. (1995), Engineering behavior of urban refuse, compaction control and slope stability analysis, *Proc. GREEN, Waste Disposal by Landfill*, R. W. Sarsby ed. Bolton, UK, A. A. Balkema Publ., Rotterdam, pp. 47–72.

Fang, H. Y. (1997), *Introduction to Environmental Geotechnology*, CRC Press, Boca Raton, FL, 652p.

Fang, H. Y. (2002), Radioactive nuclear wastes, *ASCE Practice Periodical of Hazardous, Toxic and Radioactive Waste Management*, v. 6, no. 2, pp. 102–111.

Fang, H. Y. and Chen, W. F. (1971), New method for determination of tensile strength of soils, *Highway Research Record* no. 354, pp. 62–68.

Fang, H. Y. and Deutsch, W. L. (1976), Discussion of behavior of compacted soil in tension, *Journal of Geotechnical Engineering Division, Proceedings, ASCE*, v. 102, no. GT5, pp. 569–570.

Fang, H. Y. and Evans, J. C. (1988), Long-term permeability tests using leachate on a compacted clayey liner material, *ASTM STP 963*, pp. 397–404.

Fang, H. Y. and Fernandez, J. (1981), Determination of tensile strength of soils by unconfined-penetration test, *ASTM Special Technical Publication (STP) 740*, pp. 130–144.

Fang, H. Y. and Hirst, T. J. (1973), A method for determining the strength parameters of soils, *Highway Research Record* no. 463, pp. 45–50.

Fang, H. Y. and Koerner, R. M. (1977), An instrument for measuring in situ soil-structure response during dynamic vibration, *Proc. 14th Annual Meeting of Society of Engineering Science*, pp. 1171–1180.

Fang, H. Y. and Mikroudis, G. K. (1987), Discussion of load factor versus environmental factor design criteria, *Proc. 1st Intern. Symposium on Environmental Geotechnology*, v. 1, pp. 340–342.

Fang, H, Y. and Mikroudis, G. K. (1991), Stability of earth slopes, Chapter 10, *Foundation Engineering Handbook*, 2nd ed. Kluwer Academic Publishers, Boston, MA, pp. 379–409.

Fang, H. Y. and Schaub, J. H. (1967), Analysis of the elastic behavior of flexible pavement, *Proceedings, 2nd International Conference on the Structural Design of Asphalt Pavements*, v. 1, pp. 719–729.

Fang, H. Y., Daniels, J. L., and Work, D. V. (1998) Soil contamination and decontamination mechanisms under wet-dry and freeze-thaw conditions. *Proceedings, 4th International Symposium on Environmental Geotechnology and Global Sustainable Development*, Danvers, MA, pp. 1158–1171.

Fang, H. Y., Chaney, R. C., and Pandit, N. S. (1981), Dynamic shear modulus of soft silt, *Proceedings, International Conference on Recent Advances in Geotechnical Earthquake Engineering and Soil Dynamics*, University of Missouri-Rolla, v. 2, pp. 575–580.

Fang, H. Y., Daniels, J. L., and Kim, T. H. (2004) Pollution intrusion on soil-pavement system *ASCE Journal of Transportation Engineering*, v. 130, no. 4, pp. 526–534.

Farouki, O. T. and Winterkorn, H. F. (1964), Mechanical properties of granular systems, *Highway Research Record* no. 52, pp. 10–42.

Fellenius, B. H. (1991), Pile foundations, chapter 13, *Foundation Engineering Handbook*, 2nd edn, Kluwer Academic Publishers, Boston, MA, pp. 511–536.

Fellenius, W. (1927), *Erdstatische Berechnungen* (calculation of stability of slopes), W. Ernst und Sohn. Berlin (revised edition, 1939).

Fellenius, W. (1938), Calculation of the stability of earth dams, *Transactions 2nd Congress on Large Dams* (1936), Washington, DC, v. 4, pp. 445–462.

Fenske, C. W. (1954), Influence Charts for Vertical Stress Distribution by Westergaad's Equation, *Circular* no. 21, Bureau of Engineering Research, The University of Texas, Austin, TX, 10p.

Fernandez, F. and Quigley, R. B. (1985), Hydraulic conductivity of natural clays permeated with simple liquid hydrocarbons, *Canadian Geotechnical Journal*, v. 22, pp. 205–214.

Fernando, J., Smith, R., and Arulanandan, K. (1975), New approach to determination of expansion index, *Journal of the Geotechnical Engineering Division Proceedings, ASCE*, v. 101, no. GT9, pp. 1003–1008.

Fetter, C. W. (2001), *Applied Hydrogeology*, 4th ed., Prentice Hall, Upper Saddle River, New Jersey, 598p.

Fleischer, R. L. (1983), Theory of alpha-recoil effects on radon release and isotopic disequilibrium: *Geochim. Cosmochi. Acta*, v. 47, pp. 779–784.

Fowkes, R. S. and Fritz, J. F. (1974), Theoretical and experimental studies on the packing of solid particles: a survey, *Bureau of Mines Information Circular*, IC no. 8623, 30p.

Fredlund, D. G. and Rahardjo, H. (1993), *Soil Mechanics for Unsaturated Soils*, John Wiley & Sons, New York, 517p.

Freeze, R. A. and Cherry, J. A. (1979), *Groundwater*, Prentice Hall, Englewood Cliffs, NJ.

Freundlich, H. (1935), *Thixotropy*, Hermann et Cie, Paris.

Friels, D. R. (1979), Pile capacity in cohesive soils computed from SPT N-values, Engineering and Testing, Inc. Memphis, TN, 5p.

Gazetas, G. (1991), *Foundation Engineering Handbook*, chapter 15 2nd edn, Kluwer Academic Publishers, Boston, MA, pp. 553–593.

Geuze, E. C. W. and Tan, T. T. (1953), Hallow cylinder triaxial test on clay, *Proc. 2nd Intern. Congress on Reheology*, Harrison, Ed. Oxford.

Gibson, R. E. and Morgenstern, N. R. (1962), A note on the stability of cuttings in normally consolidated clays, *Geotechnique*, v. 12, no. 3, pp. 212–216.

Gilboy, G. (1936), Improved soil testing methods, *Engineering News Record* (ENR), May 21.

Giroud, J. P. (1982), Filter criteria for geotextiles, *Proceedings of the 2nd International Conference on Geotextiles*, Las Vegas, v. 1, pp. 103–108.

Goldberg, G. D., Lovell, C. W., Jr. and Miles, R. D. (1979), Use the geotechnical data bank, *Transportation Research Record* 702, pp. 140–146.

Grant, R., Christian, J. T., and Vanmarcke, E. H. (1974), Differential settlement of buildings, *Journal of the Geotechnical Engineering Division, Proceedings, ASCE*, v. 100, no. GT9, pp. 973–991.

Gray, D. H. and Kashmeeri, N. A. (1971), Thixotropic behavior of compacted clays, *Journal of the Soil Mechanics and Foundations Division, Proceedings, ASCE*, v. no. pp. 193–207.

Gray, D. H. and Leiser, A. T. (1982), *Biotechnical Slope Protection and Erosion Control*, Van Nostrand Reinhold Co., New York.

Gray, H. (1936), Progress report on research on the consolidation of fine-grained soils, *Proc. 1st Intern. Conf. on Soil Mechanics and Foundation Engineering*, v. 2, pp. 138–141.

Gray, W. A. (1968), *The Packing of Solid Particles*, Chapman & Hall, London, 134p.

Griffith, A. A. (1921), The phenomena of rupture and flow in solids, *Transactions of the Royal Society of London*, v. 221, pp. 1–163.

Grim, R. E. (1968), *Clay Mineralogy*, McGraw-Hill Book Co., New York, 384p.

Gouy, G. (1910), Sur la constitution de la charge electrique a la surface d'un electrolyte, *Journal de* Physique, series 4, v. 9, p. 457.

Gupta, S. N. (1970), Discussion of sand densification by piles and vibroflotation, by C. E. Basore and J. D. Boitano, *Journal of the Soil Mechanics Foundation Division Proceedings, ASCE*, v. 96, no. SM4, pp. 1473–1475.

Habibagahi, K. (1976), Temperature effects on primary consolidation, *Geotechnical Engineering, Journal of Southeast Asian Society of Soil Engineering*, Bangkok, v. 7, no. 2, pp. 95–108.

Hadding, A. (1923), Eine rontgenographische Methode kristalline und kryptpkristalline Substanzen zu identifizieren, *Z. Krist.* v. 58, pp. 108–112.

Haines, W. B. (1923), The volume changes associated with variations of water content in soils, *Journal of Agricultural Science*, v. 13, p. 296.

Handy, R. L. and Fenton, T. E. (1977), Particle size and mineralogy in soil taxonomy, *Transportation Research Record* 642, pp. 13–19.

Hanna, A. M. and Meyerhof, G. G. (1980), Design charts for ultimate bearing capacity of foundations on sand overlying soft clay, *Canadian Geotechnical Journal*, v. 17, pp. 300–303.

Hardin, B. O. and Black, W. L. (1968), Vibration modulus of normally consolidated clay, *Journal of the Soil Mechanics and Foundations Division, Proceedings, ASCE*, v. 94, no. SM2, pp. 353–369.

Hardin, B. O. and Drnevich, V. P. (1972), Shear modulus and damping in soils, design equations and curves, *Journal of the Soil Mechanics and Foundations Division, Proceedings, ASCE*, v. 98, no. SM7, pp. 667–692.

Hausmann, H. (1990), *Ground Improvement Engineering*, McGraw-Hill Book Co., New York.

Hendron, D. M., Fernandez, G., Prommer, P. J., Giroud, J. P., and Orozco, L. F. (1999), Investigation of the cause of the 27 September 1997 slope failure at the Dona Juana Landfill, *Proceedings of the Seventh Waste Mangement and Landfill Symposium (Sardinia '99)*, Environmental and Sanitary Engineering Centre, Cagliari, Sardinia.

Hennes, R. G. (1953), The strength of gravel in direct shear, *ASTM Symposium on Direct Shear Testing of Soils*, Philadelphia, PA, pp. 1–43.

Higgins, C. M. (1969), Pressuremeter correlation study, *Highway Research Record* no. 284, pp. 37–62.

Hilf, J. W. (1991), Compacted fills, chapter 8, *Foundation Engineering Handbook*, 2nd edn, Kluwer Academic Publishers, Boston, MA, pp. 249–316.

Hillel, D. (1998), *Environmental Soil Physics*, Academic Press, San Diego, CA, 771p.

Hirsch, T. J., Lowery, L. L., Coyle, H. M., and Samson, C. H. (1970), Pile-driving analysis by one-dimensional wave theory, state-of-the-art, *Highway Research Record* no. 333, pp. 33–54.

Hogentogler, C. A. (1937a), *Engineering Properties of Soil*, McGraw-Hill Book Co. New York.

Hogentogler, C. A. (1937b), Essentials of soil compaction, *Proceedings, Highway Research Board*, v. 16, pp. 309–316.

Holtz, R. D. (1991), Stress distribution and settlement of shallow foundations, chapter 5, *Foundation Engineering Handbook*, 2nd edn, Kluwer Academic Publishers, Boston, MA, pp. 166–223.

Holtz, R. D. and Kovacs, W. D. (1981), *An Introduction to Geotechnical Engineering*, Prentice-Hall Co., Englewood Cliffs, NJ, 733p.

Hough, B. K. (1957), *Basic Soils Engineering*, Ronald Press Co., New York, 513p.

HRB (1952), Construction Equipment, *Highway Research Board Bulletin 58*, 73p.

HRB (1957), Glossary: pedologic (soils) and landform terminology, *Highway Research Board Special Report 25*, 32p.

HRB (1962a), The AASHO Road Test Report 2, materials and construction, *Highway Research Board Special Report 61B*, 173p.

HRB (1962b), The AASHO Road Test Report 6, special studies, *Highway Research Board Special Report 61F*, 127p.

Hu, X. H. (1981), *Technical Report on Low-cost Construction Materials*, Tongji University, Shanghai, China, 12p (in Chinese).

Huang, Y. H. (1980), Stability charts for effective stress analysis of nonhomogeneous embankments, *Transportation Research Record 749*, pp. 72–74.

Hvorslev, M. J. (1949), *Subsurface Exploration and Sampling of Soils for Civil Engineering Purposes*, Waterways Experiment Station, Vicksburg, Mississippi, 520p.

ICE (1961), *Pore Pressure and Suction in Soils*, Institution of Civil Engineers, Butterworths, Inc. London, 151p.

Ingles, O. G. (1968), Soil chemistry relevant to the engineering behavior of soils, chapter 1, *Soil Mechanics, Selected Topics*, I. K. Lee (ed.), American Elsevier Publishing Co., New York, pp. 1–57.

Inyang, H. I. (1992), Aspects of landfill design for stability in seismic zones, *Journal of Environmental Systems*, v. 21, no. 3, pp. 223–235.

Inyang, H. I. and Bergeson, K. L. (eds) (1992), *Utilization of Waste Materials in Civil Engineering Construction*, ASCE, New York, 347p.

Irwin, G. R. (1960), *Fracture Mechanics*, Pergamon Press, London.

Isenberg, R. H., Law, J. H., O'Neill, J. H., and Dever, R. J. (2001), Geotechnical aspects of landfill bioreactor design: is stability the fatal flaw? *Proceedings of the 6th Annual Landfill Symposium*, San Diego, CA, pp. 51–62.

Israelsen, C. E., Clyde, C. G., Fletcher, J. E., Israelsen, E. K., Haws, F. W., Packer, P. E., and Farmer, E. E. (1980), Erosion Control During Highway Construction: Research Report, *NCHRP Report 220*, Transportation Research Board, National Research Council, Washington, DC, 30p.

ISRM (1979), Description of discontinuities in rock masses, *International Journal of Rock Mechanics and Mining Science*, v. 15, pp. 71–89.

Janbu, N., Bjerrum, L., and Kjaernsli, B. (1956), Veiledning ved losning av fundamentering-soppgaver, *Norwegian Geotechnical Institute Publication* no. 16, 93p.

Jarquio, R. (1981), Total lateral surcharge pressure due to strip load, *Journal of Geotechnical Engineering Division, Proceedings, ASCE*, v. 107, no. GT10, pp. 1424–1428.

Johnson, A. W. and Sallberg, J. R. (1962), Factors influencing compaction test results, *Highway Research Board Bulletin 319*, National Research Council, Washington, DC, 148p.

Joisel, A. (1948), Crushing and fragmentation of rocks, *Ann. Inst. Techn. du Batiment et des Travaux*, Publics no. 26, Paris.

Jones, P. C. T. (1955), Microbiological factors in soil stabilization, *Highway Research Board Bulletin 108*, pp. 81–95.

Jumikis, A. R. (1977), *Thermal Geotechnics*, Rutgers University Press, New Brunswick, NJ, 375p.

Juran, I. and Elias, V. (1991), Ground anchors and soil nails in retaining structures, chapter 26, *Foundation Engineering Handbook*, 2nd edn, Kluwer Academic Publishers, Boston, MA, pp. 868–906.

Karpoff, K. P. (1953), Stabilization of fine-grained soils by electroosmotic and electrochemical methods, *Proceedings, Highway Research Board*, v. 32, p. 526.

Kaya, A. and Fang, H. Y. (1995), Determination of liquefaction potential using dielectric concept, *Proceedings, 3rd International Conference on Recent Advances in Geotechnical Earthquake Engineering and Soil Dynamics*, University of Missouri-Rolla, v. 3, pp. 73–76.

Kaya, A. and Fang, H. Y. (1997), Identification of contaminated soils by dielectric constant and electrical conductivity, *Journal of Environmental Engineering, ASCE*, v. 123, no. 2, pp. 169–177.

Kaya, A. and Fang, H. Y. (2000), The effects of organic fluids on physicochemical parameters of fine-grained soils, *Canadian Geotechnical Journal*, v. 37, pp. 943–950.

Kehew, A. E. (1988), *General Geology for Engineers*, Prentice Hall, Englewood Cliffs, NJ, 447p.

Kerisel, J. and Quatre, M. (1968), Settlements under foundations, calculation using the triaxial apparatus, *Civil Engineering*, London, May, and June.

Kersten, M. S. (1949), Thermal properties of soils, *Engineering Experiment Station Bulletin* no. 28, University of Minnesota, 225p.

Kersten, M. S. and Cox, A. E. (1951), The effect of temperature on the bearing value of frozen soils, *Highway Research Board Bulletin* 40, pp. 32–38.

Kezdi, A. (1964), Discussion on mechanical properties of granular systems, *Highway Research Record* no. 52, pp. 42–58.

Kinson, K., Lloyd, C. P., and Eadie, G. R. (1983), Steel piling in Australia, *Australian Geomechanics*, vol. 6, no. 12, pp. 36–45.

Ko, H. Y. and Scott, R. F. (1967), A new soil testing apparatus, *Geotechnique*, v. 17, no. 1, pp. 40–57.

Koerner, R. M. (1970), Effect of particle characteristics on soil strength, *Journal of the Soil Mechanics and Foundations Division, Proceedings, ASCE*, v. 96, no. SM4, Paper 7393, pp. 1221–1234.

Koerner, R. M. (1991), Geosynthetics in geotechnical engineering, Chapter 22, *Foundation Engineering Handbook*, 2nd ed. Kluwer Academic Publishers, Boston, MA, pp. 796–813.

Koerner, R. M. (1998), *Designing with Geosynthetics*, 4th edn, Prentice Hall, Englewood Cliffs, NJ.

Koerner, R. M., Lord, A. E., Jr., and McCabe, W. M. (1976), Acoustic emission monitoring in concrete nd foundation soils, chapter 21, *Analysis and Design of Building Foundations*, Envo Publishing Co., Bethlehem, PA, pp. 637–654.

Koppula, S. D. (1981), Statistical estimation of compression index, *ASTM Geotechnical Testing Journal*, v. 4, no. 2, pp. 68–73.

Koppula, S. D. (1984), Pseudo-static analysis of clay slopes subjected to earthquakes, *Geotechnique*, v. 34, no. 1, pp. 71–79.

Kurata, S. and Kitajima, S. (1967), Design method for cellular bulkhead made of thin steel plate, *Proc. 3rd. Asian Regional Conf. on Soil Mechanics and Foundation Engineering*, Haifa, v. 1, pp. 215–219.

Ladd, C. C. (1977), Stress-deformation and strength characteristics, state-of-the-art, *Proceedings, 9th International Conference of the Soil Mechanics and Foundation Engineering*, v. 2, pp. 421–494.

Lade, P. V. and Duncan, J. M. (1975), Elastoplastic stress-strain theory for cohesionless soil, *Journal of the Geotechnical Engineering Div., Proc. ASCE*, v. 101, no. GT10, pp. 1037–1053.

LaGrega, M. D., Buckingham, P. L., and Evans, J. C. (2001), *Hazardous Waste Management*, McGraw-Hill Book Co., New York.

Laguros, J. G. (1969), Effect of temperature on some engineering properties of clay soils, *Highway Research Board, Special Report* 103, National Research Council, Washington, DC, pp. 186–203.

Lambe, T. W. (1958), The engineering behavior of compacted clay, *Journal of the Soil Mechanics and Foundations Division Proceedings, ASCE*, v. 84, no. SM2, pp. 1–35.

Lambe, T. W. and Whitman, R. V. (1979), *Soil Mechanics, SI*, 2nd Ed. John Wiley & Sons, New York, 553p.

Lane, K. S. and Washburn, D. E. (1946), Capillarity tests by capillarimeter and by soil filled tubes, *Proceedings, Highway Research Board*, v. 25, pp. 460–473.

LaPlante, C. M. and Zimmie, T. F. (1992), Freeze-thaw effects on the hydraulic conductivity of compacted clays, *Transportation Research Record* no. 1369, pp. 126–129.

Ledbetter, R. H. (1985), Improvement of liquefiable foundation conditions beneath existing structures, *Technical Report*, REMR-GT-2, US Army Corps of Engineers, Washington, DC.

Lee, K. L. and Seed, H. B. (1967), Cyclic stress conditions causing liquefaction of sand, *Journal of the Soil Mechanics and Foundations Div., Proc. ASCE*, v. 93, no. SM1, pp. 47–70.

Lees, G. (1964), The measurement of particle shape and its influence in engineering materials, *Journal British Granite and Whitestone Federation*, v. 4, no. 2, pp. 125–141.

Leonards, G. A. (1962), Engineering properties of soils, chapter 2, *Foundation Engineering*, MaGraw-Hill Book Co., New York, pp. 66–240.

Leonards, G. A. and Ramiah, B. K. (1959), Time effects in the consolidation of clays, *ASTM Special Technical Publication* (STP) 254, pp. 116–130.

Leonards, G. A., Cutter, W. A., and Holtz, R. D. (1980), Dynamic compaction of granular soils, *Journal of Geotechnical Engineering Division Proceedings, ASCE*, v. 106, no. GT1, pp. 35–44.

Leshchinsky, D. and Perry, E. B. (1987), A design Procedure for geotextile-reinforced walls, *Proceedings, of the Geosynthetics '87 Conference*, New Orleans, v. 1, pp. 95–107.

Lew, H. S., Leyemdecker, E. V., and Dikkers, R. D. (1971), Engineering Aspects of the 1971 San Fernando Earthquake, *Building Science Series* 40, National Bureau of Standards, US Department of Commerce, Washington, DC.

Li, C. Y. (1956), Basic concepts on the compaction of soil, *Journal of the Soil Mechanics and Foundations Division, Proceedings, ASCE*, v. 82, no. SM1, Paper 862, 20p.

Liu, C. and Evett, J. B. (2003), *Soil Properties: Testing, Measurement and Evaluation*, 5th edn, Prentice Hall, Englewood Cliffs, NJ.

Lo, K. Y. (1961), Secondary compression of clays, *Journal of Soil Mechanics Foundation Division, ASCE Proc.* v. 87, no. SM4, pp. 61–66.

Lo, K. Y. (1965), Stability of slopes in anisotropic soils, *Journal of the Soil Mechanics and Foundations Division, Proceedings, ASCE*, v. 91, no. SM4, pp. 85–106.

Lovell, C. W. (1957), Temperature effects on phase composition and strength of partially-frozen soil, *Highway Research Board Bulletin* 168, pp. 74–95.

Low, P. F. (1968), Mineralogical data requirements in soil physical investigations, *SSSA Special Publication Series* no. 3, pp. 1–34.

Lowe, J. III and Zaccheo, P. F. (1991), Subsurface explorations and samplings, chapter 1, Foundation *Engineering Handbook*, 2nd edn, Kluwer Academic Publishers, Boston, MA, pp. 1–71.

Lu, N. and Likos, W. J. (2004), Unsaturated Soil Mechanics, John Wiley and Sons, New York, 584p.

Lukas, R. C. (1980), Densification of loose deposits by pounding, *Journal of Geotechnical Engineering Division, Proceedings, ASCE*, v. 106, no. GT4, pp. 435–446.

Lumb, P. (1962), The properties of decomposed granite, *Geotechnique*, v. 12, no. 3, pp. 226–243.

Marcuson, W. F. and Bieganousky, W. A. (1977), SPT and relative density in coarse sands, *Journal of the Geotechnical Engineering Division, Proceedings, ASCE*, v. 103, no. GT11, pp. 1295–1309.

Marek, C. R. (1991), Basic properties of aggregate, chapter 3, *The Aggregate Handbook*, R. D. Barksdale (ed.) National Stone Association, Washington, DC, pp. 3–74.

Margason, E. (1977), Earthquake effects on embedded pile foundations, *Piletips* Seminar on Current Practices in Pile Design and Installation, Associated Pile & Fitting Corp. Clifton, NJ, pp. 65–88.

Martin, T. R. (ed.) (1998), Stability of Natural Slopes in the Coastal Plain, *ASCE Geotechnical Special Publication* no. 77, ASCE, New York, 88p.

Martinez, J. D., Johnson, K. S., and Neal, J. T. (1998), Sinkhole in evaporite rocks, *American Scientist*, v. 86, January–Febrruary, pp. 38–51.

Massarsch, K. R. and Broms, B. B. (1976), Lateral earth pressure at rest in soft clay, *Journal of the Geotechnical Engineering Division, Proceedings, ASCE*, v. 102, no. GT10, pp. 1041–1047.

Matsuo, S. (1957), A study of the effect of cation exchange on the stability of slopes, *Proc. 4th Intern Conf. Soil Mechanics Foundation Engineering*, v. 2, pp. 330–333.

Mattson, S. (1932), The laws of soil colloidal behavior, *Soil Science*, v. 34, pp. 330–333.

Mayne, P. W. and Kulhawy, F. H. (1982), K_0-OCR Relationships in soil, *Journal of the Geotechnical Engineering Division*, vol. 108, no. SM5, pp. 63–91, ASCE.

Mayne, P. W., Holtz, R. D., and Tumay, M. T. (1995), US State of the Practice in sampling and strength testing of overconsolidated clays, *Transportation Research Record* no. 1479, pp. 1–6.

Meegoda, J. N., Ezeldin, A. S., Fang, H. Y., and Inyang, H. I. (2003), Waste immobilization technologies, *ASCE Practice Periodical* of *Hazardous, Toxic and Radioactive Waste Management*, v. 7, no. 1, pp. 46–58.

Menard, L. F. (1956), An apparatus for measuring the strength of soils in place, *Master of Sciences Thesis*, Department of Civil Engineering, University of Illinois, Urbana, IL, 46p.

Menard, L. F. and Broise, Y. (1975), Theoretical and practical aspects of dynamic consolidation, *Geotechnique*, v. 25, no. 1, pp. 3–18.

Meyerhof, G. G. (1951), The ultimate bearing capacity of foundations, *Geotechnique*, v. 2, no. 3, pp. 301–332.

Meyerhof, G. G. (1953), The bearing capacity of foundations under eccentric and inclined loads *proc. 3rd international conference on soil mechanics*, zurich, v. 1, pp. 440–445.

Meyerhof, G. G. (1955), Influence of roughness of base and groundwater conditions on the ultimate bearing capacity of foundations, *Geotechnique*, v. 5, no. 3, pp. 227–242.

Meyerhof, G. G. (1963), Some recent research on the bearing capacity of foundations, *Canadian Geotechnical Journal*, v. 1, no. 1, pp. 16–26.

Meyerhof, G. G. (1976), Bearing capacity and settlement of pile foundations, *Journal of Geotechnical Engineering Division, Proceedings, ASCE*, v. 102, no. GT3, pp. 197–228.

Meyerhof, G. G. (1984), Safety factors and limit state analysis in geotechnical engineering, *Geotechnique*, v. 21, pp. 1–7.

Meyerhof, G. G. and Hanna, A. M. (1978), Ultimate bearing capacity of foundations on layered soils under inclined load, *Canadian Geotechnical Journal*, v. 15, pp. 565–572.

Miller, C. E. and Turk, L. M. (1943), *Fundamentals of Soil Science,* John Wiley & Sons, NY.

Miller, C. J. and Lee, J. Y. (1997), Depth of frost penetration in landfill cover systems, *Environmental Engineering Science*, v. 14, no. 1, pp. 67–72.

Mitchell, J. K. (1969), Temperature effects on the engineering properties and behavior of soils, *HRB Special Report* 103, pp. 9–28.

Mitchell, J. K. (1993), *Fundamentals of Soil Behavior*, 2nd edn (1st edn, 1976), John Wiley & Sons, New York.

Mitchell, J. K. and Kao, T. C. (1977), Measurement of soil thermal resistivity. *Journal of Geotechnical Engineering Division, Proceedings, ASCE*, v. 104, no. GT10, pp. 1307–1320.

Mitchell, J. K., Seed, R. B. and Seed, H. B (1990), Kettleman Hills waste landfill slope failure. I. Liner-system properties, *ASCE Journal of Geotechnical Engineering*, v. 116, no. 4, pp. 647–668.

Mitsch, W. J. and Gosselink, J. G. (1993), *Wetlands*, 2nd ed. Van Nostrand Reinhold, NY. 722p.

Mononobe, N. and Matuo, H. (1929), On the determination of earth pressure during earthquakes, *World Engineering Congress*, Tokyo, v. 9, pp. 176–182.

Moretto, O. (1948), Effect of natural hardening on the unconfined compressive strength of remolded clays, *Proceedings 2nd International Conference of the Soil Mechanics and Foundation Engineering*, v. 1, pp. 137–144.

Morgenstern, N. R. (1963), Stability charts for earth slopes during rapid drawdown, *Geotechnique*, v. 13, pp. 121–131.

Morgenstern, N. R. and de Matos, M. M. (1975), Stability of slopes in residual soils, *Proceedings, 5th Panamerican Conference on Soil Mechanics and Foundation Engineering*, v. 3, pp. 367–383.

Moulton, L. K. (1991), Aggregate for drainage, filtration, and erosion control, chapter 12, *The Aggregate Handbook*, R. D. Barksdale (ed.) National Stone Association, Washington, DC, pp. 12.2–12.83.

Murayama, S. (1969), Effect of temperature on elasticity of clays, *Highway Research Board Special Report* 103, National Research Council, Washington, DC, pp. 194–203.

Murthy, V. N. S. (2002), *Geotechnical Engineering, Principles and Practices of Soil Mechanics and Foundation Engineering*, Marcel Dekker, New York, 1029p.

Nagaraj, T. and Murty, B. R. S. (1985), Prediction of the preconsolidation pressure and recompression index of soils, *ASTM Geotechnical Testing Journal*, v. 8, no. 4, pp. 199–202.

Naik, D. (1986), Effect of temperature and pore fluid on shear characteristic of clay, *Proceedings, 1st International Symposium on Environmental Geotechnology*, v. 1, pp. 382–390.

Nalezny, C. L. and Li, M. C. (1967), Effect of soil structure and thixotropic hardening on the swelling behavior of compacted clay soils, *Highway Research Record* no. 209, pp. 1–22.

Narain, J. and Rawat, P. C. (1970), Tensile strength of compacted soils, *Journal of the Soil Mechanics and Foundations Division, Proceedings, ASCE*, v. 96, no. SM6, pp. 2185–2190.

Newmark, N. M. (1942), Influence charts for computation of stresses in elastic soil, University of Illinois, *Engineering Experiment Station, Bulletin Series* no. 338.

Nicastro, D. H. (ed.) (1997), Failure mechanisms in building constructure, ASCE, Reston, VA, 116p.

NISEE (1994), University of California-Berkeley's National Information Service for Earthquake Engineering, NISEE and The Regents of the University of California, http://nisee.berkeley.edu/ (accessed 1/15/2004).

Nyhan, J. W., Schofield, T.G., and Salazar, J. A. (1997), A water balance study of four landfill cover designs varying in slope for semiarid regions. *Proceedings of the International Containment Technology Conference*, St. Petersburg, FL, pp. 262–269.

Obern, L. (1941), Use of Subaudible Noises for Prediction of Rockbursts, US Department of Mines, Rept. Invrs. 3555.

Ohsaki, Y. (1970), Effects of sand compaction on liquefaction during the Tokachioki earthquake, *Soils and Foundations*, v. 10, no. 2, pp. 112–128.

Okabe, S. (1924), General theory of earth pressure and seismic stability of retaining walls and dams, *Journal of the Japanese Society of Civil Engineers*, London, v. 12, no. 1, pp 104–113.

O'Neill, M. W., Brown, D. A., Townsend, F. C., and Abar, N. (1997), Innovative load testing of deep foundations, *Transportation Research Record* no. 1569, pp. 17–25.

Osterberg, J. O. (1957), Influence values for vertical stresses in semi-infinite mass due to embankment loading, *Proceedings, 4th International Conference on Soil Mechanics and Foundation Engineering*, v. 1, pp. 393–396.

Paaswell, R. E. (1967), Thermal influence on flow from a compressible porous medium, Water Resources Research, v. 3, no. 1, pp. 271–278.

Paaswell, R. E. (1973), Causes and mechanisms of cohesive soil erosion: the-state-of-the-art, *Highway Research Board Special Report* 135, pp. 52–74.

Pamukcu, S. and Topcu, I. B. (1991), Study of causes of random cracking of solidified sludge reused as capping material, Fritz Engineering Laboratory Report, Dept. of Civil Engineering, Lehigh University, Bethlehem, PA, 33p.

Pamukcu, S. and Fang, H. Y. (1989), Development of a chart for preliminary assessments in pavement design using some in-situ soil parameters, *Transportation Research Record* no. 1235, pp. 38–44.

Patterson, D. (1958), Pole embedment to resist lateral load, *Civil Engineering*, ASCE, July, p. 69.

Patterson, K. E. (1955), The early history of circular sliding surfaces, *Geotechnique*, London, v. 5, pp. 275–296.

Pauling, L. (1960), *The Nature of the Chemical Bond*, 3rd edn, Cornell University Press, Ithaca, NY.

PCA (1951), *Concrete Piles, Design, Manufacture, Driving*, Portland Cement Association, IL. 80p.

PCA (1992), *PCA Soil Primer*, Portland Cement Association, IL. 40p.

Peck, R. B., Hanson, W. E., and Thornburn, T. H. (1974), *Foundation Engineering*, 2nd edn. John Wiley & Sons, New York, 410p.

Perloff, W. H. (1975), Pressure distribution and settlement, Chapter 4, *Foundation Engineering Handbook*, 1st ed. Van Nostrand Reinhold, Co., NY, NY, pp. 148–196.

Perloff, W. H., Baladi, G. Y., and Harr, M. E. (1967), Stress distribution within and under long elastic embankment, *Highway Research Record* no. 181, pp. 12–40.

Phukan, A. (1991), Foundations in cold regions, chapter 19, *Foundation Engineering Handbook*, 2nd edn, Kluwer Academic Publishers, Boston, MA, pp. 735–749.

Plum, R. L. and Esrig, M. I. (1969), Some temperature effects on soil compressibility and pore water pressure, *Highway Research Board Special Report* 103, National Research Council, Washington, DC, pp. 231–242.

Podolay, W., Jr and Cooper, J. D. (1974), Toward an Understanding of Earthquakes, *Highway Focus*, March, v. 6, no. 1, US Department of Transportation (DOT), Federal Highway Administration (FHA), 108p.

Post, J. L. and Paduana, J. A. (1969), Soil stabilization by incipient fusion, Highway *Research Board Special Report* 103, National Research Council, Washington, DC, pp. 243–253.

Potyondy, J. G. (1961), Skin friction between cohesive granular soils and construction materials, *Geotechnique*, v. 11, no. 4, pp. 339–353.

Prandtl, L. (1920), Uber Die Haerte Plastischer Korper, Nuchrichten Von Der Koeniglichen Gesellschaften Der Wissenschaften Zu Geottingen, Mathematisch-physikalische Klasse, pp. 74–85.

Proctor, R. R. (1933), Fundamental principles of soil compaction, *Engineering News Record* (ENR), August 31, v. 111, pp. 9–13.

Qian, X., Koerner, R. M., and Gray, D. H. (2002), *Geotechnical Aspects of Landfill Design and Construction*, Prentice Hall, Upper Saddle River, 717p.

Rankine, W. J. M. (1857), On the stability of loose earth, *Philosophical Transactions Royal Society*, London, v. 147, pp. 928–935.

Rao, S. N. (1979), The influence of fabric on the shrinkage limit of clay, *Geotechnical Engineering*, Southeast Asian Society of Soil Engineering, Bangkok, v. 10, no. 2, pp. 243–251.

Raymond, G. P. and Wahls, H. E. (1976), Estimating 1-dimensional consolidation, including secondary compression of clay loaded from overconsoldated to normally consolidated state, *Transportation Research Board Special Report* 163, pp. 17–25.

Reddi, L. N. and Inyang, H. I. (2000), *Geoenvironmental Engineering: Principles and Applications*, Marcel Dekker Inc., New York, 494p.

Rendon-Herrero, O. (1980), Universal compression index equation, *Journal of the Geotechnical Engineering Division, Proceeding ASCE*, v. 106, no. GT11, pp. 1179–1200.

Reno, W. H. and Winterkorn, H. F. (1967). Thermal conductivity of kaolinite clay as a function of type of exchange ion, density and moisture content, *Highway Research Record*, no. 209, pp. 79–85.

Resal, J. (1910), *Poussee des Terres*, Paris.

Richart, F. E., Jr. (1977), Field and laboratory measurements of dynamic soil properties, *Proceeding, Conference on Dynamical Methods in Soil and Rock Mechanics*, Karlsruhe, v. 1, pp. 3–36.

Richter, C. F. (1958), *Elementary Seismology*, W. H. Freeman & Co., San Francisco & London.

Richter, C. F. (1959), Seismic regionalization, *Bulletin, Seismological Society of America*, v. 49, pp. 123–162.

Richter, C. F. and Gutenberg, B. (1954), *Seismicity of the Earth*, 2nd edn, Princeton University Press, Princeton, NJ, 310p.

Ridgeway, H. H. (1976), Infiltration of water through the pavement surface, *Transportation Research Record* no. 616, National Research Council, Washington, DC, pp. 98–100.

Ring, G. W. III (1966), Shrink swell potential of soils, *Highway Research Record* no. 119.

Road Research Laboratory (1952), *Soil Mechanics for Road Engineers*, Her Majesty's Stationery Office, London, 541p.

Robbins, N. G. (1957), Piers supported by passive earth pressure, *Civil Engineering, ASCE*, April, p. 276.

Robertson, P. K. and Campanella, R. G. (1983), Interpretation of cone penetration tests; Part I and II, *Canadian Geotechnical Journal*, v. 20, no. 4, pp. 718–745.

Rodebush, W. and Buswell, A. M. (1958), Properties of water substance, *Highway Research Board Special Report* 40, pp. 5–13.

Roscoe, K. H., Schofield, A. N., and Wroth, C. P. (1958), On the yielding of soils, *Geotechnique*, v. 8, no. 1, pp. 22–53.

Rosenqvist, I. T. (1959), Physicochemical properties of soils: soil–water systems, *Journal of the Soil Mechanics and Foundation Engineering Divisions, Proceedings, ASCE*, v. 85, no. SM2, p. 31.

Rutledge, P. C. (1940), Neutral and effective stresses in soils, *Proceedings Purdue Conference on Soil Mechanics and its Applications*, Purdue University, pp. 174–190.

Sanglerat, G. (1972), *The Penetrometer and Soil Exploration*, Elsevier Publishers, The Netherlands, 464p.

Sansalone, M. and Carino, N. J. (1986), Impact-Echo: A Method for Flaw Detection in Concrete Using Transient Stress Waves, NBSIR 86-3452, September.

Satyanarayana, B. and Satyanarayana Rao, K. (1972), Measurement of tensile strength of compacted soil, *Geotechnical Engineering, Proceedings of the Institution of Civil Engineers*, v. 3, no. 1, pp. 61–66.

Scawthorn, C. (2003), Earthquakes, seismogenesis, measurement, and distribution, chapter 4, *Earthquake Engineering Handbook*, CRC Press, Boca Raton, FL, pp. 1–58.

Schlosser, F. and Bastick, M. (1991), Reinforced Earth, chapter 21, *Foundation Engineering Handbook*, 2nd edn. Kluwer Academic Publishers, Boston, MA, pp. 778–795.

Schmertmann, J. H. (1953), The undisturbed consolidation behavior of clay, *Transactions of the ASCE*, v. 120, pp. 1201.

Schmertmann, J. M. (1955), The undisturbed consolidation of clay, *Trans. ASCE*, v. 120, p. 1201.

Schofield, R. K. (1935), The pF of the water in soil, *Transactions of the 3rd International Congress of Soil Science*, v. 2, pp. 37–48.

Schofield, A. N. and Wroth, C. P. (1968), *Critical State Soil Mechanics*, McGraw-Hill Book Co., New York.

Shibuya, T. (1973), Geological study of landslide clay, KICT Report No. 10, Kajima Institute of Construction Technology, Tokyo, July, 37p.

Seed, H. B. (1975, 1991), Earthquake effects on soil-foundation systems, chapter 16, Part I, *Foundation Engineering Handbook*, 2nd edn, Kluwer Academic Publishers, Boston, MA, pp. 594–623.

Seed, H. B. (1976), Evaluation of soil liquefaction effects on level ground during earthquakes, Problems in Geotechnical Engineering, *ASCE National Convention*, Philadelphia, PA Preprint 2725, pp. 1–104.

Seed, H. B. and Lee, K. L. (1966). "Liquefaction of saturated sands during cycling loading." *Proc. ASCE*, v. 92, no. SM6, pp. 105–134.

Seed, H. B. and Peacock, W. H. (1971), Test procedures for measuring soil liquefaction characteristics, *Journal of the Soil Mechanics and Foundations Division, Proceedings, ASCE*, v. 97, no. SM8, Paper 8330, pp. 1099–1119.

Seed, H. B., Arango, I., and Chan, C. K. (1975), *Evaluation of Soil Liquefaction Potential During Earthquakes*, Report no. EERC75-28, Earthquake Engineering Research Center, University of California, Berkeley, CA.

Seed, H. B., Idriss, I. M., and Arango, I. (1983), Evaluation of liquefaction potential using field performance data, *Journal of the Geotechnical Engineering Division, Proceedings, ASCE*, v. 111, no. 12, pp. 458–482.

Seed, H. B., Chaney, R. C., and Pamukcu, S. (1991), Earthquake effects on soil-foundation systems, Chapter 16, *Foundation Engineering Handbook*, 2nd ed. Kluwer Academic Publishers, Boston, MA, pp. 594–672.

Seed, R. B., Mitchell, J. K. and Seed, H. B. (1990), Kettleman Hills Waste landfill slope failure. II. Stability analysis, *ASCE Journal of Geotechnical Engineering*, v. 116, no. 4, pp. 669–690.

Seifert, R., Ehrenberg, J., Tiedemann, B., Endell, K., Hofmann, J., and Wilm, D. (1935), Relation between landslide slope and the chemistry of clay soils, *Mitt. Preuss. Versuchanstalt f. Wasserbau u. Schiffbau*, v. 20, p. 34.

Shi, Y. W. (1981). A study on maximum density of large size sandy gravels, *Proceedings, 10th International Conference on Soil Mechanics and Foundation Engineering (ICSMFE)*, v. 1, pp 1–5.

Shook, J. F. and Fang, H. Y. (1961), Cooperative materials testing program at the AASHO Road Test, *Highway Research Board Special Report 66*, pp. 59–102.

Sih, G. C. and Liebowitz, H. (1968), Mathematical theories of brittle fracture, *Mathematical Fundamental of Fracture*, Academic Press, New York.

Simons, N. E. (1974), Normally consolidated and lightly overconsolidated cohesive materials, *Proceedings Conference on Settlement of Structures*, Session 2, Cambridge, England.

Skempton, A. W. (1953), The colloidal activity of clays, *Proceedings 3rd International Conference of the Soil Mechanics and Foundation Engineering*, v. 1, p. 57.

Skempton, A. W. (1954), The porewater pressure coefficients A and B, *Geotechnique*, v. 4, no. 4, pp. 143–147.

Skempton, A. W. (1964), Long-term stability of clay slopes, *Geotechnique*, v. 14, no. 2, pp. 77–102.

Skempton, A. W. and Bjerrum, L. (1957), A contribution to the settlement analysis of foundations on clay, *Geotechnique*, v. 7, no. 4, pp. 168–178.

Skempton, A. W. and Northey, R. D. (1952), The sensitivity of clays, *Geotechnique*, v. 3, no. 1, pp. 30–53.

Smith, E. A. L. (1962), Pile driving analysis by the wave equation, *Transactions of the ASCE*, v. 127, part 1, pp. 1145–1193.

Snitbhan, N. (1976), Foundation stability-numerical examples, chapter 4, *Analysis and Design of Building Foundations*, Envo Publishing Co., Bethlehem, PA, pp. 103–118.

Snitbhan, N., Chen, W. F., and Fang, H. Y. (1975), Slope stability analysis of layered soils, *Proceeding, 4th Southeast Asian Conference on Soil Engineering*, pp. 26–29.

Soil Survey Division Staff (1993), *Soil survey manual*. Soil Conservation Service. US Department of Agriculture Handbook 18.

Sokolovskii, V. V. (1965), *Static of Granular Media*, Pergamon Press, New York.

Sowers, G. F. (1962), Shallow Foundations, chapter 6, *Foundation Engineering*, G. A. Leonards (ed.) McGraw-Hill Book Co., New York, pp. 525–641.

Sowers, G. F. (1973), Settlement of waste disposal fills, *Proceedings, 9th International Conference Soil Mechanics and Foundation Engineering*, Moscow, v. 4, pp. 207–210.

Sowers, G. F. and Vesic, A. B. (1962), Vertical stresses in subgrades beneath statically loaded flexible pavements, *Highway Research Board Bulletin* no. 342.

Sridharan, A. and Rao, G. V. (1972), Surface area determination of clays, *Geotechnical Engineering*, v. 3, no. 2, pp. 127–132.

Sun, Z. C. (1989), The effects of acids on physicochemical properties of laterite soil, *Chinese Journal of Geotechnical Engineering*, v. 11, no. 4, pp. 89–93 (in Chinese with English Summary).

Tamez, E. (1957), Some factors affecting the dynamic compaction test, *ASTM Special Technical Publication* (STP) 232, pp. 54–66.

Tammemagi, H. (1999), *The Waste Crisis: Landfills, incinerators and the search for a sustainable future*. Oxford: Oxford University Press, 269p.

Tan, T. K. (1957), *Secondary Time Effects and Consolidation of clays*, Academia Sinica, Harbin, China.

Taylor, D. W. (1937), Stability of earth slopes, *Journal of the Boston Society of Civil Engineers*, v. 24, p. 197.

Taylor, D. W. (1942), *Research on the Consolidation of Clays*, Department of Civil and Sanitary Engineering, Series 82, Massachusetts Institute of Technology, 147p.

Taylor, D. W. (1948), *Fundamentals of Soil Mechanics*, John Wiley & Sons, New York, 700p.

Taylor, D. W. and Merchant, W. (1940), A theory of clay consolidation accounting for secondary compression, *Journal of Mathematics and Physics*, v. 19, pp. 167–185.

Teitgen, F. C. and Fiedler, D. R. (1973), Effect of point surcharge on retaining walls, *ASCE Civil Engineering*, Engineer's Notebook, November, p. 82.

Teng, W. C. (1962), *Foundation Design*, Prentice-Hall, Englewood Cliffs, NJ, 466p.

Terzaghi, K. (1925) *Erdbaume auf bodenphysikalischer Grundlage*, Deuticke, Leipzig und Wein.

Terzaghi, K. (1936), Stability of slopes of natural clay, *Proceedings, 1st International Conference of the Soil Mechanics and Foundation Engineering*, v. 1, p. 161.

Terzaghi, K. (1942), Soil moisture and capillary phenomena in soils, *Hydrology*, O. E. Meinzer (ed.) McGraw-Hill Book Co., New York, pp. 331–363.

Terzaghi, K. (1943), *Theoretical Soil Mechanics*, John Wiley & Sons, New York, 510p.

Terzaghi, K. (1954), Anchored bulkheads, *Transactions of the ASCE*, v. cxix, p. 1243.

Terzaghi, K. and Peck, R. B. (1967), *Soil Mechanics in Engineering Practice*, 2nd edn, John Wiley & Sons, New York, 529p (1st Edition 1948).

Theis, C. V. (1935), The relation between the lowering of the piezometric surface and the rate and duration of discharge of a well using ground water storage, *Transactions of the American Geophysical Union*, v. 16, pp. 519–524.

Thibodeaux, L. J. (1996), *Environmental Chemodynamics*, John Wiley & Sons. New York.

Thiem, G. (1906), *Hydrologische Methoden*, Gebhart, Leipzig, 56p.

Timoshenko, S. (1934), *Theory of Elasticity*, McGraw-Hill, New York, pp. 104–108.

Timoshenko, S. and Goodier, J. N. (1951), *Theory of Elasticity*, 2nd edn, McGraw-Hill Book Co., New York.

Tomlinson, M. J. (1971), Some effects of pile driving on skin friction, *Proceedings, Conference on Behavior of Pile*, Institute of Civil Engineering, London, pp. 179–194.

Torrance, J. K. (1975), On the role of chemistry in the development and behavior of the sensitive marine clays of Canada and Scandinavia, *Canadian Geotechnical Journal*, v. 12, no. 3, pp. 312–325.

Townsend, F. C. and Gilbert, P. A. (1973), Tests to measure residual strength of some clay shales, *Geotechnique*, v. 23, no. 2, pp. 267–271.

TRB (1978), Effect of Weather on Highway Construction, *NCHRP Synthesis of Highway Practice* 47, Transportation Research Board, 29p.

TRB (1980), Erosion Control During Highway Construction, *NCHRP Report* 222, Transportation Research Board, 30p.

TRB (1995), Engineering Properties and Practice in Overconsolidated Clays, *Transportation Research* no. 1479, 112p.

Tschebotarioff, G. P. (1973), *Foundations and Earth Structures*, 2nd ed. McGraw-Hill Book Co., New York, 642p.

Turner, A. K. and Schuster, R. L. (ed.) (1996), Landslide investigation and mitigation, *Transportation Research Board Special Report 247*, 673p.

US Army (1993), *Design of Pile Foundations*, US Army Corps of Engineers no. 1, 106p.

US Army (1994), *Settlement Analysis*, US Army Corps of Engineers no. 9, 144p.

US Army (1996), *Design of Sheet Pile Walls: Engineering and Design*, US Army Corps of Engineers no. 15, 72p.

US Army Corps of Engineers (1998), Geophysical Exploration for Engineering and Environmental Investigations (Technical Engineering and Design Guides As Adapted from the US Army Corps of Engineers), American Society of Civil Engineers, New York, 204p.

US Department of Navy (1962), Design manual soil mechanics, foundations, and earth structures: Bureau of Yards and Docks, NAVDOCKS DM 7.

US EPA (1997), *Cleaning Up the National's Waste Sites: Markets and Technology Trends*, 1996 edn, Executive Summary, EPA Report no. EPA 542-R-96-005A, 24p.

USBR (1974), *Earth Manual*, 2nd edn., US Bureau of Reclamation, CO.

USDA (1938), *Soils and Man*, Yearbook of Agriculture, US Department of Agriculture Washington, DC.

USDA, (1969), *Wildlife Habitat Improvement Handbook*, Forest Service, U. S. Department of Agriculture, Washington, DC.

USDA (1993), *Soil Survey Manual*, Soil Conservation Service, Soil Survey Division Staff, US Department of Agriculture Handbook 18.

US DOE (1997), *Linking Legacies Connecting the Cold War Nuclear Weapons Production Processes on Their Environmental Consequences*, US. Department of Energy, Office of Environmental Management, Washington, DC.

US EPA (1986), *A Citizen's Guide to Radon*: What it is and what to do about it, EPA-86-004, Washington, DC, 13p.

US EPA (1990), EPA Report on Definition, Potentially Incompatible Wastes and Hazardous Waste Determination, Washington, DC.

US EPA (1991), Design and Construction of RCRA/CERCLA Final Covers. Seminar Publication, US Environmental Protection Agency, Office of Research and Development. EPA/625/4-91/025 Washington, DC.

US EPA (1994), Design, Operation and Closure of Municipal Solid Waste Landfills, Seminar Publication, US Environmental Protection Agency, EPA/625/R-94/008. Washington, DC.

US EPA (2003), Municipal Solid Waste Generation, Recycling, and Disposal in the United States: Facts and Figures for 2003, US Environmental Protection Agency, Washington, DC.

US Government (1957), *Principles of Modern Excavation and Equipment*, US Government Technical Manual TM 5–520 and US Government Technical Manual TM-5-9500.

US NRC (1982), 10 CFR Part 61, Licensing requirements for land disposal of radioactive waste. US Nuclear Regulatory Commission, Federal Register, 47,248,57446-57482.

Van Houten, F. B. (1955), Inheritance factor in origin of clay minerals in soil, *Highway Research Board Bulletin* 108, pp. 25–28.

Van Olphen, H. (1977), *An Introduction to Clay Colloidal Chemistry for Clay Technologists, Geologists and Soil Scientists*, 2nd ed. John Wiley & Sons, NY.

Van Rooyen, M. and Winterkorn, H. F. (1959), Structural and textural influences on the thermal conductivity of soils, *Proceedings, 38th Annual Meeting, Highway Research Board*, pp. 576–621.

Vees, E. and Winterkorn, H. F. (1967), Engineering properties of several pure clays as functions of mineral type, exchange ions, and phase composition, *Highway Research Record* no. 209, pp. 55–65.

Velde, B. (1992), *Introduction to Clay Minerals, Chemistry, Origins, Uses and Environmental Significance*, Chapman & Hall, London.

Vesic, A. S. (1970), Load transfer in pile-soil systems, *Proceedings of the Design and Installation of Pile Foundations and Cellular Structures*, Envo Publishing Co. Inc., Bethlehem, PA, pp. 47–73.

Vesic, A. S. (1975), Bearing capacity of shallow foundation, chapter 4, *Foundation Engineering Handbook*, 1st edn,. Van Nostrand Reinhold, New York, pp. 121–147.

Vidal, H. (1969), The principle of reinforced earth, *Highway Research Record* no. 282, pp. 1–16.

Vijayvergiya, V. N. and Focht, J. A. Jr. (1972), A new way to predict capacity of piles in clay, *Proceedings, 4th Annual Offshore Technology Conference*, Paper No. OTC Paper 1718, Houston, TX.

Voight, B. (1973), Correlation between Atterberg plasticity limits and residual shear strength of natural soils, *Geotechnique*, v. 23, no. 2, pp. 265–267.

Wadell, H. (1932), Volume, shape and roundness of rock particles, *Journal of Geology*, v. 40, pp. 443–451.

Waidelich, W. C. (1958), Influence of liquid and clay mineral type on consolidation of clay-liquid systems, *Highway Research Board Special Report* 40, pp. 24–42.

Westergaard, H. M. (1933), Water pressure on dams during earthquakes, *Transactions of the ASCE*, v. 98, pp. 418–472.

Westergaard, H. M. (1938), A problem of elasticity suggested by a problem in soil mechanics: soft material reinforced by numerous strong horizontal sheets, in Contribution to the Mechanics of Solids, S. *Timoshenko 60th Anniversary Volume*, Macmillan Co., New York, pp. 268–277.

Williamson, D. A. (1980), Uniform rock classification for geotechnical engineering purposes, *Transportation Research Record* no. 783, pp. 9–14.

Wilson, D. G. (ed.) (1977), *Handbook of Solid Waste Management*, Van Nostrand Reinhold, New York.

Winterkorn, H. F. (1937), The application of base exchange and soil physics to problems of highway construction, *Proc. Soil Science Society of America*, v. 1, pp. 93–99.

Winterkorn, H. F. (1942), Mechanism of water attack on dry cohesive soil system, *Soil Science*, v. 54, pp. 259–273.

Winterkorn, H. F. (1953), Macromeritic liquids, *ASTM Special Technical Publication* (STP) 156, pp. 77–89.

Winterkorn, H. F. (1955), The science of soil stabilization, *Highway Research Board Bulletin* 108, pp. 1–24.

Winterkorn, H. F. (1958), Mass transport phenomena in moist porous systems as viewed from the thermodynamics of irreversible processes, *Highway Research Board Special Report* 40, pp. 324–338.

Winterkorn, H. F. (1970a), Analogies between macromeritic and molecular liquids and the mechanical properties of sand and gravel assemblies, *Chemical Dynamics* (papers in honor of Henry Eyring), Wiley-Interscience, NY, pp. 751–766.

Winterkorn, H. F. (1970b), Suggested method of test for thermal resistivity of soil by the thermal probe, Special Procedures for Testing Soil and Rock for Engineering Purposes, *ASTM Special Technical Publication* (STP) 479, 5th edn, pp. 264–270.

Winterkorn, H. F. and Baver, L. D. (1934), Sorption of liquids by soil colloids, I, *Soil Science*, v. 38, no. 4, pp. 291–298.

Winterkorn, H. F. and Fang, H. Y. (1975), Soil technology and engineering properties of soils, *Foundation Engineering Handbook*, 1st ed. Chapter 2, Van Nostrand Reinhold Co., NY, pp. 67–120.

Winterkorn, H. F. and Fang, H. Y. (1991), Soil technology and engineering properties of soil, chapter 3, *Foundation Engineering Handbook*, 2nd edn, Kluwer Academic Publishers, Boston, MA, pp. 88–143.

Winterkorn, H. F. and Moorman, R. B. B. (1941), A study of changes in physical properties of Putnam soil induced by ionic substitution, *Proceedings, 21st Annual Meeting, Highway Research Board*, pp. 415–434.

Winterkorn, H. F. and Tschebotarioff, G. P. (1947), Sensitivity of clay to remolding and its possible causes, *Proc. Highway Research Board*, v. 23, pp. 432–435.

Wohlbier, H. and Henning, D. (1969), Effect of preliminary heat treatment on the shear strength of kaolinite clay, *Highway Research Board special Report* 103, National Research Council, Washington, DC, pp. 287–300.

Wood, H. O. and Neumann, F. (1931), Modified Mercalli intensity scale, *Bulletin, Seismological Society of America*, v. 21, pp. 277–283.

Wu, J. T. H. (1994), *Design and Construction of Low Cost Retaining Walls-the Next Generation* in *Technology*, Colorado Transportation Institute and the University of Colorado at Denver, Report no. CTI-UCD-1-94, February, 152p.

Yamanouchi, T. (ed.) (1977), *Engineering Properties of Organic Soils in Japan*, Research Committee on Organic Soils, Japanese Society of Soil Mechanics and Foundation Engineering, Tokyo, Japan, 97p.

Yegian, M. K. and Whitman, R. V. (1978), Risk analysis for ground failure by liquefaction, *Journal of Geotechnical Engineering Division, Proceedings, ASCE*, v. 104, no. GT7, pp. 921–938.

Yen, B. C. and Scanlon, B. (1975), Sanitary landfill settlement rates, *Journal of the Geotechnical Engineering Division, Proceedings, ASCE*, v. 101, no. GT5, pp. 475–487.

Yong, R. N. and Townsend, F. C. (ed.) (1981), *Laboratory Shear Strength of Soil*, ASTM Special Technical Publication (STP) 740, 717p.

Yong, R. N. and Warkentin, B. P. (1966), *Introduction to Soil Behavior*, The Macmillan Co., New York.

Youd, T. L. (1988), Recognizing liquefaction hazard, *Proceedings, Symposium on Seismic* Design and *Construction of Complex Engineering Systems*, ASCE, Saint Louis, pp. 16–29.

Zeevaert, L. (1983), *Foundation Engineering for Difficult Subsoil Conditions*, 2nd Van Nostrand Reinhold Co., New York.

Zeevaert, L. (1991), Foundation problems in earthquake regions, chapter 17, *Foundation Engineering Handbook*, 2nd edn, Kluwer Academic Publishers, Boston, MA, pp. 673–678.

Zeevaert, L. (1996), The seismic-geodynamics in the design of foundations in difficult subsoil conditions, *Proceedings, 3rd International Symposium on Environmental Geotechnology*, Technomic Publishing. Co., Lancaster, PA, pp. 19–69.

Zimmie, T. F. and Riggs, C. O. (ed.) (1981), *Symposium on Permeability and Groundwater Contaminant Transport*, ASTM Special Technical Publication (STP) 746, 245p.

Zingg, Th. (1935), Beitrag zur Schotteranalyse, Schweizer, min. pet. Mitt, v. 15, pp. 39–140.

Index